Methods in Enzymology

Volume 92
IMMUNOCHEMICAL TECHNIQUES
Part E
Monoclonal Antibodies and
General Immunoassay Methods

METHODS IN ENZYMOLOGY

EDITORS-IN-CHIEF

Sidney P. Colowick   Nathan O. Kaplan

Methods in Enzymology

Volume 92

Immunochemical Techniques

Part E
Monoclonal Antibodies and General Immunoassay Methods

EDITED BY

John J. Langone

DEPARTMENT OF MEDICINE
BAYLOR COLLEGE OF MEDICINE
HOUSTON, TEXAS

Helen Van Vunakis

DEPARTMENT OF BIOCHEMISTRY
BRANDEIS UNIVERSITY
WALTHAM, MASSACHUSETTS

1983

ACADEMIC PRESS

A Subsidiary of Harcourt Brace Jovanovich, Publishers

New York   London
Paris   San Diego   San Francisco   São Paulo   Sydney   Tokyo   Toronto

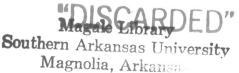

ACADEMIC PRESS, INC.
111 Fifth Avenue, New York, New York 10003

United Kingdom Edition published by
ACADEMIC PRESS, INC. (LONDON) LTD.
24/28 Oval Road, London NW1 7DX

Library of Congress Cataloging in Publication Data

Main entry under title:

Monoclonal antibodies and general immunoassay methods.

(Methods in enzymology ; v. 92, pt. E)
Bibliography: p.
Includes indexes.
1. Antibodies, Monoclonal. 2. Immunoassay--
Methodology. I. Langone, John J. (John Joseph),
Date. II. Van Vunakis, Helen, Date.
III. Series.
QP601.M49   vol. 92, pt. E 574.19'25s   82-22786
[QR186.85]   [591.19'285
ISBN 0-12-281992-2

PRINTED IN THE UNITED STATES OF AMERICA

83 84 85 86   9 8 7 6 5 4 3 2 1

Table of Contents

Section I. Hybridoma Technology

A. Production of Monoclonal Antibodies with Selected Applications

B. Detection and Assessment of Monoclonal Antibodies

Section II. Immunoassay of Antigens and Antibodies

A. Labeling of Antigens and Antibodies

B. Separation Methods in Immunoassay

C. Immunoassay Methods

D. Data Analysis

Contributors to Volume 92

Article numbers are in parentheses following the names of contributors.
Affiliations listed are current.

N. LEIGH ANDERSON (16), *Molecular Anatomy Program, Division of Biological and Medical Research, Argonne National Laboratory, Argonne, Illinois 64039*

GIULIA C. B. ASTALDI (5), *Istituto di Genetica Biochimica ed Evolutionistica C.N.R. di Pavia, 27100 Pavia, Italy*

STRATIS AVRAMEAS (38), *Département d'Immunologie, Institut Pasteur, 75724 Paris, Cedex 15, France*

WILLIAM F. BALE* (22), *School of Applied Biology, Georgia Institute of Technology, Atlanta, Georgia 30332*

RICHARD B. BANKERT (15, 21), *Department of Molecular Immunology, Roswell Park Memorial Institute, Buffalo, New York 14263*

ROBERT C. BAXTER (44), *Department of Endocrinology, Royal Prince Alfred Hospital, Sidney, New South Wales 2050, Australia*

ELSA H. BERENSTEIN (2), *Clinical Immunology Section, Laboratory of Microbiology and Immunology, National Institute of Dental Research, National Institutes of Health, Bethesda, Maryland 20205*

J. BRIGGS (36), *SYVA, Palo Alto, California 94303*

JOSEPH P. BROWN (12), *Program in Tumor Immunology, Fred Hutchinson Cancer Research Center, Seattle, Washington 98104*

ROBERT T. BUCKLER (32), *Ames Research and Development Laboratories, Miles Laboratories, Inc., Elkhart, Indiana 46515*

GÉRARD BUTTIN (14), *Unité de Génétique Somatique, Bâtiment Metchnikoff, Institut Pasteur, 75724 Paris, Cedex 15, France*

MICHAEL CAIS (25, 35), *Department of Chemistry, Technicon, Israel Institute of Technology, Haifa 32000, Israel*

M. CHRÉTIEN (23), *Protein and Pituitary Hormone Laboratory, Clinical Research Institute of Montreal, Montreal H2W 1R7, Canada*

M. ANGELES CONTRERAS (22), *School of Applied Biology, Georgia Institute of Technology, Atlanta, Georgia 30332*

ANGEL L. DE BLAS (4), *Department of Neurobiology and Behavior, State University of New York at Stony Brook, Stony Brook, New York 11794*

MARLENE DELUCA (33), *Department of Chemistry, University of California, San Diego, La Jolla, California 92093*

J. Y. DOUILLARD (13), *Institut National de la Sante et de la Recherche Medicale (INSERM) U211, Unité d'enseignement et de Recherche (UER) Medecine, 44035 Nantes, Cedex, France*

MATTHEW J. DOYLE (34), *Department of Chemistry, University of Cincinnati, Cincinnati, Ohio 45221*

V. B. ELINGS (36), *Department of Physics, University of California, Santa Barbara, California 93106*

J. FEINGERS (28), *Ames-Yissum Product Research and Development Laboratories, Jerusalem, Israel*

HELMAR FIEBACH (18), *Department of Experimental and Clinical Immunology, Central Institute of Cancer Research, Academy of Sciences of the G.D.R., DDR-1115 Berlin-Buch, German Democratic Republic*

PHILIP C. FOX (2), *Clinical Immunology Section, Laboratory of Microbiology and Immunology, National Institute of Dental Research, National Institutes of Health, Bethesda, Maryland 20205*

G. M. FULLER (42), *Department of Human Biological Chemistry and Genetics, Division of Human Genetics, The University*

* Deceased.

of Texas Medical Branch, Galveston, Texas 77550

ADRIAN P. GEE (31), Department of Pediatrics, University of Florida, Gainesville, Florida 32610

V. GHEŢIE (40), Department of Immunology, Victor Babeş Institute, 76201 Bucharest, Romania

JEAN-LUC GUÉSDON (38), Département d'Immunologie, Institut Pasteur, 75724 Paris, Cedex 15, France

H. BRIAN HALSALL (34), Department of Chemistry, University of Cincinnati, Cincinnati, Ohio 45221

BRUNO HANSEN (24), Hagedorn Research Laboratory, 2820 Gentofte, Denmark

P. HÄRING (20), Research Department, F. Hoffmann-La Roche and Co., Ltd., CH-4200 Basel, Switzerland

WILLIAM R. HEINEMAN (34), Department of Chemistry, University of Cincinnati, Cincinnati, Ohio 45221

INGEGERD HELLSTRÖM (12), Program in Tumor Immunology, Fred Hutchinson Cancer Research Center, and Department of Microbiology/Immunology, University of Washington, Seattle, Washington 98104

KARL ERIK HELLSTRÖM (12), Program in Tumor Immunology, Fred Hutchinson Cancer Research Center, and Department of Pathology, University of Washington, Seattle, Washington 98104

STEVEN H. HERRMANN (8), Department of Pathology, Harvard Medical School, Boston, Massachusetts 02115

T. HOFFMAN (13), Division of Biochemistry and Biophysics, Office of Biologics, FDA, National Center for Drugs and Biologics, Bethesda, Maryland 20205

D. INBAR (28), Ames-Yissum Product Research and Development Laboratories, Jerusalem, Israel

YI-HER JOU (21), Department of Molecular Immunology, Roswell Park Memorial Institute, Buffalo, New York 14263

DOMINIQUE JUY (14), Unité d'Immunochi-

mie Analytique, Bâtiment Metchnikoff, Institut Pasteur, 75724 Paris, Cedex 15, France

HENRY S. KAPLAN (1), Department of Radiology, Cancer Biology Research Laboratory, Stanford University School of Medicine, Stanford, California 94305

UWE KARSTEN (18), Department of Experimental and Clinical Immunology, Central Institute of Cancer Research, Academy of Sciences of the G.D.R., DDR-1115 Berlin-Buch, German Democratic Republic

KANEFUSA KATO (26), Department of Biochemistry, Institute for Developmental Research, Aichi Prefectural Colony, Kasugai, Aichi 480-03, Japan

JOHN J. LANGONE (31), Department of Medicine, Baylor College of Medicine, Houston, Texas 77030

PIERRE LEGRAIN (14), Unité de Génétique Somatique, Bâtiment Metchnikoff, Institut Pasteur, 75724 Paris, Cedex 15, France

ÅKE LERNMARK (24), Hagedorn Research Laboratory, 2820 Gentofte, Denmark

SUSANNE LINDE (24), Hagedorn Research Laboratory, 2820 Gentofte, Denmark

BO MATTIASSON (39), Pure and Applied Biochemistry, Chemical Center, University of Lund, S-220 07 Lund, Sweden

GEORGE L. MAYERS (21), Department of Molecular Immunology, Roswell Park Memorial Institute, Buffalo, New York 14263

PARK K. MAZZAFERRO (21), Department of Molecular Immunology, Roswell Park Memorial Institute, Buffalo, New York 14263

RONALD MCKAY (10), Cold Spring Harbor Laboratory, Cold Spring Harbor, New York 11724

MATTHEW F. MESCHER (8), Department of Pathology, Harvard Medical School, Boston, Massachusetts 02115

BURKHARD MICHEEL (18), Department of Experimental and Clinical Immunology,

Central Institute of Cancer Research, Academy of Sciences of the G.D.R., DDR-1115 Berlin-Buch, German Democratic Republic

V. MIGGIANO (3, 20), Research Department, F. Hoffmann-La Roche and Co., Ltd., CH-4200 Basel, Switzerland

I. MORARU (40), Department of Immunology, Victor Babeş Institute, 76201 Bucharest, Romania

DAVID L. MORRIS (32), Ames Research and Development Laboratories, Miles Laboratories, Inc., Elkhart, Indiana 46515

JAMES E. MOSIMANN (4), Laboratory of Statistical and Mathematical Methodology, Division of Computer Research and Technology, National Institutes of Health, Bethesda, Maryland 20205

ROLF MÜLLER (43), Tumor Virology Laboratory, Salk Institute, San Diego, California 92138

PETER J. MUNSON (41), Endocrinology and Reproduction Research Branch, National Institutes of Child Health and Human Development, National Institutes of Health, Bethesda, Maryland 20205

J. M. NICKERSON (42), Laboratory of Molecular Genetics, National Institutes of Health, Bethesda, Maryland 20205

J. C. NICOLAS (27), INSERM Unité 58, Montpellier 34100, France

D. F. NICOLI (36), Department of Physics, University of California, Santa Barbara, California 93106

KATHLEEN O'DAY (19), School of Veterinary Medicine, Oregon State University, Corvallis, Oregon 97331

LENNART OLSSON (1), Department of Medicine A, State University Hospital, Copenhagen 2100, Denmark

PETER PARHAM (9), Department of Structural Biology, Stanford University School of Medicine, Stanford, California 94305

TERRY W. PEARSON (16), Department of Biochemistry and Microbiology, University of Victoria, Victoria, British Columbia V8W 2Y2, Canada

JOSE M. PERALTA (29, 30), Instituto de Microbiologia, Universidade Federal do Rio de Janeiro, Rio de Janeiro 21491, Brasil

A. J. PICK (28), Ames-Yissum Product Research and Development Laboratories, Jerusalem, Israel

M. A. PIZZOLATO (17), Department of Clinical Chemistry, Faculty of Biochemistry, University of Buenos Aires, Buenos Aires 1113, Argentina

MAKARAND V. RATNAPARKHI (4), Department of Mathematics and Statistics, Wright State University, Dayton, Ohio 45435

D. G. RITCHIE (42), Department of Human Biological Chemistry and Genetics, Division of Cell Biology, The University of Texas Medical Branch, Galveston, Texas 77550

THOMAS J. ROGERS (19), Department of Microbiology and Immunology, Temple University School of Medicine, Philadelphia, Pennsylvania 19104

N. G. SEIDAH (23), Protein and Pituitary Hormone Laboratory, Clinical Research Institute of Montreal, Montreal H2W 1R7, Canada

ETHAN M. SHEVACH (7), Laboratory of Immunology, National Institute of Allergy and Infectious Diseases, National Institutes of Health, Bethesda, Maryland 20205

A. RAY SIMONS (29), Laboratory Training and Consultation Division, Centers for Disease Control, Atlanta, Georgia 30333

REUBEN P. SIRAGANIAN (2), Clinical Immunology Section, Laboratory of Microbiology and Immunology, National Institute of Dental Research, National Institutes of Health, Bethesda, Maryland 20205

IRVING L. SPAR (22), Department of Radiation Biology and Biophysics, University of Rochester, Rochester, New York 14627

TIMOTHY A. SPRINGER (11), Laboratory of Membrane Immunochemistry, Sidney Farber Cancer Institute, Boston, Massachusetts 02115

TH. STAEHELIN (3, 20), *Research Department, F. Hoffmann-La Roche and Co., Ltd., CH-4200 Basel, Switzerland*

C. STÄHLI (3, 20), *Research Department, F. Hoffmann-La Roche and Co., Ltd., CH-4200 Basel, Switzerland*

KATHRYN C. STALLCUP (8), *Department of Pathology, Harvard Medical School, Boston, Massachusetts 02115*

J. STOCKER (20), *Research Department, F. Hoffmann-La Roche and Co., Ltd., CH-4200 Basel, Switzerland*

CATHLEEN P. SULLIVAN (8), *Department of Pathology, Harvard Medical School, Boston, Massachusetts 02115*

B. TAKÁCS (20), *Research Department, F. Hoffmann-La Roche and Co., Ltd., CH-4200 Basel, Switzerland*

Y. TAMIR (28), *Ames-Yissum Product Research and Development Laboratories, Jerusalem, Israel*

B. TEROUANNE (27), *INSERM Unité 58, Montpellier 34100, France*

VICTOR C. W. TSANG (29, 30), *Helminthic Diseases Branch, Division of Parasitic Diseases, Centers for Disease Control, Atlanta, Georgia 30333*

AMAR S. TUNG (6), *Department of Immunology, Merck Sharp and Dohme Research Laboratories, Rahway, New Jersey 07065*

AARON P. TURKEWITZ (8), *Department of Pathology, Harvard Medical School, Boston, Massachusetts 02115*

D. B. WAGNER (28), *Becton Dickinson Research Center, Research Triangle Park, North Carolina 27709*

JON WANNLUND (33), *Analytical Luminescence Laboratory, San Diego, California 92121*

KENNETH R. WEHMEYER (34), *Department of Chemistry, University of Cincinnati, Cincinnati, Ohio 45221*

O. WEISS (28), *Ames-Yissum Product Research and Development Laboratories, Jerusalem, Israel*

BRITT C. WILSON (30), *2238 Lyle Road, College Park, Georgia 30337*

LEON WOFSY (37), *Department of Microbiology and Immunology, University of California, Berkeley, California 94720*

STEPHEN E. ZWEIG (7), *Department of Pharmacology, Baylor University College of Medicine, Houston, Texas 77030*

Preface

Several papers in this volume deal with advances in hybridoma technology and complement the chapter by Galfrè and Milstein that appeared in Volume 73 of this series. Methods to enhance the efficiency of producing desired fusion products, to simplify the basic procedures involved, and to assess the specificity and other properties of monoclonal antibodies are discussed. Representative examples illustrate how a battery of monoclonal antibodies with different specificities can be used to study various immunochemical and biochemical problems. Other chapters cover additional procedures that can be used to label antigens and antibodies, separate antigen–antibody complexes, and process data. Some papers classified under one heading or contained in another volume of the series include methods that can be applied in other specific areas. For example, cytotoxicity tests described in Volume 93 may be used to screen supernatant fluids in hybridoma experiments if cytotoxic activity is a desirable property of the monoclonal antibody.

The contents of these volumes continue to reflect our intention of presenting useful methods as they are developed. Thus, several section headings may be similar to those in other Immunochemical Techniques volumes. We believe that such a system makes it easier to follow recent advances. It can be particularly advantageous to investigators whose major interests may lie outside the realm of immunology but who realize that answers to their scientific problems can often be obtained by the judicious use of innovative immunochemical techniques.

Again we thank the contributors whose consistently fine efforts make these volumes possible; Carla Langone for managing the correspondence; and Nathan Kaplan, Sidney Colowick, and our colleagues at Academic Press for their continued enthusiastic support and invaluable assistance.

JOHN J. LANGONE
HELEN VAN VUNAKIS

METHODS IN ENZYMOLOGY

EDITED BY

Sidney P. Colowick and Nathan O. Kaplan

VANDERBILT UNIVERSITY
SCHOOL OF MEDICINE
NASHVILLE, TENNESSEE

DEPARTMENT OF CHEMISTRY
UNIVERSITY OF CALIFORNIA
AT SAN DIEGO
LA JOLLA, CALIFORNIA

METHODS IN ENZYMOLOGY

EDITORS-IN-CHIEF

Sidney P. Colowick   Nathan O. Kaplan

VOLUME 87. Enzyme Kinetics and Mechanism (Part C: Intermediates, Stereochemistry, and Rate Studies)
Edited by DANIEL L. PURICH

VOLUME 88. Biomembranes (Part I: Visual Pigments and Purple Membranes, II)
Edited by LESTER PACKER

VOLUME 89. Carbohydrate Metabolism (Part D)
Edited by WILLIS A. WOOD

VOLUME 90. Carbohydrate Metabolism (Part E)
Edited by Willis A. Wood

VOLUME 91. Enzyme Structure (Part I)
Edited by C. H. W. HIRS AND SERGE N. TIMASHEFF

VOLUME 92. Immunochemical Techniques (Part E: Monoclonal Antibodies and General Immunoassay Methods)
Edited by JOHN J. LANGONE AND HELEN VAN VUNAKIS

VOLUME 93. Immunochemical Techniques (Part F: Conventional Antibodies, Fc Receptors, and Cytotoxicity) (in preparation)
Edited by JOHN J. LANGONE AND HELEN VAN VUNAKIS

VOLUME 94. Polyamines (in preparation)
Edited by HERBERT TABOR AND CELIA WHITE TABOR

VOLUME 95. Cumulative Subject Index Volumes 61–74 and 76–80 (in preparation)
Edited by EDWARD A. DENNIS AND MARTHA G. DENNIS

VOLUME 96. Biomembranes (Part J: Membrane Biogenesis: Assembly and Targeting (General Methods; Eukaryotes)) (in preparation)
Edited by SIDNEY FLEISCHER AND BECCA FLEISCHER

VOLUME 97. Biomembranes (Part K: Membrane Biogenesis: Assembly and Targeting (Prokaryotes, Mitochondria, and Chloroplasts)) (in preparation)
Edited by SIDNEY FLEISCHER AND BECCA FLEISCHER

Section I

Hybridoma Technology

A. Production of Monoclonal Antibodies with Selected Applications
Articles 1 through 10

B. Detection and Assessment of Monoclonal Antibodies
Articles 11 through 20

[1] Human–Human Monoclonal Antibody-Producing Hybridomas: Technical Aspects

By LENNART OLSSON and HENRY S. KAPLAN

Human monoclonal antibodies have potential applications in a broad spectrum of biomedical areas.[1] Since our initial report of the successful production of monoclonal antibody-producing human–human hybridomas,[2] other groups have also succeeded in establishing antibody-producing human hybridomas,[3,4] encouraging the hope that human monoclonal antibodies of clinical interest will become available in increasing abundance in the years ahead.

Human–human hybridoma methodology differs in several respects from the conventional mouse hybridoma procedures.[5,6] This chapter describes our present techniques for making human hybridomas as well as some of the experience we gained in 1980–1981. Those methodological aspects that are similar to mouse–mouse (or mouse–rat) hybridoma technology are mentioned only briefly, since they have been detailed in several review papers.[7,8]

Methodology

Antigen-Primed Lymphoid Parental Fusion Partner Cells

Successful rodent hybridomas are highly dependent on the immunization procedures used and on the time of harvest of antigen-primed spleen lymphocytes. Mice and rats can be injected on any desired schedule with an essentially unlimited spectrum of antigens, including infectious agents and cancer cells, and their spleens can then be removed when they have been effectively immunized. Obviously, ethical considerations preclude

---

[1] H. S. Kaplan, L. Olsson, and A. Raubitschek, *in* "Monoclonal Antibodies in Clinical Medicine" (A. McMichael and J. Fabre, eds.), p. 17. Academic Press, New York, 1982.

[2] L. Olsson and H. S. Kaplan, *Proc. Natl. Acad. Sci. U.S.A.* **77**, 5429 (1980).

[3] C. M. Croce, A. Linnenbach, W. Hall, Z. Steplewski, and H. Koprowski, *Nature (London)* **288**, 488 (1980).

[4] K. Sikora, T. Alderson, J. Phillips, and J. V. Watson, *Lancet* **1**, 11 (1982).

[5] G. Köhler and C. Milstein, *Nature (London)* **256**, 495 (1975).

[6] G. Köhler and C. Milstein, *Eur. J. Immunol.* **6**, 522 (1976).

[7] J. W. Goding, *J. Immunol. Methods* **39**, 285 (1980).

[8] V. T. Oi and L. A. Herzenberg, *in* "Selected Methods in Cellular Immunology" (B. B. Mishell and S. M. Shiigi, eds.), p. 351. Freeman, San Francisco, California, 1979.

these procedures in human subjects. Peripheral blood lymphocytes (PBL) from individuals naturally immunized to a limited group of antigens (Rh, hepatitis B virus surface antigen, tetanus toxoid, etc.) can be used in place of spleen lymphocytes, but the numbers of specifically immune B lymphocytes in the circulation at time intervals long after immunization are likely to be very low. Lymphocytes can also be obtained from lymph nodes or other tissue specimens at the time of resection of autologous tumors.[4] These exceptions aside, it is clear that priming of human lymphocytes with most antigens will have to be carried out outside the human body. Thus, antigen priming of human lymphocytes is one of the main obstacles to the generation of human–human hybridomas. Table I gives the principal methods by which antigen-primed human lymphocytes may be obtained *in vitro* and in mouse–human chimeras.

Lymphoid Cells Primed in Vivo. Fusion of PBL without *in vitro* stimulation is the least fruitful of the procedures listed in Table I. The yield of

TABLE I

METHODS OF PRIMING HUMAN LYMPHOCYTES WITH ANTIGEN OUTSIDE THE HUMAN BODY

Lymphocytes	Site of antigen priming	Treatment prior to fusion	Yield of hybrids[a]
Spleen or PBL[b]	*In vitro*	None	Low
		PWM[c]	High
		Exposure to antigen recall	Low
		Exposure to antigen + PWM + HAT[d]	High
Spleen or PBL	*In vitro*	Complete *in vitro* priming	Low
		Complete *in vitro* priming + PWM + HAT	High
Spleen or PBL antigen-primed in a mouse-human chimera	*In vivo* (in the murine host)	None	High
		PWM + HAT	High

[a] Low: hybrid growth in less than 10% of wells; High: hybrid growth in more than 50% of wells.

[b] PBL, peripheral blood lymphocytes.

[c] Pokeweed mitogen stimulation (25 μg/ml) for 5–7 days.

[d] HAT: medium with 3×10^{-4} M hypoxanthine, 8×10^{-6} M thymidine, and aminopterin in the lowest concentration that kills 100% of the parental tumor line (in our experience in the range of 10^{-7} M to 10^{-8} M).

viable hybrids is low (<10%), and hybrids with the desired specificity are rarely obtained. Of 12 consecutive fusions, only 4 yielded a significant number of hybrids (visible cell growth in 30–50% of the wells), whereas the remaining 8 fusions resulted in hybrid growth in 0–30% of the wells. Among the wells with hybrid growth, only 20–50% exhibited immunoglobulin (Ig) production. These experiments were carried out in a search for antibodies against HL-A related antigens and the D blood type antigen (Rh antigen), using PBL from patients that had been naturally immunized against such antigens. None of the 12 fusions resulted in hybrids producing antibodies of the desired specificity.

Normal spleen lymphoid cell populations are more heterogeneous than PBL and contain a greater proportion of cells in the various stages of B-lymphocyte differentiation. Such populations therefore contain more blast cells preferentially capable of fusing with the malignant parental fusion partner.[9] Rodent hybridoma production is routinely carried out with spleen lymphoid cells, and our first human–human hybridoma fusions were also carried out with spleen cells; this was made possible by particular clinical circumstances described elsewhere.[10,11] Mouse hybridoma studies have also demonstrated that the time of spleen harvest after the last antigen injection is crucial for successful fusion. Even with the use of spleen cells, therefore, human–human fusions cannot be expected to be as efficient as mouse–mouse fusions, except in those rare cases in which a patient receives an antigen boost a few days prior to splenectomy; for most antigens such optimal timing is not ethically or logistically feasible.

Mitogen Stimulation

Since blastic B lymphocytes have a higher fusibility than mature B lymphocytes, human PBL were stimulated with lymphocyte mitogens and tested for fusibility with SK0-007 myeloma cells at various times thereafter. Figure 1 shows the results with pokeweed (PWM) or lipopolysaccharide (LPS) as mitogens. In striking contrast to its effect on mouse B lymphocytes,[9] LPS stimulation (at 50 μg/ml) had no effect on the fusibility of human PBL. Stimulation of PBL with PWM (at 25 μg/ml) gave a

---

[9] J. W. Paslay and K. J. Roozen, in "Monoclonal Antibodies and T-Cell Hybridomas: Perspectives and Technical Advances" (G. J. Hämmerling, U. Hämmerling, and J. F. Kearney, eds.), p. 551. Elsevier/North-Holland, Amsterdam, 1981.

[10] H. S. Kaplan and L. Olsson, in "Monoclonal Antibodies and T-Cell Hybridomas: Perspectives and Technical Advances" (G. J. Hämmerling, U. Hämmerling, and J. F. Kearney, eds.), p. 427. Elsevier/North-Holland, Amsterdam, 1981.

[11] H. S. Kaplan and L. Olsson, in "Hybridomas in Cancer Diagnosis and Treatment" (M. S. Mitchell and H. F. Oettgen, eds.), p. 113. Raven, New York, 1982.

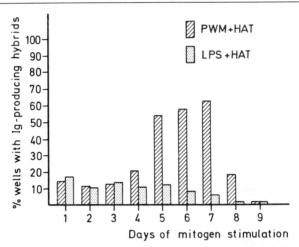

FIG. 1. Differential effects of stimulation with pokeweed mitogen (PWM) vs lipopolysaccharide (LPS) in the presence of hypoxanthine-aminopterin-thymidine (HAT) on human hybridoma yield.

significantly higher yield of hybrids at days 5, 6, and 7 after mitogen stimulation. However, we have observed that the beneficial effect of PWM for hybridoma generation depends on a well-functioning T-lymphocyte system. Pokeweed mitogen-stimulated PBL from patients with impaired T-cell function or decreased T-cell number give poor hybridoma yields.

Pokeweed mitogen-stimulated spleen cells have not been tested owing to the difficulty of obtaining fresh human spleen tissue on a regular basis. However, it can be expected that the kinetics for splenic lymphoid cells will be similar to those of PBL.

Antigen Priming of Human Lymphoid Cells in Vitro. Several reports have described the successful priming of human lymphoid cells with antigen *in vitro*.[12-16] The culture conditions of the lymphoid cells during antigen priming are very important. Even with optimal culture conditions, however, it has hitherto not been possible to obtain a secondary immune response after *in vitro* priming. This is a serious difficulty, since secondary immune responses are essential to obtain B lymphocytes producing high-affinity IgG antibodies. Encouraging, though preliminary, progress

[12] R. E. Callard, *Nature (London)* **282**, 734 (1979).
[13] N. Chiorazzi, S. M. Fu, and H. G. Kunkel, *Immunol. Rev.* **45**, 219 (1979).
[14] J.-F. Delfraisy, P. Galanaud, J. Dormont, and C. Wallon, *J. Immunol.* **118**, 630 (1977).
[15] H. M. Dosch and E. W. Gelfand, *J. Immunol. Methods* **11**, 107 (1976).
[16] M. K. Hoffmann, *Proc. Natl. Acad. Sci. U.S.A.* **77**, 1139 (1980).

has been made in our laboratory in defining the conditions for obtaining IgG responses after antigen priming of human lymphocytes *in vitro*.[17] Considerable efforts are now being expended on the development of better *in vitro* priming systems, since it can be anticipated that such systems will be the major source of antigen-primed lymphocytes for use in the human–human hybridoma system.

Mouse–Human Chimeras. The current inadequacy of *in vitro* priming methods has led us to study the possibility of using immunosuppressed mice as temporary hosts for human lymphocytes during antigen priming. Immunosuppression of young adult mice was induced by alternating exposures to whole-body X-irradiation (2×150 rad on days 0 and 7) and cyclophosphamide (2×300 mg/kg on days 4 and 11). 2,4,6-Trinitrophenyl (TNP) conjugated to keyhole limpet hemocyanin (KLH) was used as the antigen. Antigen injections were started 1 day after intrasplenic inoculation of 10^8 human PBL depleted of T lymphocytes by mass sheep red blood cell (SRBC) rosetting. Two antigen injections (100 μg each; the first in 0.2 ml emulsion of 50% phosphate-buffered saline (PBS) and 50% complete Freund's adjuvant; the second in 0.2 ml of PBS alone) were given on days 14 and 28. Four days later, the human cells were recovered from pooled mouse spleens and freed of contaminating mouse cells with anti-H-2 antibody plus complement. Recovery of human cells from BALB/c mice ranged from 21 to 29% at 3 weeks. The human lymphocytes were then fused with either human myeloma cells or human lymphoma cells (see below) and seeded in 96-well Microtiter plates. The unfused lymphocytes recovered from their murine hosts were shown to be producing specific anti-TNP antibody by plaque assay with TNP conjugated to SRBC; 1.9×10^4 direct (IgM) and 7.3×10^4 indirect (IgG) plaque-forming cells were detected per 10^8 human B lymphocytes. In the seeded plates, 87% of wells yielded viable hybridomas, and 6% produced specific anti-TNP antibody. A total of 17 stable antibody-producing clones were obtained, of which 3 produced IgM and 14 produced IgG. The human origin at the Ig products of these hybrids was confirmed by sodium dodecyl sulfate–polyacrylamide gel electrophoresis (SDS–PAGE) and isoelectric focusing (IEF), and the hybrid cells were shown to contain aneuploid, near-tetraploid numbers of human chromosomes.[18] Although these results are very encouraging, this immunosuppression procedure has the drawback that mice become highly susceptible to infection, resulting in a significant mortality (20–25%) prior to the time of harvest of the human cells. Other immunosuppression schedules are therefore being investigated to

[17] M. Bieber *et al.* In preparation.
[18] L. Olsson, C. Honsik, and H. S. Kaplan, Human–human hybridomas using lymphocytes primed with antigen in human–mouse chimeras. (Submitted for publication.)

make the procedure more practical; tolerization of newborn mice to human antigens is another alternative being studied.

Malignant Parental Fusion Partner Cells

Several murine myeloma cell lines with different Ig secretory characteristics have been selected from among the very large number of available lines for use in making mouse–mouse hybridomas.[19] In contrast, there are only a few human myeloma cell lines in existence (Table II).[20–24] We used the U-266 human myeloma line[24] for selection of a HAT-sensitive mutant that could be fused with human B lymphocytes. Many human tumor cell lines, including some myelomas and lymphomas, have a very low cloning efficiency in semisolid media. We initially attempted to select an 8-azaguanine-resistant mutant of U-266 in semisolid medium. When these first efforts failed, we developed the alternative liquid culture selection procedure outlined in Table III. The mutant cell line thus obtained (U-266AR$_1$, later designated SKO-007) had a population doubling time of 18–24 hr, but has slowed to about 48 hr. It was subsequently discovered that despite an initially negative broth culture, the U-266 cell line had been infected with mycoplasma, as was its derived mutant, SK0-007. Both lines were freed of infection by treatment with heat and antibiotics and subcloning. The mycoplasma-free U-266 cell line could readily be cloned in agarose, with a cloning efficiency of 1–5%. By plating ~10^7 cells in agarose containing, per milliliter, 20 μg of 8-azaguanine or 6-thioguanine, HAT-sensitive mutants could be obtained in a single-step procedure (J. Reese and H. S. Kaplan, unpublished results).

The U-266 myeloma line (as well as its HAT-sensitive mutants isolated to date) produces and secretes IgE(λ). Hybridomas resulting from fusions of such myeloma cells with Ig-producing lymphocytes can therefore be expected to produce variously permuted Ig molecules. In fact, a number of hybrids made with this line secrete relatively large amounts of the parental λ light chain, with which the B-lymphocyte-encoded heavy chain is extensively associated. Our efforts to obtain a non-Ig-producer variant

[19] G. J. Hämmerling, U. Hämmerling, and J. F. Kearney, eds., "Monoclonal Antibodies and T-Cell Hybridomas: Perspectives and Technical Advances." Elsevier/North-Holland, Amsterdam, 1981.

[20] K. H. Burk, B. Drewinko, J. M. Trujillo, and M. J. Ahearn, *Cancer Res.* **38**, 2508 (1978).

[21] V. Diehl, M. Schaadt, H. Kirchner, K.-P. Hellriegel, F. Gudat, C. Fonatsch, E. Laskewitz, and R. Guggenheim, *Blut* **36**, 331 (1978).

[22] M. E. Jobin, *In Vitro* **10**, 333 (1974).

[23] G. E. Moore and H. Kitamura, *N.Y. State J. Med.* **68**, 2054 (1968).

[24] K. Nilsson, H. Bennich, S. G. O. Johansson, and J. Pontèn, *Clin. Exp. Immunol.* **7**, 477 (1970).

TABLE II
Cell Lines Available for Fusion

Derivation	Modal No. of chromosomes	Ig class	Parental cell line
Mouse			
X63-Ag8	65	γ_1, κ	MOPC-21 myeloma
NSI-Ag4/1	65	κ intracellular	X63 myeloma
MPC11-45.6TG1.7	62	γ_{2b}, κ	BALB/c myeloma
X63-Ag8.653	58	None	X63-Ag8 myeloma
SP2/0-Ag14	72	None	(X63-Ag8 × BALB/c hybridoma
F0	72	None	Clone of SP2/0-Ag14
S194/5XXO.BU.1		None	BALB/c myeloma
Rat			
210.RCY3.Ag1.2.3	39	-, κ	Lou rat myeloma
Human			
U-266AR$_1$ (SKO-007)	44	ε, λ	U-266 myeloma
GM1500 GTG AL$_2$	—	γ_2, κ	GM1500 "myeloma"[a]
RPMI-8226-AR	—	-, λ	RPMI-8226 myeloma
LICR-LON-HMy2	—	α, κ	ARH-77 "myeloma"[a]
RH-L4-AR1	50	γ, κ[b]	RH-L4 lymphoma

[a] EBNA-positive; probably an Epstein-Barr virus (EBV)-transformed lymphoblastoid line.

[b] Cell surface marker; not secreted.

TABLE III
Selection of Hypoxanthine-Aminopterin-Thymidine (HAT)-Sensitive Variants of Human Myeloma/Lymphoma Cell Lines

1. Incubation for 5–7 days in RPMI-1640 + 15% FCS plus 20 μg of 8-azaguanine per milliliter[a]
2. Isolation of viable cells on Ficoll–Hypaque gradient and reseeding of cells at high density (1 to 2 × 10^5 cells/ml)
3. Cultivation of viable cells in 8-azaguanine (5 μg/ml) for 7 days
4. Isolation of viable cells on Ficoll–Hypaque gradient; then:

5a. Clone in semisolid medium (methylcellulose 1%; agarose 0.3%; agar 0.3%) with 8-azaguanine (30 μg/ml), followed by

5b. Selection of rapidly growing clones

or 6a. Expand liquid cultures with increasing concentrations (5 μg/ml per week to 30–40 μg/ml) of 8-azaguanine; then

6b. Clone in liquid culture by limiting dilution; select fast growing clones

7. Maintain cultures in 8-azaguanine (20–40 μg/ml)

[a] 8-Azaguanine may be substituted by other purine or pyrimidine analogs, e.g., 5-bromo-deoxyuridine, 6-thioguanine, 2,6-diaminopurine.

of SK0-007 cells have included repeated killing of λ light chain-positive cells with rabbit anti-λ light chain antibody plus complement, repeated fluorescence-activated cell sorter (FACS) selection of SKO-007 cells labeled with fluorescein-conjugated rabbit anti-λ light-chain antibody, and combinations of these procedures. The consistent outcome of these experiments has been that apparently non-Ig-secreting myeloma cells could be isolated, but after 3–5 weeks in culture these cells invariably again became positive for λ light chain.

We therefore decided to test other human tumor cell lines, ontogenetically related to B lymphocytes. Using 8-azaguanine in the liquid culture system outlined in Table III, a HAT-sensitive mutant was selected from a human lymphoma cell line that had been shown to express very low amounts of γ,κ, but did not secrete detectable amounts of Ig into the culture fluids. This HAT-sensitive cell line, RH-L4, has now been tested in 12 consecutive fusions with PWM-stimulated human PBL, and has yielded Ig-producing hybrids in all 12 fusions.[25] The population doubling time of this line is about 30 hr.

Human hybridomas are isolated from the parental myeloma cells, as in the case of most murine hybridomas, by selection in HAT medium.[26] It is desirable to use the lowest concentration of aminopterin that can be shown to kill 100% of the malignant fusion partner cells. Figure 2 shows such curves for 2 HAT-sensitive cells lines, SKO-007 and RH-L4, at two different aminopterin (A) concentrations, 2×10^{-7} M and 2×10^{-8} M. Differences have been found not only between myeloma and lymphoma lines, but also between sublines of the SKO-007 line. The concentrations of hypoxanthine and thymidine for optimal cell growth should also be estimated for each HAT-sensitive mutant. For our currently used subclone of SKO-007, these values are 3×10^{-4} M for hypoxanthine and 8×10^{-6} M for thymidine. We have observed occasional changes in aminopterin sensitivity of the myeloma lines; the aminopterin sensitivity of the cells is therefore checked at periodic intervals. At an aminopterin concentration of 2×10^{-7} M, the mycoplasma-free J3 subclone of SKO-007 has a low but not insignificant ($\sim 10^{-5}$) frequency of revertants to HAT resistance.

Fusion Procedures

Somatic cell hybridization between lymphocytes and myeloma-lymphoma cells is carried out using polyethylene glycol (PEG) as the fusogen, as in the case of mouse hybridomas. We have obtained equally good

[25] L. Olsson *et al.* Submitted for publication.
[26] J. W. Littlefield, *Science* **145**, 709 (1964).

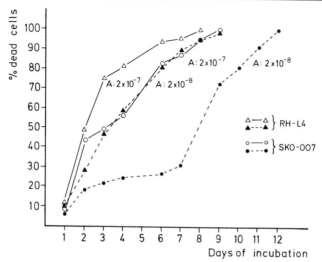

FIG. 2. Kinetics of cell death of two hypoxanthine-aminopterin-thymidine (HAT)-sensitive human cell lines at two different concentrations of aminopterin (A).

results using PEG of molecular weight (M_r) 1000 (37% w/v; Baker) or PEG M_r 1540 (38% w/v; Baker); preliminary experiments indicate that the fusion yield may be further improved by using PEG M_r 4000. No experiments have been carried out with Sendai virus as the fusogen; it is possible that this agent would also yield good results.

Fusions are performed by mixing equal numbers (usually 2×10^7 each) of SK0-007 or RH-L4 cells and antigen-primed lymphocytes, washing twice in serum-free RPMI-1640 medium, and then fusing in 2.0 ml of PEG in Ca^{2+}-free serum-free medium at 37° for 1 min. The PEG is then very gently diluted with Ca^{2+}-free serum-free medium, prewarmed to 37°. The PEG-treated cells are washed twice in warm (37°) serum-free medium, suspended at a concentration of 10^6 cells/ml in RPMI-1640 + 15% fetal calf serum (FCS), and seeded at 0.2 ml per well in flat-bottom 96-well Microtiter plates.

Cultivation and HAT Selection of Hybrids

The hybrids are very fragile in the first 7–14 days after fusion, and the cultures accordingly have to be treated with great care. Our very first experiments indicated, in keeping with experience with mouse hybridomas, that human hybridoma growth is highly cell density-dependent. This is important both during the initial growth of hybrids and in limiting-dilution cloning procedures. Human hybrids typically grow more slowly

TABLE IV
RELATIONSHIP BETWEEN INITIAL CELL CONCENTRATION AND GROWTH OF
FUNCTIONAL HYBRIDS

Number of cells seeded per well	Percentage of wells with viable cells growing at day 10 after fusion			Percentage of wells producing IgG at day 21 after fusion		
1×10^6	0,	0,	0^a	—		
5×10^5	8,	3,	4	2,	0,	0
2×10^5	54,	64,	38	29,	36,	18
1×10^5	29,	32,	14	9,	11,	6
5×10^4	32,	17,	9	8,	3,	1

^{a} Data of three successive experiments with hybridized spleen cells from three patients with Hodgkin's disease.

than mouse hybrids, and visible growth is often not observed until 3 weeks or more after seeding. Table IV shows the yield of viable hybrids as a function of the number of cells seeded per well in 96-well Microtiter plates.

In the mouse system, thymocytes, spleen macrophages, or peritoneal macrophages are routinely used as feeder cells for the freshly seeded hybrids.[7] Similar feeder cells of human origin may be difficult to obtain on a regular basis. We therefore routinely use monocytes from peripheral blood. These cells are readily isolated by virtue of their reversible adherence to plastic[27]; after incubation for 2 hr at 37° in RPMI-1640 medium with 10% FCS and 10% human AB serum, the nonadherent cells are removed, and the adherent cells are reincubated overnight, at which time they can readily be recovered by pipetting with ice-cold PBS. More recently, good results have also been obtained using gelatin columns,[28] from which the adherent cells are removed by trypsinization. Monocytes harvested by either of these methods are seeded into wells to which the hybrid cultures are later added. Typically, 5×10^4 adherent cells are seeded into each well of a 96-well plate, 2×10^5 adherent cells into each well of a 24-well plate. Plates seeded with feeder cells are conveniently prepared the day before fusion. Table V shows the significant beneficial impact of feeder cells on the yield of human hybridomas. Human thymocytes may also be used as feeder cells, when available; we use 5×10^5

[27] A. Treves, D. Yagoda, A. Haimovitz, N. Ramu, D. Rachmilewitz, and Z. Fuks, *J. Immunol. Methods* **39,** 71 (1980).
[28] T. Brodin, H. Sjögren, and L. Olsson, *J. Immunol. Methods* (in press).

TABLE V

EFFECT OF HUMAN MONOCYTE FEEDER CELLS ON GROWTH OF HUMAN MYELOMA AND
LYMPHOMA CELLS

Cell line	Number of cells seeded per well	Number of monocytes seeded per well	Number of viable cells after 14 days	Feeder: no feeder ratio
SKO-007 myeloma	10^3	None	4×10^3	~8
		5×10^4	30×10^3	
SKO-007 myeloma	10^2	None	2×10^3	~3
		5×10^4	6×10^3	
RH-L4 lymphoma	10^3	None	14×10^3	~39
		5×10^4	540×10^3	
RH-L4 lymphoma	10^2	None	2×10^3	~30
		5×10^4	59×10^3	

cells per well. Feeder cells are normally added every 7–10 days during the first 4–5 weeks after fusion and also during the first 4 weeks after cloning (see below).

In a number of fusions, we have tested the influence of preincubation in HAT medium on the fusibility of human PBL. The cells were stimulated 5 days with PWM in the presence of hypoxanthine ($3 \times 10^{-4} M$), thymidine ($8 \times 10^{-6} M$), and aminopterin at the concentration found to kill 100% of the HAT-sensitive malignant partner cells (usually 2 to $5 \times 10^{-7} M$). The idea was to select B lymphocytes with high levels of HGPRT prior to fusion. Stimulation of human PBL by PWM in the presence of HAT medium reduced [^{3}H]TdR uptake and total viable cell counts to about 10–25% of control PWM stimulation levels.

Selection of fused hybrids in HAT medium should not be started until about 48 hr after fusion. Approximately half of the RPMI-1640 + 15% FCS medium is initially removed and replaced by HAT medium. Supplementation with fresh HAT medium is repeated every 2–3 days. All the unfused myeloma and lymphoma cells are usually dead by about 14–21 days after fusion, and HAT medium can then be replaced by HT (omitting aminopterin). Simultaneously, large, round hybrid cells can be seen to grow out of the cell debris (Fig. 3). The hybrids have variable growth rates; a few grow to populations of 100 cells or more within 14–21 days after fusion, but most take 4–5 weeks to reach the same population size. Often, however, the most rapidly growing hybrids are unstable and exhibit little or no Ig production.

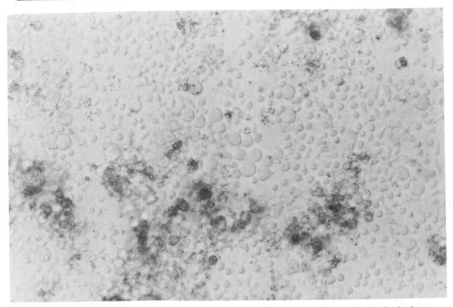

Fig. 3. Early outgrowth of a human–human hybridoma as a focus of relatively large, round lymphoid cells in a background of macrophage feeder cells, debris, and dying unfused lymphocytes.

Cloning of Human Hybridomas

The cloning step is very similar to the mouse hybridoma procedure[7]; either limiting dilution or cloning in semisolid culture medium can be used. It is crucial for successful cloning that feeder cells be used in both procedures. Human peripheral blood monocytes are used in liquid cultures at a density of 2 to 5×10^4 cells per well of 96-well Microtiter plates and added every 7–10 days during the first 3–4 weeks of cloning. In semisolid media, the concentration must be in the order of 10^6 feeder cells per milliliter. Human thymocytes, when available, are used at a concentration 2–3 times that of monocytes. Cloning should be performed as early as possible after the detection of specific Ig secretion by viable hybridomas. Human–human hybrids, though substantially less prone to chromosomal instability and loss than mouse–human hybrids,[29] may nonetheless exhibit substantial variation of chromosome number within the first 1–2 months (Fig. 4) and may lose the chromosomes bearing the genes for Ig heavy or light chain during this time. Thus, Ig secretion should be monitored repeatedly and cells be recloned at frequent intervals. Aliquots

[29] C. M. Croce and H. Koprowski, *Science* **184,** 1288 (1974).

should also be frozen in the presence of 10% DMSO at $-1°$/min and stored in liquid nitrogen to avoid permanent loss of a producer hybridoma through chromosomal change or infection.

Screening for Antibody Production

Screening for Ig production by hybrids can be started as early as 12–14 days after fusion by either solid-phase radioimmunoassay (RIA) or enzyme-linked immunosorbent assay (ELISA). Typically, spent culture supernatants are harvested and tested for their Ig content. Since fresh culture medium is simultaneously added to the cultures, the test cannot be repeated until 3–4 days later. Despite the fact that ELISA assays are more convenient than RIA, they are currently somewhat less sensitive. Other useful screening procedures involve the use of the fluorescence-activated cell sorter (FACS) to detect fluorescent antibody binding to target cell antigens and of microcytotoxicity assays for cell-surface antigens. When screening for antibody against cell surface antigens, we suggest the use of four tests: (a) ELISA for Ig production; (b) antigen-binding ELISA for specificity; (c) microcytotoxicity assay for cell-surface antigen specificity; and (d) FACS analysis for specificity. We require two of these four tests to be positive in order to continue work with a given culture.

Future Improvements

We have already mentioned that further improvement in the quality of malignant fusion partner cells can be expected. It will be necessary to

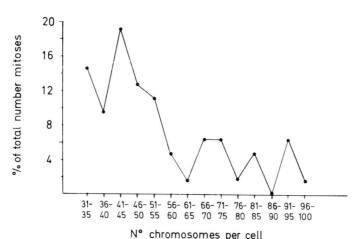

FIG. 4. Variation in chromosome numbers among cells of a noncloned human–human hybridoma established about 6 weeks earlier.

develop rapidly growing, non-Ig-producing malignant partner cells (either myelomas or lymphomas) with good fusion properties. A crucial aspect of the human hybridoma system is the effective antigen priming of human lymphocytes. The mouse–human chimera model described above seems presently to be a broadly applicable method. However, the system requires considerable resources and labor. Efforts must therefore be made to develop better *in vitro* priming procedures that will permit secondary immune responses and high-affinity IgG antibodies to be obtained.

Hybrid selection in HAT medium is a tedious procedure that probably results in loss of substantial numbers of hybrids. The selective culture medium was developed to select hybrids on the basis of their differential capacity for proliferation. However, it should be possible to select and clone Ig-producing hybrid cells on the basis of other markers, such as the antigen specificity of hybrid Ig. The development of such methods would be of major importance to human–human hybridoma technology.

Conclusions

The various steps in generating human–human monoclonal antibody-producing hybridomas have been described. We have not attempted to present an excessively detailed procedure, since this technology is still evolving rapidly and experience with mouse hybridoma technology has already shown that almost every laboratory introduces its own small technical modifications. However, the various steps outlined in this chapter should provide a workable foundation for investigators trying to establish human–human hybridomas. It is highly likely that more simplified systems will be developed in the not-distant future, but the present "state of the art" already provides us with a technique that has broad applicability in a number of biomedical areas.

Acknowledgment

Work from our laboratories described herein was supported in part by Grant CA-29876 from the National Cancer Institute, National Institutes of Health.

[2] Methods of Enhancing the Frequency of Antigen-Specific Hybridomas

By REUBEN P. SIRAGANIAN, PHILIP C. FOX, and ELSA H. BERENSTEIN

The hybridization of spleen cells from immunized mice with plasmacytoma cell lines is a powerful method for the production of monoclonal antibodies.[1] In a previous volume in this series, an extensive discussion of this technique is presented by Galfrè and Milstein.[2] However, the development of monoclonal antibodies has been hampered by the low frequency of antigen-specific hybridomas recovered after cell fusion. In an effort to improve the yield of hybridomas, we used two techniques to select and expand the number of spleen cells producing specific antibody prior to fusion.[3] The first method is the adoptive transfer of spleen cells from immunized animals to X-irradiated syngeneic recipients followed by *in vivo* antigen boosting. In the second method, spleen cells from immunized mice were cultured with antigen before fusion. The use of either of these methods resulted in a 10-to 50-fold increase in the percentage of antigen-specific antibody-secreting hybridomas. These methods for enhancing the yield of antigen-specific hybridomas and the techniques used for cell fusion and cloning by limiting dilution are described in this chapter.

Immunization

BALB/c mice, 6–8 weeks old, are commonly used for immunization. The spleen cells from immunized mice are then fused to plasmacytoma cell lines derived from this mouse strain. Hybridomas produced by the fusion of spleen cells of other strains and a BALB/c myeloma require the use of F_1 (BALB/c × immunized mouse strain) for the *in vivo* growth of the hybridomas to produce ascitic fluid. Enhancement of hybridoma formation will be effective only with adequately primed donor spleen cells, therefore the use of animals with high serum antibody titers is crucial. The injection schedules and the selection of adjuvant should be based on prior experience of immunization with a particular antigen. One effective method of immunization is the intramuscular injection at multiple sites of antigen emulsified in complete adjuvant with *Mycobacterium tuberculosis*

[1] G. Köhler and C. Milstein, *Nature (London)* **256**, 495 (1975).
[2] G. Galfrè and C. Milstein, this series, Vol. 73, p. 3.
[3] P. C. Fox, E. H. Berenstein, and R. P. Siraganian, *Eur. J. Immunol.* **11**, 431 (1981).

METHODS IN ENZYMOLOGY, VOL. 92

(H37 Ra-Difco) followed by booster injections every 4 weeks with the antigen in incomplete adjuvant (Difco). Animals should be bled individually 6–10 days after a third or fourth injection, and specific serum antibody titers determined. The spleens from mice with the highest titers are removed 10 days after the last antigen boost for either adoptive cell transfer or *in vitro* culture.

Preparation of Single-Cell Suspensions from Spleens

Materials

    Plastic petri dishes (100 × 20 mm)
    Plastic pipettes (10 ml)
    Plastic tubes (50 ml)
    Sterilized dissection instruments
    Dulbecco's modified Eagle medium with 15 mM HEPES (DMEM-HEPES) (for preparation of medium see section on hybridization technique)
    CO_2 chamber (an ice bucket containing dry-ice and water beneath an insulating layer can be used)

Procedure

1. It is essential that the entire procedure be done aseptically.
2. The mice are killed by cervical dislocation or by placing in a CO_2 chamber.
3. The animal is immobilized on a dissection board and the abdomen is wiped with alcohol. The skin and muscles are incised in the midline and stripped laterally to expose the peritoneal membranes. This area is wiped with an alcohol-soaked gauze sponge.
4. The spleen is removed and placed in a 100 × 20 mm plastic petri dish containing 10 ml of cold Dulbecco's modified Eagle medium with 15 mM HEPES (DMEM-HEPES). All procedures beyond this point are carried out in a laminar flow cabinet.
5. The spleen is divided at the midpoint with scissors. A 12-ml plastic syringe with 25-gauge needle is filled with cold DMEM-HEPES. The needle is inserted into the uncut end of the spleen capsule, and the medium is slowly injected while squeezing the capsule with a broad forceps. This expresses most of the spleen cells.
6. The cells and medium are transferred with a plastic pipette to 50-ml plastic tubes.
7. The suspension is repeatedly drawn up into a 10-ml plastic pipette to break up cell clumps, and then the tubes are placed in ice.

8. After 10 min. the supernatant is transferred to another 50-ml plastic tube, leaving the large clumps of debris that have settled out.

9. The cells are washed twice with 50 ml of cold DMEM-HEPES by centrifugation at 4° for 10 min at 300 g.

10. After the second wash, the cell pellet is resuspended in 2 ml of cold DMEM-HEPES. It is important that the cell pellet be vigorously resuspended, as clumping can occur. In general, ~100 × 10^6 cells are harvested from a single mouse spleen. Viability as measured by Trypan Blue dye exclusion is routinely >95%.

Method I: Adoptive Spleen Cell Transfer

1. Recipient syngeneic mice are X-irradiated with 550 rad 24 hr prior to donor spleen cell transfer. After irradiation, the drinking water is acidified (0.001 N HCl).

2. Spleen cells (prepared as described in the section on single-cell suspensions from spleens) are injected intravenously via a lateral tail vein. The 30 to 50 × 10^6 cells in ~0.25 ml DMEM are injected per mouse using a 27-gauge needle and a 1-ml plastic syringe. In general, a single donor spleen will be transferred into two mice. The use of heat and xylene applied topically to dilate the tail veins will facilitate injection. A new needle is used for each injection. The technique of tail vein injection should be mastered prior to actually transferring cells.

3. Immediately after the cell transfer, the mice are given an intraperitoneal injection of the immunizing antigen in 0.5 ml of 0.15 M saline. Note that adjuvant is not used at this time. The concentration of the antigen is the same as that used previously for boosting of the donor animal.

4. Four days after cell transfer, the spleens are removed aseptically from the recipients and single-cell suspensions are prepared. The number of viable cells recovered averages 27% of the number of spleen cells initially transferred.

Method II: Secondary Culture of Spleen Cells from Immunized Mice

As an alternative to adoptive transfer, spleen cells may be cultured in the presence of the immunizing antigen prior to fusion. This technique is best suited for purified soluble antigens.

Materials

Tissue culture plates (6 wells; Costar No. 3506)
Plastic tubes, 50 ml
Plastic pipettes, 2 ml

Gas mixture: 83% nitrogen, 7% oxygen, 10% CO_2
Dulbecco's modified Eagle medium with 15 mM HEPES (DMEM-HEPES) and hybridoma complete medium (HCM) (for preparation of media see the section on hybridization technique)

Procedure

1. Spleen cells (prepared as described) are placed in 6-well tissue culture plates, 2×10^7 cells per well in 1 ml of HCM. Lower cell densities in culture are not as effective in increasing the yield of antigen-specific hybridomas.
2. The immunizing antigen, diluted in HCM, is added to a final concentration of 1–50 μg per culture. The further addition of 10 μg of dextran sulfate or 1–10 μg of lipopolysaccharide per culture does not enhance the yield of antigen-specific hybridomas.
3. Cultures are placed on a rocking platform at 37° in a humidified gas mixture of 83% N_2, 7% O_2, and 10% CO_2.
4. After 4 days the culture plate is scraped with a sterile rubber policeman, the cells and medium are transferred into a 50-ml plastic tube, and the plate is washed with 10 ml of DMEM-HEPES.
5. The cells are washed twice with 50 ml of DMEM-HEPES at room temperature by centrifugation for 10 min at 300 g. The cells are resuspended in 2 ml of DMEM-HEPES, and the cell number and viability are determined. In general, ~15% of the number of cells initially placed in culture are recovered.

Hybridization Technique

A number of successful techniques for cell fusion have been described.[4,5] The following procedure is an adaptation of the method of Gefter *et al.*[6]

Materials

Dulbecco's modified Eagle medium with 4.5 g of glucose per liter without sodium pyruvate or L-glutamine (DMEM)
NCTC 109 medium with Earle's balanced salt solution and L-glutamine (NCTC 109, M.A. Bioproducts)

[4] "Lymphocyte Hybridomas" (F. Melchers, M. Potter, and N. Warner, eds.), *Curr. Top. Microbiol. Immunol.* **81** (1978).
[5] "Monoclonal Antibodies. Hybridomas: A New Dimension in Biological Analyses" (R. H. Kennett, T. J. McKearn, and K. B. Bechtol, eds.). Plenum, New York, 1980.
[6] M. L. Gefter, D. H. Margulies, and M. D. Scharff, *Somatic Cell Genet.* **3,** 231 (1977).

L-Glutamine (200 mM, GIBCO)

Antibiotic-antimycotic mixture with penicillin 10,000 U/ml, fungizone 25 μg/ml, and streptomycin 10,000 μg/ml (GIBCO)

Nonessential amino acid mixture (10 mM of each amino acid; M.A. Bioproducts)

Sodium pyruvate (100 mM; M.A. Bioproducts)

Oxaloacetic acid (Sigma). A 100 mM stock solution is prepared in distilled water, sterilized by passage through a 20-μm pore filter, and stored at $-20°$

2-Mercaptoethanol (Sigma). A 0.1 M stock solution is prepared by adding 0.1 ml to 14.1 ml of DMEM.

Insulin (U 100, regular Iletin; Lilly)

8-Azaguanine (Sigma). A 1000 $\times$ stock solution is prepared by dissolving 30.0 mg in 2.0 ml of dimethyl sulfoxide (Sigma).

Aminopterin (Sigma). A 100 $\times$ stock solution (4 $\times$ 10^{-5} M) is prepared by dissolving 1.76 mg in 0.1 N NaOH. Adjust the pH to 7.2 with HCl and bring the volume to 100 ml with distilled water. Store at $-20°$.

Hypoxanthine and thymidine (Sigma). A 100 $\times$ stock solution (hypoxanthine 1 $\times$ 10^{-2} M, thymidine 1.6 $\times$ 10^{-3} M) is prepared by dissolving 136 mg of hypoxanthine and 38.7 mg of thymidine in 0.1 N NaOH. The pH is adjusted to 9.0 with HCl, and the volume is brought to 100 ml with distilled water.

HEPES (Calbiochem). A 1.0 M stock solution, pH 7.2, is prepared and then stored at $-20°$.

Fetal calf serum. Serum lots vary widely in their capacity to maintain the growth of plasmacytoma cell lines. Individual lots of sera are prescreened to find those that best support the growth of the parental cell line. The cloning efficiency of the parental plasmacytoma cell line in different sera is determined by seeding 1–10 cells into each well of a 24-well plate. There is a 10–25% cloning efficiency with a good serum in the absence of a feeder layer.

Polyethylene glycol (PEG, 1000 M_r, Koch-Light, Research Products International Corp., Elk Grove, Illinois) PEG is autoclaved in a sterile glass bottle for 15 min and kept in a 56° water bath. A 35% solution (v/v) is prepared in prewarmed DMEM-HEPES (see below), and then kept at 37°. Depending on the lot of PEG or the cell lines, other concentrations of PEG may be required for optimal cell fusion. When hybridizing an unfamiliar combination of parental plasmacytoma and immune spleen cell, we routinely screen for the appropriate concentration of PEG by using several dilutions (e.g., 30, 35, 40, 45, and 50% PEG).

Plastic tubes (15 ml, round bottom; Falcon No. 2001).

Tissue culture plates (24 wells, Costar No. 3524; 6 wells, Costar No. 3506) and flasks (75 ml; Corning No. 25116).

Media

Hybridoma complete medium (HCM): To the Dulbecco's modified Eagle medium with 4.5 g of glucose per liter, without sodium pyruvate or L-glutamine, the following additions are made (final concentrations): 20% fetal calf serum, 10% NCTC 109, 1.0 mM pyruvate, 1.0 mM oxaloacetic acid, 2.0 mM L-glutamine, 0.1 mM each of a mixture of nonessential amino acids, 4×10^{-5} M 2-mercaptoethanol, 0.2 units of bovine insulin per milliliter, and 1:100 antibiotic–antimycotic mixture. The additives are filtered through a 0.45-μm filter prior to addition to the DMEM.

Hybridoma complete medium plus HAT (hypoxanthine-aminopterin-thymidine): To the hybridoma complete medium is added hypoxanthine (1×10^{-4} M), aminopterin (4×10^{-7} M), and thymidine (1.6×10^{-5} M). HCM-2×-HAT is hybridoma complete medium with double the concentration of hypoxanthine, aminopterin, and thymidine.

Dulbecco's modified Eagle Medium with HEPES (DMEM-HEPES): To Dulbecco's modified Eagle Medium with 4.5 g of glucose per liter, without L-glutamine or sodium pyruvate, add HEPES (final concentration 15 mM) and antibiotic–antimycotic mixture (final concentration 1:100).

Procedure

1. The plasmacytoma cell lines are maintained in HCM with the fetal calf serum reduced to 10% and the addition of an appropriate selective drug (e.g., 8-azaguanine). The cells should be in the logarithmic growth phase for optimal results. The cells are subcultured at 2 to 4×10^5 cells per milliliter in fresh medium at 4 days, and again 2 days prior to their use for fusion. Sublines from a single plasmacytoma cell line vary in their capacity to form antibody-secreting hybrids. An efficient cell line should be used soon after initiation from frozen stock.

2. The plasmacytoma cells are washed twice and resuspended in DMEM-HEPES. The cell number and viability are determined.

3. Spleen cells and myeloma cells are mixed at a 1:1 ratio of viable cells in a 15-ml round-bottom plastic tube. Between 5 and 25×10^6 spleen cells may be fused in a single tube. The procedure is done at room temperature.

4. Centrifuge the tube for 8 min at 300 g and then pour off all the medium. Be certain that all the medium is removed.

5. Lightly tap the tube on a table top to disrupt the cell pellet. Add 0.2 ml of PEG solution slowly (drop by drop) while gently shaking and mixing the contents of the tube.

6. Immediately centrifuge the tube for 6 min at 200 g.

7. Add 5 ml of DMEM-HEPES medium slowly (drop by drop) and leave the tube undisturbed for 2 min.

8. Add another 5 ml of DMEM-HEPES medium and swirl the contents of the tube gently.

9. Centrifuge the tube for 5 min at 300 g, then pour off all the medium.

10. Add 5 ml of HCM and leave undisturbed for 20–30 min.

11. Gently mix the cells and transfer to a 50-ml plastic tube. Add HCM to bring the concentration of spleen cells to 0.5×10^6/ml. If thymocytes are used as a feeder layer, they should be added at a concentration of 1×10^6 cells/ml in HCM (for preparation of thymocytes see the next section).

12. Mix the cells well, and place 1 ml of the suspension in the wells of a 24-well tissue culture plate.

13. Place the plates in a humidified 37° incubator with a 5% CO_2-containing atmosphere.

14. After 24 hr, 1 ml of HCM-2×-HAT is added to each well.

15. The wells are then fed on days 5, 8, 10, and 12 by removing 1 ml from each well and replacing with 1 ml of HCM-HAT.

16. During the third week, the cultures are fed twice with HCM with added hypoxanthine and thymidine *without* aminopterin.

17. Starting on the fourth week, the cultures are fed twice a week with HCM.

18. In general, actively growing hybrids can be seen by day 10. Supernatants can be tested for antibody activity at 2 and 3 weeks.

19. Cultures are expanded when >25% of the surface area of the well is covered by cells. Contents of a single well are transferred to one well of a 6-well tissue culture plate. The original well of the 24-well plate is refed in case the expansion does not survive.

20. Supernatants from the 6-well plate are retested for antibody activity; if positive, the cells are cloned. The cells are further expanded to the second and third wells of the 6-well plate and then to tissue-culture flasks. Cells from positive flasks are frozen as soon as possible.

Cloning by Limiting Dilution on Thymocyte Feeder Layers

Hybridoma cells can be cloned in semisolid agar or by limiting dilution. Both techniques require a feeder layer. The semisolid agar method has been described in detail in this series.[2] Thymocytes are an effective feeder layer. They are short-lived in culture but provide the necessary factors during the critical early stages of the growth of clones.[7,8]

Materials

Sterile sieve and holder (300–400 mesh; Scientific Products)
Plastic petri dish (100 × 20 mm)
Plastic pipettes (10-ml)
Plastic tubes (50-ml)
Sterilized dissection instruments
Hybridoma complete medium (HCM)
Tissue culture plates (96 wells, Costar No. 3596).
CO_2 chamber

Procedures

Harvesting and Preparation of Thymocytes

1. All procedures must be done aseptically.
2. BALB/c mice, 5–7 weeks old, are sacrificed by CO_2 narcosis. The mouse is immobilized on a board, and the ventral surface is wiped with alcohol.
3. The skin over the thorax is peeled away from the rib cage and up to the head. Pin the skin to the board away from the body.
4. The rib cage is cut and retracted laterally. The thymus is located in the upper mediastinum. It is large and bilobed at this age.
5. The thymus is removed and placed in a plastic petri dish containing 10 ml of HCM. In a laminar-flow cabinet, the thymus is transferred to a sieve sitting in a petri dish with 10 ml of HCM. With a curved tweezer, the thymus is pressed against the sieve to express all the cells.
6. The sieve is washed with 10 ml of HCM, and the media and cells from the petri dish are transferred to a 50-ml plastic tube.
7. The tube is centrifuged at room temperature for 10 min at 200 g.
8. Aspirate the supernatant and resuspend the cell pellet in 20 ml of HCM.
9. Determine the cell number. A single mouse, 5–7 weeks old, usually provides 1 to 2×10^8 thymocytes. Older mice have smaller thymuses and should not be used.

[7] J. Andersson, A. Coutinho, W. Lernhardt, and F. Melchers, *Cell* **10**, 27 (1977).
[8] W. Lernhardt, J. Andersson, A. Coutinho, and F. Melchers, *Exp. Cell Res.* **111**, 309 (1978).

Cloning by Limiting Dilution

1. In general, a minimum of 1 to 5×10^5 hybridoma cells are required for successful cloning. Therefore, the cells cannot be cloned before they are growing well in at least a single well of a 6-well plate.

2. The hybrid cells are fed 24–48 hr before cloning by replacing most of the medium with fresh HCM.

3. On the day of cloning, remove and discard most of the medium from the plate. This contains mainly dead cells. Then vigorously pipette off the attached cells with fresh HCM. Transfer the cells to a 50-ml plastic tube.

4. Wash the cells twice by centrifugation with HCM at room temperature.

5. Cell number and viability (by Trypan Blue exclusion) are determined, and the cell number is adjusted to 1×10^5 viable cells per milliliter. Most successful cloning is achieved when cell viability is >70%.

6. Two serial 10-fold dilutions are made (10^4, 10^3 per milliliter).

7. In a 50-ml tube, add 0.45 ml of the 10^3 hybridoma cells per milliliter and 68×10^6 thymocytes, and sufficient HCM to a volume of 45 ml.

8. Add 0.2 ml of the cell suspension to each well of two 96-well tissue culture plates. The final concentrations are 2 hybridoma cells and 3×10^5 thymocytes per well. Leave the plates undisturbed in a 37° incubator.

9. At 14 days, the 96-well plates are examined using an inverted microscope at low magnification. The wells with growing clones are marked to indicate the number of clones. Certain wells will contain more than a single clone of cells, and ~30% of the wells will have no hybridoma growth. The cloning efficiency of the hybridoma cells is ~50% under these conditions.

10. During the fourth week, wells with single clones are transferred to 24-well tissue culture plates in 0.5 ml of HCM. Routinely, 24 clones are selected from each 96-well plate. The 24-well plates are fed with 0.5 ml of HCM on days 5 and 10, and the supernatant is sampled and assayed for specific antibody activity on day 14. Positive wells are further expanded, first to 6-well plates and then to 75-ml flasks. It is usual to expand 6–10 positive clones to the flask stage and then choose the one with the highest antibody titer for further expansion and injection into animals for the production of ascitic fluid.

Remarks

Although the initial culture after fusion and cell cloning is in medium containing 20% fetal calf serum, established cultures in flasks may be maintained in medium containing 10% serum. Fetal calf serum or a 1:1

mixture of calf serum and horse serum may be used. Different lots of horse serum and sera mixtures should be screened for cloning efficiency of the parental cell line.

A 1:1 spleen cell-to-plasmacytoma cell ratio and the use of a thymo-cyte feeder layer following fusion have given good results. With the de-scribed adoptive transfer or *in vitro* culture techniques we have been successful in producing monoclonal antibodies against cell-surface recep-tors (e.g, the IgE-Fc receptor on basophils), bacteria (*Actinomyces, Cy-tophaga* sp.), haptens (2,4-dinitrophenol and benzylpenicilloyl), lympho-kines (Interleukin 1 and 2), and immunoglobulins (human and rat IgE).

Although both adoptive cell transfer and *in vitro* cell culture are effec-tive in increasing the number of antigen-specific hybridomas formed, we prefer to utilize adoptive transfer in most cases. This method allows anti-gen stimulation with whole cells, a higher percentage of viable cells are recovered, and the results are more consistent. A wider spectrum of hybridoma antibody isotypes have also been recovered with this tech-nique.

The choice of assay for specific antibody in supernatants depends on the nature of the antigen and its availability in pure form. Numerous assays have been described, and those utilizing the enzyme-linked im-munosorbent assay (ELISA) method have widespread applicability. A prime consideration in the choice of any assay is the ease, rapidity, and reproducibility for handling large numbers of samples.

[3] Spleen Cell Analysis and Optimal Immunization for High-Frequency Production of Specific Hybridomas

By C. Stähli, Th. Staehelin, and V. Miggiano

After the publication of Köhler and Milstein[1] reporting on a stable hybridoma secreting antibodies to SRBC,[2] many investigators reported similarly successful experiments.[3] Not only did they obtain large numbers

[1] G. Köhler, and C. Milstein, *Nature (London)* **256**, 495 (1975).

[2] Abbreviations: SRBC sheep red blood cells; BSA, bovine serum albumin; SE, specific efficiency of hybridization = number of hybridomas specific for immunizing antigen rela-tive to total number of hybridomas; KLH, keyhole limpet hemocyanin; PBS, phosphate-buffered saline without calcium or magnesium; MUHI, multiple high-dose immunizations.

[3] Compare many contributions in "Lymphocyte Hybridomas" presented at a 1978 Hybridoma Conference and published in *Curr. Top. Microbiol. Immunol.* **81** (1978).

of hybridomas from fusion experiments with mouse spleens (high efficiency of hybridization), but they also found that a high proportion of their hybridomas secreted the desired antibodies specific for the immunizing antigen [high specific efficiency (SE)]. In many of the experiments, in which mostly cells or viruses were used as antigens, specific efficiencies >10% were observed (personal communications by G. Köhler, M. H. Schreier, H. Forster, and B. Takács). In contrast, with soluble antigens such as proteins, many laboratories including our own initially had very poor success.[4] While fusion efficiencies were normal, specific efficiencies were close to zero. For instance, in five early hybridization experiments with the protein human chorionic gonadotropin (hCG, M_r 50,000) we screened over 1500 hybridomas without finding a specific one (SE < 0.1%). This failure was not the result of poor immunogenicity of hCG. These hybridizations had been carried out with spleen cells of mice selected for high serum antibody titers after three courses of immunization.

A series of hybridization experiments by Köhler in which the time between immunization and fusion was varied showed that specific efficiency increased up to 3 days and then decreased again as the interval increased further (G. Köhler, personal communication). This led to the assumption that antibody-secreting hybridomas are the result of fusion between myeloma cells and B-lymphocyte blast cells, since resting B cells are upon stimulation transformed into rapidly proliferating blasts, most of which quickly differentiate further into plasma cells.[5] Accordingly, to obtain a high specific efficiency one would need a high proportion of antigen specific B-lymphocyte blast cells relative to the total number of B blasts in the spleen at the time of fusion. Thus, B-cell activation should be synchronous and vigorous. In agreement with these considerations we found specific efficiency to be crucially dependent on the procedure of the final immunization(s) immediately preceding fusion (Table I), and we are routinely using an optimized method for the final immunizations (MUHI) described below.[4] The MUHI immunization procedure, developed with hCG as antigen (see below), has since been applied with equal success in hybridization experiments with many different purified proteins in our laboratory as well as in others.

By size analysis of the spleen cells, we were able to measure directly the increase of the numbers of blasts cells in the spleen of an immunized mouse over the background levels[4] (Table I; see also Fig. 3). Furthermore, this increase correlates linearly with specific efficiency (see Fig. 3)

[4] C. Stähli, T. Staehelin, V. Miggiano, J. Schmidt, and P. Häring, *J. Immunol. Methods* **32**, 297 (1980).

[5] A. Fagraeus, Thesis, pp. 51–53, State Bacteriological Laboratory, Stockholm, Sweden.

TABLE I

IMMUNIZATION SCHEDULES AND THEIR EFFECTS ON FREQUENCY OF STIMULATED
SPLEEN CELLS AND ON FREQUENCY OF SPECIFIC HYBRIDOMAS[a]

	Fusion:	H-9	H-10	H-11	H-12	H-14
	Mouse number:	407	412	414	413	460
Preimmunizations[b]						
1		50	10	10	10	0.5
2		50	10	10	10	5
3		10	10	10	10	50
Final immunizations[b]						
Day −7 (i.p.)		1	1	1	—	1
Day −4 (i.p.)		200	400[c]	50	200	200
Day −3 (i.p. and i.v.)		400	—	—	400	400
Day −2 (i.p. and i.v.)		400	—	—	400	400
Day −1 (i.p.)		400	—	—	400	400
Size 71–99 cells[d] above background[e]		1.1%	0.2%	0%	0.5%	0.5%
Wells with growth[f]		40	46	46	46	39
Clones per well[g]		1.8	3.2	3.2	3.2	1.7
Positive wells[f]		25	10	1	21	11
Positive clones per well[g]		0.73	0.24	0.02	0.58	0.26
Specific efficiency[g]		41%	7.5%	0.6%	18%	15%

[a] Fusion and cell culture were carried out according to standard procedure (Stahli et al.[4]).

[b] Immunizing doses are all expressed as micrograms of human chorionic gonadotropin.

[c] I.p. + i.v.

[d] Fraction of cells in channels 71–99 as percentages of nucleated cells (channels 14–99).

[e] Background: 0.3% (see Fig. 3).

[f] Out of a total of 48 wells.

[g] From the fraction of wells with growth and the fraction of positive wells (out of 48), the mean numbers of clones and of positive clones per well, respectively, were calculated by the use of Poisson statistics.

of the subsequent hybridization experiment in agreement with the hypothesis that the fusing cells are recruited from the pool of blast cells. Cell size analysis by Coulter counter has since been used as a routine method to monitor the effectiveness of a final immunization for a hybridization experiment. In the case of an inadequate B-cell response it would allow one to restart the experiment by using another mouse and possibly another immunization procedure and would save one the effort of culturing and screening only to find no specific hybridomas.

Although we have carried out a few hybridization experiments with antigenic preparations in which the protein of interest was <50% pure, we now consider effort better expended on further antigen purification than on searching for a few specific hybridomas among the (potentially) large number of those secreting antibodies against contaminating proteins. A typical hybridization experiment carried out according to the methods described later requires at least 100 μg, if possible, a few milligrams, of protein antigen. If it is impractical to purify such large amounts, it is still helpful to purify 20–50 μg (e.g., by polyacrylamide gel electrophoresis[6]) to be used in preimmunizations and assays, and the final immunization (MUHI), which requires larger amounts, is carried out with partially purified antigen.[6]

Nonimmunogenic peptides (e.g., the 28 amino acid thymosin α_1 peptide, see below) can be coupled to carriers such as keyhole limpet hemocyanin (KLH) or BSA. The major immune response may then be directed against the carrier. Again one may then have to search for very few peptide specific hybridomas among carrier specific ones, and in some cases none may be found (see below). We have developed some strategies to improve the frequency of peptide specific hybridomas.

Our standard procedure before fusion consists of the following steps, which will be described in detail: preimmunization (2–3 injections); determination of serum titer and establishment of test for screening hybridomas; final immunization; spleen cell size analysis (optional).

Preimmunization

Mice (usually BALB/cJ) are injected with 10–100 μg of protein (per immunization) in adjuvants. In the first injection, 250 μl of antigen [consisting of an emulsion,[7] prepared from 125 μl of antigen in aqueous solution supplemented with 10^9 Bordetella pertussis bacteria (Eidg. Serum und Impfinstitut, Bern, Switzerland) and 125 μl of complete Freund's adjuvant (Difco Laboratories, Detroit, Michigan)] are injected in five equal portions (5 × 50 μl) in four subcutaneous sites on the ventral surface and intraperitoneally (i.p.). Between immunizations we routinely wait at least 4 weeks, since too rapid successions of immunization may result in lower antibody titer and affinity.[8] In the second and a possible

[6] T. Staehelin, B. Durrer, J. Schmidt, B. Takács, J. Stocker, V. Miggiano, C. Stähli, M. Rubinstein, W. P. Levy, R. Hershberg, and S. Pestka, Proc. Natl. Acad. Sci. U.S.A. 78, 1848 (1981).

[7] B. A. L. Hurn and S. M. Chantler, this series, Vol. 70 [5].

[8] W. J. Herbert, in "Handbook of Experimental Immunology" (D. M. Weir, ed.), 3rd ed., Vol. 3. Blackwell, Oxford, 1978.

third immunization, the same amount of antigen as in the first immunization is injected i.p. in 100 μl of emulsion prepared from equal volumes of protein in aqueous solution and incomplete Freund's adjuvant (Difco). We routinely immunize groups of 3–5 mice and have found marked variability in response among individual mice.

Determination of Serum Titer

We use a solid-phase antibody binding assay (SABA). It has provided a successful assay system for all proteins and peptides we have tried, and it is in addition well suited for screening hybridoma supernatants. Because antibodies have several combining sites, which in this test can attach simultaneously to antigens adsorbed to the plastic surface,[4] SABA measures antibody concentrations independently of affinity down to low affinities (at least to 10^6 M^{-1}).

Preparation of Plates. SABA is carried out in 96-well flexible plastic U-shaped microtiter plates (Dynatech Laboratories, Alexandria, Virginia). Antigen is adsorbed to the surface of wells by placing 50 μl of antigen solution containing per milliliter 1–10 μg of pure antigen in PBS in each well for at least 2 hr at room temperature (RT). Most proteins and peptides adsorb easily and bind firmly, but a few, such as carcinoembryonic antigen (CEA) or prostatic acid phosphatase (PAP) do not. They may be adsorbed into wells that have been precoated with nitrocellulose in the following way. Four nitrocellulose filters (Millipore) about 5 cm in diameter are dissolved in 10 ml of acetone. This solution is then placed into a well, using a Pasteur pipette, and withdrawn again immediately and completely to avoid dissolving the plate. After evaporation of the remaining fluid, a film of nitrocellulose forms on the surface. Antigen can then be added as above. Remaining protein binding sites on the plate are saturated by a second 2-hr period at RT in which the wells are completely filled with 3% BSA. Control wells are coated with BSA alone to determine unspecific antibody binding. These plates, when wells are filled with PBS + 0.1% sodium azide, may be stored at 4° for months.

Assay. SABA is carried out by placing 50 μl of a test solution (antiserum diluted in 0.5% BSA, hybridoma supernatants) into duplicate wells. Unspecific binding is determined in wells coated with BSA only. The plates are held overnight at RT. The fluid is then removed from the plate by vigorous flicking, and the plate is washed by pouring PBS over it and flicking again. The wells are then incubated for 4–5 hr with 50 μl of ^{125}I-labeled sheep-anti-mouse-Ig antibodies (sαm) containing ~50,000 cpm. This sαm is prepared by iodination of 0.5 mg of affinity-purified antibodies with 2 mCi by the chloramine-T method.[9] Finally, after aspiration of sαm, plates are washed 3 times, cut, and counted.

Instead of ^{125}I-labeled sαm, an enzyme-labeled second antibody recognizing mouse Ig may be used. Such anti-mouse-Ig antibodies coupled with horseradish peroxidase or alkaline phosphatase can be obtained from many suppliers (e.g., New England Nuclear, Boston, Massachusetts). Wells are incubated with this antibody for 4–5 hr at RT as with the iodine-labeled sαm (see above). After washing 3 times with PBS, wells are further incubated for 0.5 hr at RT with substrate (provided by the same suppliers) and stopped with 4 N HCl. A qualitative reading may be taken by eye. A quantitative determination requires reading absorbances in a spectrophotometer or by an enzyme-linked immunosorbent assay (ELISA) reader. Besides the obvious advantages of enzyme labels compared to radioactivity (stable, nonradioactive reagent which can easily be purchased; no cutting up of plates), these also have a number of disadvantages (smaller measuring range; lower sensitivity; additional incubation step with substrate; carcinogenicity of substrates).

Figure 1 shows titrations in SABA of sera of mice immunized with hCG. The titers of these sera (inverse of serum dilution at which signal decreases by half) are around 10^4. Table II shows similar titers of mice immunized with thymosin α_1 (see below). In spite of similar, good titers of these mice immunized with hCG (Fig. 1) and thymosin α_1 (Table II), subsequent fusions varied greatly in SE, some yielding no single specific hybridoma. Thus, a good serum titer after the preimmunizations alone is not a sufficient precondition for a high specific efficiency (Tables I and II). A high SE depends even more on an adequate final immunization (see below). However, we consider it to be a desirable precondition, and we usually try to reach good titers with additional preimmunizations if necessary, since low titers must likely reflect low numbers of memory cells.

Specific Efficiency (SE) and Counting of Hybridomas

A hybridization experiment is successful if one obtains many hybrids and if a high proportion of these hybrids secrete specific antibodies. The methods described in this chapter have a bearing only on the proportion of specific hybrids, on specific efficiency (SE). The numbers of all and of those of the specific hybridomas, respectively, can be counted directly as long as no culture wells contain more than 1 hybridoma. This is the case

[9] H. C. Greenwood, W. M. Hunter, and J. S. Clover, *Biochem, J.* **89,** 114 (1963).

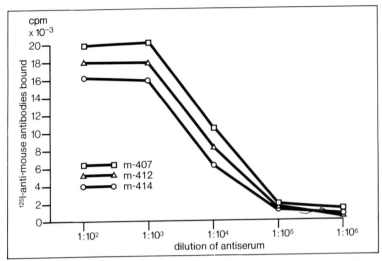

Fɪɢ. 1. Titrations in solid-phase antibody binding assay (SABA) of the sera of three mice preimmunized three times with human chorionic gonadotropin as shown in Table I. The mice were bled 14 days after the third immunization. SABA was carried out as described in the text. In the final incubation, 50 μl of sheep anti-mouse-Ig antibodies containing 66,000 cpm were added to each well.

as long as the proportion of culture wells with growth does not exceed ~20%. In this case, this proportion is equal to the average number of hybridomas per well. Above ~0.2 the proportion of culture wells with growth becomes significantly lower than the average number of hybridomas per well, because of increasing numbers of wells with more than 1 hybridoma. According to Poisson statistics, at average numbers of hybridomas per well of 0.3, 0.5, 0.7, 1, 2, 3, and 4, the proportion of wells with growth is 26%, 39%, 50%, 63%, 86%, 95%, and 98%, respectively. Conversely, by this correlation from the percentage of culture wells with growth (or of wells containing specific hybridomas), the average numbers per well can be approximately determined. For instance, if 85% of wells contain hybridomas (i.e., average of 2 hybridomas per well), and if about half of these wells, e.g., 39%, contain specific hybridomas (i.e., average of 0.5 specific hybridomas per well), SE is 0.25 (not 0.5). This is possible only as long as there are some wells without growth. For this and other reasons (with polyclonal wells there is danger of overgrowth of specific hybridomas by unspecific ones and immediate need for cloning), we try to distribute the product of fusion at a dilution that yields only ~60% of wells with growth (average of ~1 hybridoma per well).

TABLE II
IMMUNIZATION SCHEDULES AND SPECIFIC EFFICIENCIES IN A
PEPTIDE-CARRIER SYSTEM[a]

	Fusion: Mouse number:	T-1 953	T-9 954 + 955	T-10 966
Preimmunizations				
Antigen (μg, type)		160, K	160, K	160, B
Age (weeks)		5, 13	5, 12, 34	5, 11
Serum titer (SABA)		3×10^4	10^4	3×10^4
Final Immunizations: antigen (μg, type)				
Day -4 only		—	200, K	20, B
Days $-4, -3, -2$		200, K	—	160, K
Age (weeks)		21	60	31
No. of hybridomas tested		300	160	71
No. of specific hybridomas		0	5	21
Specific efficiency		<1%	3%	25%

[a] Mice were immunized with the antigens K (KLH-thymosin α_1) and B (BSA-thymosin) polymerized with glutaraldehyde. In the course of preimmunization the same amount and type of antigen was injected twice (3 times for m-954 and m-955) at the indicated ages of the mice. After the last preimmunizatin, the serum titer was determined. In the final immunizations in the last 4 days before fusion, antigen was injected according to the multiple high-dose immunization procedure or as a single high dose (see text). Fusions were carried out as described elsewhere.[4]

Final Immunizations

Final immunizations during the last days before fusion are optimally carried out according to the MUHI protocol, provided the required large amounts of antigen ($\geq$600 μg) are available. Antigen (200–400 μg in 200 μl of 0.9% NaCl) is injected on each of three successive days beginning i.v. at day -4 before fusion and i.p. on days -3 and -2. Occasionally mice die after the second or third immunization. If antigen is limiting, it is given in a single i.v. injection on day -3, however at a significant cost to specific efficiency (Table I). The SE is also very strongly related to the antigen dose of the injection (Table I). Thus, a final immunization with 50 μg of hCG, a dose perfectly adequate for obtaining good serum titers, yields an SE that is one to two orders of magnitude lower than the SE after a high dose immunization or after MUHI (Table I). Finally, as mentioned above, SE is very low when antigen is given in adjuvants for the final immunization before fusion. Thus, we have never found a specific hybridoma after a final immunization with adjuvants in a number of experiments with hCG or CEA.

A Peptide-Carrier System

Low molecular weight peptides are usually not immunogenic. Thymosin α_1[10] is a 28 amino acid peptide produced by the thymus. To make it immunogenic, 5 mg of thymosin α_1 were attached to 25 mg of either KLH or BSA (Roche, Nutley, New Jersey). The immunogenicity in mice of BSA-thymosin α_1, which was still poor, was further enhanced by gentle polymerization with glutaraldehyde: 2 mg of BSA-thymosin α_1 per milliliter were supplemented with glutaraldehyde to 0.04% and left at room temperature for over 6 hr (no visible precipitate was observed). Preimmunizations with these two carrier-peptide preparations gave good serum titers (Table II). Final immunization according to the MUHI protocol resulted, however, in an unexpected failure to obtain a single thymosin α_1 specific hybridoma (fusion 1, Table II). We then speculated that, if final MUHI were carried out with the same carrier–peptide combination that was used in the preimmunizations, such a vigorous response against the very immunogenic carrier would be stimulated as to interfere with the response to the peptide. Accordingly, the problem was solved by reducing the final immunization to a single injection (fusion 9, Table II) or, more effectively, by switching the carrier [fusion 10, Table II; note that a *small* amount of the original carrier (BSA) was also injected once in the final immunization with the intention to restimulate T-cell help].

Cell Size Analysis

Under the microscope, blast cells can easily be distinguished from normal lymphocytes by their size,[4] and their increase in numbers after MUHI over those in an unimmunized mouse is striking. A more quantitative measure of this increase can be obtained by size analysis in a Coulter counter equipped with a Channelizer (Coulter Electronics, Harpenden, U. K.). "Amplification" and "Aperture-Current" are both set to 1 on the Coulter counter; "Base-Channel-Threshold" to 2 and "Window-Width" to 100 on the Channelizer. Cells are scaled into 100 channels as shown in Fig. 2. After clumps have been disrupted by vigorous pipetting with a 100-μl Eppendorf pipette, cells are diluted to about 30,000 cells in 30 ml to avoid coincidences. Small lymphocytes are scaled into channels 10–15 to 30–40; plasma cells into 20–30 to 40–60; and blasts into 40–60 to >100 (Fig. 2).[4] The increase in the ratio of cells in channels 71–99 (uncontaminated blasts) to all cells excluding red blood cells (= cells in channels

[10] T. L. K. Low, G. B. Thurmann, M. McAdoo, J. McClure, J. L. Rossio, P. H. Naylor, and A. L. Goldstein, *J. Biol. Chem.* **254**, 981 (1979).

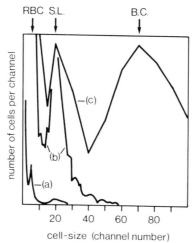

FIG. 2. Cell-size analysis of spleen cells from an unimmunized mouse (curves a and b) and of mitogen-stimulated spleen cells (curve c). Full scales are 10,000 cells (a and c) and 400 cells (b) per channel. Mitogen stimulation: 10^6 spleen cells were incubated for 4 days in 1 ml of RPMI-1640 medium containing 50 μg of lipopolysaccharide and 5×10^{-5} M 2-mercapto-ethanol; curve (c) was copied pointwise from a figure with a different scale of the x axis. Red blood cells (RBC), small lymphocytes (S.L.), and blast cells (B.C) are indicated by arrows.

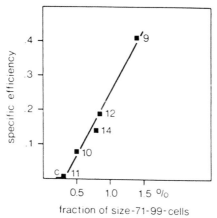

FIG. 3. Plot of specific efficiency vs fraction of large spleen cells of sizes 71–99 after final immunization with human chorionic gonadotropin. Points are labeled with the numbers of the respective fusion experiments (see Table I). The bar (c) spans the range of cell sizes of four unimmunized control mice.

14–99) is a good measure of the increase in blasts after the final immunization and correlated precisely with SE of the subsequent fusion (Fig. 3).[4] We have routinely based our decision to fuse on this size analysis. However, when we employed MUHI, we never had to abort a fusion.

Acknowledgment

We gratefully acknowledge the help of Dr. J. W. Stocker in revising the manuscript.

[4] Estimation of the Number of Monoclonal Hybridomas in a Cell-Fusion Experiment

By Angel L. De Blas, Makarand V. Ratnaparkhi, and James E. Mosimann

Köhler and Milstein[1] developed the techniques for the generation of hybridomas secreting monoclonal antibodies. Today many investigators have incorporated these techniques in their respective areas. The homogeneity (monoclonality) of the cells growing in each hybridoma culture becomes very important in order successfully to recover and maintain the hybridomas of interest. In polyclonal cultures the clone of interest may be overgrown by others. Thus, soon after the fusion the investigator has to pursue an optimal experimental strategy aiming at the safe recovery of the hybridomas of interest. Therefore it is important at that point to have an estimate of the number of the hybridoma cultures that are monoclonal.

Here we present a simple method for the estimation of the proportion of the cultures with dividing cells that are monoclonal. This method is applicable after the population of fused cells has been fractionated into a large number of cultures by limiting dilution.[2,3] Our method is based on the Poisson probability model and assumes that the only information available to the investigator is the number of culture wells with dividing hybridoma cells of the total number of wells planted.

[1] G. Köhler and C. Milstein, *Nature (London)* **256**, 495 (1975).
[2] G. Galfrè and C. Milstein, this series, Vol. 73, p. 3.
[3] A. L. De Blas, M. V. Ratnaparkhi, and J. E. Mosimann, *J. Immunol. Methods* **45**, 109 (1981).

Procedure

The clonal growth in a culture well depends on the number of hybrid cells surviving in that dish. Therefore, it is appropriate to consider a probability model for the number of surviving hybridomas in a culture well and use the same for the estimation procedure. Such a probability model, owing to the typical nature of the available date (only a single observation representing the number of wells with and without hybridoma growth is available) has to be based on the knowledge of the experimental units (hybrid cells in a suspension and their survival process). In what follows, we consider the Poisson distribution as a possible model and use it for the estimation of the number of monoclonal cultures.

Distribution of Surviving Hybridomas in Culture. For each culture well planted with an aliquot of the original volume of the cell suspension, we assume the points below.

1. The total number (X) of hybrid cells initially present has a Poisson distribution with parameter α. After the fusion, the cells must be dispersed (this should be verified by microscopical observation) and randomly distributed in the original cell suspension by thorough mixing.
2. The surviving process of the hybrid cells follows a binomial distribution, i.e., the number of surviving cells (Y), given that $X = x$ cells are plated, has the binomial distribution with parameters x and π, where π is the survival probability of each hybrid cell.

These assumptions appear to be valid for bacterial suspensions and their survival.[4,5] Under these assumptions, the marginal distribution of the number of surviving cells (Y) is Poisson with parameter $\lambda = \alpha\pi$:
Since

$$P_Y(y) = \sum_{x=y}^{\infty} P_X(x) \, P_{Y|X}(Y|x)$$

substituting for $P_X(x)$ and $P_{Y|X}(y|x)$, we have

$$P_Y(y) = \sum_{x=y}^{\infty} \frac{e^{-\alpha}\alpha^x}{x!} \binom{x}{y} \pi^y(1 - \pi)^{x-y}$$

which after simplification becomes

$$P_Y(y) = \frac{(\alpha\pi)^y e^{-\alpha\pi}}{y!} = \frac{\lambda^y e^{-\lambda}}{y!}$$

Hence the result.

[4] F. M. Wadley, *Ann. Appl. Biol.* **36**, 196 (1949).
[5] D. J. Finney, "Statistical Methods in Biological Assay." Griffin, London, 1964.

Estimation of the Number of Monoclonal Hybridomas in a Fusion Experiment. A monoclonal culture is the result of a single surviving cell, i.e., $Y = 1$. With our model, the probability of monoclonal hybridoma cultures among the culture wells that exhibit growth is

$$P_1 = \text{Prob } (Y = 1|\text{growth}) = e^{-\lambda}\lambda/(1 - e^{-\lambda})$$

and the required estimate of P_1 is

$$\hat{P}_1 = \frac{(1 - u) \, [-\ln(1 - u)]}{u}, \qquad 0 < u < 1$$

where u = (number of culture wells with growth)/(total number of culture wells planted), is an estimate of $(1 - e^{-\lambda})$, the probability of growth.

Similarly, the respective estimates of $\hat{P}_2$, $\hat{P}_3$, and $\hat{P}_4$ of the proportion of the surviving hybridoma cultures containing two, three, and four different clones are

$$\hat{P}_i = \frac{(1 - u) \, [-\ln(1 - u)]^i}{i! \, u} \tag{1}$$

where $i = 1, 2, 3,$ or 4 and ln denotes a natural logarithm.

A graphical representation of Eq. (1) for $\hat{P}_i$, $i = 1, 2, 3,$ or 4, is given in Fig. 1. These graphs provide the necessary estimates, directly from the observed values of u.

The following are two examples where Fig. 1 and Eq. (1) are applied. If dividing hybridoma cells are present in 600 wells of 700 originally planted, then the expected percentage of the hybridoma cultures contain-

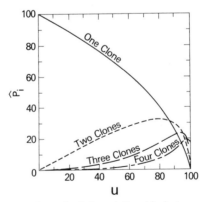

FIG. 1. Computer-generated graph of the relationship between $\hat{P}_i$, the estimate of the percentage of hybridoma cultures with i growing clones ($i = 1, 2, 3,$ or 4) and u, the percentage of wells with dividing hybridoma cells. This figure is based on the Poisson model.

ing one, two, three, or four clones would be 32.4, 31.6, 20.5, and 10.0, respectively. Subsequently the expected number of wells with one, two, three, or four different clones, respectively, would be 195, 189, 123, and 60, respectively. However, if only 100 of the 700 wells planted show growth, then the expected percentage for wells with one, two, three, or four clones would be 92.5, 7.1, 0.4, and 0.0. Thus about 93 of the 100 cultures are expected to be monoclonal, the remaining seven containing two clones each.

The practical consequences of Eq. (1) go beyond the estimation of the monoclonality of the hybridoma cultures. Further, using the above model we have shown that the probability of hybridoma survival decreases with postfusion cell dilution even in the presence of a constant number of feeder cells.[3]

[5] Use of Human Endothelial Culture Supernatant (HECS) as a Growth Factor for Hybridomas

By GIULIA C. B. ASTALDI

The growth of hybridoma cells is improved by the presence of feeder cells, particularly when the hybridoma cells are not yet adapted to the culture conditions. This is especially so for cultures at low numbers of cells.[1-4] We have previously reported that the human endothelial culture supernatant (HECS) contains factor(s) that promote the growth and the stability of hybridoma cells and in these respects is superior to the commonly used feeder cells.[5-7]

[1] G. Köhler and C. Milstein, *Eur. J. Immunol.* **6**, 511 (1976).

[2] G. J. Hämmerling, H. Lemke, U. Naüerling, C. Höhmann, R. Wallich, and K. Rajewsky, *Curr. Top. Microbiol. Immunol.* **81**, 100 (1978).

[3] W. Lernhardt, J. Andersen, A. Coutinho, and F. Melchers, *Cell Res.* **111**, 309 (1978).

[4] S. Fazekas de St. Groth and D. Scheidegger, *J. Immunol. Methods* **35**, 1 (1980).

[5] G. C. B. Astaldi, M. C. Janssen, P. M. Lansdorp, C. Willems, W. P. Zeijlemaker, and F. Oosterhof, *J. Immunol.* **125**, 1411 (1980).

[6] G. C. B. Astaldi, M. C. Janssen, P. M. Lansdorp, W. P. Zeijlemaker, and C. Willems, *Protides Biol. Fluids* **28**, 443 (1980).

[7] G. C. B. Astaldi, M. C. Janssen, P. M. Lansdorp, W. P. Zeijlemaker, and C. Willems, *J. Immunol.* **126**, 1170 (1981).

METHODS IN ENZYMOLOGY, VOL. 92

Culture Media

Human endothelial cells (HEC) were cultured at 37° in an atmosphere of 5% CO_2 in endothelial cell growth medium (ECGM). This consisted of: 35% HEPES-buffered RPMI-1640 medium (Gibco, New York), 35% medium 199 containing Hank's salts (Gibco), 30% pooled human serum (obtained from 12–18 donors and tested for its capacity to support the growth of HEC), 2 mM L-glutamine, 100 U of penicillin per milliliter, and 100 μg of streptomycin and 2.5 μg of fungizone per milliliter. In our laboratory, the combination of cultured media enhanced endothelial cell growth more strongly than either RPMI-1640 or medium 199 alone (C. Willems, personal communication). Hybridoma cells were cultured at 37° in an atmosphere of 6–7% CO_2 in hybridoma cell culture medium (HCCM). This consisted of HEPES-buffered RPMI-1640 medium (Flow Laboratories, Irvine, Scotland), 10% fetal calf serum (FCS; Seralab., Crawley Down, Sussex, U.K.), 4 mM L-glutamine, 2 mM sodium pyruvate (Gibco), 2.5 μg of fungizone per milliliter (Squibb, Rijswijk, The Netherlands), 50 μg of gentamycin per milliliter (Schering Corp., Bloomfield, New Jersey) and 5 × 10^{-5} M mercaptoethanol.

Isolation and Culture of Human Endothelial Cells

Human endothelial cells were isolated from umbilical cord veins (within 24 hr after delivery) by the method of Jaffé et al.[8] with minor modifications.[5,9] Umbilical cord veins were cannulated and washed with phosphate buffer (138 mM NaCl, 4.1 mM KCl, 0.5 mM $Na_2HPO_4 \cdot 7 H_2O$, 0.15 mM KH_2PO_4, 11.1 mM glucose, pH 7.4) to remove traces of blood. Subsequently, the vessel was filled with 0.25% trypsin-EDTA (Gibco) dissolved in phosphate buffer. The cannulas were clamped at both ends and incubated for 15 min at 37°. Next, the incubation medium was collected and the vessel was filled again with phosphate buffer. After gentle traumatization of the vessel walls, the buffer solution was collected and pooled with the first effluent (both effluents contained endothelial cells). To this cell suspension, which generally consisted of 1 × 10^6 endothelial cells, 5 ml of ECGM was added to inhibit the trypsin activity. The cells were centrifuged for 5 min at 200 g, resuspended in 10 ml of ECGM and inoculated into 75-ml tissue culture flasks that had been precoated with

[8] E. A. Jaffé, R. L. Nachmann, C. G. Becker, and C. R. Minick, J. Clin, Invest. 53, 2745 (1973).
[9] C. Willems, W. G. van Aken, E. M. Peuscher-Prakke, J. A. van Mourik, C. Dutilh, and F. ten Hoor, J. Mol. Med. 3, 195 (1978).

partially purified fibronectin (this coating procedure enhanced the adhesion of endothelial cells to the flasks). The ECGM was refreshed after 24 hr and again after 3 days. When, after 5–6 days, a monolayer of cells was obtained (1 to 2×10^5 cells/cm^2), the endothelial cells were either maintained in culture (not longer than 10 days, in which the ECGM was refreshed twice) or harvested, by incubation with trypsin-EDTA for 3–5 min at room temperature. After collection and washing as described above, new cultures were set up (split ratio 1 : 4). The cultures could be split 8 to 10 times. They were identified as endothelial cells by the criteria that (a) they showed a density-dependent regulation of growth; (b) the presence of factor VIII-related antigen could be demonstrated by immunofluorescence; and (c) the presence of Weibel–Palade bodies could be confirmed using electron microscopy.[8]

Production of HECS and Assay of Its Activity on Hybridoma Cells

Cultures of HEC were maintained in ECGM as described above. Generally, confluent monolayers of HEC were used for the production of HECS, although the supernatant of semiconfluent monolayers also contained some hybridoma cell growth-promoting activity.[7] Twice weekly the supernatant was collected to be used as HECS and replaced by fresh ECGM. The HECS was filtered through a 0.22 μm Millipore filter to remove any contaminating cellular material and stored at $+4°$ or at $-20°$. The activity of HECS was measured by its ability to increase the incorporation of [^{3}H]thymidine ([^{3}H]TdR) into hybridoma cells. The hybridoma cells were cultured in the continuous presence or absence of HECS in HCCM, in flat-bottom Microtiter tissue culture plates, with an initial concentration of 100 cells/well. Twenty-four hours before harvesting, the cultures were labeled with 0.8 μCi of [^{3}H]TdR (Radiochemical Centre, Amersham, U.K.: specific activity 400 mCi/mmol). Before being used to test HECS activity, the hybridoma cells were weaned from HECS for at least a week.

As shown in Fig. 1, the supernatant of a confluent culture of endothelial cells already secreted detectable growth-promoting activity after 2 hr of culture, and this activity increased linearly with time up to 72 hr.[7]

A confluent monolayer of HEC could be obtained only if the ECGM contained pooled human serum. However, once confluent, the endothelial cell culture could also be maintained in medium containing FCS. Such cultures also showed production of hybridoma growth-promoting activity (HECS). Similarly, endothelial cells in serum-free medium also produced HECS activity. This indicates that the production of the HECS activity

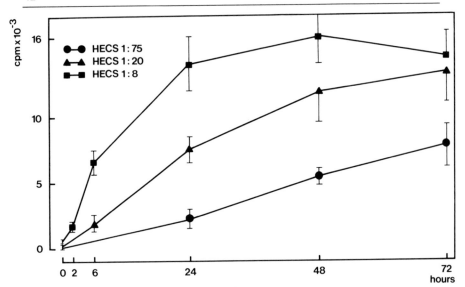

FIG. 1. Time course of production of hybridoma growth-promoting activity by confluent human endothelial cells (human endothelial culture supernatant; HECS) as a function of time. Endothelial cells were washed and incubated at 37° in growth medium. Samples of the supernatant were collected from the culture at times indicated on the abscissa, and the activity was determined by the thymidine incorporation into hybridoma cells, as described in the text. Each point represents the mean counts per minute (cpm) of six determinations ± SD.

did not depend upon interaction of HEC and human serum components. However, serum-free HECS was found to be much more unstable during storage than was HECS produced in the presence of serum.

Effects of HECS on Hybridoma Cells

HECS was found to improve the growth of hybridoma cells at all stages of the hybridoma cell production.[5-7] The details of the methods used in our laboratory to produce hybridoma cells have been reported elsewhere.[5-7] In brief, hybridoma cells were obtained by fusing human or murine lymphocytes with the Sp 2/0-Ag 14 cell line, kindly given to us by Dr. G. Köhler, using the method of Galfrè et al.,[10] with minor modifications.[5,7] After fusion, the cells were dispensed into 96-well Microtiter plates containing different feeder cells or HECS. As shown in Table I,

[10] G. Galfrè, S. C. Howe, C. Milstein, C. W. Butcher, and J. C. Howard, *Nature (London)* **266,** 550 (1977).

TABLE I[5]
Effect of Different Systems on the Recovery of Hybrids after Fusion[a]

Feeder system	Fusion 1	Fusion 2	Fusion 3	Fusion 4	Fusion 5
None					
a	96	96	96	96	96
b	40	28	5	0	15
%	42	29	5	0	16
Mouse spleen cells					
a	96	ND[b]	96	ND	ND
b	42		16		
%	44		17		
Mouse macrophages					
a	96	ND	96	ND	ND
b	44		18		
%	46		19		
HECS 1:5					
a	96	48	96	48	48
b	70	38	48	38	17
%	73	79	50	79	35

[a] Results are expressed as number of wells containing growing hybrid clones (b) over the total number of wells (a) and percentage = $b/a \times 100$. Fusions 1, 2, and 3 were of mouse myeloma cells with mouse spleen cells; fusions 4 and 5 were of mouse myeloma cells with human tonsillar lymphocytes. The same number of cells (1.5×10^5) had been added to each well.

[b] ND, not done.

HECS was superior to feeder cells (commonly used to improve hybridoma cell growth) in increasing the recovery of hybrid cells after fusion. In these experiments, the growth of the hybridoma cells was assessed after 2 weeks of culture. Moreover, HECS could substitute for feeder cells during the cloning of the hybridoma cells, leading to a cloning efficiency approaching 100% (Table II).[5] In contrast, none of the supernatants of the other feeder cells was able to improve the growth of hybridoma cells at the single-cell level.

As mentioned above, the activity of HECS was usually assayed by its capacity to increase incorporation of [³H]thymidine into hybridoma cells. As shown in Fig. 2, spleen cell feeders were found to be less efficient in this system as well. In other experiments, we could show that HECS improved the proliferation of hybridoma cells regardless of their stage of adaptation to culture conditions[5] and at all concentrations of hybridoma cells tested.[7] The effect of HECS on the hybrid cells was found to be dose-dependent (Fig. 1).[7]

TABLE II[5]

EFFECT OF VARIOUS FEEDER SYSTEMS ON THE GROWTH OF HYBRID CELLS AT
THE SINGLE-CELL LEVEL[a]

Feeder system	Number of wells containing hybrid clones[b]	Total number of wells
None	0.4 ± 0.2	36
Mouse spleen cells	12.4 ± 3.6	36
Mouse spleen cells supernatant 1 : 5[c]	0	36
Mouse thymocytes	13.4 ± 3.2	36
Mouse thymocytes supernatant 1 : 5[c]	0	36
Human endothelial cells	21.4 ± 4.2	36
HECS 1 : 5	20.0 ± 2.7	36
ECGM 1 : 5	0.5 ± 0.1	36

[a] One hybrid cell, obtained after fusion of human lymphocytes or mouse spleen cells with mouse myeloma cells, was added to each well at the initiation of the culture. After 12 days, clones of hybrids were visible in those wells where growth was successful.

[b] Results are expressed as means ± SE of the number of wells containing growing hybrid clones, from five different experiments.

[c] One single experiment.

To grow hybridoma cells, we now routinely use HECS at a concentration of 10% (v/v) in HCCM. At this concentration, it is possible to use serum-free HCCM to grow the hybridoma cells, because the serum and the factor(s) present in 10% HECS are sufficient to allow optimal growth (Table III).

TABLE III

EFFECT OF HUMAN ENDOTHELIAL CELL SUPERNATANT (HECS) ON THE GROWTH OF
HYBRIDOMA CELLS IN HCCM WITH OR WITHOUT FCS[a]

% of FCS in HCCM	Percentage of HECS in HCCM (v/v)			
	0	2	10	20
0	131 ± 78	4745 ± 1813	13 589 ± 2049	10 297 ± 2918
5	387 ± 142	15 214 ± 6380	17 032 ± 4581	18 317 ± 3814
10	2683 ± 1544	11 874 ± 1878	13 152 ± 3340	16 334 ± 3433

[a] One hundred hybridoma cells were added to each well at the start of the culture, in hybridoma cell culture medium (HCCM) supplemented with various concentrations of fetal calf serum (FCS) and HECS. Incorporation of [³H]TdR was measured after 6 days of culture. Results are expressed as mean counts per minute ± SD of six determinations.

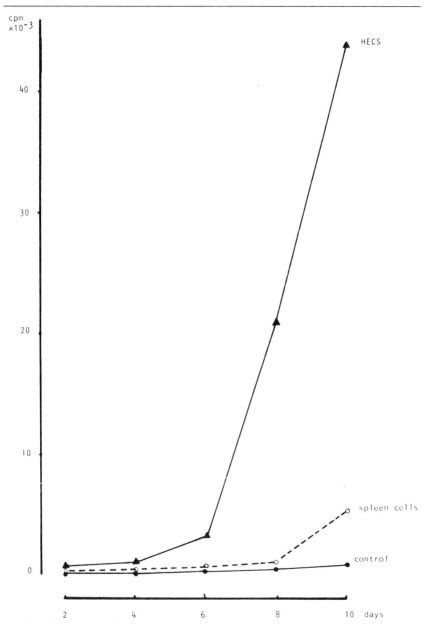

FIG. 2. Effect of HECS on thymidine incorporation into hybridomas. One hundred cells were added to each well at the start of the culture in the presence of HECS or spleen cells as feeders. Each point represents the mean counts per minute (cpm) of six determinations.[6]

It was found that endothelial cells increased the stability of human–mouse hybridoma cells producing human immunoglobulins, as compared with spleen cells as feeders.[5,6] Thus, it was possible to maintain human–mouse hybridoma cells producing anti-Rh[+]-erythrocyte antibodies in culture with HECS for at least 5 months without losing their capacity to produce the antibodies, although the proportion of clones producing the antibodies gradually decreased.[11,11a]

The HECS was found to be adsorbed by hybridoma cells but not by thymocytes[7]; furthermore, hybridoma cells that had been in contact with HECS continued to proliferate even when HECS was removed.[7] These findings suggest that the action of HECS on hybridoma cells involves binding to the membrane and delivery of a growth-promoting signal.[7]

Other Factors Promoting the Growth of Hybridoma Cells

Because the supernatant of endothelial cells was found to contain bone-marrow colony-stimulating factor(s) (CSF),[12] we investigated whether CSF from other sources would stimulate the growth of our hybridoma cells. Of these, giant cell tumor supernatant (Gibco) was ineffective, but conditioned medium from human placenta[13] as well as the supernatant of mouse peritoneal macrophages[14] did indeed promote growth, although to a lesser extent than did HECS at all the different concentrations tested. Furthermore, a lectin-free, partially purified preparation of human T-cell growth factor (TCGF), obtained from the supernatant of concanavalin A-stimulated human lymphocytes, was tested. This factor stimulates the growth of mature T and B lymphocytes[15] and was also found to have activity in bone-marrow cultures (R. du Bois, personal communication). The TCGF did show some growth-promoting activity on the hybridoma cells but, again, it was less efficient than HECS.[16] Conversely, HECS had no effect at all on the growth of T lymphocytes (L. Aarden, personal communication). Therefore, it seems likely that the HECS activity is distinct from that of TCGF. To establish the nature of the factor(s) in HECS responsible for its effect on hybridoma cell growth, further investigation of its chemical properties is required.

[11] G. C. B. Astaldi, C. Willems, W. Alleyne, P. M. Lansdorp, W. P. Zeijlemaker, and M. C. Janssen, *Protides Biol. Fluids.* **29,** 609 (1981).

[11a] G. C. B. Astaldi, E. P. Wright, C. Willems, W. P. Zeijlemaker, and M. C. Janssen, *J. Immunol.* **128,** 2539 (1982).

[12] P. J. Quesenberry and M. A. Gimbone, Jr., *Blood* **56,** 1060 (1980).

[13] A. W. Burgess, E. M. A. Wilson, and D. Metcalf, *Blood* **49,** 573 (1977).

[14] A. C. Eaves and W. R. Bruce, *Cell Tissue Kinet.* **7,** 19 (1974).

[15] G. Möller, *Immunol. Rev.* **51,** (1980).

[16] G. C. B. Astaldi, M. C. Janssen, C. Willems, W. P. Zeijlemaker, and P. M. Lansdorp, *Proc. Int. Leukocyte Culture Conf.* **14,** 650 (1981).

Properties of HECS

The active principle(s) of unpurified HECS was found to be stable over time upon storage at +4° or at −20°, either as liquid or a lyophilized powder, for a period of at least 4 months. However, if no serum was present (either after production in serum-free medium or after gel filtration), the activity was very unstable, being completely abolished after 1 week at +4° or −20°. Such preparations were, however, stable in the presence of 0.01% polyethylene glycol 4000 (PEG) when stored at −196°. In the presence of serum, HECS was stable at pH levels ranging from 5 to 10 for at least 1 hr at 37° and for 1 month at +4°. The HECS activity was resistant to treatment at 56° for 30 min, but was abolished after incubation for 2 min at 80° or after treatment with perchloric acid or periodate. The HECS activity was precipitated with 80% ammonium sulfate, but not with 50%. The active principle(s) of HECS were found to bind to DEAE-cellulose, but not to CM-cellulose, at pH 7.2. Upon gel filtration on an ACA-54 column, the activity of HECS was eluted as a single peak corresponding to an apparent molecular weight of about 33,000.[17]

Acknowledgments

I am grateful to Drs. Charles Willems and Jan van Mourik of the Department of Blood Coagulation of the Central Laboratory of the Netherlands Red Cross Blood Transfusion Service, Amsterdam, for their expertise in endothelial cell culture and critical discussion of this work.

[17] G. C. B. Astaldi et al., manuscript in preparation.

[6] Production, Purification, and Characterization of Antigen-Specific Murine Monoclonal Antibodies of IgE Class

By Amar S. Tung

Since the discovery of immunoglobulin E (IgE) in humans by Ishizaka,[1] significant progress has been made in the understanding of immediate hypersensitivity reactions mediated by reaginic antibodies. Unlike most other immunoglobulin classes, IgE is normally found in extremely low concentrations in serum. Hence the purification and characterization of antibodies of this class of immunoglobulin has been extremely difficult.

[1] K. Ishizaka, T. Ishizaka, and M. M. Hornbrook, J. Immunol. 97, 75 (1966).

The availability of a few IgE-producing myelomas of human[2] and rat[3] origin has contributed significantly to our current knowledge of physical and biochemical properties and effector functions of IgE; however, none of the available IgE myeloma proteins possess a known antigenic specificity. Therefore in all the studies with myeloma IgE for triggering of mast cell or basophil degranulation leading to mediator release, aggregation of surface-bound IgE was achieved by anti-IgE antibodies rather than by the antigen that is the physiological signal for triggering. Studies of murine IgE have therefore been limited owing to the lack of availability of IgE-producing mouse myelomas.

The development of somatic cell fusion techniques to immortalize antibody-producing lymphocytes by Köhler and Milstein,[4] has resulted in the production of IgE-producing hybridomas against specific protein and hapten determinants. These hybridomas include IgE antibodies with specificity for ovalbumin,[5] dinitrophenol (DNP),[6,7] and trinitrophenol (TNP).[8] In experiments where splenic lymphocytes from conventionally immunized mice have been fused with drug-sensitive mouse myeloma cells, IgE-producing hybridomas have not been detected whereas hybridomas making antibodies of other immunoglobulin classes do appear.

The IgE antibody response displays certain unique characteristics.[9,10] For example, it is highly antigen and adjuvant dependent, is of lower magnitude, is transient in nature, and in normal circumstances is localized in the respiratory and gastrointestinal mucosa as well as in the regional lymph nodes. This is also due to the fact that the frequency of IgE-producing B lymphocytes is very small both in the spleen and lymph nodes of both high- and low-responder inbred mice.[11] Therefore a different method must be employed to immunize mice for enrichment of IgE-producing B lymphocytes suitable for generating antibody-producing hybridomas.

Various aspects of monoclonal antibodies have been reviewed elsewhere.[12] The primary objective here is to discuss the practical aspects of

[2] S. G. O. Johansson and H. Bennich, *Immunology* **13,** 381 (1967).

[3] H. Bazin, A. Beckers, C. Deckers, and M. Moriame, *J. Natl. Cancer Inst.* **51,** 1359 (1973).

[4] G. Köhler and C. Milstein, *Nature (London)* **256,** 495 (1975).

[5] I. Bottcher, G. Hämmerling, and J.-F. Kapp, *Nature (London)* **275,** 761 (1978).

[6] Z. Eshhar, M. Ofarim, and T. Waks, *J. Immunol.* **124,** 775 (1980).

[7] F.-T. Liu, J. W. Bohn, E. L. Ferry, H. Yamamoto, C. A. Molinaro, L. A. Sherman, N. R. Klinman, and D. H. Katz, *J. Immunol.* **124,** 2728 (1980).

[8] A. K. Rudolph, P. D. Burrows, and M. R. Wabl, *Eur. J. Immunol.* **11,** 527 (1981).

[9] *Immunological Reviews*, Vol. 41 (G. Möller, ed.). Munksgaard, Copenhagen, 1978.

[10] D. H. Katz, "Lymphocyte Differentiation, Recognition, and Regulation." Academic Press, New York, 1977.

[11] J. M. Teale, F.-T. Liu, and D. H. Katz, *J. Exp. Med.* **153,** 783 (1981).

[12] A. S. Tung, *Annu. Rep. Med. Chem.* **16,** 243 (1981).

raising antigen-specific monoclonal IgE antibody, its purification and characterization, and the production of rabbit antibodies specific for epsilon (ε) heavy chain.

Selection of Mice

Mice of an inbred strain that are known to be high responders for the desired immunogen should be chosen. Mice of the A, BALB/c, and CAF$_1$ strains are suitable for generating IgE-producing hybridomas as well as hybridomas of other immunoglobulin classes against classical protein antigens, e.g., keyhole limpet hemocyanin (KLH), *Limulus polyphemus* hemocyanin (Hy), ovalbumin and their dinitrophenol conjugates made by reacting fluorodinitrobenzene (FDNB) or dinitrobenzenesulfonic acid by standard methods.[13]

Choice of Immunogen

Any one of the following immunogens can be used to raise antigen-specific monoclonal antibodies of IgE class (optimal doses for eliciting IgE antibody response for each immunogen are given in parentheses): DNP$_{2.1}$-*Ascaris*[7] (10 μg), DNP$_8$-KLH of DNP$_8$-Hy (2 μg), ovalbumin (0.2 μg), KLH or Hy (2 μg).

Subscripts indicate the number of moles of DNP per mole of KLH and Hy; in case of DNP-*Ascaris* it refers to moles of DNP $\times$ 10^{-7}/mg of *Ascaris* protein. It should be borne in mind that heavily substituted hapten-carrier conjugates (large number of hapten molecules per mole of carrier protein) are very poor immunogens in eliciting IgE antibody responses. Large doses of immunogen also induce a poor IgE antibody response. In our studies we have used DNP$_8$-Hy as the immunogen and DNP$_{25}$-BSA as the test antigen for eliciting passive cutaneous anaphylaxis reaction (PCA) and solid-phase radioimmunoassay (RIA).

Choice of Adjuvant and Its Preparation

Complete Freund's adjuvant (CFA) is not a suitable adjuvant for eliciting IgE antibody responses. Aluminum hydroxide gel, commonly known as alum, is a very strong adjuvant for the induction of IgE as well as IgE antibody responses. It is prepared by the method described in Table I.

---

[13] H. N. Eisen, *Methods Med. Res.* **10**, 94 (1964).

TABLE I
PREPARATION OF ALUMINUM HYDROXIDE GEL (ALUM) ADJUVANT

Reagents: A. Aluminum sulfate, $Al_2(SO_4)_3 \cdot 18\ H_2O$
           B. 1 N NaOH
           C. Sodium chloride (crystals)
           D. Distilled water

Reaction: $Al_2(SO_4)_3 + 6\ NaOH \rightarrow 2\ Al\ (OH)_3 + 3\ Na_2SO_4$

Procedure: 1. In a 500-ml graduated cylinder, add 33.3 g of $Al_2(SO_4)_3$ to 250 ml of distilled water. This does not go into solution well but will form a paste.
           2. Slowly add 1 N NaOH solution with continuous stirring, bringing total volume of mixture to 350 ml.
           3. After the contents are completely dissolved and stirred, transfer to a 2-liter graduated cylinder and wash vigorously with 1500 ml of distilled water 6 times, allowing alum to settle out completely each time before aspirating off the supernatant fluid, which contains NaOH and Na_2SO_4. After completion of washes, resuspend to 750 ml in H_2O.
           4. Transfer the suspended alum to a Waring blender and blend for about 3–5 min.
           5. Wash with 1 liter of distilled water and allow to settle completely before aspirating.
           6. Measure the total volume of alum and add enough NaCl to make final suspension in 0.85% NaCl (isotonic).
           7. Lyophilize 1.0 ml of above suspension to determine the dry weight of alum. All dose calculations are made on the basis of dry weight.

Immunization and Bleeding of Mice

The following protocol is designed to give reproducible results in generating IgE-producing hybridomas against a desired antigen. This procedure evolves from the consideration of two important observations made during studies on the regulation of IgE antibody synthesis.

1. Mice of both high- and low-responder phenotype have nonspecific suppressor T (thymus-derived) cells that dampen the IgE antibody response.[14–16]

2. The adoptive secondary IgE antibody responses as determined by PCA are 4- to 8-fold higher when antigen-primed cells are transferred into a syngeneic irradiated host followed by boost immunization with the homologous antigen as compared to a secondary immunization of the original host.[10] The enrichment of antigen-specific reaginic antibody-forming cells in the spleens of these adoptive recipient mice is 100-fold higher than

[14] N. Watanabe, S. Kozima, and Z. Ovary, *J. Exp. Med.* **143,** 833 (1976).
[15] N. Chiorazzi, D. Fox, and D. H. Katz, *J. Immunol.* **117,** 1629 (1976).
[16] A. S. Tung, N. Chiorazzi, and D. H. Katz, *J. Immunol.* **120,** 2050 (1978).

TABLE II

IMMUNIZATION PROTOCOL FOR AMPLIFICATION OF IgE-PRODUCING B CELLS

Day 0: Using a gamma irradiator (Gamma Cell 40, Atomic Energy of Canada) irradiate mice with 250 rads (group I). Immediately thereafter inject 2 μg of DNP$_8$-Hy in 4 mg of aluminum hydroxide gel (alum) i.p. using a 23-gauge ⅝-inch needle in a total volume of 0.5 ml per mouse. Volume is made by addition of normal saline. The recommended procedure is first to dilute stock solution of antigen in saline and then add alum with immediate mixing. Since, with time, alum with adsorbed antigen will settle to the bottom of the vial, mix immediately prior to injecting the recipient mice.

Day 10: Repeat immunization procedure with the same dose of antigen in 2 mg of alum.

Day 24–28: Bleed the mice through the retroorbital venous plexus using Nattleson heparinized capillary tubes (Fisher Scientific) and isolate plasma using a Beckman or Eppendorf microfuge. Test the plasma for IgE titer by passive cutaneous anaphylaxis reaction (PCA). The procedure for PCA appears under assay for IgE positive clones.

Day 35: Boost the mice i.p. with same dose of antigen in saline, total volume of 0.5 ml per mouse.

Day 42–63: Irradiate syngeneic mice with sublethal dose of 700 rads (group II). Four hours later, sacrifice group I mice by cervical dislocation, remove spleens, and make a single-cell suspension in RPMI or MEM containing 10 mM HEPES buffer. Wash the cells once with the same medium, and resuspend the cell pellet in serum-free medium at a cell concentration of 10^8 cells/ml and inject 0.5 ml (5×10^7 cells) i.v. to each of the irradiated recipients (group II). Immediately inject i.p. 2 μg of DNP$_8$-Hy in 2 mg of alum. In working with weak immunogens, individual mice will vary immensely with regard to their serum antibody titer. If this is the case, it is recommended that only spleen cells from those mice (group I) having the highest titer of antibody (as determined by PCA or radioimmunoassay) on day 24 or 28 be adoptively transferred into group II recipients.

The mice should be given antibiotic-containing water for 4 days following treatment or injected with 0.5 ml of 1 mg/ml of gentamycin solution (Schering Corp., Kenilworth, New Jersey) i.p. on days 1 and 2 following cell transfer. After 6–7 days, the recipient mice are bled using Nattleson heparinized capillary tubes, and the plasma is separated as described above. At this time the mice are sacrificed and their spleens are removed for cell fusion.

the original immunized host as determined by heterologous adoptive cutaneous anaphylaxis.[17]

A detailed immunization protocol is described in Table II. In this protocol naturally occurring suppressor T cells are abrogated by low-dose whole-body irradiation prior to the immunization of mice (group I). After an appropriate boosting schedule, spleen cells from these mice are transferred into sublethally irradiated syngeneic recipients (group II), which are then boosted with the homologous antigen in order to amplify IgE-producing B lymphocytes. Cyclophosphamide (Mead Johnson, Evansville, Indiana) treatment of mice 3 days prior to immunization also en-

[17] N. Chiorazzi, Z. Eshhar, and D. H. Katz, *Proc. Natl. Acad. Sci. U.S.A.* **73**, 2091 (1976).

hances IgE antibody responses.[15] Therefore, if irradiation facilities are not available, a cyclophosphamide dose of 50–100 mg per kilogram of body weight should be given intraperitoneally (i.p.) 3 days prior to the first immunization. Both of these maneuvers result in 4- to 8-fold enhancement of IgE antibody titer (PCA) in treated mice over untreated controls.[15]

Generation of Hybridomas

Six or seven days after immunization of adoptive recipients, the mice are sacrificed by cervical dislocation, the spleens are removed aseptically, and a single-cell suspension is prepared using sterile techniques. Normally 15 to 20 $\times$ 10^6 viable cells are obtained from each spleen. The standard procedures for cell fusion have been described.[18,19] A detailed review describing the practical aspects of cell fusion to generate monoclonal antibodies has appeared in this series.[20] The single most important point in cell fusion is the growing stage of myeloma cells. These cells should be in a logarithmic phase of growth at the time of fusion and more than 95% viable.

Several companies now supply prescreened batches of fetal calf serum and polyethylene glycol (PEG), which work effectively in generating somatic cell hybrids (MA Bioproducts, Walkersville, Maryland). In our hands, Fisher PEG-1000 (Fisher Scientific) used at 35% concentration[18] or E Merck PEG-4000 (supplied by ACE Scientific) used at 50% concentration in serum-free medium, have been used reproducibly to obtain positive fusions. The E Merck PEG-4000 takes longer to melt and requires a higher temperature than does Fisher PEG. If PEG-4000 is used as an agent for cell fusion the following method is recommended. Wash the mixture of 10^8 spleen cells and 2 $\times$ 10^7 SP2/0 or X63/Ag 8.653 tumor cells in serum-free medium in a Falcon (Becton Dickinson, Rutherford, New Jersey) 15-ml round-bottom tube. Centrifuge at 1000 rpm for 10 min, and aspirate the supernatant completely. Add 0.5 ml of 50% PEG solution, made with serum-free Dulbecco's Modified Essential Medium (DMEM) and kept at 37°, drop by drop over a period of 1 min while mixing the cell pellet. After 30 sec more, add 10 ml of serum-free medium (kept at 37°) drop by drop over a period of 5 min. Centrifuge at 800 rpm for 10 min. Aspirate the supernatant fluid, gently break the cell pellet by tapping the

[18] R. H. Kennett, K. A. Dennis, A. S. Tung, and N. R. Klinman, *in* "Lymphocyte Hybridomas" (F. Melchers, M. Potter, and N. L. Warner, eds.), p. 77. Springer-Verlag, Berlin and New York, 1978.

[19] S.-P. Kwan, D. E. Yelton, and M. D. Scharff, *in* "Genetic Engineering" (J. K. Setlow and A. Hollaender, eds.), Vol. 2, p. 31. Plenum, New York, 1980.

[20] G. Galfrè and C. Milstein, this series, Vol. 73, p. 3.

tube, and resuspend the cells in 30 ml of hybridoma culture medium. The culture medium is made with high glucose DMEM (MA Bioproducts or GIBCO) containing 20% FCS, 10% NCTC 109 (MA Bioproducts), 2 mM L-glutamine, 50 μg of gentamycin per milliliter, 1 mM oxaloacetate, 0.45 mM pyruvate, and 0.2 unit of bovine insulin per milliliter (Sigma Chemical Co., St. Louis, Missouri). Littlefield's[21] concentrations of thymidine ($1.6 \times 10^{-5}M$) and hypoxanthine ($1 \times 10^{-4}M$) (Sigma Chemical Co. or MA Bioproducts) are added in addition to the concentrations of these two bases that are already present in NCTC 109. Using a 5-ml disposable glass pipette with pipette-aid (Drummond Scientific Co., Bromall, Pennsylvania) 1 drop (~50 μl) of the cell suspension is added per well to six 96-well flat-bottomed Microtiter plates (Linbro, Hamden, Connecticut) that have already been seeded with 1 drop of 2×10^6/ml irradiated and washed thymocytes suspended in culture medium. Thymocytes are prepared in a sterile manner, and the irradiation dose given is 2500 rads, using Gamma Cell 1000 (Atomic Energy of Canada). We have used both syngeneic and allogeneic thymocyte feeder layers with essentially the same results. The plates with feeders are prepared 2–24 hr prior to fusion and are kept at 37° in an 8–10% CO_2 and humidified atmosphere. The plates with feeders are effectively used up to a week after seeding with thymocytes. At the later times, however, the medium is aspirated and replaced with fresh medium. Four- to six-week-old mice are used as donors of thymocytes.

On the next day 1 drop (~50 μl) of the above culture medium containing aminopterin is added to give a final concentration of 4×10^{-7} M in the wells. Alternatively, fused cells can be plated immediately in culture medium containing aminopterin after fusion with the same results. A week later, the cells are fed with 1–2 drops of culture medium. Macroscopic colonies of hybrid cells will appear around day 10. The cultures are fed with 1–2 drops of culture medium at intervals of 3–4 days. Before feeding most of the supernatant is aspirated using a sterile Pasteur pipette. When the supernatants from wells with positive growth start turning yellow, about 150 μl of supernatant fluid are collected for assay of antibody.

The experimental protocol for production of antigen-specific hybridoma of the IgE class is schematically illustrated in Fig. 1. The entire procedure is divided into four parts: A, optimal immunization of mice to amplify IgE-producing B lymphocytes; B, PEG-mediated cell fusion of spleen cells from these mice with myeloma cells to generate hybridomas; C, expansion of IgE-producing hybridoma cells *in vitro* and *in vivo*; and D, affinity purification of hybridoma IgE from culture supernatant fluids and/or hybridoma ascitic fluids.

[21] J. W. Littlefield, *Science* **145,** 709 (1964).

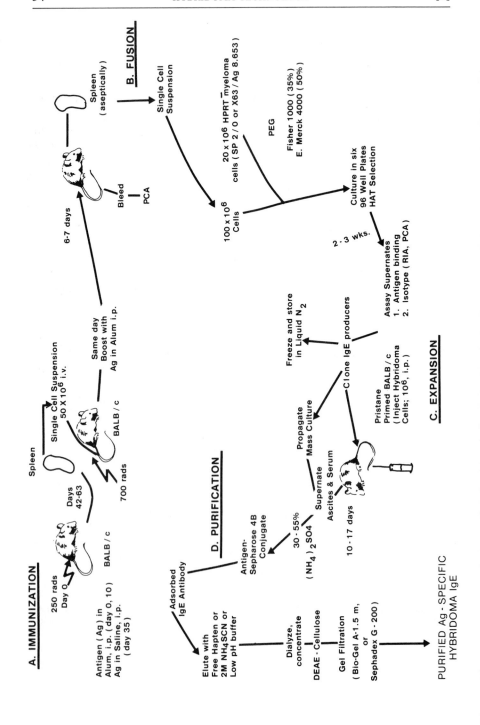

Assay for IgE-Positive Clones

Screening for the antibody-producing colonies can be done by PCA,[22,23] solid-phase RIA,[24] or enzyme-linked immunoabsorbent assay (ELISA).[25,26] The PCA assay will identify only IgE-positive clones (≥ 4 hr sensitization period), whereas the RIA and ELISA can detect clones producing all isotypes of antibody. Several chapters in Volumes 70, 73, and 74 in this series describe various methodological aspects of RIA and ELISA.[26] In addition, the use of a sensitive ELISA using galactosidase–antibody conjugate for quantitation of antigen-specific mouse IgE has been published.[27] In this paper all three of these methods are described.

The first task is to determine which colonies from the 96-well culture plates are making antigen-specific antibody. If the investigator has no preference between anti-hapten or anti-carrier IgE antibody, it is suggested that the supernatants from growing colonies be screened against both the hapten and the carrier protein. The supernatants from antibody-producing colonies should then be checked for the presence IgE antibody by any of the methods described in detail below. For RIA and ELISA, a rabbit or goat antibody monospecific for mouse epsilon (ε) heavy chain and standard IgE antibody of known titer are needed.

Screening of Antigen-Specific Hybridoma Products by Solid-Phase RIA (Fab Assay)

Collect the supernatants (150 μl) from wells with positive growth and assay for anti-DNP and anti-Hy antibody by the following procedure. A manifold dispenser with 8 or 12 delivery tips attached to a 1-ml Cornwall glass syringe (Dynatech Laboratories, Alexandria, Virginia) is conve-

[22] Z. Ovary, *J. Immunol.* **81**, 355 (1958).
[23] I. Mota and D. Wong, *Life Sci.* **8**, 813 (1969).
[24] N. R. Klinman, *J. Exp. Med.* **136**, 241 (1972).
[25] A. M. W. M. Schuurs and B. K. Van Weemen, *Clin. Chim. Acta* **81**, 1 (1977).
[26] J. E. Butler, this series, Vol. 73, p. 482.
[27] P. N. Hill and F.-T. Liu, *J. Immunol. Methods* **41**, 51 (1981).

FIG. 1. Experimental protocol for production of antigen-specific hybridoma-secreting antibodies of IgE class. A. Immunization: BALB/c mice are given 250 rad of whole body irradiation on day 0 and immunized i.p. with 2 μg of DNP$_8$-Hy on days 0, 10, and 35 as indicated. Two weeks later their spleen cells are adoptively transferred into sublethally irradiated BALB/c recipients, which are immediately boosted with antigen. B. Cell fusion: Six to seven days later spleen cells of these mice are fused with X63/Ag 8.653 myeloma cells, and colonies producing IgE antibody are cloned. C. Expansion: IgE antibody-producing clone is propagated in tissue culture and ascites. D. Purification: The IgE-enriched immunoglobulin fraction from culture supernatants and ascites is affinity purified, characterized as described, and used to raise IgE heavy chain-specific antisera in rabbits.

niently used to dispense antigen, washing solutions, and labeled antibody. Briefly, polyvinyl assay plates (Dynatech Laboratories) are coated with 100 μl of 0.1–1.0 mg of antigen (DNP$_{25}$-BSA or Hy) per milliliter for 4 hr at room temperature or overnight at 4°.

The plates are washed with PBS and the nonspecific sites are saturated by incubating the plates with 0.2 ml of 10% γ-globulin-free horse serum (HS) for 30 min. The plates are washed once with 0.1 ml of PBS containing 1.5% horse serum (PBS-HS). Twenty-five microliters of hybridoma supernatant fluid are added to the wells, and the plates are covered and incubated for 2 hr at room temperature. The plates are washed twice with 0.1 ml of PBS-HS, and 0.1 ml of ^{125}I-labeled affinity-purified rabbit anti-mouse antibody (Fab) are added.[28] This antibody recognizes both κ and λ light chains and therefore detects all antibody-producing clones. Approximately 20,000 cpm of the label are added per well. The plates are incubated at 37° for at least 4 hr or overnight at 4° in a humidified atmosphere. The unbound radioactivity is removed using a 96-channel aspirator attached to a vacuum with a charcoal filter. The plates are then thoroughly washed with tap water at least 10 times either manually or with a 96-channel washing device. The washed plates are dried and cut with a plate cutter; individual wells are counted in a gamma scintillation counter. The 96-channel aspirator, washing device, and plate cutter have been developed by Dr. A. Pickard of Department of Pathology of the University of Pennsylvania School of Medicine, Philadelphia.

The antibody-producing wells are identified by the amount of radioactivity above background and upon comparison to standard antibody run routinely at concentrations of 10, 20, 40, 80, 200, 400, and 800 ng/ml. This assay determines all the antibody-producing clones regardless of the heavy-chain class.

Identification of Heavy Chain

The supernatant fluids from antibody-producing wells are then subjected to assay for isotype identification. The principle of the assay is as described for the Fab assay with the requirement for one additional step. After the culture supernatant fluids have reacted with the antigen, rabbit antibody monospecific for mouse IgM, IgG1, IgG2a, IgG2b, IgG3, IgA, and IgE heavy chains is added. After a 2-hr incubation, the plates are washed with PBS-HS and labeled by addition of 100 μl of affinity-purified ^{125}I-labeled goat antibody against rabbit immunoglobulins. The plates are incubated for 4 hr at 37° or overnight at 4° and individual wells are counted as described in the Fab assay. The specificity of the rabbit antibodies for a particular mouse heavy chain is determined by running standard antigen-

[28] P. J. Gearhart, N. H. Sigal, and N. R. Klinman, *J. Exp. Med.* **145,** 876 (1977).

specific murine hybridomas or antigen-binding myeloma immunoglobulins possessing different heavy chains. Rabbit antibodies monospecific for murine IgM, IgG1, IgG2a, IgG2b, IgG3, and IgA heavy chains were obtained from Litton Bionetics (Rockville, Maryland), and rabbit antisera specific for mouse IgE heavy chain (RAME) used in the earlier studies was a generous gift of Dr. Zelig Eshhar of Weizmann Institute of Science, Rehovot, Israel.

Expansion of IgE-Producing Hybridoma Cells

The IgE antibody-producing colonies are cloned as soon as possible by limiting dilution either in agar or single-cell culture.[18–20] For cloning hybridoma cells in soft agar, we routinely use 3T3 fibroblasts as feeders; and for cloning by single-cell culture, 1 drop of cell suspension (20 cells/ml) is added to each well of 96-well plates previously seeded with 1 drop of 2×10^6 irradiated thymocytes per milliliter. Propagate the clones making antibodies of IgE class in tissue culture or *in vivo* by injecting 10^6 hybridoma cells i.p. into pristane-primed syngeneic mice to obtain IgE antibody-containing ascitic fluids. A single injection of 0.5 ml of pristane (Aldrich Chemicals, Milwaukee, Wisconsin) given i.p. 1–5 weeks earlier adapts mice for growth of tumor cells in ascites form. Details on the labeling of antibody with ^{125}I, cloning, and growing the hybridoma cells to obtain large amounts of monoclonal antibodies are given in this series[20] and in other papers on monoclonal antibodies in this volume.

ELISA for IgE Anti-DNP Antibody

1. Using 96-well Falcon polystyrene plates (Becton-Dickinson Inc., Rutherford, New Jersey), add 0.1 ml of 100 μg of DNP$_{25}$-BSA per milliliter in phosphate-buffered saline (PBS) containing 0.02% NaN$_3$. Cover the plates and incubate for 4 hr at room temperature or overnight at 4°.

2. Remove the DNP$_{25}$-BSA solution for reuse, wash the wells twice with PBS, and add 0.2 ml of PBS containing 0.04% NaN$_3$ and 10% γ-globulin-free horse serum (HS) (GIBCO) to block nonspecific protein-binding sites on the surface of the wells. Incubate for 1 hr at room temperature. γ-Globulin-free HS is used throughout this procedure.

3. Remove PBS-HS (10%) and save for reuse; wash the plates with 0.1 ml of PBS-HS (1%). Discard the wash and add 0.1 ml of test sample as well as several concentrations covering a wide range (1–800 ng/ml) of standard IgE anti-DNP antibody solution to individual wells. Cover the plates and incubate for 2 hr at room temperature or 4 hr at 4°.

4. Aspirate the liquid from the wells and rinse the plates 4 times with 0.1 ml of PBS and add 0.1 ml of affinity-purified RAME at a concentration of 1.0 μg/ml in PBS-HS (1%). The method for obtaining affinity-purified

RAME is described under Purification of Rabbit Anti-Mouse IgE. Incubate the plates for 2 hr at room temperature or overnight at 4°. Discard the RAME solution, and wash the wells 4 times with PBS as before.

5. Add 0.1 ml of a working dilution of β-galactosidase conjugate of goat anti-rabbit IgG.[27] The conjugate is diluted in PBS containing 1% HS. β-Galactosidase is obtained from Boehringer-Mannheim Biochemicals, Indianapolis, Indiana. (Other antibody-enzyme conjugates for ELISA are commercially available.) Incubate for 4 hr at 4°. Discard the conjugate solution, and wash the wells 4 times with PBS containing 1 mM 2-mercaptoethanol (2-ME) and 0.1% glycerol.

6. Immediately after the last rinse, add 0.1 ml of assay buffer consisting of 10 mM NaCl, 10 mM MgCl$_2$, 10 mM Tris base, at pH 7.5 adjusted by addition of acetic acid, containing the enzyme substrate O-nitrophenyl-β-D-galactoside (Calbiochem-Behring Corp., La Jolla, California) at a final concentration of 0.7 mg/ml and 0.1% (v/v) freshly prepared 2-ME. Cover the plates and incubate at 37° for 4 hr.

7. Read the absorbance of individual wells using Titertek Multiscan Automated 96-well plate reader equipped with a 414-nm filter (Flow Laboratories, McLean, Virginia). Determine the sensitivity of the assay by plotting probits of absorbance at 414 nm versus log IgE concentration. The procedure is same for other antigens except that DNP-BSA solution is replaced by appropriate antigen solution.

Determination of IgE Antibody by PCA[23]

Physiological Basis for PCA

In the mouse both IgE and IgG1 possess the ability to bind to specific receptors on homologous and, in certain cases, heterologous tissue mast cells through the Fc portion of the antibody molecule. The CH$_2$ domain of the ε heavy chain is known to bind to the mast cell receptor. IgE binds to its own receptor with greater avidity than IgG1. Mast cells bearing IgE and IgG1 molecules are referred to as sensitized or armed. The bridging of the resulting antibody–receptor complex by cross-linking of either the Fab regions of IgE molecules by di- or polyvalent antigens or the Fc regions by anti-IgE antibody molecules results in the degranulation of sensitized cells and to the release of vasoactive compounds such as histamine, serotonin, slow-reacting substance of anaphylaxis (SRS-A) [also known as leukotriene C$_4$ (LTC$_4$), eosinophil chemotactic factor of anaphylaxis (ECF-A), a platelet activating factor (PAF), and prostaglandin-generating factor of anaphylaxis (PGF-A)].[29] The symptoms of immedi-

[29] L. Steel and M. Kaliner, *J. Biol. Chem.* **256,** 12692 (1981).

ate-type hypersensitivity are a direct result of the interactions of the released mediators of anaphylaxis with their target cells. Upon release of the vasoactive amines, the surrounding blood vessels become more porous and hence allow the transudation of intravascular fluids and proteins resulting in a localized area of induration. This local reaction can be visualized by the use of a dye that binds to albumin and is therefore specifically localized to areas of increased vascular permeability.

Materials Required for PCA

Culture supernatants, serum or ascitic fluids from IgE hybridoma-bearing mice

CD albino rats, retired breeders, preferably males (Charles River Laboratories, Wilmington, Massachusetts)

Normal rat serum, 1%, as diluent

Syringes, 1 ml with 27-gauge needles (Becton-Dickinson, Rutherford, New Jersey)

Appropriate test antigen or highly substituted hapten-carrier conjugate, e.g., Hy and DNP_{25}-BSA at a concentration of at least 5 mg/ml

Evans blue dye, 5%, in saline

Nembutal sodium (Abbott Laboratories, North Chicago, Illinois)

Preparation of Rats for PCA

Retired breeders CD rats are suitable for this assay, and 60–65 individual samples can be assayed on the back of a single rat weighing about 600 g. Weigh the rat and inject 1 ml of Nembutal solution for every 250 g of body weight. Nembutal solution is prepared by mixing together 12 ml of Nembutal sodium, 9 ml of absolute ethanol, and 79 ml of normal saline. After about 5 min, shave the hair from the back of the rat and draw squares with a marking pen to keep a record of test sample injection sites.

PCA Assay[23]

The concentration of mouse IgE can easily be determined in a semiquantitative fashion in rats after a 4–24-hr sensitization period. IgE remains bound to the mast cells via its receptor *in situ* for a period exceeding 7 days. At first, assay the culture supernatants of antibody-producing wells without further dilution. For the first dilution of ascites or serum, use 1 : 320 as the lowest dilution for assay (5 μl of test serum or ascitic fluid plus 1.6 ml of diluent). Add 0.2 ml of diluent to a set of tubes and make twofold serial dilutions by transferring 0.2 ml of diluted test serum or ascitic fluid. Keep the dilutions at 4°. Inject 0.1 ml of test dilutions intradermally (i.d.) into the shaved dorsal skin of the assay animal. Always inject 3–5 dilutions of standard antibody with known PCA titer. This is done to normalize the PCA results obtained when more than one rat is used in a given experiment. After 4 hr or on the next day, challenge the assay animal intravenously with a solution containing 1 mg of antigen per milliliter and 1% Evans blue dye diluted in saline. The recommended amount is 1.0 ml of diluted solution per 250 g of shaved body weight of the assay animal. If the antigen used to elicit a PCA reaction is too costly, then inject 1 μg of the antigen in 0.1 ml of saline i.d. into the same site as the original test sample. Immediately thereafter inject 1% Evans blue dye i.v. Inject female rats in the tail vein and male rats in the tail or penal vein. Rats are lightly anesthetized with ether or Metofane (Pittman-Moore, Inc., Washington Crossing, New Jersey) at the time of i.v. injection of the antigen. Thirty minutes later, sacrifice the animal by exposure to ether or Metofane and read the PCA titer. The PCA titer is recorded as the reciprocol of the highest serum or ascites dilution that gives a 5-mm blue spot 30 min after challenge with antigen.

The PCA titer of culture supernatants from IgE-producing cultures is usually 2580, whereas hybridoma ascites show a positive PCA reaction even at a dilution greater than one million (PCA titer $>10^6$). Hill and Liu[27] have observed that a PCA titer of 1280 is equivalent to a concentration 2.9 μg of IgE per milliliter.

Criteria for IgE Anti-DNP Antibody

1. The skin-sensitizing activity, demonstrated by PCA reaction, is abolished by heat treatment of the sample at 56° for 4 hr.

2. The skin-sensitizing activity persists at the injected sites for periods longer than 72 hr.

3. This activity is adsorbed by passing culture supernatants or serum or ascitic fluids from IgE-producing hybridoma-bearing mice over anti-rat IgE immunoadsorbent. Rat and mouse IgE show strong cross-reactivity. The activity is not adsorbed by immunoadsorbents made with goat or rabbit antisera monospecific for murine IgM, IgG1, IgG2a, IgG2b, IgG3, and IgA.

4. Intravenous injection of the antigen (DNP_{25}-BSA) into hybridoma-bearing mice causes anaphylactic shock leading to death.

5. Murine IgE antibodies bind to rat basophilic leukemia cells and compete with rat myeloma IgE for binding to the IgE receptor.

Purification of Monoclonal IgE

After collecting hybridoma ascitic fluid,[28] filter it through nylon mesh and centrifuge at 2500 rpm for 15 min. Remove the fatty layer at the top either by aspiration or by using cotton-tipped applicators. Add cold saturated ammonium sulfate solution (SAS), pH 7.2, to the ascitic fluids to 30% saturation with continuous stirring. Leave it for 1 hr at 4°, and then centrifuge at 15,000 rpm for 20 min. Collect the supernatant (discard the pellet), and add cold ammonium sulfate solution at a saturation of 55% with continuous stirring at 4°. To the culture supernatants from mass culture, add cold SAS to 55% saturation in a single step. After 1 hr, centrifuge at 15,000 rpm for 20 min at 4°. Discard the supernatant, and wash the IgE-containing precipitate with cold 55% ammonium sulfate solution using one-half volume of the starting volume of ascitic fluid. Centrifuge again at 15,000 rpm for 20 min. Discard the supernatant, and dissolve the precipitate in a minimal volume of distilled water (one-fifth or less of the starting volume of ascitic fluid). Dialyze the precipitate twice against PBS.

For hapten-specific (anti-DNP) hybridoma IgE, specific purification can be carried out by affinity chromatography. Apply the dialyzed IgE antibody to a DNP_{25}-BSA–Sepharose 4B immunoadsorbent column prepared by reacting 5–8 mg of DNP_{25}-BSA per gram of CNBr-activated Sepharose using the manufacturer's procedure (Pharmacia Fine Chemicals, Piscataway, New Jersey). When using for the first time, always prewash the packed immunoadsorbent column with the eluting agent. Apply 1–2 mg of antibody per milligram of immobilized antigen. The first two procedures given below are carried out at room temperature or at 4°, and the low pH elution is always performed at 4°.

Elution with Hapten

Remove the unbound material by washing the immunoadsorbent column with 0.05 M, pH 8.0, N-carbobenzoxyglycine (Sigma Chemical Co.). When the eluate shows no trace of protein, start eluting IgE anti-DNP with 0.02 M DNP-glycine, pH 8.0. Follow the elution of bound antibody by testing a drop of eluate for trichloroacetic acid precipitability. The volume of each fraction collected will vary with the size of the immunoadsorbent column employed. Concentrate the eluted antibody by ultrafiltration, and dialyze against PBS extensively to remove from DNP-glycine.

Elution with Ammonium Thiocyanate (NH_4SCN)

After IgE anti-DNP antibody is bound to the DNP-BSA immunoadsorbent, wash the column with PBS until the absorbance at 280 nm of the eluate is less than 0.01. Apply 2 M NH_4SCN solution (prepared in distilled H_2O) and read the absorbance of eluted antibody fractions at 280 nm. Pool the antibody-containing fractions and dialyze immediately against PBS.

Concentrate the purified antibody by ultrafiltration using a vacuum flask or Millipore immersible CX ultrafiltration units (Millipore Corp., Bedford, Massachusetts).

Elution with Low-pH Buffer

The procedure is same as above except that 0.1 M glycine-HCl buffer, pH 2.3, is used as the eluting agent. The antibody-containing fractions are immediately neutralized by addition of 1 M Tris followed by dialysis against PBS. Concentration of purified IgE is determined by reading the absorbance at 280 nm ($E_{1\,cm}^{1\%} = 16$).

Further Purification

For analytical studies, further purification of IgE antibody is carried out by ion-exchange chromatography followed by gel filtration.[7] Two milligrams of antibody protein are applied per milliliter of packed DEAE-cellulose column, (DE-52, Whatman, Clifton, New Jersey) and is eluted with a linear gradient of 0.01 M Tris-HCl, pH 8.0, to 0.01 M Tris-HCl, pH 8.0, containing 0.15 M NaCl. IgE is eluted at 0.01 M Tris-HCl, pH 8.0, at 0.04–0.07 M NaCl. The IgE peak is pooled, concentrated, and further purified by gel filtration using an appropriate sized column of any one of the following gels: Sephadex G-150 or G-200, Sepharose 6B, Sephacryl-300 (Pharmacia Fine Chemicals) or BioGel-A, 0.5 m or 1.5 m (Bio-Rad Laboratories, Richmond, California). All purified antibodies are stored frozen at $-70°$ or in the presence of 0.02% NaN_3 at $4°$.

Physical and Chemical Characterization of Hybridoma IgE Antibodies

The physicochemical characteristics of a DNP-specific hybridoma IgE (clone SPE-7),[6] derived from fusion of C57BL/6 spleen cells with NS1 plasmacytoma cells, are given in Table III. In SDS–PAGE this purified antibody displayed a single band and indicated an apparent molecular weight of 200,000 for the intact molecule and 90,000 for the ε chain. This IgE bound to the rat basophilic leukemia cells and competed with the same efficiency as the rat IgE myeloma protein (1R-162) for the IgE receptors present on the surface of these cells.

In the extensive study of another DNP-specific hybridoma IgE (clone H1 DNP-ε-26),[7] the molecular weights of the intact IgE molecule, its heavy and its light chain were estimated to be 184,000 ± 3000, 82,200 ± 2900, and 22,900 ± 500, respectively. The estimated molecular weight for the intact IgE molecule was based on a partial specific volume ($\bar{V}$) calculated from the amino acid and carbohydrate composition, $\bar{V} = 0.715$. The molecular weights of heavy and light chains were estimated by comparing

TABLE III
PHYSICHOCHEMICAL CHARACTERISTICS OF
DNP-SPECIFIC MURINE HYBRIDOMA IgE[a]

Molecular weight	187,000[b]
$s_{20,\omega}$	7.8[c]
$D_{20,\omega}$	3.5 × 10[7] cm[2]/sec[c]
Association constant	6.5 × 10[7] M^{-1} [d]

[a] Data given in the table are from Eshhar et al.[6]
[b] Molecular weight determined from Svedberg equation by using partial specific volume ($\bar{V}$) of 0.713 cm[3]/g.
[c] Sedimentation velocity measurements made in a Model E Spinco ultracentrifuge operated at 56,000 rpm at 20°.
[d] Determined by fluorescence quenching method.

their migration to that of standard proteins in SDS–PAGE analysis. This monoclonal IgE had a carbohydrate content of 13.3% and it mediated antigen-induced triggering of rat basophilic leukemia cells to release incorporated [3H]serotonin. The sedimentation coefficient and $E_{1\,cm}^{1\%}$ for this molecule were determined to be 8.2 and 16.2, respectively. In this study, binding affinity was determined by equilibrium dialysis against [3H]DNP-lysine, and the association constant was calculated to be 1.4 × 10[8] M^{-1} at 25° and 7.1 × 10[7] M^{-1} at 37°. The IgE used in this study was isolated by affinity chromatography using DNP-BSA–Sepharose immunoadsorbent and further purified by DEAE-cellulose ion-exchange chromatography followed by gel filtration over Sephadex G-200.

Production of Antibodies Specific for Mouse ε Heavy Chain

New Zealand red rabbits (Three Springs Kennels, Inc., Jackson Center, Pennsylvania) weighing 6–8 pounds are preferred over white rabbits because they respond with a relatively higher antibody titer. Three injections of IgE are given at intervals of 2 weeks. A fourth injection is given a month after the third injection. The dose per rabbit is 2 mg of purified IgE for the first injection, 1 mg for second and third, and 500 μg for the fourth injection. The adjuvant used is CFA for first two immunizations and incomplete Freund's adjuvant (IFA) for the third and fourth immunizations. The total volume of inoculation for each rabbit is 3.0 ml containing equal volumes of antigen solution and adjuvant. Emulsions are made using Spex Mixer Mill (Spex Industries, Metuchen, New Jersey). The emulsion is injected intradermally and subcutaneously at multiple sites (about 15) on the back of the rabbit. The investigator might find it convenient to shave the fur of the rabbit's back and proximal limbs before immunization. Rabbits are bled from the ear 3 weeks after the third injection and weekly thereafter (bleeding should precede the booster injections).

A quick test of the strength of antisera is done by using 50–100 μl of rabbit serum from a given bleed and adding 2–20 μl of IgE hybridoma ascites. The volume and nature of the precipitate is directly indicative of the antibody titer. When the titer has reached a plateau value, usually after 3–4 months, the rabbit may be bled to death in order to collect a large amount of serum. Alternatively, the animal can be bled weekly and boosted at regular intervals using 100 μg of IgE in 10 mg of alum in a total volume of 3 ml i.p. for the first boost and IFA for subsequent injections spaced at intervals of 2–3 months using 100 μg of IgE for each inoculation.

We have also made rabbit antisera specific for IgE by using isolated heavy chains as the immunogen. Heavy chains are prepared from the whole molecule by a modification of the method of Bridges and Little.[30] This method involves reduction of S—S bonds with 0.01 M dithiothreitol and alkylation of free sulfhydryl (—SH) groups with 0.022 M iodoacetamide. Dialysis of the reduced and alkylated molecules is done against 1 M propionic acid containing 6 M urea, and chains are separated by gel filtration over Sephadex G-100 equilibrated in 1 M propionic acid and eluted with the same solution. When the isolated ε heavy chain is used as the immunogen (100-μg dose for each immunization) or for the boost immunization of the rabbits originally immunized with the whole IgE molecule, the antiserum produced by this approach, is less laborious to adsorb to render it monospecific for ε heavy chain.

Purification of Rabbit Anti-Mouse IgE

Preparation of Sepharose 4B–Mouse γ-Globulin (MGG) Conjugate

Prepare MGG from normal mouse serum or CFA-induced ascitic fluids.[31] The MGG is precipitated from the serum or ascitic fluid by 18% Na_2SO_4 at room temperature (2.57 ml of Na_2SO_4 per milliliter of serum or ascitic fluid). After 1 hr, centrifuge at 15,000 rpm for 20 min, discard the supernatant, and wash the precipitate in 18% Na_2SO_4 solution. Recentrifuge at 15,000 rpm for 20 min, dissolve the precipitate in a minimal volume of water, and dialyze against coupling buffer (0.1 M $NaHCO_3$, 0.5 M

[30] S. H. Bridges and J. R. Little, *Biochemistry* **10**, 2525 (1971).
[31] A. S. Tung, this series, Vol. 93 [2].

NaCl, pH 8.3). Normally 1 ml of serum or ascitic fluid yields 2–5 mg of MGG. The concentration of MGG is estimated by reading the absorbance at 280 nm ($E_{1\,cm}^{1\%} = 14$). Prepare Sepharose 4B–MGG conjugate using 5–8 mg of MGG per milliliter of packed Sepharose according to manufacturer's protocol (Pharmacia Fine Chemicals).

Purification Procedure

The rabbit antiserum is passed over Sepharose 4B–MGG conjugate. The capacity of Sepharose 4B–MGG conjugate to adsorb anti-mouse Ig activity (with the exception of ε heavy chain) can be determined in a trial run. Sepharose 4B–MGG conjugate adsorbed rabbit antiserum shows strong reactivity only for murine IgE antibody, and the same dilution (1 : 500 or higher) of this antiserum shows reactivity close to background when antigen-binding hybridoma or myeloma immunoglobulins of classes other than IgE are tested. Prepare a rabbit γ-globulin (RGG) fraction from the absorbed rabbit antisera by slowly adding 25% Na_2SO_4 solution at room temperature with stirring to a final concentration of 18% as described before. Wash the precipitate once with 18% Na_2SO_4 solution and centrifuge at 15,000 rpm for 20 min. Discard the wash and dissolve the precipitate in one-fifth the starting volume of rabbit antiserum, and dialyze against PBS. Affinity-purify the rabbit antibody specific for ε heavy chain by passing the RGG fraction of adsorbed antiserum over Sepharose 4B–mouse IgE immunoadsorbent. This immunoadsorbent is prepared by conjugating 5–8 mg of affinity-purified antigen-specific IgE per gram of CNBr-activated Sepharose. Wash the column with PBS until the absorbance of the eluate at 280 nm drops to less than 0.01. Now elute ε chain-specific rabbit antibodies by using 2.5 M NH_4SCN solution. Read the absorbance at 280 nm to determine the antibody-containing fractions. Pool the desired fractions, dialyze against PBS, and concentrate by ultrafiltration as described under purification of antigen-specific IgE. Determine the quantity of affinity-purified rabbit antibody by reading the absorbance at 280 nm ($E_{1\,cm}^{1\%} = 14$). Ouchterlony tests of affinity-purified rabbit anti-mouse IgE showed a single precipitin line against IgE, but showed no precipitin line against purified hybridoma or myeloma proteins belonging to other classes of murine immunoglobulins. Store rabbit antiserum, RGG fraction of adsorbed rabbit antiserum, and affinity-purified antibody at −70° or in the presence of 0.02% NaN_3 at 4°.

Conclusions

A method is described to obtain murine antigen-specific monoclonal antibodies of IgE class. The availability of large quantities of murine monoclonal IgE antibody will permit studies aimed at delineating the mechanisms involved in the regulation of IgE synthesis and its role in allergic diseases. Antigen-specific IgE as well as anti-IgE antibodies will be greatly useful in studies designed to investigate the dynamics of IgE-receptor aggregation on the surface of basophils and tissue mast cells and the triggering events that lead to their degranulation. Purified IgE and anti-IgE antibodies should also prove to be useful in the purification and characterization of different populations of cells involved in allergic reactions.

Acknowledgments

The author wishes to thank Drs. Norman Klinman, Alan Rosenthal, and Hans Zweerink for their enthusiasm and scientific support. Dr. Regina Skelly for her sincere suggestions and critical review of the manuscript, Drs. Roger Kennett and Kathie Dennis for introducing me to the practical world of making hybridomas, and Eileen Frees for typing this manuscript.

[7] Production and Properties of Monoclonal Antibodies to Guinea Pig Ia Antigens

By Stephen E. Zweig and Ethan M. Shevach

This chapter is primarily concerned with describing the selection procedures that are used to generate monoclonal antibodies specific for guinea pig Ia antigens and the experimental procedures used to define the properties of the guinea pig Ia antigenic determinants that the antibodies recognize. Monoclonal antibodies[1] have proved to be very useful tools for the in depth characterization of the role that Ia antigens play in the immune response. Their extreme specificity allows the Ia antigen system to be dissected at a much higher level of resolution than was previously possible using conventional alloantisera. Using this technology, it is possible to distinguish different epitopes within a single Ia molecule[2,3] and to define new Ia subclasses and modification states.[4-6] Rigorous character-

[1] G. Köhler and C. Milstein, *Nature (London)* **256**, 495 (1975).
[2] R. Burger and E. M. Shevach, *J. Exp. Med.* **152**, 1011 (1980).
[3] H. Lemke, G. J. Hämmerling, and U. Hämmerling, *Immunol. Rev.* **47**, 175 (1979).
[4] L. A. Lampson and R. Levey, *J. Immunol.* **125**, 293 (1980).
[5] D. A. Shackelford, L. A. Lampson, and J. L. Strominger, *J. Immunol.* **127**, 1403 (1981).
[6] V. Quaranta, L. E. Walker, M. A. Pellegrino, and S. Ferrone, *J. Immunol.* **125**, 1421 (1980).

ization of the binding properties of a given set of antibodies is very important and, in fact, is often more work than the actual generation and isolation of the antibodies themselves. In this review, particular attention will be directed to this characterization process. The hybridoma production process that we use is a standard one and is therefore described in less detail. For a more complete discussion of the general aspects of the hybridoma production process, see the review by Drs. Galfrè and Milstein in this series (Vol. 73 [1]).

General Considerations

The two inbred guinea pig strains that are most commonly used in this work are designated strain 2 and strain 13. These two strains have similar class I major histocompatibility complex (MHC) antigens, but differ in their class II MHC (Ia) antigens. Using serological techniques, the guinea pig *I*-region has been subdivided into subregions coding for different molecules.[7] Inbred strain 13 guinea pigs have been shown to have the phenotype Ia.1, 3, 5, 6, 7, and the *I* region of strain 13 is composed of three subregions. The Ia.1 and Ia.6 determinants are borne on a molecule comprising one or two chains of 26,000 daltons each. The Ia.3 and Ia.5 antigens are borne on a molecule that apparently comprises two noncovalently associated proteins of 33,000 and 25,000 daltons; the Ia.7-bearing molecule is a 58,000-dalton protein having two chains of 33,000 and 25,000 daltons bound by sulfhydryl linkage. Strain 2 guinea pigs express Ia.2, 4, 5, 6, and the *I* region of strain 2 guinea pigs can be broken down into two or possibly three subregions. The Ia.2 determinant is on a molecule of 33,000- and 25,000-dalton chains bound by noncovalent forces, while the Ia.4, 5 antigens are on a molecule of comparable size, but having sulfhydryl-linked chains. The molecule bearing the Ia.6 determinant in strain 2 animals has not been identified, and the existence of a third *I* subregion in strain 2 animals is inferred from the data obtained from strain-13 animals.

The assignment of an immune response (Ir) gene to a given region is based on an association of that Ir gene with an Ia antigen in outbred populations, the ability of anti-Ia sera to block T-cell responses controlled by that Ir gene, and the association between Ir genes and Ia antigens in macrophage–T cell interactions.[8] A diagram of the guinea pig Ia system is shown in Fig. 1.

---

[7] B. D. Schwartz, A. M. Kask, W. E. Paul, A. F. Geczy, and E. M. Shevach, *J. Exp. Med.* **146**, 547 (1977).

[8] E. M. Shevach, M. L. Lundquist, A. F. Geczy, and B. D. Schwartz, *J. Exp. Med.* **146**, 561 (1977).

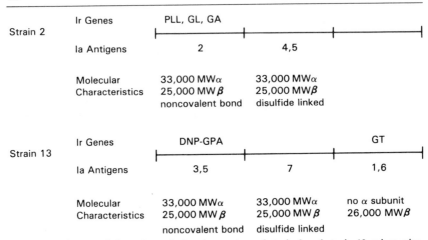

FIG. 1. Diagram of the guinea pig *I* region system of strain 2 and strain 13 guinea pigs. Serological analysis has shown that the guinea pig *I* region is composed of a number of subregions, each consisting of Ia antigens with distinct molecular properties. The Ir genes that control immune system recognition of a number of synthetic antigens have been shown to be associated with certain of these subregions. MW, molecular weight.

Production of Monoclonal Antibodies

Immunization of Mice

Ia antigens are highly immunogenic,[9] and, as a result, immunization is easily accomplished. Owing to the extreme specificity of monoclonal antibodies, it is unnecessary to immunize using purified Ia antigens. We have successfully used unfractionated spleen cells, a leukemic B cell line (EN-L2C), and even T cells for this purpose.[10] Typically, BALB/c mice are injected with 20 million cells (suspended in 0.5 ml of PBS) intraperitoneally at intervals of 3 weeks. Generally, two injections are given. Three days before the fusion, the mice are given a final boost using 10 million cells given intravenously via the tail vein.

Myeloma Cell Lines

We have had experience with two myeloma fusion partners: NS-1 and its variant, SP2/0. We have found that SP2/0 is a superior partner because SP2/0 does not produce any immunoglobulin molecules constitutively.

[9] L. T. Clement, A. M. Kask, B. D. Schwartz, and E. M. Shevach, *Transplantation* **27**, 397 (1979).
[10] R. Burger, L. Clement, J. Schroer, J. Chiba, and E. M. Shevach, *J. Immunol.* **126**, 32 (1981).

Although NS-1 does not constitutively produce an immunoglobulin heavy chain, it does produce a κ light chain and after the fusion hybrid antibody molecules will be obtained.

Production of Monoclonal Antibodies

We had good results using the polyethylene glycol fusion procedure of Galfrè et al.[11] as modified by Oi and Herzenberg.[12] We have not made any substantial modifications to their procedure. For further information, the reader is referred to the chapter by Galfrè and Milstein (this series, Vol. 73 [1]).

Selection

Given a successful fusion, large colonies grow and can be tested for antibody production after 2–3 weeks in culture. For the preliminary screening, we use binding and/or cytotoxicity testing. In the guinea pig system, we are fortunate to have both an Ia-positive leukemia cell line, EN-L2C, and an Ia-negative variant of this cell line, BZ-L2C,[13] for use in selection. If such lines are not available, selection can be done using lymphoid cells, which are generally Ia positive, and nonlymphoid cells, which are generally Ia negative. In this later case, however, the danger of isolating monoclonal antibodies that are directed against a molecule other than Ia will be greater. In the guinea pig system, we test for reactivity toward EN-L2C and lack of reactivity toward BZ-L2C. Since EN-L2C is of strain 2 origin, this procedure serves as a screen for both antibodies to allotypic determinants of strain 2 Ia as well as for antibodies to framework Ia determinants. Antibodies specific for allotypic determinants of strain 13 Ia can be screened by tests for differential reactivity between strain 2 spleen cells and strain 13 spleen cells. Since the MHC of these two strains differs only in the *I* region, most antibodies obtained should be directed against Ia antigens.

Binding Assay. We use a radioimmunoassay, using the binding of iodinated F(ab')₂ rabbit anti-mouse Ig (Cappel Labs, Cochranville, Pennsylvania) to detect the mouse antibodies. This second antibody is iodinated according to the standard chloramine-T method.[14]

[11] G. Galfrè, S. C. Howe, C. Milstein, G. W. Butcher, and J. C. Howard, *Nature (London)* **266,** 550 (1977).
[12] V. T. Oi and L. A. Herzenberg, *in* "Selected Methods in Cellular Immunology" (B. B. Mishell and S. M. Shiigi, eds.), Chapter 17. Freeman, San Francisco, 1980.
[13] G. Forni, E. M. Shevach, and I. Green, *J. Exp. Med.* **143,** 1067 (1976).
[14] F. C. Greenwood, W. M. Hunter, and J. S. Glover, *Biochem. J.* **89,** 114 (1963).

The binding assay is done using 96-well polyvinyl plates (Cooke No. 220-24). Cells are placed in the Microtiter wells along with the antibody containing hybridoma cell culture supernatants. The antibodies are allowed to bind to the cells, and the cells are washed by centrifuging in the Microtiter plates using microplate carrier centrifuge holders (Beckman).[10] The radiolabeled second antibody (2×10^6 cpm/ml) is then added and allowed to bind; the cells again washed. Finally, the plate is dried; the wells are cut out using a puncher built for the purpose (Division of Research Services, NIH) and then counted in a gamma counter. The details of the assay follow.

Cells are washed by centrifugation and resuspended in PBS containing 5% fetal calf serum (FCS), 0.02% sodium azide, to a final concentration of 16 million cells/ml. Then steps 1–15 are carried out.

1. Plate 25 μl of cells in each well.
2. Add 20 μl of hybridoma supernatant to each well.
3. Mix by gentle vortexing (Dynatech Micro-Shaker II).
4. Let stand on ice for 30 min.
5. Mix again by gentle vortexing.
6. Let stand on ice an additional 15 min.
7. Centrifuge at 1000 rpm (200 g) for 3 min.
8. Remove supernatant; do not disturb cell pellet.
9. Add 150 μl of the assay buffer to each well, vortex, and wash the cells by repeating steps 7–9 twice.
10. To the final pellet, add 50 μl of the second antibody (100,000 cpm).
11. Mix; let stand on ice for 30 min.
12. Mix again and let stand an additional 15 min.
13. Add 150 μl of assay buffer.
14. Wash the cells as in steps 7–9 for a total of three centrifugations. Note that the supernatant should be disposed of in radioactive waste containers.
15. Allow plate to dry; cut and count as described above.

An example of the binding results that are obtained from this assay using a monoclonal reagent that recognizes an alloantigenic determinant on strain 13 Ia is shown in Fig. 2.

Cytotoxicity Testing. The binding assay is the method of choice for preliminary screening because it can detect all subclasses of mouse Ig. However, for cases in which complement-binding cytotoxic antibodies are specifically desired, a standard chromium release assay can be used.[15] As in the binding assay, the Ia-positive cell line EN-L2C and the Ia-

---

[15] E. M. Shevach, D. L. Rosenstreich, and I. Green, *Transplantation* **16**, 126 (1973).

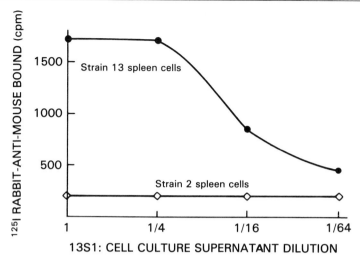

FIG. 2. Binding of serial dilutions of the strain 13 specific monoclonal antibody 13S1 to strain 2 spleen cells and strain 13 spleen cells. Even when the binding to strain 13 spleen cells is saturated, there is essentially no binding to strain 2 spleen cells.

negative cell line BZ-L2C, together with strain 2 and strain 13 spleen cells, are used. The details of this assay follow.

Absorption of complement. We use commercially available rabbit complement (Gibco, 5 ml of rabbit complement, lyophilized). This must first be preabsorbed with guinea pig spleen cells.

1. Prepare guinea pig spleen cells from one spleen.
2. Resuspend rabbit complement in 5 ml of RPMI-1640 (Grand Island Biological Co., Grand Island, New York).
3. Add complement to spleen cells. Vortex.
4. Incubate on ice for 1 hr.
5. Centrifuge at 2800 rpm (1700 g) for 5–10 min. Remove supernatant; discard pellet.
6. Complement is ready to use. It may be stored in small aliquots at $-70°$ for later use.

Chromium labeling of cells. The technique essentially consists of incubating cells together with chromium, followed by washing to remove excess chromium.

1. Suspend cells at 50 million/ml in 0.9 ml of RPMI 5% FCS.
2. Add 0.1 ml (100 μCi) of ^{51}Cr (sodium chromate, NEZ-030, New England Nuclear).

3. Incubate for 30–45 min in CO_2 incubator at 37°.
4. Wash cells 3 times with Hanks' buffer to remove excess chromium.
5. Resuspend cells to 5 ml.

Cytotoxicity assay. This assay is done using 96-well nonsterile Titertek/Linbro U round-bottom plates, (Flow Laboratories, Inc., No. 76-311-05), and the Titertek supernatant collection system. This set of components is designed to soak up the supernatant at the end of the assay and put it into containers suitable for counting, while leaving the remaining cells behind. The set is composed of harvesting frames (catalog No. 78-201-05), transfer tube frames (No. 78-211-05), harvesting press (No. 78-201-00), alignment rack (No. 78-202-00), transfer fork (No. 78-203-00), and transfer tube adaptor (No. 78-204-99). All these components are available from Flow Laboratories. The assay consists of the following steps.

1. Plate 25 μl of cells/well.
2. Add 50 μl of hybridoma cell supernatant. Vortex.
3. Add 50 μl of the preabsorbed rabbit complement, first diluting the complement 1 : 3 before use. Vortex.
4. Cover plate and incubate in 37° CO_2 incubator for 30–45 min. Plate covers are No. 76-311-05.
5. Harvest supernatant (free from cells) and count.

Controls for cell lysis in the absence of antibody are performed by using a normal myeloma cell supernatant in one or more wells. Controls to determine the number of counts in the supernatant that represent total cell lysis are done by adding 50 μl of 0.1% Nonidet P-40 (NP-40) to several wells in place of the hybridoma supernatant. (Note that this should be added before adding complement.) It is also a good idea to include, as a positive control, anti-Ia antibodies of known activity. In this way, the activity of the hybridoma antibodies can be expressed as a percentage of the lysis of a known anti-Ia antibody.

Because the desired clone is usually growing in the presence of nonproducing clones, it must be isolated by recloning as soon as possible. A decision as to which clones to save and which clones to discard should be made within a maximum of 6 weeks from the initial cloning.

Immunoprecipitation Studies

Molecular characterization is the final step in the selection procedure. The definitive proof that an anti-Ia antibody has been made is the demonstration that the antibody will immunoprecipitate an Ia-like molecule from a detergent-solubilized[16] cell extract of a radiolabeled cell preparation. We

---

[16] S. E. Cullen and B. D. Schwartz, *J. Immunol.* **117**, 136 (1976).

have used the detergent NP-40, which is made by Shell and can be obtained from Bethesda Research Laboratories, (catalog No. 5500UA). Since Ia antigens are integral membrane proteins, they are too hydrophobic to be soluble in solution in the absence of the NP-40 detergent.

Both external labeling[17] and internal labeling[18] procedures are useful, and each procedure has its unique advantages and disadvantages. External labeling by lactoperoxidase-catalyzed iodination is a rapid and convenient method for labeling cell-surface molecules and has the further advantage that iodine is easily detected. However, it has the disadvantage that not all parts of the Ia molecule are equally susceptible to iodination.[17] Alternatively, Ia antigens can be internally radiolabeled with [³H]leucine or [³⁵S]methionine, followed by purification of the cell extract on a lentil lectin column. Lentil lectin specifically selects for the glucose and mannose side chains of glycoproteins, and thus preferentially binds cell-surface molecules, many of which are glycosylated. Lectin purification is generally required when Ia antigens are biosynthetically labeled, because these antigens must be separated from other internally labeled proteins. Certain Ia antigens, however, do not adhere well to lentil lectin. Thus, to get a complete picture, more than one method of labeling should be used.

The choice of cells to be used in labeling studies is also an important consideration. Reports have suggested that differential glycosylation may occur in different cell types,[19] and our recent work suggests that this occurs also in the guinea pig system. Thus standard sources of cells should be used to prevent confusion.

Lactoperoxidase Iodination

1. Fresh iodine is essential for this procedure. Preferably, it should be used within a week of delivery. Iodine used is carrier-free high-concentration sodium [¹²⁵I]iodine, 17 Ci/mg, 522 mCi/ml (New England Nuclear).
2. Make up
   Lactoperoxidase 1.5 mg/ml in 1 ml of PBS. The lactoperoxidase used is 86 units/mg solid (Sigma L-2005).
   Diluted hydrogen peroxide: 25 μl of stock H_2O_2 in 25 ml of cold PBS. The hydrogen peroxide stock is 30% (Fisher Scientific).
3. Obtain guinea pig lymph node cell suspension.
4. Centrifuge cells and resuspend to a concentration of 100 million cells/ml in PBS.

[17] B. D. Schwartz, E. S. Vitetta, and S. E. Cullen, *J. Immunol.* **120,** 671 (1978).
[18] B. D. Schwartz, A. M. Kask, W. E. Paul, and E. M. Shevach, *J. Exp. Med.* **143,** 541 (1976).
[19] S. E. Cullen, C. S. Kindle, D. C. Shreffler, and C. Cowing, *J. Immunol.* **127,** 1478 (1981).

5. Note that precautions should be taken when working with [125]I. Wear double gloves, work in efficient fume hood, and check for thyroid uptake at routine intervals. With reasonably good technique, no detectable radioiodine should be taken up by the thyroid.

6. Dilute iodine with distilled water so that it is at a concentration of 1 mCi/10 μl.

7. The following steps are carried out at room temperature. Place the cells in a 50-ml conical centrifuge tube. The following steps assume 100 million cells in 1 ml of PBS. If more cells are used, scale up accordingly.

8. Iodination: Add 2 mCi of ^{125}I to cells. Add 100 μl of lactoperoxidase and 25 μl of diluted H_2O_2. Agitate gently for 3 min. Then add 50 μl of lactoperoxidase and stir gently for 2 min. After this, add 25 μl of the diluted H_2O_2 again and stir for 3 min more. Finally, add an additional 25 μl of H_2O_2 and stir gently for 4 min more.

9. Washing: Fill the tube with cold PBS to nearly the top. Cap the top securely. Centrifuge the tube at 1200 rpm (300 g) for 5 min. Discard the supernatant in radioactive waste. Repeat this washing step twice more.

10. Place the cells on ice. Resuspend the final cell pellet in 0.5 ml of tris-buffered saline (abbreviated TBS; this is 0.01 M Tris, 0.15 M NaCl, pH 7.4). Add 0.5 ml of 1% NP-40 TBS; let sit on ice for 30 min.

11. Centrifuge the extract at 20,000 rpm using a Beckman Ti-40 rotor. Save the supernatant; discard the pellet.

12. Removal of free iodine: Free iodine is most easily removed by running the supernatant through a small Sephadex G-25 column previously equilibrated with TBS 0.3% NP-40.

Immunoprecipitation. Generally, the best immunoprecipitation technique is the solid-phase method of Kessler.[20] The first antibody is bound to formalin-fixed *Staphylococcus aureus* bacteria either directly via the bacteria's protein A receptor (which binds the Fc portion of immunoglobulins), or indirectly via *S. aureus* coated with a rabbit-anti-mouse immunoglobulin serum. We have found that this latter technique is superior to the alternative technique of precipitating the monoclonal antibody by using the second antibody directly, since an immune precipitation lattice is not formed using *S. aureus*. As a result, the chance of spurious proteins being nonspecifically bound by the immune complex is much less, and the efficiency of washing is much greater. *Staphylococcus au-*

[20] S. W. Kessler, *J. Immunol.* **115,** 1617 (1975).

reus can be commercially obtained under various names (Immuno-precipitin, Bethesda Research Laboratories; Pansorbin, Calbiochem). The preparations tend to deteriorate upon prolonged storage at refrigerator temperatures, and freezing of aliquots is recommended for prolonged use. The commercial preparations are washed in TBS 0.3% NP-40 buffer several times before use to remove any nonsedimentable components.

In order to precoat the bacteria with anti-mouse immunoglobulins (rabbit anti-mouse-IgG serum works well), the bacteria are pelleted and resuspended in a 2× volume of anti-mouse immunoglobulin serum for 1 hr. After the incubation, the bacteria are washed 4–5 times in TBS 0.3% NP-40 before use. Since the titer of the anti-mouse immunoglobulin sera may vary, the immunosorbent's capacity to bind hybridoma immunoglobulins is evaluated in a competition binding experiment in which increasing amounts of unlabeled ascites are used to compete with a small amount of radiolabeled hybridoma antibody for binding to the immunosorbent. We have found that 100 μl of a 10% suspension of rabbit anti-mouse Ig-coated *S. aureus* has the capacity to bind about 1 μl of ascites fluids.

Preabsorptions. A certain number of components of a cell extract tend to bind nonspecifically to *S. aureus;* it is therefore necessary to preabsorb the cell extract with *S. aureus* and additionally to preabsorb with anti-immunoglobulin coated *S. aureus* if this latter reagent is to be used. A 1-hr preabsorption on ice with a 10% volume suspension of *S. aureus* is usually sufficient to remove the nonspecific background binding.

Immunoprecipitation

1. Count radioactivity of cell extract. As a rule of thumb, use 1 million cpm per immunoprecipitate. This varies with the antibody/antigen used, however. Sometimes as much as 10 million cpm is required.
2. Add 1 μl of ascites or 25 μl of cell culture supernatant to the cell extract. This, and the following steps, are all done on ice.
3. Wait 45 min. This allows the antibody to bind to its antigen free from interference by the *S. aureus.*
4. Add *S. aureus* immunosorbent using 100 μl of a 10% suspension per tube.
5. Wait for 45 min, swirling tube occasionally.
6. Centrifuge. The Microfuge B (Beckman) works well for this procedure. Save the supernatant if a sequential precipitation is going to be done later. Wash the pellet by resuspending in 1 ml of cold TBS NP-40 and centrifuging again. Wash three times. At the last resuspension, transfer contents to another tube to avoid the background of labeled proteins absorbed nonspecifically to the first tube.

7. Store pellets in the cold. It is recommended that the pellets be dissolved in sodium dodecyl sulfate (SDS) gel dissociation buffer to avoid possible proteolysis.

The immunoprecipitates are then analyzed by conventional SDS–polyacrylamide gel electrophoresis. This can be done using either 10% tube gels or 7 to 15% gradient slab gels. Tube gels can be analyzed by running the gel through a gel slicer (electrophoresis gel slicer, Model 190, Bio-Rad) and counting each section. Slab gels are analyzed by autoradiography. Greater sensitivity in autoradiography can be obtained by the use of intensifying screens, such as the DuPont Chronex screens, and by exposure at −70°.

Electropherograms of typical gels of a hybridoma (22C4) directed against both the Ia.2-bearing and the Ia.4, 5-bearing molecules and of a hybridoma (27E7) directed against the Ia.2-bearing molecule alone are shown in Fig. 3.

Interpretation of Immunoprecipitation Results. In the guinea pig system, the different Ia molecules can usually be detected by their different molecular weights.[7] In strain 2 guinea pigs, the α and β chains of the Ia.4, 5-bearing molecule are linked by disulfide bridges, whereas the α and β

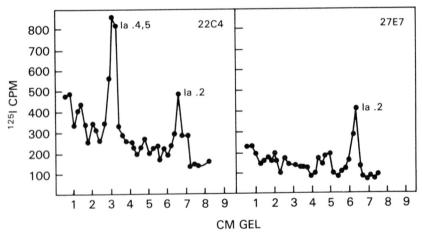

FIG. 3. Immunoprecipitation of lactoperoxidase iodinated strain 2 guinea pig Ia antigens using two different monoclonal antibodies. The gels are run under nonreducing conditions, hence the disulfide-linked Ia.4, 5-bearing molecule runs as a single chain (M_r 58,000), while the Ia.2-bearing molecule dissociates into its subunits and runs at lower molecular weight. Using lactoperoxidase, only the M_r 25,000 subunit of Ia.2-bearing molecule is iodinated. Antibody 22C4 precipitates the Ia.2- and Ia.4, 5-bearing molecules. Antibody 27E7 precipitates only the Ia.2-bearing molecule.

chains of the Ia.2-bearing molecule are associated noncovalently. Thus, by running the gels under nonreducing conditions (generally done by simply omitting the 2-mercaptoethanol from the SDS dissociation buffer), the two different Ia molecules can be distinguished. Similarly, in strain 13 guinea pigs, the two chains of the Ia.7-bearing molecule are also linked by a disulfide bridge, while the two chains of the Ia.3, 5-bearing molecule are noncovalently associated.

Sequential precipitation studies are often needed to answer definitively exactly what subpopulation of Ia molecules is recognized by a hybridoma. The Ia antigen preparation is first exhaustively depleted of all molecules reactive with one antibody, followed by a second precipitation with a different antibody. Normally, two cycles of immunoprecipitation by one antibody suffice to deplete essentially all molecules reactive with that antibody. Each immunoprecipitation should be followed by incubation with S. aureus alone to prevent a buildup of unbound antibody in the supernatant. In the immunoprecipitation with a second antibody, a crucial control is a sample in which the depleted cell extract is again immunoprecipitated with the first antibody. It must be shown that all material reactive with the first antibody has been removed.

The interpretation of sequential precipitation studies with monoclonal antibodies may not be straightforward. Sequential precipitation may be used to prove that a hybridoma recognizes the same species of Ia antigen seen by a conventional anti-Ia serum. A cell extract is first depleted with the alloantisera and a control serum. If the molecules capable of subsequently reacting with the hybridoma are depleted by the alloantiserum, but not by the control serum, it is likely that the hybridoma recognizes some of the same Ia molecules seen by the serum. The interpretation of sequential precipitation studies involving two hybridoma reagents may be more difficult because it has been found that two hybridomas, apparently directed against the same Ia molecule, actually distinguish different populations of Ia molecules in the sequential precipitation analysis. In some cases, this has been shown to be the result of one antibody preferentially binding to a biosynthetic Ia precursor molecule.[5] In other cases, it has been suggested that new, previously undetected, Ia subregions may be involved.[4] Further studies using two-dimensional gel electrophoresis or peptide mapping may be needed to give definitive answers.

Epitope Binding Assay

Once it has been determined that two or more monoclonal antibodies bind to the same Ia molecules, it becomes of interest to determine whether the antibodies are binding to the same site on the Ia molecule or to different sites. Studies of this type are important in the correlation of the structure of the Ia molecule with its immunoregulatory functions.[21] Competition binding assays are performed between a radiolabeled monoclonal antibody on the one hand and excess amounts of an unlabeled competitor antibody on the other.[22] This assay is thus a modification of the binding assay discussed previously. The principal difference is that the unlabeled competitor antibody is added first and allowed time to bind to the cells; the labeled tracer antibody is then added and also given time to bind. The cells are then washed and counted. If the unlabeled competitor antibody does not block the binding of the labeled tracer antibody, then it can be assumed that the two bind to different sites. Conversely, if the unlabeled competitor antibody does block the binding of the tracer antibody, it is likely that the two do bind to the same site.

Although a monoclonal antibody in ascites form can be purified by ammonium sulfate precipitation, iodinated by the chloramine-T method (or by gentler methods), and used as a tracer antibody in this assay, it is generally better to use [³H]leucine internally labeled antibodies produced by hybridoma cells in culture. Antibodies labeled in this way keep for very long periods of time and in this assay give results superior to those with iodinated antibodies (lower background and less scatter). Internally labeled antibodies can be produced by the following procedure.

Internal Labeling of Antibody

1. Hybridoma cells are grown to late log phase in culture. In our hands this is approximately 1 million cells/ml. This protocol assumes that amount.
2. Cells (5 ml in medium) are spun down, and resuspended in leucine (−) minimal essential medium without sera, with 1% 100 × glutamine, and 1% Gibco antibiotic–antimycotic mixture (filter sterilize before use). Resuspend cells in 2 ml of medium. Add 0.2 ml (200 μCi) of [³H]leucine (Amersham TRK.510, 136 Ci/mmol) to the cells.
3. Put the cells in a sterile petri dish and incubate in a CO_2 incubator over night.
4. Centrifuge cells. Save the supernatant and discard the pellet.
5. Run the supernatant through a small G-25 column to separate antibodies from free leucine. The column should be equilibrated and

[21] F. C. Grumet, D. J. Charron, B. M. Fendly, R. Levey, and D. B. Ness, *J. Immunol.* **125,** 2785 (1980).

[22] J. C. Howard, G. W. Butcher, G. Galfrè, C. Milstein, and C. P. Milstein, *Immunol. Rev.* **47,** 139 (1979).

run in PBS. The labeled antibodies can be frozen or stored for extended periods of time in the refrigerator with the addition of 0.04% sodium azide.

Binding Assay

The cells and assay conditions are set up exactly as in the binding assay discussed previously. The ^{3}H tracer antibody solution is diluted in the PBS, 5% fetal calf serum (FCS), 0.04% sodium azide solution so that each milliliter contains 2 million cpm of the antibody. Serial dilutions are made of the competitor antibody ascites; as a control, similar dilutions are made of the (unlabeled) tracer antibody ascites and of a control ascites. These serve as positive and negative controls for this assay. Dilutions should be in steps of 10, starting with 1 : 10 dilution of the ascites and going down to a 1 : 1,000,000 dilution. If the amount of ascites is limiting, a higher dilution, such as a 1 : 20, is usually safe to use as a first step instead of the 1 : 10 dilution. The assay conditions follow.

Competition Binding Assay

1. Plate 25 μl of cell suspension per well.
2. Add 50 μl of the competitor ascites per well; vortex.
3. Let sit for 30 min on ice.
4. Without washing the cells, add 50 μl of tracer antibody per well. Be careful not to contaminate the tracer antibody solution with the competitor ascites.
5. Vortex cells; let sit an additional 30 min on ice.
6. Wash cells three times.
7. Cut out wells. Place in vials with scintillation fluid to count.

A typical result from such a binding assay is shown in Fig. 4. The particular antibodies in this example were previously shown to bind to the Ia.2 bearing molecule of the I region of strain 2. By combining results from immunoprecipitation data and competition binding assays, a map of the different epitopes on Ia molecules recognized by these different antibodies can be drawn. This is shown for the Ia.2-bearing molecule in Fig. 5.

It should be noted that the interpretation of binding assays is occasionally not straightforward. In the case in which the two different antibodies do not compete with each other for binding, it is clear that the two have different binding sites. However, in the case in which two antibodies do compete with each other, there are two possibilities. One is that the two antibodies share a binding site (or that the binding sites are close enough that steric interference prevents the two antibodies from binding side by

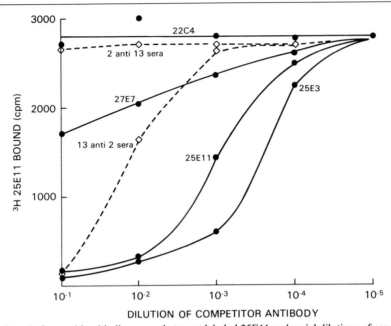

FIG. 4. Competition binding assay between labeled 25E11 and serial dilutions of several competitor ascites. Antibody 25E3 totally competes with 25E11 for binding. Antibody 27E7 shows partial competition. Antibody 22C4 shows no competition. Strain 13-anti-2 alloantiserum totally competes with 25E11 for binding, while the control, strain 2-anti-13 alloantiserum shows no competition. Thus, the monoclonal antibody 25E11 recognizes a determinant also seen by conventional anti-Ia alloantisera.

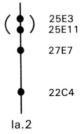

FIG. 5. Diagram of the Ia.2-bearing molecule which shows one possible scheme for interpreting the data shown in Fig. 4. Antibodies 25E3 and 25E11 may bind to neighboring epitopes, while 27E7 may be somewhat more distant. 22C4 may bind to a totally different part of the molecule. Although this is the simplest interpretation, alternative schemes in which 27E7 binds to a totally different location and inhibits the binding of 25E11 by producing a conformational shift in the Ia molecule are possible.

side). The other possibility is that one antibody, although bound to a different part of the molecule, has somehow changed the conformation of that molecule in such a way that the second antibody no longer recognizes its binding site. Occasionally, the reverse effect is seen in which an antibody binding to one site can actually enhance the binding of the "tracer" antibody to its site.[23] Despite these restrictions, it has been found that binding studies such as these are useful to put comparative data from a number of antibodies into perspective and allow one to define regions of a target molecule that appear to have different biological roles.

Fluorescence-Activated Cell Sorter Analysis

An understanding of the cellular distribution of the molecules recognized by a particular monoclonal antibody is very useful in studies that attempt to assign a particular function to that molecule. Here, fluorescence-activated cell sorter (FACS) analysis is the method of choice.[24] Studies by Burger and Shevach,[25] for example, have indicated that in contrast to the findings obtained by cytotoxicity studies in the human and murine systems, the majority of nonactivated guinea pig T cells and thymocytes bear Ia antigens. (This has also been seen by other workers.) This is shown in Fig. 6. In other systems, B cells and macrophages strain strongly for Ia,[26] but most unactivated T cells stain weakly, if at all. The implications of this finding to the regulation of the guinea pig immune system are still under investigation.

Procedure for FACS Analysis

1. Hanks' balanced salt solution containing 0.1% NaN_3 and 0.1% bovine serum albumin is used throughout.
2. Lymphoid cells to be tested (2–3 million) are incubated as a pellet at 4° for 30 min with 25 μl of a 1 : 50 dilution of ascitic fluid. This proved to be a saturating amount of antibody.
3. The suspension is then washed twice and incubated at 4° for 45 min with 25 μl of a fluorescein-conjugated F(ab')$_2$ rabbit anti-mouse Ig (Cappel Laboratories, Cochranville, Pennsylvania).
4. The cells are washed twice and resuspended in 0.7 ml of the Hanks' solution. They are then kept on ice until analyzed on the FACS.

---

[23] J. C. Howard, G. W. Butcher, G. Galfrè, C. Milstein, and C. P. Milstein, *Immunol. Rev.* **47**, 139 (1979).
[24] M. R. Loken and L. A. Herzenberg, *Ann. N. Y. Acad. Sci.* **254**, 163 (1975).
[25] R. Burger and E. M. Shevach, in preparation.
[26] C. G. Fathman, J. L. Cone, S. O. Sharrow, H. Tyrer, and D. H. Sachs, *J. Immunol.* **115**, 584 (1975).

CELL NUMBER

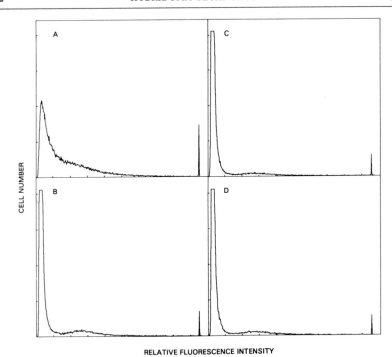

RELATIVE FLUORESCENCE INTENSITY

Fig. 6. Fluorescence-activated cell sorter analysis of the binding of antibody 25E11 to lymph node cells of strain 2 and strain 13 guinea pigs. Panels A and B: strain 2 cells; panels C and D: strain 13 cells. 25E11 is used in panels A and C, and a control ascites is used in panels B and D. Antibody 25E11 binds only to strain 2 cells (note increase in panel A over panel C). Note also that the majority of lymph node cells, which are primarily T cells, are labeled by the anti-Ia antibody.

5. Flow microfluorometry is performed using an FACS (FACS II, Becton-Dickinson). Data on individual cells are collected, stored, and analyzed using a PDP-11/40 computer (Digital Equipment Corporation, Marlboro, Massachusetts) interfaced to the FACS II. Data are collected on 5×10^4 viable cells, as determined by forward light scatter intensity, and are displayed as immunofluorescence profiles in which increasing fluorescence intensity is plotted on the x axis and cell number on the y axis.

Use of Monoclonal Anti-Ia Antibodies to Inhibit T-Lymphocyte Proliferation

The role that Ia antigens play in the immune response has been studied in antigen-induced T-cell proliferation assays for many years with the use of alloantisera directed against Ia.[27] These studies have shown that, in the initial stages of the immune response, antigens are taken up by antigen-presenting cells (macrophages) and are presented, in the context of the Ia molecule, to antigen-specific T cells. The T cells appear to recognize antigen and Ia simultaneously, both being required to initiate the T-cell response to antigen. In earlier studies, alloantisera directed against distinct Ia specificities were used to block this T-cell response to antigen in *in vitro* assays.[27] The alloantisera are believed to act by coating the Ia molecule with antibody, thus preventing the T-cell receptor for Ia from binding to the antigen-presenting cell. Monoclonal antibodies offer a way to extend these studies at a higher level of resolution. In alloantisera, antibodies directed against a number of Ia epitopes are present and the functional effects of the alloantisera are due to the sum of the blocking of each of these individual epitopes. By contrast, when monoclonal antibodies are used, the functional significance of a single Ia epitope can be assessed.[2]

Immunization of Guinea Pigs. A T-cell proliferation assay starts with the primary immunization of guinea pigs with the antigens that will later be used to stimulate proliferation in the assay.[28] Approximately 2 weeks after the immunization, the immunized guinea pigs are injected in the peritoneal cavity with 25 ml of sterile mineral oil (Marcol 52; Humble Oil and Refining Co., Houston, Texas).[29] This oil induces peritoneal exudate cells to accumulate in the peritoneal cavity. These cells are subsequently removed, and after column purification, serve as a source of T cells for the antigen-induced proliferation assay. At the same time that the immunized guinea pigs are injected with mineral oil, unimmunized animals of the same strain should also be injected. Peritoneal exudate cells from these animals will serve as a source of antigen presenting cells for the proliferation assay.

Preparation of Monoclonal Antibodies. Monoclonal antibodies are used at a relatively high concentration in the proliferation assay. The monoclonal antibody ascites is diluted 1 : 10 in Hanks' and filter-sterilized by passage through a small Millex HA filter (Millex-HA 0.45 μm filter unit, Millipore Corp., Bedford, Massachusetts).

Proliferation Assay. Sterile round-bottom Microtiter plates (Linbro Chemical Co., New Haven, Connecticut) and cover (Falcon No. 3041 lid) are used. Macrophages are prepared as described[28] and pulsed with a number of soluble protein antigens [for example, the copolymer L-glutamic acid, L-lysine (GL), ovalbumin (OVA), or purified protein deriva-

[27] E. Shevach, *Springer Semin. Immunopathol.* **1**, 207 (1978).
[28] E. M. Shevach, *J. Immunol.* **116**, 1482 (1976).
[29] D. L. Rosenstreich, J. T. Blake, and A. S. Rosenthal, *J. Exp. Med.* **134**, 1170 (1971).

tive of tuberculin (PPD)] for 60 min at 37° and then washed extensively to remove unbound antigens. In each well of the Microtiter plates, 0.1 ml of the pulsed macrophages (1 × 10⁶/ml) are mixed with 0.1 ml of purified T lymphocytes (1 × 10⁶/ml); 20 μl of the diluted sterile ascites are then added to each well. Each data point is done in triplicate. Important controls for this experiment are an unpulsed macrophage control, which measures the baseline of T-cell thymidine incorporation in the absence of antigen, a control ascites to measure any nonspecific cytotoxic effects, and a positive control using a known anti-Ia antibody. The plates are then incubated for 4 days. Eighteen hours before harvesting, each well is given 1.0 μCi of tritiated thymidine (6.7 Ci/mmol; New England Nuclear Corp.). The plates are harvested with a semiautomated harvesting device (mini-MASH, M.A. Bioproducts). This device draws up the contents of each well, deposits it on filter paper (mini-MASH glass-fiber filter No. 23-994,

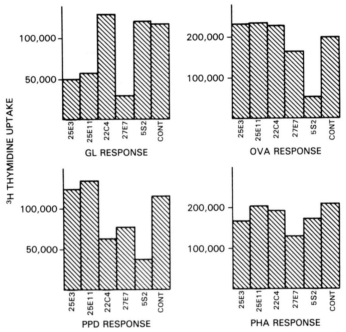

FIG. 7. Inhibition by monoclonal antibodies of the T-cell proliferative response to macrophages pulsed with three different antigens: GL, OVA, PPD, and the mitogen PHA. A given monoclonal antibody will selectively inhibit the proliferative response, lowering the response to some antigens, but not to others. Thus, 25E3 and 25E11 lower the response to GL, but not to OVA. Conversely, 5S2 lowers the response to OVA but not to GL. CONT, control.

M.A. Bioproducts) and washes it with distilled water. The pieces of filter paper are then put into counting vials with scintillation fluid (National Diagnostics Ultra Fluor) and counted.

An example of the results that can be obtained from such an assay are shown in Fig. 7. In this figure it can be seen that antibody 25E3, which specifically binds to the Ia.2-bearing molecule, blocks the response to GL but not the response to OVA. Conversely, the antibody 5S2, which binds to the Ia.4, 5-bearing molecule but does not bind the Ia.2-bearing molecule, blocks the response to OVA but not the response to GL. By contrast, 22C4, which binds both the Ia.2- and Ia.4, 5-bearing molecules, does not inhibit either the proliferative response to GL or OVA appreciably, but does inhibit the response to PPD. The anti-Ia antibodies have only a minimal effect on the proliferative response induced by macrophages pulsed with the mitogen phytohemagglutinin (PHA). Studies of this type suggest that different epitopes on individual Ia molecules may play unique roles in the activation of T-cell proliferation to different protein antigens.

Summary

Although the use of monoclonal antibodies to characterize the Ia antigen system is still in its infancy, the usefulness of this approach is clearly apparent. Since xenogeneic immunization can be used to generate these antibodies, the chances are good that a greater variety of Ia types and determinants will be detected compared to those detected with conventional alloantisera. In addition, the advantages of unlimited quantity and constant specificity allow a much greater degree of standardization than was previously possible. Finally, monoclonal antibodies open up lines of investigation, such as the examination of correlations between the monoclonal antibody's inhibition of T-cell proliferation and the antibody's corresponding Ia epitope-binding characteristics, that it would be impossible to approach using conventional alloantisera.

Acknowledgment

S. E. Z. is a fellow of the Arthritis Foundation.

[8] Purification of Murine MHC Antigens by Monoclonal Antibody Affinity Chromatography

By Matthew F. Mescher, Kathryn C. Stallcup,
Cathleen P. Sullivan, Aaron P. Turkewitz, and
Steven H. Herrmann

The growing understanding of the role of major histocompatibility complex (MHC) antigens in transplantation and immune system recognition has resulted in considerable interest in the functional and structural characteristics of these cell-surface glycoproteins. Primary structural analysis of the murine class I (H-2K, D and L) and class II (Ia) antigens has been largely done by radiochemical microsequencing of antigens obtained by immunoprecipitation.[1,2] Relatively large amounts of H-2 antigens have been purified by conventional methods,[3-6] but the large amounts of starting material required, the lengthy procedure, and the low yields obtained ($<10\%$) have prohibited the general use of this approach to obtain material for structural and functional studies. The availability of monoclonal antibodies (MAbs) specific for the H-2 and Ia antigens has permitted development of affinity purification procedures that are rapid and result in high yields (70–90%) of serologically and biologically active antigens. Given the high yields, starting material sufficient for purification of milligram amounts of antigen is readily obtainable from tumor cells grown as ascites in 20–100 mice. Large-scale monoclonal antibody affinity purification of murine MHC antigens was first done for H-2K^{k}[7] using the 11-4.1 antibody developed by Oi et al.,[8] and has subsequently been extended to use of other MAbs to purify H-2K and D(L) antigens[9] from several haplotypes and Ia antigens[10] of the *H-2^{k}* haplotype. Monoclonal antibody affinity chromatography has also proved to be useful for purifi-

[1] S. G. Nathenson, H. Uehara, B. M. Ewenstein, T. J. Kindt, and J. E. Coligan, *Annu. Rev. Biochem.* **50**, 1025 (1981).

[2] E. S. Vitetta and J. D. Capra, *Adv. Immunol.* **26**, 148 (1978).

[3] A. Shimada and S. G. Nathenson, *Biochemistry* **8**, 4048 (1969).

[4] O. Henriksen, E. A. Robinson, and E. Appella, *Proc. Natl. Acad. Sci. U.S.A.* **75**, 3322 (1978).

[5] J. H. Freed, D. W. Sears, J. L. Brown, and S. G. Nathenson, *Mol. Immunol.* **16**, 9 (1979).

[6] M. J. Rogers, E. A. Robinson, and E. Appella, *J. Biol. Chem.* **254**, 11126 (1979).

[7] S. H. Herrmann and M. F. Mescher, *J. Biol. Chem.* **254**, 8713 (1979).

[8] V. T. Oi, P. P. Jones, J. W. Goding, L. A. Herzenberg, and L. A. Herzenberg, *Curr. Top. Microbiol. Immunol.* **81**, 115 (1978).

[9] K. C. Stallcup, T. A. Springer, and M. F. Mescher, *J. Immunol.* **127**, 924 (1981).

[10] A. P. Turkewitz, C. P. Sullivan, and M. F. Mescher, manuscript submitted.

METHODS IN ENZYMOLOGY, VOL. 92

cation of human MHC antigens, as described by Parham in this volume [9].

A. General Considerations

The general scheme used for the purification of murine MHC antigens by monoclonal antibody affinity chromatography is shown in Fig. 1. The details of materials and methodologies employed at each step depend on a variety of considerations, and these are discussed in this section. Section B gives detailed protocols for purification of specific antigens.

1. Antibody Choice, Purification, and Affinity Column Preparation

A large number of monoclonal antibodies specific for murine MHC antigens are now available from individual investigators, commercial sources, and the American Type Culture Collection (Rockville, Maryland) (which now distributes cell lines formerly available from the Cell Distribution Center of the Salk Institute). Choice of the MAb to be used for affinity purification is dictated by specificity for the antigen of interest, affinity of the antibody, and ease of obtaining and purifying sufficient amounts of the antibody (5–20 mg).

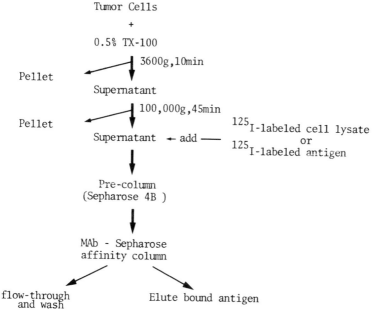

FIG. 1. General scheme for the purification of murine major histocompatibility complex antigens by affinity chromatography on monoclonal antibody (MAb)–Sepharose columns.

Once an antibody of appropriate specificity has been obtained, we have used the criteria of immunoprecipitation[7] to determine whether it is of sufficiently high affinity to be useful for affinity purification. Spleen (or tumor) cells bearing the appropriate antigens are surface labeled by lactoperoxidase (LPO)-catalyzed iodination (see Section B) and lysed with 0.5% Triton X-100 (Sigma Chemical Co.) or Nonidet P-40 (NP-40; Particle Data Laboratories Ltd., Elmhurst, Illinois) in 15 mM sodium phosphate, pH 7.3, at 5×10^7 cells/ml. The lysate is allowed to stand on ice for 10 min and then centrifuged at 1000 g for 10 min to pellet the nuclei. The supernatant is transferred to another tube, 2–20 μg of MAb are added per 0.2 ml of lysate, and incubation is done for 60 min at 4°. A second reagent is then added to precipitate the MAb–antigen complexes. We have routinely used protein A (Sigma Chemical Co.) or rabbit anti-mouse Ig coupled to Sepharose 4B (Pharmacia) at 2 mg of protein per milliliter of beads as the second reagent. Fifty-microliter aliquots (wet settled volume) of beads are added to lysate containing the MAb and to control lysate samples lacking the MAb, and the samples are incubated for 60 min at 4° with shaking to keep the beads suspended. The beads are then washed three times in 0.5 ml of lysis buffer, and the bound antigen is eluted by heating in 0.1 ml of 1% sodium dodecyl sulfate (SDS) at 100° for 5 min. The beads are then pelleted, and the supernatant containing the labeled antigen is removed. We have found that the amount of radioactivity eluting from the control beads is usually comparable to that from the beads which contained the MAb (presumably due to binding of free iodine or radiolabeled lipid present in the lysate). It is therefore necessary to analyze the eluates by SDS–polyacrylamide gel electrophoresis (see below) to determine whether specific precipitation of the antigen has occurred. While this procedure is only qualitative, specific precipitation under these conditions is adequate to demonstrate that the MAb is of sufficiently high affinity for use in antigen purification. Failure to obtain precipitation under these conditions does not necessarily preclude the possibility that the antigen will bind sufficiently well under the conditions used for affinity chromatography to allow purification. We have not attempted to use nonprecipitating MAbs for purification.

The MAbs are obtained from hybridoma culture supernatants or from the ascites fluid of mice injected intraperitoneally with the hybridomas. The antibody can be purified by conventional immunoglobulin purification schemes, although the distinctive properties of a given MAb can result in somewhat different behavior during purification than that seen for most of the immunoglobulin in a polyclonal mixture. Parham (this volume [9]) has discussed these considerations and outlined a general strategy for the purification of MAbs. Alternatively, chromatography on protein

A–Sepharose provides a rapid and convenient method for purifying those MAbs that bind to protein A.[11] Binding of immunoglobulins to protein A varies with the species of origin and the class and subclass of Ig. Murine IgG2a, 2b, and 3 bind, as do some IgG1s.[12] Use of protein A–Sepharose for immunoglobulin purification has been reviewed elsewhere,[11,13] and a detailed protocol is given in Section B.

The monoclonal antibody, having been selected and purified, is then coupled to a solid support for use in affinity purification of the antigen. We have used Sepharose 4B as the solid support. The purified MAb is dialyzed into 0.1 M NaHCO$_3$, 0.15 M NaCl, pH 8.3, adjusted to a concentration of approximately 1 mg/ml, and mixed with cyanogen bromide-activated Sepharose 4B for coupling. A detailed protocol is given in Section B. We routinely obtain greater than 90% coupling using this procedure. The MAb–Sepharose can be qualitatively assessed for activity by using 50 μl of beads for immunoprecipitation as described above.

MAb–Sepharose columns of up to 5 ml packed volume are made in plastic syringes plugged with glass wool. Larger columns are made in 10-ml plastic disposable pipettes. When not in use, the columns are stored at 4° in neutral pH buffer containing 0.02% sodium azide to prevent bacterial growth. Binding capacities of the columns have not been accurately determined in most cases. Columns prepared from the 11-4.1 MAb coupled to Sepharose at 2.2 mg/ml have a capacity of about 0.15 mg of H-2K^k per milliliter. We have used columns repeatedly over a period of more than a year. In fact, it appears that the purity of the antigens obtained from a column is sometimes not as good for the first one or two uses as for subsequent preparations. Repeated use sometimes results in a greatly reduced flow rate, presumably due to packing, accumulation of particulate material, or both. This can normally be corrected by removing the MAb–Sepharose from the column, washing it in bulk, and preparing a new column.

2. Antigen Source

The ideal starting material for MHC antigen purification is a cell that expresses a large amount of antigen and can be readily obtained in high numbers. In the case of human MHC antigens, Epstein–Barr virus (EBV)-transformed B lymphoblastoid cell lines are the source of choice. They express large amounts of antigen relative to normal lymphocytes

[11] H. Hjelm and J. Sjöquist, Scand. J. Immunol. 3, 51 (1976).
[12] J. Goudswaard, J. A. VanDerDonk, A. Noordzij, R. H. VanDam, and J.-P. Vaerman, Scand. J. Immunol. 8, 21 (1978).
[13] J. W. Goding, J. Immunol. Methods 20, 241 (1978).

and are readily grown *in vitro*.[14] We have identified one murine tumor cell line that expresses comparably high amounts of an MHC antigen. RDM-4, a lymphoma, was found to yield approximately 150 μg of H-2K^k per 10^9 cells and is estimated to have about 2×10^6 H-2K^k molecules per cell.[7]

In an effort to identify other murine tumors expressing high levels of H-2, several cell lines were examined by fluorescent staining with M1/42, a rat anti-mouse H-2 monoclonal antibody that appears to react with most, if not all, H-2s.[9,15] Cells were treated with M1/42 followed by fluorescein isothiocyanate (FITC)-coupled rabbit anti-rat Ig [F(ab)$_2$, absorbed with mouse Ig] under saturating or near saturating conditions and analyzed on a fluorescence-activated cell sorter (FACS) (Table I). Fluorescence intensities obtained were directly proportional to the number of antibodies bound per cell. T-cell and B-cell blasts and EL-4 (H-2^b), P388D (H-2^d), P815 (H-2^d), and YAC (H-2^a) cell lines all expressed 2.3- to 3.5-fold more H-2 than normal spleen cells. As expected, RDM-4 (H-2^k) cells expressed about 10-fold more H-2 than normal spleen cells. Light-scattering measurements on the FACS (Table I) and measurement of cell size by microscopy indicated that, with the exception of RDM-4, the increased expression of H-2 on the tumor and blast cells in comparison to spleen cells could be accounted for by the larger size of these cells. RDM-4 is similar (or smaller) in size than the other tumor cells examined, indicating that it has a higher surface density of H-2. Yields obtained from affinity purification indicate that only H-2K^k is expressed in unusually high amounts on these cells.[9] Expression of D^k (surface density) is about normal. The mutant cell line RIE/TL8X.1 had previously been shown to lack H-2 and Tla antigens[16] and was included as a negative control in this experiment.

Although the experiment shown in Table I failed to identify lines other than RDM-4 that express high amounts of H-2, it did demonstrate the usefulness of this approach in searching for such lines. It also demonstrated that the cell lines examined are the preferable source of starting material in comparison to spleen or blast cells bearing the same antigens. There is more H-2 present on a per cell basis owing to their larger size, and those that can be grown *in vivo* (P815, EL-4, YAC, and RDM-4) can be readily and inexpensively obtained in large numbers. Typically, syngeneic mice injected intraperitoneally with 1 to 2×10^7 tumor cells yield 3 to

[14] J. M. McCune, R. E. Humphreys, R. R. Yocum, and J. L. Strominger, *Proc. Natl. Acad. Sci. U.S.A.* **72,** 3206 (1975).
[15] T. Springer, G. Galfrè, D. S. Secher, and C. Milstein, *Eur. J. Immunol.* **8,** 539 (1978).
[16] R. Hyman and I. Trowbridge, *Cold Spring Harbor Symp. Quant. Biol.* **41,** 407 (1977).

TABLE I
QUANTITATION OF H-2 EXPRESSION USING M1/42 AND FACS

Cell	H-2 (spleen cell relative units)[a]	Scatter intensity (mean channel no.)
Spleen	(1)	52
Concanavalin A blasts[b]	2.3	134
Lipopolysaccharide blasts[b]	3.1	155
RIE/TL8X.1	0	106
RDM-4	9.8–12.7[c]	97
YAC	3.2	106
P815	3.5	140
P388D$_1$	2.1	127
EL-4	3.5	108

[a] Cells were labeled with saturating M1/42 MAb or irrelevant MAb, then fluorescein isothiocyanate (FITC)-anti-rat IgG and analyzed on the fluorescence-activated cell sorter (FACS) as described [T. A. Springer, G. Galfrè, D. S. Secher, and C. Milstein, Eur. J. Immunol. 9, 301 (1979)]. The mean fluorescence intensity of each cell type was calculated by integration, corrected for background labeling, and expressed relative to the intensities of C57BL/6J splenic nucleated cells.
[b] Scatter was gated to include only blasts.
[c] Twofold more FITC anti-rat IgG gave 1.3-fold brighter labeling.

5×10^8 cells 1 week later. Thus, as an MHC antigen source, murine tumor cells compare favorably, in terms of labor and expense, with human EBV-transformed cell lines that must be grown in vitro. A large number of murine tumor cell lines of different MHC haplotypes are available through individual investigators or the American Type Culture Collection. It is likely that many will prove to be as good a source of H-2 antigen as those we have examined and that other lines expressing high levels, as does RDM-4, may be found.

Tumor cell lines that bear Ia antigens and can be grown intraperitoneally are less readily available than those bearing H-2 antigens. We have used CH1,[17] an IgM-bearing B-cell lymphoma, as a source of I-A^k and I-E^k antigens. It appears to have amounts of these antigens comparable to normal B cells but less than lipopolysaccharide-induced spleen cell blasts. Again, the availability of large numbers of CH1 cells by in vivo growth makes them the preferable starting material for purification.

[17] M. A. Lynes, L. L. Lanier, G. F. Babcock, P. J. Wettstein, and G. Haughton, J. Immunol. 121, 2352 (1978).

3. Solubilization and Radiolabeling

H-2K and D antigens can be solubilized by treatment of cells or membranes with papain to release a water-soluble fragment that remains associated with β_2-microglobulin and retains serological activity.[1,18] Alternatively, the intact molecule can be solubilized by detergent treatment. We have been unable to obtain reproducibly good yields of the papain cleavage product by treatment of cells or membranes with the protease. However, good yields of the cleavage product can be obtained by treatment of the intact, affinity-purified H-2 with papain, and conditions for this are described in Section B. In the case of the Ia antigens, detergent treatment is the only method available for effectively.solubilizing the antigens in serologically active form.

Detergent solubilization of the MHC antigens in preparation for affinity chromatography has been done either by direct lysis of the cells in the detergent or by preparing a crude membrane preparation and then treating with detergent. Direct lysis of the cells has been effective in the case of RDM-4, YAC, EL-4, and Ch1 tumor cells. Direct lysis of P815 cells results in a very viscous lysate that cannot be passed over the affinity columns without drastically decreased and unworkable flow rates. This problem can be eliminated by lysing the cells by nitrogen cavitation, centrifuging at low speed to pellet the nuclei and then at high speed to pellet membranes. The nuclear and membrane pellets can then be extracted with detergent, and the extracts used for affinity chromatography. Yields of H-2 thus obtained are comparable to those obtained from other cells by direct cell lysis, provided that the detergent extract of the nuclear pellet is also used. (The nuclear pellet contains about 30–40% of the plasma membrane,[19] probably owing to physical trapping). Although a lengthier procedure, isolation of crude membranes, and removal of cytoplasmic proteins, does allow antigen purification in good yield from cells that give highly viscous whole-cell detergent lysates.

Cell lysis (and membrane solubilization) has been done using either Triton X-100 or NP-40. No differences have been noted between the preparations. These detergents effectively solubilize the MHC antigens while leaving the nuclei sufficiently intact to allow them to be removed by centrifugation. Lysis is normally done by suspending washed tumor cells in buffer at a density of 1×10^8 cells/ml and adding an equal volume of 1% Triton X-100. Lysis has also been done at higher cell densities to minimize the volume of lysate and reduce column loading time. In these cases the concentration of detergent has been proportionately increased to maintain

[18] S. G. Nathenson and A. Shimada, *Transplantation* **6**, 662 (1968).
[19] F. Lemonnier, M. Mescher, L. Sherman, and S. Burakoff, *J. Immunol.* **120**, 1114 (1978).

the same detergent-to-protein ratio. Phenylmethylsulfonyl fluoride, 250 μM (PMSF; Sigma Chemical Co.) is included during lysis to inhibit proteolysis. After 20 min at 4° the cell lysate is centrifuged at 1000 g to pellet the nuclear debris. The supernatant is then centrifuged at 100,000 g for 45 min. The resulting pellet is enriched in actin, and omission of this centrifugation step results in actin contaminating the affinity purified antigen.[7] Solubilization of membranes with Triton X-100 (details given in Section B) also results in an actin-containing detergent-insoluble fraction that can be removed by pelleting at 100,000 g.[20] In some cases lysates have been passed through 0.45 μm filters (Nalgene filter unit; Nalge Co., Rochester, New York) to further remove particulates. Attempts to store cell lysates at 4° or −20° prior to affinity chromatography have resulted in increased viscosity and precipitation of proteins. Cells are therefore harvested and detergent-solubilized, and the affinity chromatography is done on the same day.

Antigen purification can be easily monitored if radiolabeled antigen is included in the initial cell lysate. This can be accomplished by internal labeling with ^{14}C, ^{3}H, or [35]S. Alternatively, cells can be surface labeled with ^{125}I by lactoperoxidase-catalyzed iodination[21] or using Iodogen.[22] Iodination has the advantage of allowing the radioactive column fractions, etc., to be measured without sacrificing material. We have routinely used this procedure for initial purification attempts. Once purified antigen has been obtained, small amounts of it can be iodine-labeled using Iodogen[22,23] and added to the initial cell lysate to monitor purification. Examples of each of these approaches are detailed in Section B.

4. Affinity Chromatography

Prior to using a monoclonal antibody for affinity chromatography, conditions must be found for elution of the bound antigen. If the purified antigen is to be used for serological or biological studies, it is necessary that the elution conditions be as mild as possible. A variety of elution conditions can be examined simultaneously by binding ^{125}I-labeled antigen to MAb–Sepharose beads, washing the beads, and then treating aliquots of the beads with buffers of different pH, salt concentration, detergent composition, etc. Results of this type of experiment using the 11-4.1 (anti-H-2K^k)MAb are shown in Table II.[7] High pH, high salt, and deoxycholate

[20] M. F. Mescher, M. J. L. Jose, and S. P. Balk, *Nature (London)* **289**, 139 (1981).

[21] D. B. Phillips and M. Morrison, *Biochemistry* **10**, 1766 (1971).

[22] P. J. Fraker and J. C. Speck, Jr., *Biochem. Biophys. Res. Commun.* **80**, 849 (1978).

[23] P. R. P. Salacinski, C. McLean, J. E. C. Sykes, V. V. Clement-Jones, and P. J. Lowry, *Anal. Biochem.* **117**, 136 (1981).

TABLE II
ELUTION OF ^{125}I-LABELED H-2K[k] FROM 11-4.1 SEPHAROSE[a]

0.5% detergent	15 mM buffer	pH	NaCl	Percent H-2 eluted
NP-40	Phosphate	7	—	0
NP-40	Glycine	5	—	12
NP-40	Tris	8	—	9
NP-40	Carbonate	9	—	49
NP-40	Carbonate	10	—	50
NP-40	Carbonate	11	—	94
NP-40	Phosphate	7	0.14 M	38
NP-40	Phosphate	7	0.5 M	94
NP-40	Phosphate	7	1.0 M	98
Deoxycholate	Tris	8	—	73
Deoxycholate	Tris	8	0.14 M	86
SDS	—	—	—	96

[a] Three hundred microliters of MAb–Sepharose were incubated at 4° for 1 hr with whole-cell lysate from ^{125}I surface-labeled RDM-4 cells in 0.5% Nonidet P-40 (NP-40), 15 mM phosphate buffer. Twenty-microliter aliquots of the MAb–Sepharose were added to individual tubes. Five hundred microliters of the indicated detergent solutions were added to the tubes, and the beads were suspended by shaking. After 20 min the beads were pelleted by centrifugation and the supernatant was removed. This extraction was repeated once. The remaining ^{125}I-labeled H-2K[k] bound to the beads was eluted by heating in 100 μl of 1% SDS at 100° for 5 min. The SDS eluates were analyzed on a sodium dodecyl sulfate (SDS)–12% polyacrylamide gel after reduction. Percentage of elution was calculated using the sum of the ^{125}I radioactivity in the H-2K[k] heavy chain band and the β_2-microglobulin band. Heavy chain accounted for 75–80% and β_2-m for 20–25% of the total radioactivity. Samples treated with 0.5% NP-40, 15 mM phosphate buffer were taken as 0% elution. From Herrmann and Mescher.[7]

(DOC) (at pH 8) all eluted H-2K[k] from the MAb–Sepharose beads. Similar experiments with other MAb–Sepharose beads have been done to define the minimal conditions for elution of Ia and other H-2 antigens. In each case, the conditions found to be effective for elution from the beads in suspension have also been effective for elution from affinity columns made from the same beads.

H-2 antigens bound to either of the two anti-H-2 MAbs that we have examined can be eluted with 0.5% DOC at pH 8 with moderate (0.14–0.65 M) NaCl concentration. Elution under these very mild conditions may result from an effect of DOC on the conformation of the H-2 molecule,[24] and similar conditions may prove to be effective for elution from other anti-H-2 MAbs. In contrast, Ia antigens were not eluted from affinity columns made from the 10-2.16 (anti-I-A^k)[8] or 14-4-4S (anti-I-E^k)[25] MAbs under these mild conditions. In both cases, high pH buffers gave effective elution.

The choice of buffers used for affinity purification depends upon the desired pH and the method to be used for quantitation of the purified protein (see below). We have used a variety of buffers including sodium phosphate, Tris, triethanolamine (TEA), and N-2-hydroxyethylpiperazine-N'-2-ethanesulfonic acid (HEPES) at pH 7–8.5 and sodium carbonate at higher pH. No differences in yield or serological activity of the purified antigens have been noted with the different buffers. Triton X-100 (or NP-40) is the preferred detergent for antigen solubilization from cells or membranes. Nuclei remain relatively intact in this detergent and can be removed by centrifugation, and the detergent does not interfere with antigen binding by antibodies. The choice of detergent present during elution depends upon the effect of the detergent on antigen binding (see above), on stability of the antigen in the detergent, and on the desired use of the purified antigens. We have used DOC most extensively, since it can be removed by dialysis, allowing incorporation of the purified antigens into liposomes for functional studies. H-2 has also been purified using Triton X-100 or octyl-β-glucoside in the elution buffer.

The affinity chromatography is normally done by first passing the cell lysate (containing ^{125}I-labeled antigen) through a 5–10 ml column containing Sepharose 4B to remove material that will nonspecifically bind to Sepharose. This is attached to the MAb–Sepharose column with Tygon tubing so that the lysate then passes through the affinity column. The flow rate is adjusted to 20–25 ml/hr or less. Large-scale preparations generally require 12–24 hr to pass all of the lysate over the columns. The Sepharose precolumn is then disconnected and discarded, and the affinity column is washed with 5 to 10 column volumes of lysis buffer. If possible, it is also desirable to wash the column with several volumes of a buffer of higher pH, higher NaCl concentration, or different detergent composition that will remove nonspecifically bound material but leave the antigen bound. For example, washing the 10-2.16 or 14-4-4S columns with 0.5% DOC at pH 8 leaves the Ia bound but removes contaminating proteins.[10]

After washing is complete, the wash buffer remaining on the column is removed, elution buffer is added, and elution is begun with collection of 1-

[24] S. H. Herrmann, C. M. Chow, and M. F. Mescher, *J. Biol. Chem.* (in press).
[25] K. Ozato, N. Mayer, and D. H. Sachs, *J. Immunol.* **124**, 533 (1980).

ml fractions. Failure to remove the remaining wash buffer before elution results in broadening of the elution peak and dilution of the antigen. When elution requires high pH, fractions are collected into tubes containing high molarity neutral pH buffer to minimize exposure of the antigen to high pH. Exposure of the affinity column to high pH is also minimized by washing with neutral pH buffer as soon as elution is complete. Columns are normally washed with two to three column volumes of pH 4 or pH 10 buffer (or both) after use and then reequilibrated in lysis buffer containing 0.02% azide. All the steps for affinity purification are carried out at 4°.

5. Analysis and Quantitation of the Purified Antigen

The inclusion of radioactively labeled antigen in the initial detergent lysate allows the elution profile of the affinity column to be readily determined. We have found that the profile obtained using ^{125}I-labeled antigen corresponds to that obtained by determining the protein content of each fraction. Antigen usually elutes in a volume of 1–1.5 times the column volume, and pooling the fractions to include 85–95% of the antigen results in a final protein concentration of 30–200 μg/ml. Protein concentrations in this range can be measured using Folin–Ciocalteu (phenol) reagent as described by Lowry et al.,[26] the o-phthalaldehyde method of Butcher and Lowry,[27] or by using fluorescamine.[28] Using bovine serum albumin (BSA) as the protein standard in each case, we have found that the concentration values obtained for purified H-2 by the assay of Lowry et al.[26] and the o-phthalaldehyde method are in good agreement, whereas the fluorescamine assay gives a 30–40% lower value. It is likely that the o-phthalaldehyde method provides the most accurate estimate, since this measures total primary amines after hydrolysis. The fluorescamine or o-phthalaldehyde methods can be used for samples buffered with HEPES, TEA, or carbonate. The method of Lowry et al.,[26] can be used for samples buffered with Tris or carbonate.

Purity of the isolated antigens has been assessed by SDS–polyacrylamide gel electrophoresis. It is usually necessary to concentrate the protein prior to application to the gel, and this is easily done by acetone precipitation. One volume of protein sample is mixed with six volumes of ice-cold acetone, kept at −20° overnight, and centrifuged for 10 min at 2000 rpm to pellet the protein. The supernatant is removed and discarded, and the pellet is dried under an N_2 stream. Eighty to one-hundred percent

[26] O. H. Lowry, N. J. Rosebrough, A. L. Farr, and R. J. Randall, J. Biol. Chem. 193, 265 (1951).
[27] E. C. Butcher and O. H. Lowry, Anal. Biochem. 76, 502 (1976).
[28] S. Udenfriend, S. Stein, P. Bohlen, and W. Dairman, Science 178, 871 (1972).

of the protein is precipitated under these conditions. The protein is then solubilized directly in sample application buffer for gel electrophoresis. Electrophoresis using the buffer system of Laemmli[29] on long (20 cm) gels consisting of a 5 to 15% gradient of polyacrylamide allows very high resolution. The H-2 heavy chain is well resolved from actin, and the Ia heavy and light chains are well separated. Convenient molecular weight markers include ovalbumin (M_r 45,000) and lysozyme (M_r 14,300) having molecular weights similar to H-2 heavy chain and β_2-m and glyceraldehyde-3-phosphate dehydrogenase (M_r 36,000), which is similar in size to the Ia chains. Five to ten micrograms of antigen are easily visualized by staining with Coomassie Blue, and radiochemical purity is determined by autoradiography. ^{125}I-Labeled protein can be quantitated by locating the bands by autoradiography, cutting them from the dried gel, and counting in a gamma counter.

The serological activity of the purified antigen can be assessed by determining the percentage that will rebind to the MAb–Sepharose column. Alternatively, the purified antigen can be used as an inhibitor of immunoprecipitation, cell binding, or complement-mediated cytolysis by monoclonal antibodies or alloantisera. Quantitative comparison to the inhibition obtained by lysate from a known number of the starting cells allows calculation of the yield of active antigen. The biological activity of the purified H-2 antigens can also be qualitatively assessed by incorporating them into liposomes and measuring stimulation of a secondary CTL response.[9,30] Since a number of factors, including the structure of the liposomes,[30] affect generation of a response, this assay does not allow for convenient quantitation of the yield of active antigen.

B. Specific Examples

1. Purification of H-2K^k

a. Antibody Affinity Column. The 11-4.1 monoclonal cell line[8] (available from the American Type Culture Collection) was grown *in vitro*, and the antibody was purified from the supernatant of stationary phase cultures on a protein A–Sepharose column. The culture supernatant was passed over the column at pH 7.8, and the column was then washed with 10 volumes of 10 mM sodium phosphate, pH 7, 0.14 M NaCl (phosphate-buffered saline; PBS). Bound antibody was eluted with 50 mM glycine at pH 4, and the eluate was collected in 1-ml fractions into tubes containing

[29] U. K. Laemmli, *Nature (London)* **227**, 680 (1970).
[30] S. H. Herrmann and M. F. Mescher, *Proc. Natl. Acad. Sci. U.S.A.* **78**, 2488 (1981).

0.1 ml of 0.1 M Tris, pH 8. Forty to fifty micrograms of MAb were obtained per milliliter of culture supernatant. The purified antibody was then dialyzed overnight versus 20 volumes of 0.1 M NaHCO$_3$, 0.15 M NaCl, pH 8.3, and adjusted to about 1 mg/ml by dilution or by concentration using an Amicon ultrafiltration unit (Amicon Corp., Lexington, Massachusetts).

Sepharose 4B was cyanogen bromide-activated for coupling of the protein. Activation was done at room temperature in a chemical fume hood. Ten milliliters (wet settled volume) of Sepharose 4B were washed with 15 volumes of water in a Büchner funnel, suspended in 10 ml of water, and transferred to a beaker. The beads were kept suspended by slowly stirring with a magnetic stirring bar and adjusted to pH 11 with 0.1 N NaOH. With continued stirring, 0.2 mg of CNBr (dissolved in acetonitrile at 0.67 mg/ml) was added. The pH of the suspension begins to drop upon addition of the CNBr and continues to drop during the course of the reaction. The solution was maintained at pH 11 (10.5–11.5) by adding small aliquots of 0.1 N NaOH. The reaction was allowed to proceed for 10 min, and the beads were then transferred to a Büchner funnel and washed with 15 volumes of 0.001 M HCl. The beads were then resuspended in 10 ml of 0.1 M NaHCO$_3$, 0.15 M NaCl at pH 8.3 and cooled to 4°.

Coupling of the MAb to the activated Sepharose was done by mixing the suspended beads with 25 ml of purified MAb at about 1 mg/ml. This was mixed continuously for 16–20 hr at 4° by rocking or rotating. Mixing was then stopped, and the beads were allowed to settle sufficiently to remove 0.5 ml of supernatant, which was then assayed to determine the amount of uncoupled antibody left in solution. A sufficient volume of 0.1 M glycine, pH 7, was added to the bead suspension to give a final concentration of 0.05 M, and mixing was continued at 4° for 4 hr. The beads were then pelleted by low speed centrifugation, the supernatant was removed, and the beads were washed with 15–20 volumes of 0.1 M NaHCO$_3$, 0.5 M NaCl, pH 10, followed by 15–20 volumes of 0.1 M sodium acetate, pH 4. Washing with buffers of high and low pH was repeated for a total of four cycles, and the beads were then washed 3 times with the buffer in which they were to be used. This procedure routinely results in coupling of more than 90% of the antibody.

The MAb–Sepharose column was prepared in a 10-ml plastic disposable pipette plugged at the bottom with glass wool. The bead suspension in 0.5% Triton X-100, 15 mM sodium phosphate, pH 7.2, was poured in, allowed to settle without flow for 10 min, and then packed by gravity flow and washed with several volumes of buffer.

  b. Cells and ^{125}I-Labeling. RDM-4 lymphoma cells were maintained by weekly passage (i.p.) in AKR mice. The cells grow to higher density in

(AKR × DBA/2)F$_1$ (AKD2F$_1$) mice, and these are therefore used for growing cells for large-scale preparations. Interperitoneal injection of 2 × 10^7 cells per mouse yields about 5 × 10^8 cells per mouse in 6 days. Cells from 20 mice were harvested and washed three times in PBS to yield about 10^{10} cells.

If ^{125}I-labeled cell surface proteins are to be used to monitor purification, 5 × 10^7 cells are removed and labeled by LPO-catalyzed iodination (described in Section B,3). Alternatively, previously purified H-2K^k is ^{125}I-labeled using Iodogen.[23]

Iodogen was used to ^{125}I-label[23] 25 μg of purified H-2K^k for use in monitoring the purification. Iodogen (Pierce Chem. Co., Rockford, Illinois) was dissolved in dichloromethane (at room temperature) at 0.2 mg/ml, and 0.1 ml was added to a 13 × 45 mm flat-bottom glass vial. The solvent was allowed to evaporate completely in a fume hood, and 25 μg of H-2K^k in 0.1 ml of 0.5% DOC, 10 mM Tris, 0.14 M NaCl along with 0.5 mCi of Na^{125}I were added; the reaction was allowed to proceed for 10 min at room temperature with occasional mixing. The solution was then removed from the reaction vial and mixed with 0.02 ml of 2 mM tyrosine and 0.1 ml of 2 mM KI, 0.5% hemoglobin in 0.5% DOC, 20 mM Tris, pH 8. Labeled protein was then separated from free iodine by gel filtration on Sephadex G-25 as described by Tuszynski et al.[31] The resulting ^{125}I-labeled H-2K^k sample had an activity of about 5 × 10^6 cpm per microgram of starting protein (the recovery of protein was not determined).

c. Preparation of Cell Lysate. Approximately 10^{10} washed RDM-4 cells were pelleted by centrifugation; the supernatant was removed, and the cells were resuspended at 4° in 20 ml of 10% Triton X-100, 50 mM sodium phosphate, pH 7.2, containing 2 mM PMSF. The suspension was left for 20 min at 4° with occasional mixing and then centrifuged for 10 min at 3600 g to pellet the nuclei. The supernatant was transferred to new tubes and centrifuged for 45 min at 100,000 g. The supernatant from this spin was diluted to a final volume of 100 ml with 5 mM sodium phosphate, pH 7.2, and passed through a 0.45 μm filter. ^{125}I-Labeled H-2K^k tracer was added to the lysate, and affinity chromatography was done.

d. Affinity Chromatography. A 10-ml precolumn of Sepharose 4B was prepared in a disposable plastic syringe, and the outflow was connected to the MAb–Sepharose affinity column. The detergent lysate was passed over both columns by gravity flow with the flow rate maintained at 25 ml/hr or less. After all the lysate had passed over the columns, the precolumn was disconnected and the MAb–Sepharose column was washed with 75

[31] G. P. Tuszynski, L. Knight, J. R. Piperno, and P. N. Walsh, *Anal. Biochem.* **106,** 118 (1980).

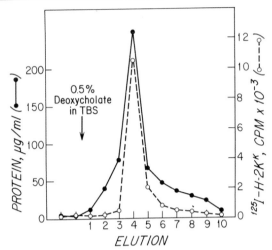

FIG. 2. Elution of H-2K^k from an 11-4.1 MAb–Sepharose column. Three-milliliter fractions were collected and assayed for protein (●——●) and ^{125}I radioactivity (○---○).

ml of 20 mM triethanolamine, 0.5% Triton X-100, pH 8. The flow was then stopped, wash buffer was removed from the column down to the level of the beads, and elution buffer was added. Elution was done with 0.5% sodium deoxycholate, 140 mM NaCl, pH 8, with gravity flow at a rate of 10–20 ml/hr. Three milliliter fractions were collected and assayed for radioactivity and protein content by the o-phthalaldehyde method of Butcher and Lowry.[27] The resulting elution profile is shown in Fig. 2.

e. Yield and Purity. Yields of H-2K^k by this procedure are typically 1–1.5 mg from 10^{10} RDM-4 cells. Coomassie Blue-stained gels of the purified material show bands having apparent molecular weights of 47,000 (heavy chain) and 12,500 (β_2-m) (Fig. 3). The only detectable contaminant is a faintly staining band of about 53,000 molecular weight seen on heavily loaded gels of some preparations.[7] About 80% of the purified antigen will rebind to the MAb–Sepharose column, and about 75% of the H-2K^k serological activity present in the whole-cell lysate is recovered with the purified protein.[7] The purified antigen retains biological activity as measured by stimulation of an allogeneic CTL response. When incorporated into liposomes having the appropriate structure, the purified H-2K^k has activity comparable to that of membranes having an equivalent amount of antigen.[30,32]

[32] S. H. Herrmann, O. Weinberger, S. J. Burakoff, and M. F. Mescher, *J. Immunol.* **128**, 1968 (1982).

RDM-4 cells express unusually high levels of H-2K^k. MAb–Sepharose affinity purification yields only 0.3 to 0.5 mg of H-2K^k from 10^{10} YAC (H-2^a) tumor cells and 0.1–0.2 mg from 10^{10} liver cells.[7] We have found that the level of H-2K^k on RDM-4 cells sometimes decreases dramatically when the cells are maintained over a prolonged period by repeated *in vivo* passage. We therefore start fresh cells from frozen stock after 10 passages (about 10 weeks).

2. Papain Cleavage and Repurification of H-2K^k

We have been unable to obtain the water-soluble papain cleavage product of H-2K^k in good yield by releasing it from cells or membranes by papain treatment. It has been possible, however, to obtain the cleavage product by papain treatment of the purified antigen.[24] Efficient cleavage

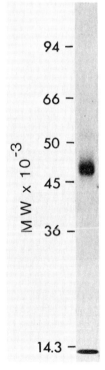

FIG. 3. Sodium dodecyl sulfate–polyacrylamide gel electrophoresis. Four micrograms of affinity-purified H-2K^k were electrophoresed on a slab gel consisting of a 5 to 15% gradient of polyacrylamide. Protein was visualized by staining with Coomassie Blue. Molecular weights are based on mobilities of standard proteins run in parallel.

and good recovery of the limited proteolysis product is dependent on the detergent composition of the solution, the concentration of reducing agent, and the time and temperature of the reaction. The effects of these parameters, as well as conditions for obtaining limited proteolysis products lacking part or all of the hydrophilic C-terminal region by treatment with chymotrypsin or trypsin, are described in detail elsewhere.[24] The procedure for obtaining an optimum yield of the papain cleavage product is described here.

Papain (Sigma) was activated by incubation at 1 mg/ml in 20 mM cysteine in 50 mM TEA, pH 8 for 30 min at 37°. Affinity-purified H-2K^k (at 50–200 μg/ml) in 0.25% Triton X-100, 0.25% DOC, 10 mM TEA, 20 mM NaCl, 5 mM cysteine, pH 8, was incubated for 30 min at 24° with 0.08 μg of activated papain per microgram of H-2. The reaction was then stopped by adding 0.2 M iodoacetamide (adjusted to pH 8) to give a final concentration of 0.01 M. The sample was then either dialyzed overnight versus 0.5% Triton X-100, 10 mM TEA, pH 8, or diluted 10-fold with this same buffer to reduce the DOC concentration sufficiently to allow rebinding to the affinity column. Repurification of the cleavage product was then done by passing the sample over a column of normal rabbit IgG coupled to CNBr-activated Sepharose at 2 mg/ml and then over the affinity column. The normal rabbit IgG–Sepharose column is necessary because the papain will bind to Ig and if bound to the MAb–Sepharose column will coelute with the H-2. The papain cleavage product was eluted from the affinity column exactly as described for the intact H-2K^k. Alternatively, it can be eluted in detergent-free buffer by first washing the column with several volumes of 10 mM TEA, pH 8, and then eluting with the same buffer containing 0.5 M NaCl. About 60% of the starting material was recovered as the papain cleavage product with an apparent molecular weight for the heavy chain of 40,600. In most preparations, no uncleaved heavy chain can be detected on SDS gels.

3. Purification of H-2K^d and D^d(L^d) Antigens Using the M1/42 Antibody

The M1/42 monoclonal antibody was obtained after immunizations of a DA rat with B10 spleen cells and fusion of the immune rat spleen cells with the mouse myeloma line NS1.[33] Binding studies and immunoprecipitation have shown that the antibody reacts with H-2 antigens from all haplotypes tested (including *a, b, d, j, k, s,* and *u*).[9] Furthermore, it appears to react with both the H-2K and D(L) products of at least the *d*

---

[33] T. A. Springer, *in* "Monoclonal Antibodies" (R. H. Kennett, T. J. McKearn, and K. B. Bechtol, eds.), pp. 185–217. Plenum Press, New York, 1980.

and k haplotypes.[9] The protocol used to purify the H-2K^d, D^d (L^d) antigens from P815 tumor cells is described.

a. Antibody Affinity Column. The M1/42 monoclonal cell line (available from the American Type Culture Collection) was grown *in vitro,* and the antibody was purified from the supernatant of stationary phase cultures. This antibody is not bound by protein A and was therefore purified by ammonium sulfate precipitation and Sephadex G-200 (Pharmacia) chromatography and coupled to CNBr-activated Sepharose 4B at 2.5–3 mg per milliliter of wet gel, as described in Section B,1.

b. Cells and ^{125}I-*Labeling.* P815 (H-2^d) tumor cells were maintained by weekly passage in (BALB/c × DBA/2)F$_1$ mice (CD2F1). Cells for large-scale antigen purification were obtained by i.p. injection of 60 mice at 1×10^7 cells per animal. Cells were harvested 1 week later and washed three times with PBS. The yield was about 5×10^8 cells per mouse.

Washed cells (5×10^6) were suspended in 0.2 ml of PBES (Earle's balanced salt solution (GIBCO, Grand Island, New York) buffered at pH 7.2 with 10 mM sodium phosphate). Ten micrograms of lactoperoxidase (Calbiochem, San Diego, California) and 0.2 mCi of carrier-free Na^{125}I (New England Nuclear, Boston, Massachusetts) were added, and the reaction was started by addition of 5 μl of 1.8 mM H$_2$O$_2$. The reaction was carried out at room temperature. Additional 5-μl aliquots of H$_2$O$_2$ were added at 3-min intervals for a total of six additions. After the final addition, the sample was left for 5 min at room temperature, and the cells were then pelleted by centrifugation and washed three times in cold PBS containing 10% fetal calf serum. Labeling by this procedure yields about 1–4 cpm per cell. The washed cells were then lysed by suspending them in 0.5% Triton X-100, 20 mM Tris, pH 8, and leaving them for 20 min at 4°. The lysate was centrifuged at 2000 rpm for 10 min to pellet the nuclei, and the supernatant containing the ^{125}I-labeled cell-surface proteins was kept on ice and added to the large-scale detergent lysate immediately before the affinity chromatography was done.

c. Preparation of Large-Scale Detergent Lysate. Direct detergent lysis of P815 cells results in highly viscous lysates unsuitable for affinity purification. Cells were therefore lysed by nitrogen cavitation and centrifuged to yield two pellet fractions, one containing nuclei and the other membranes (Fig. 4). The nuclear fraction was solubilized by addition of 0.5% Triton X-100, 20 mM Tris, pH 8 (containing PMSF) using 1 ml per 5×10^7 starting cells. After 20 min on ice with mixing, the suspension was centrifuged at 3600 g for 10 min, and the supernatant was removed and centrifuged for 45 min at 100,000 g. The membrane pellet was similarly solubilized by addition of 1% Triton X-100, 20 mM Tris, pH 8 (with PMSF) at a detergent-to-protein ratio of 5 : 1. After 10 min on ice, the

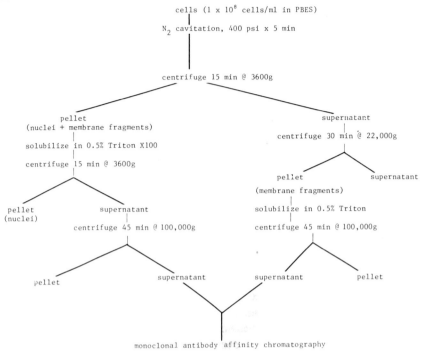

FIG. 4. Scheme for subcellular fractionation and antigen solubilization from P815 tumor cells. From Stallcup *et al.*[9]

sample was centrifuged at 100,000 *g* for 45 min, and the supernatant was collected. The supernatants obtained from detergent solubilization of the nuclear and membrane pellets were combined, and the lysate from the ^{125}I-labeled cells was added.

d. Affinity Chromatography. Antigens were purified by passing the lysate over a 1-ml precolumn containing Sepharose 4B, then a 1-ml column of M1/42–Sepharose. Loading and washing were done as described in Section B,1,d, and the bound antigens were then eluted with 0.5% DOC, 0.65 *M* NaCl, 20 m*M* Tris, pH 8 (Fig. 5). This solution is most easily prepared by addition of an appropriate volume of 5 *M* NaCl to 0.5% DOC, 0.15 *M* NaCl, 10 m*M* Tris, pH 8. Solutions of DOC in salt gel upon prolonged standing, and the elution buffer is therefore prepared immediately before use. For the same reason, the column is washed with Triton X-100 containing buffer as soon as elution is complete.

e. Yield and Purity. This procedure yields 110–180 µg of antigen from 10¹⁰ P815 cells. About 60% of the purified antigen will rebind to the affinity column, and the antigen, when incorporated into liposomes, will stimulate

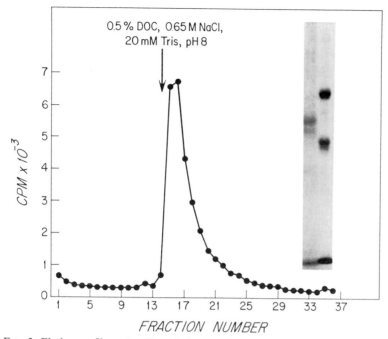

FIG. 5. Elution profile and sodium dodecyl sulfate (SDS) gel electrophoresis of H-2 antigens purified from P815 cells by affinity chromatography on M1/42–Sepharose. The left lane of the SDS gel contained purified H-2^d. The right lane had bovine serum albumin (M_r 69,000), ovalbumin (M_r 42,000), and lysozyme (M_r 14,300) as standards.

generation of a secondary allogeneic CTL response.[9] Coomassie Blue staining of the antigen on SDS gels visualizes only the heavy chains and β_2-m. H-2K^d and D^d heavy chains have slightly different molecular weights[34] and are resolved by electrophoresis on 5 to 15% polyacrylamide gradient gels in SDS. Both heavy chains are present in antigens purified from P815 cells by affinity chromatography on M1/42 Sepharose.

The M1/42 antibody reacts with the H-2K and D antigens of a variety of haplotypes. It is likely that it also reacts with L locus products, but this has not been determined. Affinity chromatography on M1/42-Sepharose has been used successfully to purify H-2^b antigens from EL-4 and H-2^k antigens from RDM-4. In tandem with the 11-4.1–Sepharose column, M1/42 has been used to obtain H-2D^k from RDM-4 (K^{k}D^k) cells and H-2D^d from YAC (K^{k}D^d) cells.

[34] T. Krakauer, T. H. Hansen, R. D. Camerini-Otero, and D. H. Sachs, *J. Immunol.* **127,** 2149 (1981).

3. Purification of I-A^k and I-E^k Antigens

a. Antibody Affinity Columns. Hybridoma cell lines 10-2.16 (anti-I-A^k)[8] and 14-4-4S (anti-I-E^k),[25] both available from the American Type Culture Collection, were grown *in vitro*. The antibodies were purified from supernatants of stationary-phase cultures using a protein A–Sepharose column as described in Section B,1. The purified antibodies were coupled to CNBr-activated Sepharose at 2 mg/ml (see Section B,1,a), and columns containing 1–2 ml of MAb–Sepharose were prepared.

b. Cells and Labeling. CH1 (H-2^a) tumor cells[17] were maintained by passage in (C57BL6 × A/J)F$_1$ mice (B6AF$_1$) or B10.A mice. Prolonged *in vivo* passage in B6AF$_1$ mice results in greatly decreased yields of tumor cells. Better yields are obtained when the cells are maintained in B10.A mice. Because they are more readily available, B6AF$_1$ mice are used for growing large numbers of cells for antigen purification. Injection of 4 × 10^7 cells i.p. yields about 3 × 10^8 cells per mouse in 8–9 days. Typically, cells from 35 mice are harvested and washed three times in PBS to yield about 10^{10} cells.

CH1 cells label poorly by lactoperoxidase-catalyzed iodination. Cells were therefore surface-labeled with ^{125}I using Iodogen.[22,35] Iodogen was dissolved in dichloromethane at 0.25 mg/ml, and 0.25 ml was added to a flat-bottom glass vial of 25-mm diameter. The solvent was removed by evaporation in a fume hood, and 5 × 10^7 cells in 1 ml of PBES were added along with 2.5 mCi of Na^{125}I in PBS. The contents were mixed and let incubate for 15 min at room temperature. The cell suspension was then transferred to a plastic tube, and the cells were washed three times with PBES at 4°. The labeled cells, having 1–2 cpm per cell, were lysed by incubation for 20 min at 4° in 0.5% Triton X-100, 20 mM Tris, pH 8, 0.14 M NaCl. The lysate was then centrifuged at 3600 g for 10 min, and the supernatant was centrifuged at 100,000 g for 45 min. The final supernatant was added to the large-scale detergent lysate immediately before loading the MAb–Sepharose columns.

c. Preparation of Large-Scale Detergent Lysate. Cells (10^{10}) were suspended in 200 ml of 0.5% Triton X-100, 10 mM Tris, pH 8 (with PMSF), 0.14 M NaCl. After 20 min on ice, the lysate was centrifuged at 3600 g for 15 min. The supernatant was transferred to new tubes and centrifuged at 100,000 g for 45 min. Supernatant from this centrifugation was passed through a 0.45-μm filter, and the ^{125}I-labeled cell lysate was added.

d. Affinity Chromatography. The cell lysate was passed through a precolumn containing 3 ml of Sepharose 4B and then through the

---

[35] J. G. Salisburg and J. M. Graham, *Biochem. J.* **194**, 351 (1981).

10-2.16–Sepharose column. The flow through was then passed through the 14-4-4S–Sepharose column. The MAb–Sepharose columns were then disconnected and washed individually with 50 column volumes of lysis buffer. Each column was then washed with 20–30 column volumes of 0.5% DOC, 20 mM Tris, pH 8, 0.14 M NaCl. This wash buffer removes nonspecifically bound material while leaving the Ia antigens bound.

After washing, the columns were eluted with high-pH buffer containing DOC. I-A^k eluted from the 10-2.16–Sepharose column at pH 11 with 0.5 M NaCl. Higher pH or salt concentration did not remove additional

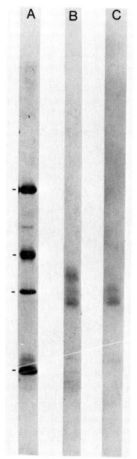

Fig. 6. Sodium dodecyl sulfate–polyacrylamide gel electrophoresis of affinity-purified I-A^k. (A) Standard proteins including bovine serum albumin (M_r 69,000), ovalbumin (M_r 42,000), G-3-P (M_r 36,000), and lysozyme (M_r 14,300). (B) I-A^k stained with Coomassie Blue. (C) Autoradiogram of ^{125}I-labeled I-A^k.

antigen. Some of the bound I-E^k eluted from the 14-4-4S–Sepharose column at pH 11, 0.14 M NaCl, but additional antigen was eluted at pH 12, 0.14 M NaCl. The basis for this heterogeneous elution is not known. All the antigen is eluted at pH 12, as demonstrated by failure to elute any additional Ia upon treatment of the MAb–Sepharose with 1% SDS at 100°. During elution, 1-ml fractions were collected into tubes containing 0.1 ml

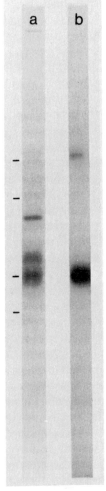

FIG. 7. Sodium dodecyl sulfate–polyacrylamide gel electrophoresis of affinity purified I-E^k. Bars at left indicate positions of protein standards run in parallel. These included bovine serum albumin (M_r 69,000), Ig heavy chain (M_r 50,000), G-3-P (M_r 36,000), and Ig light chain (M_r 25,000). (a) I-E^k stained with Coomassie Blue. (b) Autoradiogram of ^{125}I-labeled I-E^k.

of 0.4 M Tris, pH 8, to neutralize the sample. Fractions containing the eluted Ia antigen were pooled, dialyzed to reduce the concentration of salt and buffer, and analyzed.

e. Yield and Purity. Approximately 200 μg of I-A^k are obtained from 10^{10} CH1 cells. About 70% of the purified antigen rebinds to the 10-2.16–Sepharose column. Coomassie Blue staining of the antigen run on SDS gels resulted in bands with apparent molecular weights of about 39,000 and 34,000 (Fig. 6). Proteins having mobilities identical to those of actin and myosin are frequently found as contaminants in the antigen preparations, and conditions that reliably avoid the presence of these contaminants have not been found.

Purification of I-E^k has not been examined as extensively, and reliable values for yields and rebinding activity have not been obtained. Comparison to I-A^k on Coomassie Blue-stained gels indicates that similar amounts of each antigen are obtained (Fig. 7). The apparent molecular weights of the heavy and light chains of I-E^k are slightly lower than these of I-A^k. The I-E^k which elutes at pH 11 appears identical on SDS gels to that which elutes at pH 12. As in the case of I-A^k, proteins having the same mobilities as actin and myosin are frequently seen as contaminants in the I-E^k preparations (Fig. 7).

Surface-labeling with ^{125}I by LPO-catalyzed iodination labels only the light chain of I-A^k [36] (Fig. 6). This is also true for I-E^k when labeled on intact cells using Iodogen (Fig. 7). Vectorial labeling (labeling of only cell-surface proteins) by Iodogen is indicated by failure to detect label in the actin- and myosin-like proteins contaminating the I-E^k preparation.

I-A^k and I-E^k have been purified from lipopolysaccharide (LPS)-stimulated spleen cells (2.5–3.5 days) with essentially identical results to those shown here using CH1 cells.[10] Lipopolysaccharide blasts appear to have more antigen per cell, but the difficulty of obtaining large numbers of cells by *in vitro* stimulation with LPS make CH1 a more convenient source of antigen for purification.

C. Conclusions

Monoclonal antibody affinity chromatography has proved to be a rapid and efficient means of purifying relatively large amounts of both class I and class II murine MHC antigens. We have successfully used four monoclonal antibodies as affinity reagents. It is likely that the same general approach will be applicable to purification of a variety of MHC antigens using different monoclonal antibodies. The antigens purified in this way retain serological and biological activity and are proving to be useful for studying the structure and function of these molecules.

[36] B. D. Schwartz, E. S. Vitetta, and S. E. Cullen, *J. Immunol.* **120,** 671 (1978).

[9] Monoclonal Antibodies against HLA Products and Their Use in Immunoaffinity Purification

By PETER PARHAM

HLA monoclonal antibodies have been produced from mice immunized with whole cells,[1] membrane fractions,[2] and purified antigen preparations.[3] However, a comparison of the type, frequency, and specificity of antibodies obtained with different immunization protocols has yet to be made. There appear to be biases one can introduce, but in assessing the available information it is difficult to separate selectivity introduced by the assays as opposed to the immunization. When producing antibodies for a specific purpose, it is sensible to use a screening assay based on the required purpose. For example, clinical HLA typing uses an antibody-dependent, complement-mediated cytotoxicity assay, and antibodies required for immediate use as typing reagents will be successfully found by screening with the standard typing method as assay.[4]

Cells provide the most convenient source of immunogens and contain HLA products in their native form. Paradoxically, the one antibody (Q1/28) that strongly binds HLA-A,B,C heavy chain, without associated β_2-microglobulin was produced by immunization with cells.[5] Obviously, antibodies to HLA will be diluted among a general immune response to the human cell surface. Identification of HLA antibodies in such a mixture is not difficult, as there are cells that do not express HLA-A,B,C, or DR but retain many other antigens. A screen for identifiable polymorphisms will also limit the window to mainly HLA products. However, one consequence of immunization with cells is that the number and frequency of antibodies against a given subset of molecules, i.e., HLA products, will probably be less. Many reports describe single HLA antibodies arising from such fusions with only rare cases of multiple antibodies. In contrast, with purified antigen preparations we have commonly obtained up to 10 HLA antibodies from a single fusion.[6] Naturally, these considerations do

[1] M. M. Trucco, J. W. Stocker, and R. Ceppellini, *Nature (London)* **273**, 666 (1978).

[2] C. J. Barnstable, W. F. Bodmer, G. Brown, G. Galfrè, C. Milstein, A. F. Williams, and A. Ziegler, *Cell* **14**, 9 (1978).

[3] P. Parham and W. F. Bodmer, *Nature (London)* **276**, 397 (1978).

[4] F. C. Grumet, B. M. Fendly, and E. G. Engleman, *Lancet* **2**, 174 (1981).

[5] V. Quaranta, L. E. Walker, G. Ruberto, M. A. Pellegrino, and S. Ferrone, *Immunogenetics* **13**, 285 (1981).

[6] P. Parham and F. M. Brodsky, Research Monographs in Immunology (G. J. Hämmerling, U. Hämmerling, and J. F. Kearney, eds.), Vol. 3, p. 67. Elsevier/North-Holland, Amsterdam, 1981.

not apply to immunizations between inbred and congenic strains of rodents where the genetics provides an operationally pure set of MHC molecules in the form of cells. McKearn and co-workers produced numerous antibodies against MHC antigens by allogeneic cell immunization of rats and many significantly cross-react with cells of other species including man.[7] Mouse monoclonal antibodies against H-2 region products have been produced from the spleen cells of mice immunized by the methods used to generate alloantisera. The immunogen is tissue, e.g., skin graft, or spleen cell suspension, and genetic considerations determine the choice of donor and recipient strains and the potential specificities of the antibodies.

It has not been easy to generate H-2 monoclonal antibodies, as indicated by Hämmerling's[8] and Sachs'[9] groups in their first reports: "We would like to mention that in many other fusions performed under comparable conditions among thousands of hybridomas no positive ones could be obtained."[8] "Production of anti-H-2 hybridomas was at first surprisingly difficult relative to other antigenic specificities, although over the past two years several laboratories, including our own, have succeeded in producing hybridomas."[9] Nevertheless, the frequency of success has apparently been high enough to deter concerted investigation of alternative approaches. The tempo of immunization and fusion has been varied,[8,9] and differences were seen in the numbers of hybridomas obtained. However, the small sample size and large numbers of uncontrolled variables in any given fusion experiment precludes any assessment of an optimal protocol. Changes in strain combination alter both the spectrum of the antigenic disparity and the potential immune response. For example, Ozato and Sachs[10] found that the frequency of IgM hybridomas was much higher from C3H anti-C3HSW than the reciprocal combination.

It is of concern that cytotoxicity-based assays are usually used to screen for H-2 monoclonal antibodies because IgG1, a predominant class of mouse immunoglobulins, is not detected by this method. As isotype and specificity may not be completely independent variables, this method may result in a bias that may in part account for the difficulty in obtaining H-2 monoclonal antibodies. An additional inadequacy in cytotoxic assays was alluded to by Hämmerling et al.[8] They found that only the most powerful complement gave good cytotoxicity with monoclonal antibod-

[7] D. E. Smilek, H. C. Boyd, D. B. Wilson, C. M. Zmijewski, F. W. Fitch, and T. F. McKearn, J. Exp. Med. 151, 1139 (1980).
[8] G. J. Hämmerling, U. Hämmerling, and H. Lemke, Immunogenetics 8, 433 (1978).
[9] K. Ozato, N. Mayer, and D. H. Sachs, J. Immunol. 124, 533 (1980).
[10] K. Ozato and D. H. Sachs, J. Immunol. 126, 317 (1981).

ies, and this has been confirmed by Richiardi *et al.*[11] using HLA monoclonal antibodies. The implication of these results is that insufficient sites are binding antibody, because of its monoclonality, to guarantee sufficient complement fixation to produce cell lysis. The observation that two anti-rat MHC monoclonal antibodies were cytotoxic only in combination suggests this as a feasible possibility.[12] If complement fixation is limiting, it is likely that antibodies of lower avidity or with specificity against less densely distributed target molecules will be missed even though they may be of complement-fixing isotype. This problem can be reduced by using a second-stage enhancing antibody with specificity for mouse immunoglobulin.[9] Alternatively, binding assays using general anti-mouse immunoglobulin reagents can be adopted.[12a]

Ceppellini, Trucco, and colleagues have immunized mice with human peripheral blood lymphocytes and B-cell lines and produced antibodies with reactivities similar to some of our antibodies.[13,14] They reported a significant proportion of cytotoxic monoclonal IgGs, which may be a consequence of the type of antigen and the minimal immunization schedule used (one or two immunizations). We have mostly hyperimmunized mice with purified antigen preparation, and noncytotoxic IgG1 antibodies have predominated.[3,15,16] The single fusion (BB7) that produced cytotoxic HLA-A2 and monomorphic HLA-A,B,C antibodies involved hyperimmunization with HLA-B7 and a final boost with a mixture of HLA-A2 and HLA-B7, again indicative of gentle immunization for HLA-A2 possibly favoring IgG2.[15,17] Cytotoxic IgM antibodies have been produced by immunization with cells[4,18] and predominate in cross-reactive rat anti-MHC antibodies.[7] We have never obtained IgM HLA antibodies.

Membrane preparations offer little obvious advantage over cells as an immunogen. On a mass basis membranes are enriched for HLA products,

[11] P. Richiardi, A. Amoroso, T. Crepaldi, and E. S. Curtoni, *Tissue Antigens* **17**, 368 (1981).
[12] G. Galfrè, S. C. Howe, C. Milstein, G. W. Butcher, and J. C. Howard, *Nature (London)* **266**, 550 (1977).
[12a] A. F. Williams, G. Galfrè, and C. Milstein, *Cell* **12**, 663 (1977).
[13] M. M. Trucco, G. Garotta, J. W. Stocker, and R. Ceppellini, *Immunol. Rev.* **47**, 219 (1979).
[14] T. M. Neri and M. M. Trucco, Research Monographs in Immunology (G. J. Hämmerling, U. Hämmerling, and J. F. Kearney, eds.), Vol. 3, p. 77. Elsevier/North-Holland, Amsterdam, 1981.
[15] F. M. Brodsky, P. Parham, C. J. Barnstable, M. J. Crumptom, and W. F. Bodmer, *Immunol. Rev.* **47**, 3 (1979).
[16] P. Parham, *Immunogenetics* **13**, 509 (1981).
[17] P. Parham and F. M. Brodsky, *Hum. Immunol.* **3**, 277 (1981).
[18] F. C. Grumet, B. M. Fendly, L. Fish, S. Foung, and E. G. Engleman, *Hum. Immunol.* **5**, 61 (1982).

but it is not clear that the immune response they produce is less heterogeneous than with cells. An exception is if cytoplasmic parts of HLA molecules are the antigen of interest. However, despite these theoretical objections, the first and one of the most useful HLA monoclonal antibodies, W6/32, was made in this way and the method has been successful when used.[2,15]

In general a purified antigen preparation is preferable before embarking on production of monoclonal antibodies against any molecule. The immune response is restricted, and a significant proportion of positively reacting cultures will have antibodies against the component of interest. Assays to confirm the specificity of monoclonal antibodies are easier to design if purified components are available. Problems arise with the availability of such preparations if considerable effort or material is required to make them. However, there is a compromise between expenditure of resources at the stage of antigen production versus the stages of fusion, screening, and subsequent tissue culture, which must be assessed for any given antigen. With our involvement in the biochemical characterization of HLA antigens,[19] purified proteins were available and the microgram quantities required for production of monoclonal antibodies did not seriously deplete the tens of milligrams used for structural studies.[20,21] With the availability of some monoclonal antibodies against HLA antigens, purification of these molecules by immunoaffinity methods has become routine, simple, and short compared to the traditional chromatographic schemes that were used.[19,22] HLA antigens are significant components of lymphocyte membranes and can be solubilized in good yield with detergent or papain proteolysis. They can be removed selectively from the solubilized extract by passage through a column of insolubilized antibody or by immune precipitation. An antigen can then be eluted by a method that will depend on the monoclonal antibody. Change of pH, detergent, or ionic strength are commonly used. Invariably, the antigens can be recovered in an immunologically active and antigenic form. McMichael and co-workers have described production of a monoclonal antibody with specificity for HLA-B7 and B27[23] similar to that described by Grumet et al.[4] Their primary immunization was with whole cells, subsequent boosts were with single HLA antigenic specificities purified by these techniques.

[19] P. Parham, B. N. Alpert, H. T. Orr, and J. L. Strominger, *J. Biol. Chem.* **252,** 7555 (1977).

[20] J. A. Lopez de Castro, H. T. Orr, R. J. Robb, T. G. Kostyk, D. L. Mann, and J. L. Strominger, *Biochemistry* **18,** 5704 (1979).

[21] H. T. Orr, J. A. Lopez de Castro, D. Lancet, and J. L. Strominger, *Biochemistry* **18,** 5711 (1979).

[22] R. J. Robb, C. Terhorst, and J. L. Strominger, *J. Biol. Chem.* **253,** 5319 (1978).

[23] S. A. Ellis, C. Taylor, and A. McMichael, *Hum. Immunol.* **5,** 49 (1982).

For purposes of immunization this degree of purification is unnecessary and usually not worth the extra effort. If an antigen preparation is 20–50% pure, there is little to be gained and valuable material may be lost by attempting further purification.

Production of Monoclonal Antibodies against HLA Antigens[24]

Choice of Mice

Most experiments have used female BALB/c mice: females, because they are less aggressive; BALB/c because that is the strain from which the myeloma cells used as fusion partners were derived. The advantage of having both fusion partners of the same strain is that the resulting hybridoma cells can be passaged as tumors in BALB/c mice. When cells from different strains are used, then F1 hybrid mice, which are more expensive and difficult to obtain, can be used to carry the tumors. However, apart from these considerations, no important problems have been encountered with immunizing mice other than BALB/c. Indeed there is considerable logic in trying a number of strains. Genetic control of immune responses by the MHC is a well established, general phenomenon that will affect the spectrum of monoclonal antibodies obtainable from the different strains of mice. This is particularly important when one is making antibodies against MHC antigens because strain differences in MHC type will determine not only the potential immune responses, but also the immunogenicity of the antigen. The extreme example is when anti-H-2 monoclonals are required and only certain strain combinations of immunizing and immunized strain will work. The requirements when immunizing with human antigens will be less stringent but still significant, as it has been shown that murine responses to various HLA alloantigens do correlate with H-2 type.

Immunization

The purified antigen preparation is dialyzed against phosphate buffered saline (PBS) and concentrated by ultrafiltration to a concentration of ≈ 0.1 mg/ml; 0.5 ml is added to an equal volume of complete Freund's adjuvant (CFA) in a 1.5-ml Eppendorf centrifuge tube and vortexed for 5–15 min to produce a relatively liquid emulsion. If vortexing is prolonged, the emulsion takes on the consistency of well beaten egg white and becomes very difficult to manipulate and inject. In consequence, one can readily lose a significant proportion of what is often a valuable antigen preparation. Although more-liquid emulsions are considered less immu-

[24] G. Galfrè and C. Milstein, this series, Vol. 73, p. 3.

nogenic, we have generally compromised that aspect in favor of the possible short-term benefits of seeing most of the antigen end up in the mouse rather than adhering in delicate little meringues to the tube and syringe. Five mice are immunized subcutaneously on the left side of the belly with 200 μl of the emulsion so that each animal receives 20 μg. The Eppendorf centrifuge tube is short and has a wide mouth, so that antigen loss on the sides of the tube and syringe is reduced. A glass syringe with an 18-gauge needle is used to inject the antigen. After 3–4 weeks the procedure is repeated with the following changes: incomplete Freund's adjuvant (IFA) is used and mice are immunized subcutaneously on the right side of the belly. A second boost in IFA is given in the peritoneum 3–4 weeks after the first boost; 7–10 days after the second boost, the mice are bled from the tail in order to determine their level of serum antibodies against the immunogen.

Tail Bleeding

The mice are lightly warmed under a lamp and placed in a restrainer; the last 5–10 mm of the tail is chopped cleanly off with a new razor blade. One drop of blood is collected into a 1.5-ml Eppendorf centrifuge containing 100 μl of PBS and 10 μl of heparin (1000 units/ml). Bleeding is staunched with a paper pipette wiper, and the mice are individually marked with an ear punch. We try to minimize the amount of blood lost, which can be altered by varying the length of the piece of tail removed. This method was chosen because it leaves the bulk of the tail undamaged and suitable for intravenous injection at a later date.

Assay of Antibodies[24–26]

The choice of serological assay to be used throughout the process of making monoclonal antibodies should be decided at an early stage and depend upon the properties required of the monoclonal antibodies. This is important because the effector properties of different classes of antibody varies, and the range of binding energies is sufficiently large that a given antibody can perform very differently depending on the assay system. For example, in seeking for new HLA typing reagents one should use a cytotoxicity assay. If one wants a reagent for immune precipitation, it is advantageous to use a protein A-based binding assay as this interaction forms the basis of the best method of immune precipitation. Although major histocompatibility complex (MHC) antigens are a minor component

[25] R. H. Kennett, T. J. McKearn, K. B. Bechtol, eds., "Monoclonal Antibodies," p. 363. Plenum, New York, 1980.
[26] G. J. Hämmerling, U. Hämmerling, and J. F. Kearney, eds., "Monoclonal Antibodies and T-Cell Hybridomas," p. 563. Elsevier/North-Holland, Amsterdam, 1981.

of cells they are relatively easy to assay owing to their location on the outside of the plasma membrane. Cellular binding or cytotoxicity with appropriate combinations of cells provide simple assays with no requirement for a supply of purified antigens. This means that the only necessary consumption of purified antigen is during immunization. For proteins not found on the surfaces of cells or viruses, purified antigen is required for assay, usually by solid-phase radioimmunoassay, which consumes quantities of material far greater than that required for immunization. The general all-purpose screening method we use is an indirect trace binding assay using affinity-purified ^{125}I-labeled F(ab')$_2$ fragments from the IgG of a rabbit anti-mouse IgG serum (RAM) as a second-step reagent.[12a] Target cells (10^5 to 10^6) in 25 μl of PBS and 0.1% bovine serum albumin (BSA) are incubated in flexible U-bottomed Microtiter plates with 25 μl of mouse antibody at 4° with shaking for 1 hr. The cells are washed three times by centrifugation and aspiration with PBS, 0.1% BSA and then resuspended in 25 μl of the same buffer containing 20% horse serum and 3 × 10^5 cpm of RAM with a specific activity of ≈4 × 10^7 cpm/μg. After incubation for 1 hr at 4° with shaking, the cells are washed four times and assayed for bound radioactivity. To save time we use multiwell washing and aspiration devices and a hot-wire cutter to separate the Microtiter plate for assay of radioactivity. In general, 20–30% of the RAM is active, yielding maximum signals of 60,000–90,000 cpm when neither antigen nor first antibody is limiting. Under those conditions background values are 500–2000 cpm depending on the target cell. Values in excess of 5000 cpm are considered positive.

For target cells we use human lymphoid cell lines grown in RPM1-1640 medium, supplemented with penicillin, streptomycin, and 10% horse serum. The horse serum contains antibodies that bind to human cells (not to MHC antigens as far as we can tell) and cross-react weakly with our RAM preparations. It is to remove the consequent background that we add horse serum to the RAM. B-cell lines are a convenient source of targets and have an enhanced expression of HLA antigens (10 to 30-fold) when compared to normal lymphocytes. When HLA-A,B,C antigens are assayed, the B-cell line Daudi, which does not express HLA-A,B,C, is used as negative control. When HLA-DR antigens are assayed, the T-cell lines HSB-2 and MOLT 4 provide negative controls. In all situations the cell from which the immunogen was derived is used as the primary target.

When sera from mice immunized with HLA antigens are titrated in the indirect trace binding assay, positive binding is usually seen at dilutions of 500–1000 but not of 5000–10,000. Sera from unimmunized mice are used as negative serum controls. Individual mice within a group have always given similar titers. The specificity of the response depends upon the purity of the immunogen. When highly pure HLA-A2 was used, only a weak serum reaction with Daudi was seen. However, when the immune sera preparations of HLA-B40 containing small quantities of DR antigens were used, the reaction with Daudi was as strong as that seen with LB, the cell from which the material was prepared. This dual specificity was subsequently reflected in the spectrum of monoclonal antibodies produced. Polymorphic reaction patterns with panels of B-cell lines are rarely seen with these sera in binding assays.

Mice giving antibody titers <100 are either immunized again or discarded. The remaining mice are ranked according to antibody titer as candidates for fusion. They are left for a period of 1 month to 1 year, to allow their active immune response to subside before reimmunizing prior to fusion.

Only antibodies that recognize the native structures of cell-surface antigens are detected in the cellular binding assay. On occasion one wants to make antibodies against denatured forms, separated polypeptides, or fragments of the component polypeptides of MHC antigens. The assay of choice should then involve binding to the immunizing material, not to a related but different structure found on the cell surface. Most proteins and large peptides can be passively absorbed in detectable amounts to the wells of flexible Microtiter plates providing a solid-phase antigen. An indirect trace binding assay can then be carried out as described above with the difference that (a) washing does not require centrifugation, merely addition and removal of buffer to the plate; (b) horse serum is not added to the RAM. The feasibility of this approach is dependent upon the quantities, stability, and stickiness of the individual antigens. Protein concentrations of ~100 μg/ml give good passive absorption of nanogram amounts per well of the Microtiter plate. Thus the antigen solution is reusable although a threshold amount is required.

Fusion

Preparation of Myeloma Cells

Four mouse myeloma cell lines have been used in our fusion experiments and are described in the table. They are distinguished by the secretion of nonspecific antibody chains that introduce heterogeneity and reduced serological activity into the resultant monoclonal antibodies. Most of our experience has been with NS1, although SP2 and 653 are now exclusively used.

Myeloma cells, NS-1, are cultured in RPMI-1640 supplemented with penicillin, streptomycin, and 10% fetal calf serum. Two weeks before

MOUSE MYELOMA DERIVED CELL LINES USED FOR FUSION

Cell line	Shortened designation	Ig chains	Reference[a]
P3-X63-Ag8	X63	δ, κ	1
P3-NSi-Ag4/1	NS1	κ	2
P3-X63-Ag8.653	653	None	3
P-3-SP2/0-Ag14	SP2	None	4

[a] Key to references: 1, G. Köhler and C. Milstein, *Nature* (*London*) **256,** 495 (1975); 2, G. Köhler and C. Milstein, *Eur. J. Immunol.* **6,** 511 (1976); 3, J. F. Kearney, A. Radbruch, B. Liesegang, and K. Rajewsky, *J. Immunol.* **123,** 1548 (1979); 4, M. Shulman, C. D. Wilde, and G. Köhler, *Nature* (*London*) **276,** 271 (1978).

fusion they are passaged for 3 days in medium supplemented with 8-azaguanine (20 μg/ml) to ensure retention of the HGPRT minus trait and removal of revertants. They are subsequently grown in the absence of 8-azaguanine to ensure effective removal of the drug by the time of fusion. During this time an aliquot of cells is tested for susceptibility to HAT medium. If cells are either susceptible to 8-azaguanine or resistant to HAT, a fresh culture of myeloma cells is started from frozen stocks. It is of prime importance to keep the myeloma culture in logarithmic growth. Growth is best between densities of 5×10^4 and 5×10^5/ml, and the cells die rapidly if allowed to overgrow beyond $\approx 10^6$/ml. Cultures with high viability >95% live cells can be achieved, maintained, and used in fusion. Dead myeloma cells can be deceptive to the inexperienced eye as they retain significant structural integrity. They can be distinguished from viable cells by their smaller size, less distinct outline, and granular appearance and by inability to exclude Trypan Blue dye. As the doubling time of mouse myeloma cells is short compared to that for many other mammalian cells ($\approx$8–16 hr), it is common to see many attached pairs of recently divided cells in a healthy culture. This is not true for dying or dead cultures.

Preparation of Spleen Cells

Four to five days before the planned fusion, three mice are immunized with antigens (20 μg in 100 μl of PBS) in a tail vein using a 1-ml disposable plastic syringe and a 27-gauge needle. The rationale is to initiate secondary immune responses selectively in the spleen as opposed to lymph nodes. In cases where intravenous injection technically fails, the antigen is injected into the peritoneum.

The general experience has been that hybridomas are more difficult to make against soluble antigens than against cellular antigens. We found that a minority (1 in 3 or 4) of fusions gave rise to hybridomas producing antibodies of interest. This variability has been ascribed to the poorer immunogenicity and susceptibility to degradation of soluble antigens. To try and overcome these effects, protocols involving daily intravenous boosts of soluble antigen for a week prior to fusion have been developed.[27] The only apparent disadvantage is the increased consumption of antigen. We have used this method to produce antibodies against three rather conserved proteins, actin, myosin, and clathrin. The problems associated with making antibodies against MHC antigens have not been sufficiently severe to encourage the use of these protocols. Detergent solubilized antigens can be immunized subcutaneously or intraperitoneally in adjuvant without removal of detergent. Intravenous injection of detergent is not advisable, and removal of the detergent by acetone or ammonium sulfate precipitation or by dialysis prior to injection is recommended.

The day before fusion the solution of polyethylene glycol (PEG) is made. Two to four grams of PEG are autoclaved in a screw-cap vial and cooled to a temperature at which it can be held comfortably. An equal volume of RPMI-1640 medium is added and mixed thoroughly. The cap is screwed tight and kept at 37°. Before starting the fusion, a water bath at 37° is placed in the tissue culture hood and bottles of RPMI-1640 at 4° and room temperature set aside.

The three fusions are done separately between 72 and 96 hr after the final boost. Like most other aspects of hybridoma technology, there are many opinions on how long one should wait. Although almost all investigators have varied some parameters, there have been no large controlled studies from which one can distill a calming and optimal protocol. We always use a number of mice with the fusion spaced over a period of 1 day in order to hedge bets. An extreme case was the experiment to make antibodies against actin in which nine mice were separately worked up over the course of a day.

Mice are killed by cervical dislocation, the spleen is aseptically removed, trimmed of fat and excess connective tissue, and teased apart with forceps and scissors in cold (4°) RPMI-1640 medium. The spleen is first trimmed in one petri dish containing medium and then transferred to a second before making the cell suspension. The spleen is extracted with 2 × 5 ml of medium and transferred to a 15-ml centrifuge tube on ice.

[27] C. Stähli, T. Staehelin, V. Miggiano, J. Schmidt, and P. Häring, *J. Immunol. Methods* **32**, 297 (1980).

The cells are resuspended with a Pasteur pipette and allowed to stand for 1 min to let large aggregates of cells and pieces of connective tissue settle out. The cell suspension is transferred to a fresh tube and sedimented for 5 min at 1500 rpm at 4° in a Beckman TR6 centrifuge.

There is no premium on extracting every last lymphocyte from the spleen; more important is that the cell suspension be clear of debris, kept cool, and prepared quickly. The spleen cells are resuspended in 10 ml of medium, counted, and washed again. At this time myeloma cells, one-third to one-fifth the number of spleen cells, are taken from culture, centrifuged at room temperature (1500 rpm for 3 min in Beckman TR6), and washed once by centrifugation. Each cell preparation is resuspended in 10 ml of medium, combined in a 50-ml plastic centrifuge tube, and centrifuged together (1500 rpm for 5 min) at room temperature. The supernatant is completely removed by aspiration (it is preferable to lose cells than retain liquid); the pellet is loosened, and the cells are distributed around the bottom of the tube by light vortexing or vigorous flicking with the index finger. The tube is placed in the 37° water bath for 5 min, and then an estimated 1 ml of prewarmed PEG is added dropwise with a Pasteur pipette onto the cells over a 1-min period. The cell mixture is incubated at 37° with gentle swirling for 1 min. On removal from the bath ≈1 ml of RPMI-1640 at room temperature is added with a Pasteur pipette over a period of 1 min. The tube is then filled dropwise with medium, and the end of the pipette is gently rotated in order to ensure dilution of the PEG. It is crucial to treat the cellular aggregates gently and to minimize shear forces when pipetting. The cells are centrifuged (1500 rpm for 5 min), the supernatant is removed, and the pellet is gently broken up and resuspended by addition of 50 ml of RPMI-1640 supplemented with 15% FCS, penicillin, and streptomycin. The cells are *not* drawn up and down the pipette but resuspended by gently stirring the pipette in the tubes. The mix is then distributed among 96 wells of 4 × 6, 24-well Linbro plates using a 10-ml pipette and gentle suction and ejection. To enhance the survival of hybridomas, layers of feeder cells are sometimes used. Peritoneal cells from mice primed with pristane (2,6,10,14-tetramethylpentadecane, Aldrich) have been most effective. They are prepared as described[28] and set up 24 hr before fusion in order to ensure sterility; 2×10^5 cells per well in a volume of 0.3 ml of RPMI-1640 with 10% FCS are used. Fused cells are then added to these wells in a volume of 0.2 ml. The following morning 0.5 ml of medium containing 2 × HAT constituents is added to the 0.5 ml of culture in each well.

---

[28] B. Mishell and S. M. Shiigi, eds., "Selected Methods in Cellular Immunology," p. 6. Freeman, San Francisco, California.

The cultures are cursorily examined every 2 or 3 days for a period of 3 weeks to a month. Culture wells with growing cells are marked; when the colonies are either easily visible with the naked eye or cover about one-quarter of the area of the well, a few drops of supernatant are removed and assayed for specific antibody activity. Colonies are most simply identified by looking at the underface of the plate. They appear as off-white balls or smears. The first screen is only against the immunizing cell or the cell from which the immunizing antigens were derived. It identifies positive cultures that are subsequently tested against a small panel of cells aimed at discerning HLA-A,B,C, or DR specificity and the polymorphisms that are likely based on the immunization. For example, if the immunizing antigen was B27 from a cell homozygous for A2, B27, C1, DR3 we would perform a secondary screen on a panel of four cells each with one of these antigens in the absence of the others, the immunizing cell, Daudi (no HLA-A,B,C), a T-cell line, e.g., HSB-2 (no HLA-DR), and two cells of completely different HLA type. On the basis of this screen, antibodies of interest can be identified. At this point it is critical to make decisions and to embark on further tissue culture only with cells that are giving both a significant signal-to-noise ratio in the binding assay and a reaction pattern of real interest. If such decisions are not made, the process becomes labor intensive and wasteful of expensive materials, e.g., serum and plasticware. It is common for more than one hybrid clone to be present in each culture. These may often be seen as separate balls or smears of cells in the well. When a positive culture has been identified, the individual colonies are separately subcultured on mouse macrophage feeder layers (2.0 × 10⁵ macrophages per well) in 24-well plates.[28] By this simple procedure one often produces cloned hybridomas. The subclones are assayed for specific antibody production, and the strongest are selected. At this early stage the priorities are to freeze stocks of the culture and clone the line. These goals do not conflict, as the number of cells required to clone a line is small compared to those obtained from a single culture well.

Cloning Procedures

Three methods have been successfully used in our laboratory: (a) cloning in soft agar on feeder layers of irradiated human fibroblasts as described[25,29]; (b) cloning by limiting dilution on mouse macrophage feeder layers[28]; (c) single-cell cloning using the fluorescent activated cell

---

[29] R. G. H. Cotton, D. S. Secher, and C. Milstein, *Eur. J. Immunol.* **6**, 82 (1973).

sorter to distribute individual cells into the wells of Microtiter plates containing mouse macrophage feeder layers.[29,30]

The cell sorter method is the most satisfactory, providing a suitably modified instrument is available.[30] It provides a better control of the numbers of cells in each culture without generating large numbers of empty cultures and allows for selection of only the most healthy cells. When clones have grown up they are tested for antibody activity, and the clones with strongest activity are selected. Judicious selection at this stage can result in the selection of clones secreting minimal amounts of the nonspecific NS1 light chain. Many HLA antibodies with interesting polymorphic specificity have sufficiently low affinity that only bivalently bound immunoglobulin molecules are detected with serological assays that involve washing steps. Molecules having one or two NS1 light chains will be inactive, and cultures producing different amounts of that light chain can be distinguished.

Freezing Cells

It is wise to respect Murphy's laws and to freeze stocks of cells at all stages. We centrifuge cells out of culture (200 g for 2–5 min), remove the supernatant, and resuspend at 10^5 to 10^7 per milliliter in 95% heat-inactivated FCS containing 5% dimethyl sulfoxide, which is kept at 4°. Ampoules are placed in insulated containers, i.e., expanded polystyrene racks at $-70°$ for 12–48 hr, and then transferred to liquid nitrogen. Only healthy cultures are worth freezing because of the loss of viability, which is higher for less vigorous cultures, resulting from freezing and subsequent thawing when the cell is required again.

A cloned line is expanded from Microtiter culture to 24-well plates to flasks, at which point growth does not require feeder layers. Ten to twenty ampoules containing $\simeq 10^7$ cells each are frozen.

Hybridomas can then be adapted to growth in medium without HAT constituents. This is done by feeding cultures with medium containing hypoxanthine and thymidine without aminopterin for 1 week and then feeding with the normal medium used for NS1 cells. The first stage of this adaptation is usually trivial and little change in the growth rate of hybridoma cells is seen. The second stage usually involves a refractory period when growth slows and cells may die. It is best to inspect the cultures daily and feed the cultures only when normal growth has resumed.

The advantage of adapting hybridoma lines in this way is that the cells

[30] D. R. Parks, V. M. Bryan, V. T. Oi, and L. A. Herzenberg, *Proc. Natl. Acad. Sci. U.S.A.* **76**, 1962 (1979).

will grow more efficiently as tumors in mice, which do not provide a supply of HAT. However, if a relatively large number of nonadapted cells are inoculated, >10^7, tumor growth can usually be initiated though there is greater probability of the tumor not having the bulk properties of the inoculum.

Production of Ascitic Fluid Containing Monoclonal Antibodies

BALB/c mice are given 0.5 ml of pristane intraperitoneally at least a week prior to giving them 5×10^5 to 10^7 hybridoma cells by the same route. The mice are inspected every 2 days; when a swelling of the belly is seen, ascites fluids are collected by holding the mouse vertical and inserting an 18-gauge needle into the peritoneum. The fluids are collected in heparin and centrifuged at 400 g for 15 min. The supernatants are frozen, the cells are either frozen (10^7 cells/ml in 95% heat-inactivated FCS with 5% dimethyl sulfoxide) or resuspended in PBS for passage in additional mice (10^6 ascites cells per mouse). With increased passage the tumors increase in virulence and the period between detection of an ascites tumor and death decreases from 1 or 2 weeks to just a few days. As a consequence, the quantity of ascites fluid and antibody obtained per mouse decreases. We usually passage the tumors only three times and then start again from tissue culture cells if further ascites fluid is required. As the cell density in the ascites fluid is $\approx 10^8$/ml and 5–15 ml of fluid per mouse can be routinely obtained, it is not difficult to prepare liter quantities from a relatively modest number (10^7–10^8) of tissue culture cells.

Purification of Immunoglobulin[16,17]

Conventional procedures for the purification of immunoglobulin are generally based on the properties of a majority of antibody molecules in a polyclonal mixture. A given monoclonal antibody has distinctive properties that may not be those of the majority. Therefore, it is wise to monitor the purification of monoclonal antibodies with a specific serological assay. In devising a general strategy for purification, it is better to use procedures that separate on the basis of properties that do not vary greatly between different antibodies, e.g., size, rather than those that do vary greatly, e.g., charge. We will restrict our discussion to monoclonal IgM and IgG antibodies.

Step 1. Clarification of Ascites Fluid. Immunoglobulin preparations are routinely made from 50 ml of ascites fluid, which usually contains 50–500 mg of monoclonal antibody. Ascites fluid is thawed and passed through a plug of glass wool or cheesecloth to remove large aggregates of lipid and cellular debris. Centrifugation at 10,000 g for 3 min removes smaller particulate material.

Step 2. Ammonium Sulfate Precipitation. The supernatant from centrifugation is sequentially precipitated with 30% and 50% ammonium sulfate. Saturated ammonium sulfate adjusted to pH 7.4 is added dropwise and with stirring at 4° to the appropriate final concentration and left stirring for 30 min. The precipitate is removed by centrifugation at 10,000 g for 30 min at 4°, and the supernatant is submitted to further precipitation. The precipitates are redissolved in a volume of PBS one-tenth that of the original volume of ascites fluid and analyzed by serological assay and sodium dodecyl sulfate–polyacrylamide gel electrophoresis (SDS–PAGE). The active immunoglobulin is usually in the 30–50% fraction.

Step 3. Sephadex G-200 Chromatography. The dissolved precipitate is applied to a Sephadex G-200 column (2.5 × 100 cm), which is equilibrated and eluted in PBS. Three protein-containing peaks of material that absorb at 280 nm are seen: an excluded peak containing lipid and IgM, and two included peaks containing IgG and serum albumin. For ascites fluids with high concentration of specific IgG, the IgG can be >90% pure at this stage.

Step 4. Dialysis. The pooled Ig from the Sephadex G-200 column is dialyzed extensively against 5 mM Tris-HCl, pH 7.5. This causes precipitation of IgM, which is separated by centrifugation at 10,000 g for 30 min.

If the monoclonal antibody is of the IgM class, the precipitate is washed with ice-cold dialysis buffer, centrifuged, redissolved in 0.1 M Tris-HCl pH 8.6, and dialyzed against PBS. No further purification of IgM has so far been needed.

Step 5. DEAE–Cellulose Chromatography. If the monoclonal antibody is of the IgG class the supernatant from centrifugation is applied to a column (1 ml per 10 mg of IgG) of DE-52 equilibrated with 5 mM Tris-HCl, pH 7.5. The column is washed with the same buffer until the absorbance at 280 nm reaches a background value <0.01. The column is eluted with a linear gradient (40 ml) from 0 to 0.1 M NaCl in the same buffer.

The column is then purged with 0.5 M NaCl. Fractions (4 ml) are collected and assayed for absorbance at 280 nm. Those containing significant quantities of absorbing material are analyzed serologically by SDS–PAGE. Active fractions that are pure as assessed by SDS–PAGE are pooled, concentrated to 2–10 mg/ml by ultrafiltration, and stored frozen in liquid N$_2$. Most monoclonal antibodies bind to the column and elute near the beginning of the gradient. Many of the contaminating proteins including albumin are retained on the column until the high-salt purge.

Sometimes ascites fluids contain significant amounts of nonspecific IgG and IgM derived either from the serum of the mice carrying the tumor or from spontaneously arising or variant myelomas. These can be misleading if one monitors only the purification by bands on a gel. Often the specific monoclonal antibody will be separated from much of the nonspecific immunoglobulin on the DEAE column.

Additional Points

The purification scheme described above has been used successfully to purify mouse monoclonal IgM, IgG1, IgG2a, and IgG2b from ascites fluid. With some 25 different antibodies, we have always obtained purity in excess of 90%. If tissue culture supernatant is the source of immunoglobulin, the purification obtained is far less.

Columns of *Staphylococcus* protein A have been used to purify monoclonal antibodies. This method is particularly suited to antibodies of the IgG2 class, which bind tightly to protein A, and for single-step purification from small quantities of ascites fluid or tissue culture supernatant. We have not used this method because (*a*) the majority of our antibodies were of the IgG1 class, which binds poorly to protein A; (*b*) we wished to avoid stringent elution buffers that might denature the antibody; and (*c*) we were interested in relatively large-scale preparation. The reader is referred to Ey *et al.*[31]

Characterization of HLA-A,B,C Monoclonal Antibodies

This section concerns methods for distinguishing the different specificities of antibodies already identified by cell binding assays as putative anti-A,B,C. A large proportion of such antibodies recognize structures that are common to all HLA-A,B,C gene products. These are called, alternatively, species, framework, common or monomorphic antigenic determinants, and antibodies. The remaining antibodies recognize determinants that vary among different gene products and are called polymorphic or discriminatory. Some polymorphic antibodies are relatively specific, bind to a small number of gene products, and can be identified as such by cellular binding or cytotoxicity assays. Others have a broader specificity, differential affinity, react with all cells, and are difficult to distinguish from monomorphic antibodies.[32] The two groups of antigenic determinants defined by these specific polymorphic and broadly polymor-

[31] P. L. Ey, S. J. Prowse, and C. R. Jenkin, *Immunochemistry* **15**, 429 (1978).
[32] F. M. Brodsky and P. Parham, *J. Immunol.* **128**, 129 (1982).

phic antibodies correspond to the private and public H-2D, K antigens of mice and have a different evolutionary history.[33,34]

Inhibition with Solubilized Antigens

Assignment of the specificity of HLA antibodies can be confirmed by inhibiting the binding to cells with purified antigens. This is particularly convincing for polymorphic antibodies if equivalent preparations of different specificities are used.[3,15] For monomorphic HLA antibodies we compare the inhibition with HLA-A,B,C, β_2-microglobulin and HLA-DR. Inhibition can be measured using direct or indirect assays, and the procedures are similar. In order to find sensitive conditions, the antibody is titrated in the binding assay with 10^5 cells of a lymphoid line or 10^6 peripheral blood lymphocytes. A dilution of antibody that gives a measurable signal (5–10 times background), but is away from the plateau of the titration, is chosen. Dilutions of inhibitors are preincubated with the fixed quantity of antibody for 1–8 hr, cells are added, and the assay is carried out in the normal way. After incubation with antibody and inhibitor, the cells are centrifuged without addition of wash buffer and the supernatant is removed before washing three times. This prevents dilution of the inhibitor in the first wash and possible redistribution of the antibody to favor the bound form. As preparations of purified antigens are usually precious, the incubation volumes are kept low, the cells, antibody, and inhibitor each being added in 5–25 μl amounts.

Antibodies can bind multivalently to cells and only monovalently to many soluble antigens. Therefore in certain cases solubilized antigens will compete poorly with cells for antibody. For example, about 50 times the amount of solubilized HLA-A2 is required to inhibit the binding of PA2.1 antibody to cellular antigen.[3] To increase the sensitivity of inhibition assays, one can solubilize antigen in a polymeric form. A common method is to solubilize and purify in the presence of a dialyzable detergent, such as sodium deoxcholate (DOC) or octyl glucoside, and then remove the detergent by dialysis.[35,36] This produces small soluble aggregates of the membrane antigens, which can often bind antibody multivalently. Alternatively, one can design the assay so as to minimize the dissociation of soluble complexes once cells have been added to the mixture of antigen

[33] F. M. Brodsky and P. Parham, *Immunogenetics* **13**, 151 (1982).
[34] P. Ivanyi, R. A. Reisfeld, and S. Ferrone, eds., "Current Trends in Histocompatibility," p. 133. Plenum, New York, 1981.
[35] P. W. Kuchel, D. G. Campbell, A. N. Barclay, and A. F. Williams, *Biochem. J.* **169**, 411 (1978).
[36] G. Uterman and K. Simons, *J. Mol. Biol.* **85**, 569 (1975).

and antibody.[37] Reducing the time of incubation with cells as shown for BB7.2 is one way to do this.[17]

Anti-β_2-m antibodies are obtained with high frequency. They are equally inhibited by free and HLA-associated β_2-m and constitute one class of monomorphic antibodies.[38] A second group of monomorphic antibodies are equally inhibited by all HLA antigens, but not by β_2-m.[32,39,40] They define determinants on the HLA heavy chain. A third group is weakly inhibited by β_2-m, about 1000 times less strongly than by HLA, and identifies combinatorial determinants of the two chains.[32]

No polymorphic antibody that is inhibited to any extent by β_2-m has been identified. Specific polymorphic antibodies are inhibited by some HLA antigens, but not by others. Broadly polymorphic antibodies are inhibited by many HLA antigens, but to different extents, which is indicative of differential affinity. This provides one method for distinguishing monomorphic from broadly polymorphic antibodies.

In general the inhibition of polymorphic antibodies by appropriate HLA preparations has been significantly weaker (10–100 times) than for monomorphic antibodies. For a limited number of antibodies we have shown this to be a direct consequence of their relative affinities. To extrapolate these results, we conclude that monomorphic antibodies are more commonly of high affinity, and polymorphic antibodies of low affinity. Knowing that different HLA antigens are highly homologous, this does not come as a surprise. It is merely an immunochemical restatement of the fact that an HLA antigen is more similar to other HLA antigens than it is to H-2 antigens.

Somatic Cell Hybrids

Mouse–human hybrids that express only one of the human components of class I antigens are used as targets in cell-binding assays to confirm the specificity of monomorphic antibodies. Horl 9.8R 3.3 has only one human chromosome, No. 15, and expresses human β_2-m in association with H-2. CTP41.17.2 has human chromosome No. 6 but not No. 15 and expresses HLA heavy chain in association with mouse β_2-m.[41,42] β_2-m antibodies bind to Horl 9.8R 3.3, but not to CTP41.17.2. Anti-HLA bind

[37] D. W. Mason and A. F. Williams, *Biochem. J.* **187**, 1 (1980).

[38] F. M. Brodsky, W. F. Bodmer, and P. Parham, *Eur. J. Immunol.* **9**, 536 (1979).

[39] P. Parham, C. J. Barnstable, and W. F. Bodmer, *J. Immunol.* **123**, 342 (1979).

[40] P. Parham, H. T. Orr, and J. G. Golden, Houston symposium 1981 (1982) in press.

[41] E. A. Jones, D. Phil. Thesis, Oxford University, 1976.

[42] E. A. Jones, P. N. Goodfellow, R. H. Kennett, and W. F. Bodmer, *Somatic Cell Genet.* **2**(6), 483 (1975).

to CTP41.17.2, but not to Horl 9.8R 3.3. The antibodies against combinatorial determinants bind to neither hybrid, as they require that both human polypeptides be expressed.[15,32] These properties have enabled the chromosomal location of the MHC in owl monkeys to be identified.[43]

Phylogenetic Comparisons

Comparison of the binding to cells from other species often distinguishes antibodies that appear identical from their reactions with humans.[7,15,33,38,44,45] Thus, different antigenic determinants show distinguishable degrees of evolutionary conservation. In general, monomorphic antibodies exhibit monomorphic behavior in all species; i.e., all MHC products react similarly, whether positive or negative. As one moves phylogenetically away from humans, the affinity between an antibody and the various HLA homologs decreases. The point in phylogeny at which positive reactions give way to negative ones depends on the nature of the assay and the minimum antibody avidity required to give a significant signal. For example, BB7.7 reacts only with primate cells in a binding assay, whereas bovine cells were also positive as assessed by complement-mediated cytotoxicity.[45] This behavior is consistent with the nature of molecular evolution as assessed in many families of homologous molecules. It has proved to be particularly useful in distinguishing monomorphic antibodies from broadly polymorphic antibodies. In addition, identification of high-affinity cross-reactions enables isolation and characterization of MHC antigens from cross-reacting species.[46]

Affinity Chromatography

The mainstays of purification procedures for globular, water-soluble proteins, including the papain-fragment of transplantation antigens, are ion-exchange and size-exclusion chromatography; these are poorly suited for the separation of small membrane proteins in large detergent micelles.[19]

Two important breakthroughs in the struggle to isolate and characterize "native" detergent-solubilized HLA antigens were successive applications of lentil lectin[47,48] and anti-β_2-microglobulin affinity chromatogra-

[43] N. S. F. Ma, T. Simeone, J. McLean, and P. Parham, *Immunogenetics* 15, 1 (1982).
[44] P. Parham, P. W. Sehgal, and F. M. Brodsky, *Nature (London)* 279, 639 (1979).
[45] F. M. Brodsky, W. H. Stone, and P. Parham, *Hum. Immunol.* 3, 143 (1981).
[46] P. Parham and H. L. Ploegh, *Immunogenetics* 11, 131 (1980).
[47] M. J. Hayman and M. J. Crumpton, *Biochem. Biophys. Res. Commun.* 47, 923 (1972).
[48] D. Snary, P. Goodfellow, M. J. Hayman, W. F. Bodmer, and M. J. Crumpton, *Nature (London)* 247, 457 (1974).

phy.[49] Both procedures depended upon large quantities of proteins, i.e., lentil lectin and β_2-microglobulin which were discovered, characterized, and made available through efforts unconnected with research on transplantation. Their usefulness lies in the increasing selectivity of lentil lectin for a class of glycoproteins and anti-β_2-m for β_2-m associated, or class I, histocompatibility antigens. The monoclonal antibody method can provide large quantities of a variety of binding proteins of even greater selectivity and has many possible applications.

A monomorphic antibody, W6/32, was the first HLA monoclonal antibody obtained,[2] and its suitability for protein purification was investigated while experiments to produce additional monoclonal antibodies of polymorphic specificity were in progress.[39] W6/32 was coupled (10 mg of protein per milliliter of Sepharose) to cyanogen bromide-activated Sepharose 4B (25 mg of CNBr per milliliter of Sepharose). This basic method has been used exclusively with the replacement of Sepharose 4B by Sepharose CL-4B. Two methods of cyanogen bromide activation have been used. The buffer method of March et al.[49a] has been more reproducible and reliable than the older procedure involving continuous adjustment of pH.[49b] With that method inactivation of antibody due to overcoupling has been a problem.

Coupling Immunoglobulin to CNBr-Sepharose[49a]

Materials

Sepharose CL-4B (Pharmacia)
CNBr (Eastman Kodak): check that this is *white*
Filter funnel with sintered-glass bed
Side-arm flask, large
Glass dish (for ice bath), shallow
Magnetic stirrer
Water-powered vacuum pump

Procedure. Note that activation (steps 3–8) must be done in a fume cupboard.

1. Define Sepharose in distilled water (DW).
2. Make 50% slurry in DW.
3. Add 1 volume of Sepharose slurry to glass beaker containing magnetic stir bar. Place in ice bath on stirrer. Stir gently.
4. Add 1 volume of 2 M Na$_2$CO$_3$ at room temperature.

---

[49] R. J. Robb, J. L. Strominger, and D. Mann, *J. Biol. Chem.* **251,** 5427 (1976).
[49a] S. C. March, I. Parikh, P. Cuatrecasas, *Anal. Biochem.* **60,** 149 (1974).
[49b] J. Porath, R. Axèn, and S. Ernbäck, *Nature (London)* **215,** 1491 (1967).

5. Increase the stirring rate until vigorous. Now add, all at once, 0.05 volume of CNBr solution (100 mg/ml in acetonitrile).

6. Stir vigorously for 1 min.

7. Pour the slurry onto filter. Wash under (gentle) vacuum with the following ice cold solutions: $5 \times 0.1\ M$ NaHCO$_3$, pH 9.5; $5 \times$ DW; $5 \times 0.06\ M$ Na$_2$HPO$_4$, $0.04\ M$ NaH$_2$PO$_4$, pH 7.0. The volume of the wash depends on the amount of Sepharose. Use about 2–5 volumes of starting slurry for each step.

8. Finally vacuum down to a moist cake and *either* transfer this to a fresh beaker *or* reslurry in half its own volume of phosphate-buffered saline (PBS) pH 7.0, and transfer to beaker(s).

9. Add protein solution to Sepharose. Protein should be dialyzed into PBS, pH 7.0, before use. The concentraton may be varied to give the desired coupling ratio, but the final Sepharose concentration should be *50%*. Initial protein ratios of 2–10 mg per milliliter of Sepharose yield 50–70% efficiencies of coupling.

10. Stir the protein–Sepharose mixture overnight (16 hr) at 4°.

11. Centrifuge the Sepharose slurry (1000 rpm for 3 min is sufficient). Remove the supernatant and keep for analysis and/or recycling of antibody.

12. Add an equal volume of $0.2\ M$ β-alanine, $0.1\ M$ (NH$_4$)HCO$_3$. Reslurry and allow to stand for at least 30 min at room temperature.

13. Wash the coupled Sepharose on a glass filter as follows: 5×0.1 sodium acetate, $0.5\ M$ NaCl, pH 4.0; 5×0.1 NaHCO$_3$, 0.5 NaCl, pH 10.0; $5 \times$ PBS/N$_3$. Again each step of washing should be 3–5 volumes of Sepharose.

14. Slurry and store at 4° in PBS with 0.02% sodium azide. A typical coupling is with 200 mg of purified immunoglobulin and 40 ml of packed Sepharose CL-4B.

W6/32-Sepharose selectively bound small quantities of radioactive HLA. The capacity of the absorbent was not estimated owing to lack of sufficient HLA. A comparison of elution methods showed that low-pH and high-salt buffers were poor and that optimal elution was with buffers of pH >11.0.[39] This observation has held for other HLA monoclonal antibodies. A contributory factor is that HLA tends to precipitate at pH <7.0. The elution buffer of choice has been 50 mM diethylamine adjusted with HCl to pH 11.5.[50–52] Trucco and co-workers[53] found that the bind-

[50] R. S. D. Read, J. C. Cox, H. A. Ward, and R. C. Nairn, *Immunochemistry* **11**, 809 (1974).
[51] M. Letarte-Muirhead, A. N. Barclay, and A. F. Williams, *Biochem. J.* **151**, 685 (1975).
[52] W. R. McMaster and A. F. Williams, *Immunol. Rev.* **47**, 117 (1979).
[53] M. M. Trucco, S. DePetris, G. Garotta, and R. Ceppellini, *Hum. Immunol.* **3**, 233 (1980).

ing constants of an anti-β_2-microglobulin (51.26/114) and other monomorphic HLA antibodies decreased with increased temperature. They were able to bind HLA to columns of 51.26/114 at 4° and elute at 37°.[54] The temperature elution was as effective as high pH, and, in agreement with Parham *et al.*,[39] they found that low pH performed poorly. Quaranta *et al.*[55] found that 3 *M* sodium isothiocyanate was a superior eluent for HLA-DR antigens from Q5/13 antibody coupled to Sepharose 4B. Mescher and co-workers (see this volume [8]) have identified deoxycholate as a good eluting reagent for H-2 monoclonal antibody columns. This is an attractive method, as the potential for denaturation is far less than with the above-mentioned reagents. Deoxycholate was not successful in eluting affinity columns of HLA antibodies W6/32, BBM.1, PA2.1, and BB7.1, although we have found it to be useful for the purification of bovine clathrin using the monoclonal antibody CVC.7.

Monoclonal antibodies PA2.1[1] and BB7.1[15] are serologically specific for HLA-A2 and HLA-B7. This specificity was retained when these antibodies were used for immunoaffinity purification of HLA from solubilized extracts of the JY cell line, homozygous A2, B7.[56] Radioiodinated preparations of A2 bound only to PA2.1–Sepharose and of B7 only to BB7.1–Sepharose. A general method for purifying HLA-A,B,C antigens from crude papain-solubilized extracts of JY cells was developed. The method relied on using multiple monoclonal antibody columns in series, an idea also exploited by Springer[57] to characterize differentiation antigens. Five columns were used:

1. Virgin Sepharose—to trap particulate and nonspecifically adhering material. This column is renewed for each experiment.
2. Normal BALB/c IgG-Sepharose—to remove material specifically sticking to IgG, in particular actin.[58]
3. PA2.1-Sepharose—to bind HLA-A2.
4. BB7.1-Sepharose—to bind HLA-B7.
5. W6/32-Sepharose—to bind HLA-C and other uncharacterized class I antigens.

All procedures are at 4° with buffers containing 0.02% sodium azide. Columns are made in plastic syringes connected with Tygon tubing. Columns 3, 4, and 5 are individually prewashed with one column volume

[54] G. Garotta and M. Trucco, *Protides Biol. Fluids* **28**, 463 (1980).
[55] V. Quaranta, L. E. Walker, M. A. Pellegrino, and S. Ferrone, *J. Immunol.* **125**, 1421 (1980).
[56] P. Parham, *J. Biol. Chem.* **254**, 8709 (1979).
[57] T. A. Springer, *in* "Monoclonal Antibodies," (R. H. Kennett, T. J. McKearn, and K. B. Bechtol, eds.), p. 185. Plenum, New York (1980).
[58] M. Fecheimer, J. L. Daiss, and J. J. Cebra, *Immunology* **16**, 881 (1979).

of elution buffer (50 mM diethylamine–HCl, pH 11.5) to remove material that would elute nonspecifically with HLA, and then with five column volumes of buffer (0.1 M Tris-HCl pH 7.8). Columns 1 and 2 are just prewashed with five column volumes of buffer. The columns are connected in series, and the sample is applied. Samples in a variety of buffers with pH 7–8 and salt concentrations 10–200 mM have been used with no obvious differences. Papain-solubilized antigens were prepared as described[19]; the intermediate fraction from the second DEAE column, containing approximately equal amounts of A2 and B7, was initially used as a sample. Subsequent experiments used the pool from the first DEAE or Sephadex G-150 columns. Flow rates of 10 ml/hr were typically used for sample application and elution. There appeared to be no advantage in stopping the flow at various times to allow increased contact between sample and columns. Higher flow rates (up to 100 ml/hr) were used for washing columns. Sufficient buffer is then applied completely to wash the sample through all the columns. It is disadvantageous to wash the columns in series as impurities from one column will pass to the next. The columns are separated; column 1 is discarded and the rest are washed with 10 column volumes of buffer. This did not elute significant quantities of bound HLA antigens. Column 2 was put aside, and columns 3–5 were individually eluted with three column volumes of elution buffer followed directly by a 10-column volume wash of buffer to minimize antibody denaturation. When native HLA antigens were required, fractions were collected into tubes containing sufficient 1 M Tris-HCl, pH 7.8, to overcome the pH of the elution buffer. Other investigators have used solid glycine for the same purpose with less dilution.[52] When this is done, antigens can be obtained in a serologically active form. Elution was monitored spectrophotometrically, by measurement of pH (if collected fractions were not neutralized) or by SDS–PAGE. An advantage of the diethylamine elution buffer is that it can be lyophilized. If salt-free material is required, then sodium azide should be eliminated from the elution buffer.

Bleeding of bound HLA from PA2.1, BB7.1, W6/32 columns, and anti-β_2-microglobulin columns made with the BBM.1 antibody[38] is not a major problem. HLA antigens of the appropriate specificity are quantitatively bound and efficiently retained during the washing procedure. Recoveries obtained in experiments to measure the capacity of the absorbents showed that efficient elution was obtained with 50 M diethylamine–HCl, pH 11.5. In general 2–3 mg of relatively pure (>75%) IgG was coupled per milliliter of Sepharose, and the capacity was ≈0.5 mg of HLA per milliliter of Sepharose.

Immunosorbents when kept in azide-containing buffers at 4° have remained stable and active for periods in excess of 2 years. Columns can be reused many times, though some loss of activity occurs. This problem increases with the size of column, as the exposure time to high pH in an experiment will correspondingly increase. No quantitative assessment of the denaturing effect of the elution buffer or antibodies has been made.

This scheme provided a method for purification and complete separation of the HLA-A2 and B7 antigens of JY cells and for obtaining a fraction, bound to W6/32–Sepharose, that was non-HLA-A2 or -B7 and supposedly enriched for HLA-C antigens. This material was electrophoretically similar to HLA-B7 and also susceptible to cleavage by dilute acid.[59] "HLA-C" was reduced and alkylated with [³H]iodoacetic acid, and the component polypeptides were separated on a column of BioGel P300 in 6 M guanidine hydrochloride. Ten moles of each polypeptide were recovered and subjected to NH_2-terminal amino acid sequence analysis. The sequence up to position 12 was identical to HLA-B7.[20,21]

A similar approach was used to search for non-HLA-A,B class I antigens of the human thymocyte-like cell line Molt 4. Quantitative binding studies[38] showed a significant excess of BBM.1 to W6/32 binding material on the surface of human thymocytes and Molt 4 cells. An extract from Molt 4 was passed through three W6/32 columns followed by two BBM.1 columns. Each was eluted, and the eluates were compared by SDS–PAGE. Specifically eluting material was confined to the first W6/32 and first BBM.1 column, showing that all W6/32 reactive molecules, i.e., HLA-A,B,C were removed from the extract before passage through the BBM.1 columns. On analysis by SDS–PAGE the W6/32 eluate had the characterization pattern of HLA-A,B antigens. The BBM.1 eluate showed a diffuse heavy-chain band of slightly higher mobility than HLA-A,B, heavy chains. These and similar results[60] showed the existence of a set of class I antigens serologically and biochemically different from HLA-A,B,C on Molt 4. These molecules were further characterized more recently.[61,62]

Purification of detergent-solubilized HLA-A,B products was as follows. Human B lymphoblastoid cells lines were grown in bicarbonate-buffered RPMI-1640 medium, supplemented with 10% FCS, penicillin (100 U/ml), and streptomycin (100 U/ml). Growth was maintained at cell densities of 10^5 to 10^6 cells per milliliter in standing cultures of 200 ml,

[59] C. Terhorst, R. Robb, C. Jones, and J. L. Strominger, *Proc. Natl. Acad. Sci. U.S.A.* **74**, 4002 (1977).

[60] N. Tada, N. Tanigaki, and D. Pressman, *J. Immunol.* **120**, 513 (1978).

[61] C. Terhorst, A. van Agthoven, K. LeClair, P. Snow, E. Reinherz, and S. Schlossman, *Cell* **23**, 771 (1981).

[62] T. Cotner, H. Mashimo, P. C. Kung, G. Goldstein, and J. L. Strominger, *Proc. Natl. Acad. Sci. U.S.A.* **78**, 3858 (1981).

which were split 1 : 1 every 2 days until a total culture volume of 10 liters was obtained. Two days after the final feeding, cells were examined for viability by exclusion of the dye Trypan Blue. Cultures showing viability >95% were harvested by centrifugation and washed once by resuspension in PBS. The method of solubilization was different from that usually used to purify membrane proteins and was designed so that both mRNA coding for MHC products and the products themselves could be obtained. It is derived from the method of Efstratiadis and Kafatos.[63] All procedures were at 4°. Cells ($\simeq 10^{10}$) were solubilized in 60 ml of 25 mM NaCl, 5 mM MgCl$_2$, 25 mM Tris-HCl, pH 7.5, 2% Triton X-100, 1 mg of heparin per milliliter (solubilization buffer) for 30 min and centrifuged at 27,000 g for 5 min. The supernatant was added to an equal volume of the solubilization buffer containing 200 mM MgCl$_2$ and stirred for 60 min to precipitate a crude mRNA fraction. Aliquots (20 ml) were layered over 7-ml cushions of solubilization buffer without Triton X-100, or heparin containing 1 M sucrose in 30-ml Corex tubes, and centrifuged for 5 min at 27,000 g. The supernatants were decanted, pooled, and provided the source of membrane proteins. Phenylmethylsulfonyl fluoride (PMSF) 0.01 mM was added to the extract and included in all buffers used in affinity chromatography. The RNA-containing precipitate was further purified and used for cDNA cloning.[63,64] The extract contained some precipitated material, presumably a result of precipitation of membrane proteins at the interface between applied sample and the sucrose cushion, which contains no detergent. At this stage the extract was in 25 mM NaCl, 78 mM MgCl$_2$, 25 mM Tris-HCl pH 7.5, 1.5% Triton X-100, 0.75 mg of heparin per milliliter, and 0.25 M sucrose. It could be frozen and thawed with no observable deleterious consequences. Prior to affinity chromatography, the extract was centrifuged at 27,000 g for 20 min to remove the precipitate and then applied to the same series of affinity columns as used for papain-solubilized antigens. Columns of 1 ml were used, and the procedures were identical except that 0.5% Triton X-100 or NP-40 and 0.01 mM PMSF was added to all buffers.

Pober et al.[65] subsequently used PA2.1 and BB7.1 to isolate HLA-A2 and HLA-B7 from larger quantities of frozen JY cells. They made a membrane preparation that was solubilized with detergent (4% Brij, 0.1 mM PMSF, 12.5 mM Tris-HCl, pH 8.0) and applied to the immunoaffinity columns. Their initial HLA preparations contained significant actin

[63] A. Efstratiadis and F. C. Kafatos, *Methods Mol. Biol.* **8,** 1 (1976).
[64] H. L. Ploegh, H. T. Orr, and J. L. Strominger, *Proc. Natl. Acad. Sci. U.S.A.* **77,** 608 (1980).
[65] J. S. Pober, B. C. Guild, J. L. Strominger, and W. R. Veatch, *Biochemistry* **20,** 5625 (1981).

contamination (5–50%), which they eliminated by preextraction of the membranes with a buffer (2 mM Tris-HCl, pH 8.0, 200 μM NaATP, 200 μM MgCl$_2$, 500 μM dithiothreitol (DTT), 0.002% sodium azide), which provided conditions to depolymerize actin filaments. Extracted proteins were removed by centrifugation, and the membranes were then solubilized with detergent. This problem of actin contamination of HLA preparation is not so great if viable cells are extracted by the magnesium precipitation method.[56] This is probably because the high magnesium concentration causes filamentous actin to precipitate in paracrystals[66,67] and be eliminated from the extract applied to the immunoaffinity columns. In addition, lower protein concentrations and proportionately larger columns of nonspecific immunoglobulin were used by Parham[56] to remove actin that bound nonspecifically to immunoglobulin.[58]

Quaranta et al.[55] also encountered problems with actin contamination when using the Q6/13 monoclonal antibody to purify a subset of HLA-DR antigens from detergent-solubilized extracts of human B lymphoblastoid cell lines. They observed actin contamination when antibody covalently coupled to Sepharose, but not when antibody bound to protein A–Sepharose was used as the immunoabsorbent. This suggests that actin binds nonspecifically to the Fc of CNBr-coupled immunoglobulin. Actin was not observed in preparations made with Q5/13-Sepharose from a glycoprotein fraction made from cell extracts with a lentil lectin column. In contrast, Pober et al.[65] found that actin present in HLA-A,B preparations could not be separated on a lentil-lectin column.

It is worth pointing out that although human B lymphoblastoid cells lines have advantages as a source of HLA antigens—i.e., they grow indefinitely in vitro and have a high expression of HLA—there are disadvantages associated with their large actin content. The cell surface contains numerous membrane projections that are full of filamentous actin. Actin is the major cellular component as assessed by SDS–PAGE. The problem of actin contamination in immunoaffinity chromatography may be less with other types of cells, as demonstrated for the mouse cell lines used by Mescher et al. in this volume [8].

Williams and co-workers have extensively used monoclonal antibody affinity columns to purify various rat lymphocyte glycoproteins including Ia antigens.[51,52,68,69] Purification of Ia antigens from rat spleen cell membranes was accomplished using MCR OX 4 antibody coupled to CNBr-

[66] J. Hanson, Nature (London) 213, 353 (1976).
[67] J. A. Spudich, H. E. Huxley, and J. T. Finch, J. Mol. Biol. 72, 619 (1972).
[68] C. A. Sunderland, W. R. McMaster, and A. F. Williams, Eur. J. Immunol. 9, 115 (1979).
[69] W. R. Brown, A. N. Barclay, C. A. Sunderland, and A. F. Williams, Nature (London) 289, 456 (1981).

activated Sepharose 2B.[52] Membrane solubilization was with 2% deoxycholate, and a purification of 200-fold as assessed by measurement of antigenic activity was obtained. In their studies, elution with 50 mM diethylamine-HCl, pH 11.5, always proved to be effective.

Antibodies from Antibodies

One use of these purification procedures is to provide immunogen for production of more HLA monoclonal antibodies. For example, we were interested in obtaining antibodies against HLA-B27 and discovered that all the available large-scale preparations of HLA-B27 were in fact HLA-B7. This was probably due to a confusion in identification of B cell lines. We therefore grew a 10-liter culture of 23.1 cells (A2,2;B27,27). The cells were solubilized as described above, and HLA-A,B,C antigens were purified on a 1-ml column of W6/32–Sepharose preceded by 5-ml columns of Sepharose and normal BALB/c IgG–Sepharose. The bound material was eluted with 50 mM diethylamine, pH 11.5, and 0.5-ml fractions were collected and neutralized with 1.0 M Tris-HCl, pH 8.0. Aliquots (20 μl) of each fraction were removed before neutralization for analysis by SDS–PAGE. This showed four fractions to have significant quantities of HLA and β_2-microglobulin and a yield of about 0.5–1 mg. The four fractions were pooled and divided into five aliquots; four were frozen at $-20°$. The fifth aliquot was emulsified with an equal volume of complete Freund's adjuvant and used to immunize five mice. After three subsequent immunizations in incomplete Freund's adjuvant mice were given a final intravenous immunization with an acetone precipitate of the fifth aliquot of antigen. Antigen was mixed in a 1.5-ml Eppendorf centrifuge tube with four times the volume of cold ($-70°$) acetone and placed at $-70°$ overnight. After centrifugation in the Eppendorf centrifuge for 5 min, the supernatant was removed and the precipitate was taken up in 250 μl of PBS. These fusions from mice immunized with A2 and B27 gave rise to many HLA monoclonal antibodies including MA2.2,[6] which is highly specific for HLA-A2 and similar to PA2.1[3] and BB7.2.[15] Unfortunately, our goal of obtaining a B27 reagent was not realized. Ellis et al.[23] using a different strategy solubilized a mixture of HLA-A32,2,B27,17 from human spleen cell membranes and removed A2 and B17 with a column of MA2.1 antibody.[70] A32 and B27 were purified on a column of W6/32–Sepharose and used to boost mice that had been previously immunized with B27-expressing spleen cells. This resulted in production of an antibody (ME1) with specificity for B27, B7, and B22 similar to that of Grumet et al.[4] obtained by a regime involving only whole-cell immunizations.

[70] A. J. McMichael, P. Parham, N. Rust, and F. M. Brodsky, Hum. Immunol. 1, 121 (1980).

Affinity Considerations

Our initial experiments to purify HLA antigens on monoclonal antibody columns were done in the absence of any clear knowledge of the affinity of the various antibodies, W6/32, PA2.1, BB7.1, and BBM.1, used.[39,56] Their relative success suggested that the basic idea would be of general usefulness and that it would be helpful to be able to identify suitable monoclonal antibodies. Naively we reasoned that high-affinity antibodies would be difficult to elute, low-affinity antibodies would not give stable binding, and antibodies in a compromise range of affinity would be optimal. Therefore, we began to measure the affinities of our "successful" antibodies to see if this were true. The methods used were based on those of Trucco et al.[55] and Mason and Williams.[37]

Purified W6/32 and PA2.1 IgG bound to appropriate cells with binding constants of $\simeq 10^9$ M^{-1}. F(ab')$_2$ of PA2.1 bound similarly to IgG, F(ab')$_2$ of W6/32 bound significantly more strongly than IgG. Fab of PA2.1 bound with a binding constant of $\simeq 2 \times 10^7$ M^{-1}, 50-fold less than for F(ab')$_2$, whereas Fab of W6/32 bound similarly to F(ab')$_2$ and 3- to 4-fold stronger than IgG. These results showed that W6/32 has both high avidity and high affinity, and that PA2.1 has high avidity and low affinity and is dependent on bivalent attachment for stable interaction with the cell surface.[71] These results are consistent with the observations that (a) W6/32 binding to cells is inhibited $\simeq 50$ times more effectively by monomeric soluble HLA antigens than is PA2.1; (b) W6/32 efficiently immunoprecipitates solubilized HLA antigens, and PA2.1 does not. The basis for these differences is the rate of dissociation; for W6/32 Fab the half time of dissociation is $\simeq 24$ hr; for PA2.1 Fab, it is $\simeq 10$ min. Knowing these binding characteristics, we would have predicted that PA2.1 was unlikely to provide a useful immunoabsorbent. In conclusion, we can only recommend an empirical approach and suggest that no antibody be discarded from consideration as an immunoabsorbent on the basis of binding properties measured in solution. The essentially stable complex found by HLA-A2 and PA2.1 when conjugated to Sepharose could be the result of additional binding interactions between antigen and Sepharose and/or an effectively very high concentration of PA2.1 in the vicinity of the matrix. It is interesting to note that PA2.1–Sepharose will effectively immunoprecipitate HLA-A2 from solution, whereas an equivalent amount of soluble PA2.1 bound to Staphylococcus with rabbit anti-mouse IgG does not.

Analysis of the binding properties of a number of antibodies indicates that most polymorphic HLA-A, B, C monoclonal antibodies will be of

[71] P. Parham, M. J. Androlewicz, F. M. Brodsky, N. J. Holmes, and J. P. Ways, *J. Immunol. Methods* **53**, 133 (1982).

relatively low affinity and that the properties we have outlined for PA2.1 will be generalizable. It is our impression from the literature, however, that polymorphic H-2 monoclonal antibodies perform better in solution immunoprecipitation than most HLA monoclonal antibodies.

Acknowledgments

I thank F. Brodsky, N. Holmes, H. Orr, and J. Ways, whose unpublished results are referred to in this paper, and B. Buckley for preparing the manuscript. The original research described was supported by Grants NSF PCM 80-17834 and NIH 1R01 AI 17892.

[10] Immunoassay for Sequence-Specific DNA–Protein Interaction

By RONALD MCKAY

Proteins that bind to specific DNA sequences are known to play important roles in regulating DNA replication and transcription in prokaryotes and eukaryotic viruses. This knowledge suggests that specific DNA sequence binding proteins will also regulate genetic activity in eukaryotic cells. With the possible exceptions of yeast and *Drosophila*, where such regulatory proteins may be accessible by genetic techniques, the complexity of the eukaryote genome limits the value of traditional genetic approaches to such regulatory proteins. Hybridoma technology makes possible an alternative strategy. Monoclonal antibodies against the viral tumor (T) antigen of SV40 have been used to confirm that the viral protein specifically binds to a particular restriction fragment of the viral genome containing the origin of DNA replication and the promoter for T-antigen transcription.[1] This technique uses the specificity of the monoclonal antibody to make quantitative statements of binding affinity even in crude protein extracts of virus-infected cells that contain the viral T antigen. In this chapter the immunological technique is shown (*a*) to be capable of detecting the specific binding properties of SV40 T antigen in crude protein extracts of SV40-transformed cell lines, which, unlike the infected cells, contain only one copy of the viral genome; (*b*) in conjunction with a partial DNase I digestion to give information on the length of the DNA sequence protected by the DNA binding protein; and (*c*) by modification

---

[1] J. Tooze, "The Molecular Biology of Tumor Viruses, DNA Tumor Viruses," 2nd ed. Cold Spring Harbor Laboratory, Cold Spring Harbor, New York, 1980.

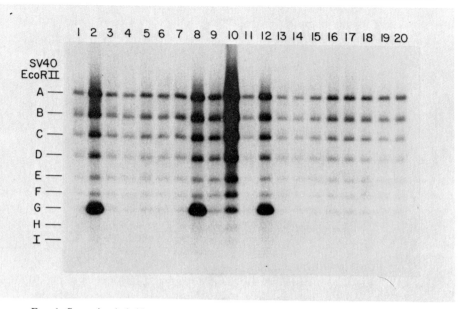

FIG. 1. Screening hybridoma supernatants for monoclonal antibodies that immunoprecipitate the regulatory region of the SV40 genome via specifically bound SV40 T antigen. Tracks 2, 8, and 12 show the specific precipitation of the SV40 control sequences in the *Eco*RII-G fragment. Track 10 shows an activity that nonspecifically immunoprecipitates DNA fragments.

of the DNase footprinting procedure to give more precise information on the specific DNA sequence bound by the protein.

Screening Monoclonal Antibodies

Monoclonal antibodies against SV40 large T antigen were obtained by immunizing BALB/c mice with the partially purified related protein synthesized by the Ad2D2 recombinant virus described by Hassel *et al.*[2] The hybridoma cell lines were screened by indirect immunofluorescence on SV40-transformed cells and by the ability of the supernatant medium specifically to immunoprecipitate the radiolabeled *Eco*RII/G restriction fragment of SV40 DNA in the presence of T antigen (Fig. 1). These antibodies have also been shown to bind directly to radiolabeled T antigen by immunoprecipitation and Western blot analysis.[3] The indirect immuno-

[2] J. A. Hassel, E. Lukanidin, G. Fey, and J. F. Sambrook, *J. Mol. Biol.* **120**, 209 (1978).
[3] H. Towbin, T. Staehelin, and J. Gordon, *Proc. Natl. Acad. Sci. U.S.A.* **76**, 4350 (1979).

precipitation of DNA through T antigen binding has been described in detail.[4] In brief, the procedure calls for radiolabeled DNA, a source of T antigen, and a specific antibody against T antigen. An aliquot of T antigen is added to the labeled DNA in binding buffer. The DNA fragments with T antigen bound are then separated from the unbound fragments by adding monoclonal anti-T antibody, rabbit anti-mouse IgG, and immunosorption to fixed *Staphylococcus aureus* fragments. The monoclonal antibody can be used as hybridoma cell supernatant. The assay procedure is compatible with many antisera containing anti-T antibodies generated in rabbits and hamsters. The fractionated DNA fragments were then released from the *S. aureus* fragments with sodium dodecyl sulfate (SDS) and analyzed by electrophoresis.

In Fig. 1 this assay was used as the primary screen of supernatants from hybridoma lines to detect monoclonal antibodies against the soluble DNA-binding form of SV40 T antigen. Five nanograms of *Eco*RII SV40 DNA fragments (nick translated to a specific activity of 7×10^7 cpm/μg) was added to 1 ml of binding buffer [20 mM sodium phosphate, pH 6.8, 2.0 mM dithiothreitol, 100 μg of bovine serum albumin per milliliter, 0.1 mM EDTA, 0.05% (v/v) Nonidet P-40 (NP-40), 3% dimethyl sulfoxide]. The DNA in binding buffer was incubated at 25° for 10 min with a protein fraction containing the SV40 T antigen-related protein derived from HeLa cells infected with the adeno-SV40 recombinant virus, Ad2+D2 (see McKay[4] for details). Supernatant (40 μl) from hybridoma cultures was added and incubated for 10 min, then followed by 30 μl of rabbit antiserum against mouse immunoglobulin for another 10 min. Eighty microliters of 10% (w/v) fixed *S. aureus* (The Enzyme Center Inc., Boston, Massachusetts) were added, and the preparation was incubated for 5 min. The immunosorbent was spun down in an Eppendorf centrifuge and washed twice with 150 mM NaCl, 10 mM Tris-HCl, pH 8.0, 0.5% NP-40. The bound fragments were released by 1% (w/v) SDS plus 10 mM EDTA and analyzed by agarose gel electrophoresis. For autoradiography the agarose gel was dried after equilibration with ethanol.

Specific Binding with Extracts of Transformed Cell Lines

The experiment described above (Fig. 1) used protein extracts of virus-infected cells that carry many copies of the gene coding for SV40 T antigen and consequently are highly enriched for the DNA binding protein. Figure 2 shows that the immunoassay procedure is sufficiently sensitive to detect the specific binding of T antigen in protein extracts of rat cell

[4] R. D. G. McKay, *J. Mol. Biol.* **145,** 471 (1981).

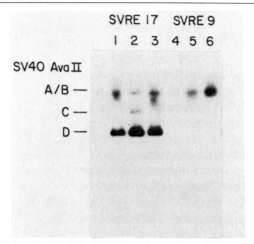

FIG. 2. Crude protein extracts of the SV40-transformed rat cell line, SVRE 17 allow the specific immunoprecipitation of the SV40 AvaII-D fragment containing the regulatory region. In contrast, similarly prepared extracts of the SVRE9 cell line known to carry a defective T antigen fail to mediate specific immunoprecipitation of the AvaII-D fragment.

lines transformed with SV40 virus. These cell lines are known to carry only one integrated copy of the viral genome.[5-7] Extracts of the two cell lines were prepared as follows: 1.5×10^7 SV40-transformed cells were scraped from tissue culture dishes, washed in phosphate-buffered saline, and swollen in 3 ml of 10 mM Tris-HCl, pH 8.0, 0.5 mM dithiothreitol (DTT), 10 mM NaCl, 1 mM EDTA, 150 μg of phenylmethylsulfonyl fluoride per milliliter for 5 min at 0°. The cells were disrupted by Dounce homogenization, and the supernatant from a 20,000 g spin was used directly in the assay.

The binding experiments were performed as described above. The protein from the cell line SVRE 17 gives specific immunoprecipitation of the appropriate SV40 AvaII-D fragment. Under the same conditions the extracts of the SVRE9 showed no specific binding to the AvaII-D fragment (Fig. 2). This difference is consistent with other information that shows the SVRE9 T antigen to be mutant and defective.[6,7] In this context this lack of specific DNA binding confirms the view that the binding to the AvaII-D fragment in the SVRE17 extract is due to the presence of functional large T antigen. One of the functions of SV40 large T antigen is to act as a repressor of the transcription of the message RNAs coding for

[5] M. Botchan, W. Topp, and J. Sambrook, Cell 9, 269 (1976).
[6] J. R. Stringer, J. Virol. 38, 671 (1981).
[7] J. R. Stringer, J. Virol. in press (1982).

large T and small t antigens. SVRE9 cells that lack functional T antigen also synthesize more large T antigen than an equivalent number of SVRE17 cells (J. Stringer and R. McKay, unpublished data). This finding suggests that in the SVRE17 cells the amount of T antigen may be regulated by the inhibitory effect of SV40 T antigen on the transcription of the single SV40 genome. Quantitative studies[4] of T antigen binding affinity suggest that the apparent equilibrium binding constant is $10^{-12} M$. We can speculate then that if other eukaryotic regulatory proteins have similar properties they will be detectable with this assay system.

DNase Protection

In addition to information on the restriction fragment containing the specific binding site for a regulatory protein, it is possible to determine the size and location of the binding site using the immunoassay. The information is obtained by including a partial DNase I digestion in the procedure. If nick-translated DNA is used as the substrate, then the binding site is protected from digestion by the bound protein. These protected fragments can then be analyzed on a polyacrylamide gel. Figure 3 shows the results of a DNA protection experiment showing a series of protected fragments.

The binding buffer was 10 mM HEPES, pH 7.75, 20 mM NaCl, 10 mM MgCl$_2$, 2.5 mM DTT; 2.5×10^5 counts of nick-translated SV40 DNA (10^7 cpm/μg) were added to the buffer, followed by 10 μl of a protein extract from HeLa cells infected with adeno-SV40 recombinant virus Ad2D2. This amount of protein extract had previously been shown to give specific immunoprecipitation of the EcoRII-G restriction fragment of SV40 DNA. After 15 min at room temperature, 10 μl of 0.1 M CaCl$_2$ were added to each 1-ml reaction. Then 1 μl and 12.5 μl of a DNase I solution (1 mg/ml) were added to tubes 2 and 3. Tube 1 was a control where no T antigen was added and 1 μl of DNase I was used. After 10 min at room temperature the reaction was stopped by the addition of 250 μl of 0.1 M EDTA, 3 M ammonium acetate, 0.1 mg of bovine serum albumin per milliliter, and 5 μl of total mouse DNA (1.5 mg/ml). The immunoprecipitation of DNA fragments resistant to digestion and bound to T antigen was as described above. Of the 2.5×10^5 input cpm, 310 cpm were immunoprecipitated in the absence of T antigen (tube 1), 4438 cpm were immunoprecipitated after treatment with 1 μl of DNase I (tube 2), 2872 cpm were immunoprecipitated after digestion with 12.5 μl of DNase I (tube 3). The size of the immunoprecipitated double-stranded DNA fragments was determined by electrophoresis on a 15% polyacrylamide gel (Fig. 3). The major fragments have a bimodal distribution around a mean of 30 and 75 base pairs,

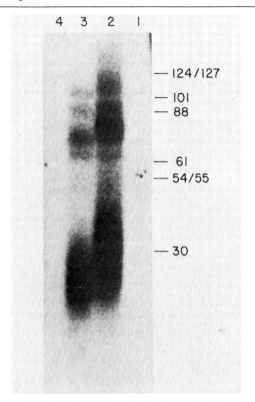

FIG. 3. DNase protection. Track 1 shows that no low molecular weight DNA is immuno-precipitated if no T antigen is present during the DNase I digestion. In contrast, in the presence of T antigen, tracks 2 and 3 immunoprecipitation reveals two major protected fragments with a double-stranded size of 30 and 75 base pairs. Track 4 shows that no fragments in this size range are seen when the immunoprecipitation is carried out in the presence of T and the absence of DNase I.

as we expect from the DNase protection experiments of Tjian.[8] In the absence of added binding protein, no DNA is immunoprecipitated; in the absence of divalent ions, all the precipitated DNA remains high molecular weight (data not shown) as the DNase is not active.

Footprinting

If the DNA substrate is end-labeled, then the immunoassay, including a partial DNase I step, gives information on the location of the specific

[8] R. Tjian, *Cell* **13**, 165 (1978).

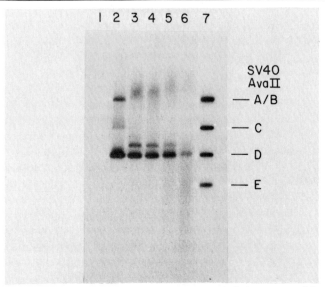

FIG. 4. Analysis of partial DNase I-digested end-labeled, immunoprecipitated AvaII SV40 DNA fragments by agarose gel electrophoresis. Track 1 shows the immunoprecipitated DNA in the absence of T antigen; track 2 shows DNA fragments immunoprecipitated in the absence of DNase I but in the presence of T antigen; tracks 3–6 show the consequences of including increasing amounts of DNase (see text for details). Track 7 shows the five large SV40 AvaII fragments.

binding site within a DNA restriction fragment. The conditions used here were modified from the DNase protection experiment as follows. SV40 DNA was cleaved with the restriction enzyme AvaII. The cleavage site on the AvaII-D fragment at 5099 base pairs was specifically labeled with the Klenow form of DNA polymerase I in the presence of radiolabeled dATP. The other end of this fragment at 538 base pairs on the SV40 genome is unlabeled by this procedure. Because of the immunoprecipitation step in this procedure, the other AvaII restriction fragments need not be separated from the specific binding D fragment. Per 50 ng of SV40 DNA, 10^5 cpm/min were added to 1 ml of binding buffer of 20 mM sodium phosphate pH 6.8, 2.0 mM DTT, 3% dimethyl sulfoxide, 10 mM sodium chloride, 1 mM calcium chloride. Crude T antigen derived from HeLa cells infected with a recombinant virus synthesizing full length SV40 T antigen[9] was added to each aliquot. After 15 min at room temperature, different amounts of DNase I were added to 1-ml aliquots of the binding solution to give a final concentration of 1 μg/ml, 10^{-1} μg/ml, 10^{-2} μg/ml,

[9] D. Solnick, Cell 24, 135 (1981).

10^{-3} μg/ml. The DNase digestion was allowed to proceed for 10 min at room temperature and terminated by the addition of 100 μl of EDTA (0.5 M, pH 7.0) and 5 μl of total mouse DNA (1.5 mg/ml). The immunoprecipitation of the DNA fragments bound by T antigen was as described above.

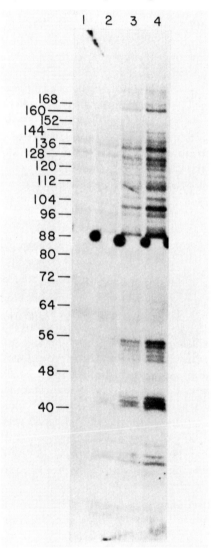

FIG. 5. A denaturing polyacrylamide gel showing the single-stranded size distribution of the DNA fragments shown in tracks 3–6 of Fig. 4. The gaps in the ladder from base pairs 56–89 and 146–160 correspond precisely to the first and second binding sites of SV40 T antigen.[10]

The immunoprecipitated DNA was released from the *S. aureus* fragments by SDS. An aliquot was run on a native agarose gel (Fig. 4) to show (*a*) that specific binding of the *Ava*II-D fragment had occurred and (*b*) which concentration of DNase I first gave partial digestion. The remainder of the immunoprecipitated DNA was ethanol precipitated, dissolved, and denatured by heating in 70% formamide, 20 mM EDTA, 0.3% xylene cyanol, and 0.3% bromophenol blue. The size distribution of these single-stranded DNA fragments was analyzed on a denaturing acrylamide gel. The double-stranded immunoprecipitated DNA after agarose gel electrophoresis is shown in Fig. 4. The major immunoprecipitated fragment is the *Ava*II-D fragment. The origin of the DNA band that runs at a slightly higher molecular weight than *Ava*II-D is not known. The appropriate DNase digestion conditions can be assessed by the appearance of DNA fragments of lower molecular weight (lanes 5 and 6). Figure 5 shows the results of running the same samples as in lanes 3–6 in Fig. 4 on a denaturing gel. The lack of fragments of 55–88 base pairs and 136–160 base pairs is the expected result from the known location of the T antigen binding sites.[10]

Summary

An immunoassay that allows a quantitative measure of the interaction of DNA and specific DNA binding proteins has been described here. A detailed quantitative description of the assay has previously been published.[4] In this chapter the assay has been used to determine the size and location of the SV40 sequences that bind the viral protein, T antigen, by new DNase protection and footprinting procedures. These results are consistent with the observations of Tjian,[8,10] who used purified protein to establish the sizes and locations of T-antigen binding sites. In addition, the assay is shown to be capable of detecting specific binding even in crude protein extracts of SV40 virus-transformed cell lines that carry only one copy of the viral genome. These results suggest that this procedure may be useful in the identification and purification of specific DNA binding proteins in eukaryotic cells.

[10] R. Tjian, *Cold Spring Harbor Symp. Quant. Biol.* **43**, 655 (1979).

[11] Quantitation of Hybridoma Immunoglobulins and Selection of Light-Chain Loss Variants

By TIMOTHY A. SPRINGER

Compared to conventional antisera, hybridoma antibodies offer many advantages including specificity, consistency, availability in large quantities, and the ability to use impure immunogens.[1] Hybridoma antibodies also differ in two other respects from conventional antibodies. First, some hybridomas secrete myeloma as well as specific antibody chains. Methods are described in Section I for selecting variant clones that secrete homogeneous immunoglobulins from mouse–rat or mouse–mouse hybrids. Second, the specific antibody component of monoclonal antibodies consists of a single heavy-chain subclass and light-chain isotype. In addition to measuring antibody activity, it is often desired to measure monoclonal antibody immunoglobulin concentration. Monoclonal antibodies express only a portion of the antigenic determinants found in whole immunoglobulins. This has important implications for the measurement of monoclonal immunoglobulin concentration by immunoassay. Section II describes methods for measuring rat or mouse monoclonal immunoglobulins derived from rat–mouse, mouse–mouse, or rat–rat hybrids.

I. Quantitation of Myeloma Light-Chain Secretion and Selection of Loss Variants Using Radioimmunoassay

In immunoglobulin-synthesizing cells, the genes for immunoglobulin heavy and light chains are expressed by one chromosome each, and the expression of allelic genes on the homologous chromosomes are excluded. In hybridoma cells, the active genes for immunoglobulin synthesis from both the myeloma and spleen cell parents continue to be expressed.[1,2] Thus, two different heavy chains and two different light chains may be made by a single hybridoma cell. Theoretically, there are nine different combinations in which these chains can be assembled into an IgG. For purposes of nomenclature, the heavy and light chains from the spleen cell parent have been designated H and L, and those from the myeloma parent have been designated G and K, respectively. In HGLK

[1] G. Galfrè and C. Milstein, this series, Vol. 73, p. 3.
[2] C. Milstein, K. Adetugbo, N. J. Cowan, G. Köhler, D. S. Secher, and C. D. Wilde, *Cold Spring Harbor Symp. Quant. Biol.* **41**, 793 (1977).

METHODS IN ENZYMOLOGY, VOL. 92

lines, i.e., hybrids secreting all four chains, hybrid molecules are assembled and secreted in which all permutations of heavy–light chain combinations occur and random heavy–heavy associations also occur unless the heavy chains differ in class.[3] If all chains are synthesized in equal amounts, then the bivalently active, specific, H_2L_2 antibody accounts for only 1/16th of the total antibody.

There are two approaches to obtaining more homogeneous and active antibody. First, variant myeloma lines are available as fusing partners that synthesize only the myeloma K chain or no chains at all.[1,4] The use of nonproducing myelomas is by far the easiest way of obtaining new hybridomas secreting homogeneous H_2L_2 antibodies. Second, from hybridomas previously prepared with myeloma-chain producing myelomas, such as P3-X63, NSI, or MPC 11, variant hybridomas may be selected that no longer secrete myeloma chains. There is one point in favor of making new hybridomas with myeloma lines, such as NSI, that make a myeloma K chain. From an HLK hybrid, HK variants as well as HL variants can be selected. The HK antibodies serve as the best possible control for the HL antibodies, since they have exactly the same heavy chain, but in association with an inappropriate light chain that in almost all cases leads to loss of specific antigen-binding capacity. Such variants are also useful for studies on idiotypes and the antigen binding site.

Hybridoma lines lose specific and myeloma chains at similar frequencies.[5] The pattern of chain loss is essentially random, except that hybridomas with an excess of expressed heavy chains over light chains are rarely seen.[5] This may be due to toxicity by free heavy chains. Loss of chains is correlated with loss of specific chromosomes, 6 and 12 for mouse kappa and heavy chains, respectively.[6] When mass cultures are screened for variants, they are usually found at frequencies of 1/50 to 1/200.[5,7]

The effects of various chain loss events in NSI × spleen cell HLK hybrids are described in the table. NSI is unusual in that it synthesizes, but does not secrete, the myeloma K chain. K chain secretion is reactivated in the presence of an H chain provided by a spleen cell[2] (see the table). The types of chains secreted by parental hybridoma lines and their variants may be confirmed by internal radiolabeling and SDS–PAGE and by comparison to the myeloma products. Chains secreted by K, L, or H chain-loss variants of HLK lines are illustrated in Fig. 1.

[3] G. Köhler and M. J. Shulman, *Curr. Top. Microbiol. Immunol.* **81**, 143 (1978).

[4] G. J. Hämmerling, U. Hämmerling, and J. F. Kearney, *in* "Monoclonal Antibodies and T-Cell Hybridomas," pp. 563–571. Elsevier/North-Holland, Amsterdam, 1981.

[5] G. Köhler, H. Hengartner, and C. Milstein, *Protides Biol. Fluids* **25**, 545 (1977).

[6] H. Hengartner, T. Meo, and E. Muller, *Proc. Natl. Acad. Sci. U.S.A.* **75**, 4494 (1978).

[7] T. A. Springer, *J. Immunol. Methods* **37**, 139 (1980).

CHAIN-LOSS VARIANTS ARISING FROM HLK NSI × SPLEEN CELL HYBRIDS

Nature of chain loss	Intracellular synthesis	Ig secretion phenotype[a]	K chain assay inhibition	Antigen binding
None	HLK	HLK	+	+
−K	HL	HL	−	+
−L	HK	HK	+	−[b]
−H	LK	L[c]	−	−

[a] NSI does not secrete its K chain in the absence of an H chain.[2]
[b] There may be rare cases in which K chain complements antigen binding activity.
[c] There may be rare cases in which H chain expression is required for L chain secretion; however, this will not affect the radioimmunoassay results.

Previous methods of screening for variants involved the internal radiolabeling and polyacrylamide gel electrophoresis (PAGE) analysis of a large number of subclones.[5,8] However, this is time consuming and expensive. Therefore, a rapid radioimmunoassay specific for idiotypic or V_κ subgroup determinants on the NSI and P3-X63 K chain was developed.[7] The assay measures the ability of hybridoma culture supernatants competitively to inhibit the binding of ^{125}I-labeled anti-P3 V_κ idiotype to HLK or HK IgG coated on plastic Microtiter wells. The target antigen (HLK or HK) and immunogen (P3, which is GK) share only the K chain, rendering the assay specific for this chain. P3 IgG, but not other myeloma proteins, inhibit, demonstrating the idiotype or V_κ subgroup specificity of the absorbed anti-P3 IgG serum (Fig. 2A). HL variants do not inhibit, but HLK monoclonal antibodies in the form of hybridoma culture supernatants give potent inhibition (Fig. 2B). In addition to screening for loss variants, the assay is also useful for typing hybrid lines and for quantitating the amount of K chain secreted (Fig. 2B).[7]

A. P3 K Chain Radioimmunoassay and Loss-Variant Selection[7]

1. Antibodies

Antibodies described here are for use with the NSI and P3-X63 myeloma lines, which were derived from P3 (MOPC 21) and hence have idiotypically identical light chains. Antibodies specific for K or G chains of nonrelated myelomas, e.g., MPC-11, could be obtained by similar procedures. P3 (MOPC 21) IgG may be obtained from serum or ascites of BALB/c mice bearing P3-X63 or MOPC 21 tumors (American Type Cul-

[8] G. Köhler and C. Milstein, *Eur. J. Immunol.* **6**, 511 (1976).

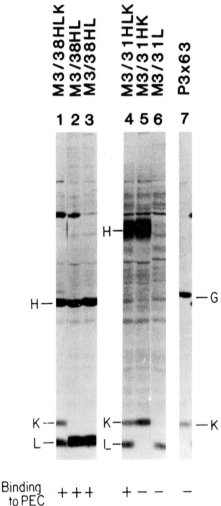

FIG. 1. SDS–PAGE of immunoglobulin chains secreted by HLK hybridoma cells and their chain-loss variants. M3/38.1.2 and M3/31.1.1.6 are NSI × rat spleen cell hybrids that secrete HLK antibodies of IgG2a and IgM subclass, respectively, which bind to the Mac-2 antigen [M.-K. Ho and T. A. Springer, *J. Immunol.* **128,** 1221 (1981)] on murine peritoneal exudate cells (PEC). Subclones (460 of each line) were tested for K chain loss by the P3 K chain radioimmunoassay and for rat L chain loss by radial immunodiffusion against rabbit anti-rat Fab. After labeling with [^{14}C]leucine, secreted products were subjected to SDS–PAGE under reducing conditions and autoradiographed. From M3/38.1.2 (lane 1) were obtained 2 HL variants (lanes 2 and 3), 2 HK variants, and 3 L variants (not shown). From M3/31.1.1.6 (lane 4) were obtained 3 HK variants (one shown in lane 5) and 2 L variants (one shown in lane 6). The L variants are presumably due to H chain loss, which secondarily leads to absence of K chain secretion.[2] The K chain secreted by P3-X63 (lane 7) serves to identify the corresponding chain in the NSI hybrids. NSI was derived from P3 and therefore has an identical light chain. Loss of H or L chain in all cases correlated with loss of antibody activity, measured in a binding assay, whereas loss of K chain did not.

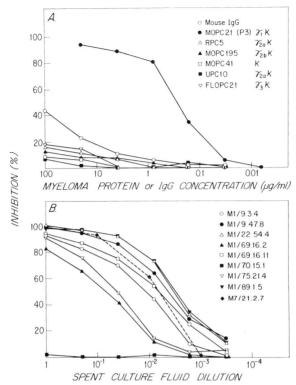

FIG. 2. Characteristics of the P3 K chain assay. (A) Inhibition by MOPC 21 (P3) IgG, other myeloma proteins containing kappa chains, and mouse IgG. (B) Inhibition by products secreted into culture medium by NSI × rat spleen cell hybrids. M1/70.15.1 is an HL line; the others are HLK lines as verified by SDS–PAGE of ^{14}C-labeled secreted products. The solid lines and dashed lines represent the assay carried out with target HLK antigen coated onto sheep red blood cells[7] or Microtiter wells, respectively. The characteristics of the assay are quite similar in both cases.

ture Collection), or from Litton Bionetics. Rabbits are immunized by three monthly injections of 1 mg of P3 IgG in complete Freund's adjuvant distributed to multiple intramuscular sites. Bleeds may be taken biweekly thereafter for several months. The antiserum is absorbed by passing 10 ml through a 4-ml column of rat or mouse serum coupled to Sepharose CL-4B (Pharmacia) (19 mg of protein per milliliter of settled beads). This absorption renders the antiserum specific for P3 IgG idiotypic determinants. Completeness of the absorption should be checked by comparing binding to normal IgG and to K chain-containing MAb in the radioimmunoassay.

2. Iodination[9,10]

Materials

Microfuge tube (Sarstedt, No. 701)
Tubing, $\frac{1}{8}$ inch i.d., $\frac{3}{16}$ inch o.d. (Tygon R-3603)
Tubing, 0.05 inch i.d., 0.09 inch o.d. (Tygon S-54-HL)
Tubing connector, $\frac{1}{16}$ inch to $\frac{1}{8}$ inch (Value Plastics, Loveland, Connecticut, No. AC)
Glass wool
Hemostat clamp
Adjustable micropipettor and disposable plastic tips
Sephadex G-25 swollen in phosphate-buffered saline (PBS)
P3 or M1/69 HK IgG, coupled at 2 mg/ml to Sepharose CL-4B
NaPO$_4$, 0.2 M, pH 7.5
PBS: 0.8% NaCl, 0.05 M NaPO$_4$, pH 7.5
Absorbed anti-P3 IgG serum (anti-P3 idiotype)
Glycine-HCl, 0.1 M, pH 2.5
Glycine-HCl, 0.1 M, pH 2.5; 1% bovine serum albumin (BSA)
Carrier-free Na^{125}I, 1–2 mCi in 1–20 μl
HCl, 0.1 N
Chloramine-T, 2 mg/ml, dissolved before use in PBS
Tris-HCl, 1 M, pH 8.4

Procedure. Anti-P3 idiotype is purified by adsorption to a P3 IgG–Sepharose CL-4B or M1/69 IgG–Sepharose CL-4B affinity matrix and is iodinated *in situ.*

1. Prepare a minicolumn by cutting off the tip of a microfuge tube, pushing a $\frac{1}{2}$ inch long piece of $\frac{1}{8}$ inch i.d. tubing over the bottom, and adding a 6 inch piece of 0.05 inch i.d. tubing with the connector. The column is mounted in one of the 96 holes formed when the top half of a hard-well Microtiter plate is sawed off, and the plate is held in a ringstand. Flow rate is regulated either by looping the outlet through a hole in the Microtiter plate and adjusting the hydrostatic head, or with a hemostat. Tamp a ball of glass wool to the bottom of the column and fill it with PBS. Add Sephadex G-25 to make a 50-μl bed, then add a 10-μl bed of P3 IgG–Sepharose CL-4B.

2. Preelute the column with 1 ml of glycine-HCl, pH 2.5, and wash with 2 ml of PBS.

3. Slowly, over 10 min, pass through sufficient anti-P3 idiotype to saturate the column (about 0.2 ml).

[9] L. E. M. Miles and C. N. Hales, *Biochem. J.* **108,** 611 (1968).
[10] L. A. Herzenberg and L. A. Herzenberg, *in* "Handbook of Experimental Immunology" (D. M. Weir, ed.), p. 12.1. Blackwell, Oxford, 1978.

4. Wash with 4 ml of PBS, and clamp off the effluent when no PBS remains above the Sepharose.

5. The iodination is carried out in a properly equipped fume hood. Prepare ^{125}I by adding 5 μl of 0.2 M NaPO$_4$, pH 7.5, then sufficient 0.1 N HCl to neutralize the NaOH in the [125]I.

6. Add 5 μl of 2 mg/ml chloramine-T to the ^{125}I, mix, immediately transfer with the micropipettor to the Sepharose bed in the column, and suspend the Sepharose layer by gentle stirring with the plastic tip for 1 min.

7. Wash with 2 ml of PBS (collect as flow-through)

8. Elute with 2 ml of 0.1 M glycine-HCl, pH 2.5, + 1% BSA into a tube containing 1 ml of 1 M Tris-HCl, pH 8.4, and mix by vortexing (collect as eluate).

9. Determine with a portable radioactivity monitor the relative proportion of counts in the flow-through, the eluate, and those remaining on the column. About 30% incorporation into the eluate is normally achieved.

10. Dialyze the eluate overnight versus 1 liter of PBS, dilute to 2000–20,000 cpm/μl with 5% BSA in PBS, and store at $-80°$ in 2-ml aliquots.

3. Subcloning

Hybridoma lines are allowed to grow for 2–4 months after the previous subcloning to allow variants to arise. Soft agar (0.3%) cloning of 1000 and 3000 cells/100-mm petri dish and growth of hybridoma cells are carried out as described elsewhere.[1] After 7–10 days, agar plugs containing single clones are transferred to 96-well microculture plates (Costar, Cambridge, Massachusetts) containing 0.2 ml of medium per well. A blank agar plug is transferred as a control. Since K chain loss variants are found at a frequency of about 10^{-2}, about 400 clones are picked. Clones grow at different rates. After some clones grow sufficiently to lower the pH of the medium (phenol red turning orange), about two-thirds of the medium in every well is aspirated and replaced with fresh medium every 2 days. At least three changes are made before assay, to allow most of the residual K chain from the agar plug to be diluted out. After this number of changes, HL variants and the control agar plug may inhibit the P3 K chain assay slightly, but will be clearly distinguishable from HLK lines. For assay, about 100 μl of culture fluid are transferred into sterile V-well Microtiter "master" plates (Flow Laboratories, No. 76-222-05) with a 12-channel, 50–200 μl adjustable micropipettor (Flow Laboratories, No. 77-889-00) equipped with autoclaved plastic tips. One set of 12 tips is used for an entire Microtiter plate, rinsing twice with sterile distilled water in a reservoir (Flow Laboratories, No. 77-824-00) and touching off residual solution onto an autoclaved paper towel between each transfer.

4. Radioimmunoassay[11,12]

Materials

[125]I-anti-P3 idiotype

Purified HK or HLK MAb, 10 μg/ml in PBS (M1/69 HK IgG may be obtained from Boehringer-Manheim, Indianapolis, Indiana, or the M1/69 HK cell line from the American Type Culture Collection, Rockville, Maryland).

Normal IgG, 10 μg/ml in PBS

Fetal calf serum (FCS), 10% in PBS

BSA, 1% in PBS

Micropipettor, 50–200 μl, 12 channel (Flow Laboratories, No. 76-222-05)

Soft-well Microtiter radioimmunoassay (RIA) plates (Costar No. 2595)

Syringe, 250 μl, with repeating dispenser for 5-μl aliquots (Hamilton, Reno, Nevada)

Hot wire Microtiter plate cutter (D Lee, 932 Kintyre Way, Sunnyvale, California)

Adhesive tape (plate sealing tape, Cooke Laboratories, Alexandria, Virginia)

Manifold, 12 needle (Cooke No. 300-5) and Cornwall syringe for delivering wash buffer. A similar 12-needle manifold for aspiration, and a Microtiter plate shaker (Cooke No. 2-225-06) are handy but optional pieces of equipment.

Procedure

1. The HLK or HK target antigen in PBS is transferred with the 12-tip pipettor to soft-well Microtiter plates (50 μl/well) and allowed to adsorb to the wells for 2 hr at room temperature or overnight at 4°. The use of an HLK or HK target antigen, rather than P3 IgG or HGLK, ensures that the subset of anti-idiotype antibodies specific for the P3 K chain, but not the G chain, are included in the assay. Wells coated with normal IgG and with 1% BSA are included as controls.

2. Wells are emptied by aspiration, and residual binding sites on the plastic are saturated with 120 μl of 1% BSA in PBS for 5 min; the wells are then washed twice more with the same solution.

[11] T. A. Springer, A. Bhattacharya, J. T. Cardoza, and F. Sanchez-Madrid, *Hybridoma* 1, 257 (1982).

[12] G. H. Parsons, Jr., this series, Vol. 77, p. 224.

3. Aliquots to be tested (50 μl) are transferred with the multichannel pipettor from the "master" Microtiter plate to the soft-well assay plate. Serial dilutions in 10% FCS–PBS of supernatant from the parent hybridoma are also tested to establish a standard curve. The diluent for the competition step contains 10% serum, which gives a lower background than 1% BSA.

4. Aliquots of 5 μl of ^{125}I anti-P3 idiotype are added with the Hamilton syringe (for highest accuracy the aliquot is added while the plate is shaking). The plate is shaken and allowed to stand ½ hr at room temperature.

5. The plates are aspirated and washed 3 times with 1% BSA–PBS. Residual liquid is removed from plates by throwing them upside down onto absorbent paper.

6. Adhesive tape is applied to the bottom of the soft-well plate, and the top is cut off by pushing the plate horizontally through the hot wire cutter, leaving the isolated wells adhering to the tape. Wells are plucked off and gamma-counted.

7. Percentage of inhibition is calculated as

$$(1 - e/t)/(1 - c/t) \times 100$$

where e = experimental cpm bound in the presence of inhibitor to antigen-coated wells, c = control cpm bound in the presence of 10% FCS–PBS to BSA-coated wells, and t = total cpm bound in the presence of 10% FCS–PBS to antigen-coated wells. Total cpm bound should be 10–30% of input cpm.

5. Further Characterization of Variant Subclones

If HK variants are also desired, parallel testing of the subclones for loss of antigen-binding can be carried out.

In testing for HL variants with the P3 K chain assay, it is common to find a large proportion of subclones that give complete inhibition (HLKs) and about 1% that give moderate to weak inhibition (HLs). The latter are transferred to 2 ml of medium in 24-well plates. True HL variants are found to give no inhibition after further growth and media changes.

Both H and K chain loss can lead to absence of K chain secretion (see the table). Therefore it is important to confirm the nature of the chain loss both by testing for antibody activity and by internal radiolabeling and PAGE of the secreted chains (Fig. 1). Specific light chains (of mouse or rat) can almost always be resolved from the myeloma K chain by discontinuous SDS–PAGE.[13] Separation may be due to charge differences, which are important in the stacking gel, as well as to size differences between different V_L regions.

[13] T. Springer, G. Galfrè, D. S. Secher, and C. Milstein, *Eur. J. Immunol.* **8**, 539 (1978).

Subclones showing the desired chain loss are expanded by further growth for collection of culture supernatants and liquid N_2 storage of frozen cells.[1]

II. Quantitation of Monoclonal Immunoglobulins

Immunoassays are useful for determining the concentration of monoclonal antibodies secreted into culture medium, ascites, or serum and to guide purification. The accurate determination of monoclonal Ig concentrations by immunoassay presents special problems. The use of anti-IgG sera is inappropriate, since such sera contain a high proportion of antibodies directed to subclass-specific determinants on the Fc region. In inhibition assays, the proportion of inhibitable anti-IgG antibodies thus depends on the subclass of the MAb being tested (Fig. 3B). This can lead to artifactually low estimates of monoclonal immunoglobulin concentrations. Anti-Fab sera show much less subclass preference. Rabbit anti-Fab sera is routinely used in this laboratory in single radial immunodiffusion (Mancini assay[14]) for the determination of mouse and rat MAb concentrations. This assay takes very little time to carry out and has the convenience that the Mancini plates may be stored for at least up to a year before being used.

Antisera to Fab fragments prepared from whole rat or mouse IgG are primarily specific for kappa chain determinants. It should be cautioned that these sera almost invariably do not react by radial immunodiffusion with lambda chain-containing monoclonal immunoglobulins. About 10% of mouse and rat immunoglobulins contain lambda chains, and we have found about the same proportion in rat anti-mouse cell surface MAb.[11] Antisera to lambda chains or the appropriate heavy-chain subclass must be used to quantitate these MAb.

In working with antisera to rat Fab or rat kappa chains, it is important to take the kappa allotype into account. Two allelic forms of rat kappa chains have been defined, RI-1a in DA, ACI, and Fisher rats, and RI-1b in Lewis, BN, LOU, and Wistar rats.[15–17] The two allotypes differ by 11 amino acids in the kappa chain constant region. This is a very large difference and is comparable to the interspecies difference between mouse and rat RI-1a kappa chains of 14 residues.[18,19] Therefore, the rat

[14] G. Mancini, A. O. Carbonara, and J. F. Heremans, *Immunochemistry* **2**, 235 (1965).

[15] G. A. Gutman and I. L. Weissman, *J. Immunol.* **107**, 1391 (1971).

[16] A. Beckers, P. Querinjean, and H. Bazin, *Immunochemistry* **11**, 605 (1974).

[17] G. A. Gutman, *Immunogenetics* **5**, 597 (1977).

[18] G. A. Gutman, *Transplant. Proc.* **13**, 1483 (1981).

[19] H. W. Sheppard and G. A. Gutman, *Proc. Natl. Acad. Sci. U.S.A.* **78**, 7064 (1981).

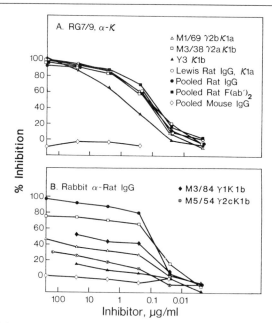

FIG. 3. Radioimmunoassay determination of rat monoclonal immunoglobulin concentrations.[11] Serial dilutions of rat monoclonal immunoglobulins were incubated with ^{125}I-labeled, monoclonal mouse anti-rat kappa chain (panel A), or affinity-purified rabbit anti-rat IgG (panel B). The residual binding capacity of the ^{125}I-labeled antibodies was then measured after transfer to rat IgG-coated soft-well plates. (A) Kappa 1a and 1b allotype-bearing monoclonal immunoglobulins and whole IgG all inhibited the monoclonal anti-rat K chain antibody identically, showing that it can be used to determine monoclonal antibody concentration. The monoclonal anti-kappa antibody was prepared against whole IgG and apparently has a lower affinity for free kappa chains. (B) Inhibition by the monoclonal immunoglobulins of ^{125}I-labeled rat IgG plateaued at submaximal levels, showing that each can inhibit only a subpopulation of the anti-rat IgG antibodies with the appropriate specificity.

IgG preparation used to construct the standard curve for radial immunodiffusion or radioimmune assays, and the samples being analyzed, should always be of the same allotype. In contrast to the rat, mouse kappa chains do not express allotype differences.

Radioimmunoassay is an alternative to single radial diffusion for determining monoclonal immunoglobulin concentrations. This laboratory has prepared a monoclonal antibody that has equal affinity for rat kappa chains of 1a and 1b allotypes.[11] It is an excellent reagent for determination of monoclonal immunoglobulin concentrations (Fig. 3A) and for use in the indirect binding assay and many other applications.[11] Monoclonal antibodies with specificity for rat IgG subclasses and kappa allotypes,[11] and with specificity for mouse kappa chains and IgG subclasses are also available.[20] Since many of the monoclonal anti-IgG reagents that have subclass specificity are directed to Fc region determinants,[11] they are also useful for monitoring or effecting the removal of Fc fragments and undigested IgG from Fab or F(ab')$_2$ preparations.

A. Single Radial Immunodiffusion Determination of Monoclonal Immunoglobulin Concentrations

1. Anti-Rat (or Mouse) Fab Serum

To purified rat (or mouse) IgG, 10–20 mg/ml in PBS, 2 mM EDTA, is added 1/100 by weight of papain (Worthington, Freehold, New Jersey) and 2-mercaptoethanol to bring to 10 mM. Digestion is carried out under N$_2$ for 18 hr at 37°, and sodium iodoacetate and Tris-HCl, pH 8.6, are added to 11 mM and 20 mM, respectively, to stop digestion. Fab and Fc fragments are separated from small peptides and any residual undigested IgG by Sephadex G-200 filtration. The Fab and Fc pool is dialyzed versus 0.05 M Tris-HCl, pH 8.0, and applied at 6 mg of protein per milliliter of bed volume to a DEAE-cellulose column (DE-52, Whatman) equilibrated with the same buffer. The column is washed with 0.05 M Tris-HCl, pH 8.0, and eluted with a gradient of 0 to 0.2 M NaCl in the same buffer. Separation of Fab from Fc is monitored by immunoelectrophoresis[21,22] or radioimmunoassay. Pure Fab should elute in the flow-through.

Antisera to Fab fragments are prepared as described in Section I,A,1.

2. Single Radial Immunodiffusion[14,22,23]

In single radial immunodiffusion, antigens in a central well diffuse into antiserum-containing agar. When equilibrium is reached, the area bounded by a precipitate ring around the well is proportional to the amount of antigen applied.

To determine the proper amount of antiserum to use, the procedure described below is scaled down for use with nonfrosted microscope slides. Each slide receives agar containing a different amount of antiserum, three 3 mm in diameter wells are punched, and 8-μl aliquots of IgG at 200, 100, and 50 μg/ml are applied. The lowest antiserum concentration

[20] D. E. Yelton, C. Desaymard, and M. D. Scharff, *Hybridoma* **1**, 5 (1981).
[21] D. R. Stanworth and M. W. Turner, *in* "Handbook of Experimental Immunology" (D. M. Weir, ed.), p. 6.25. Blackwell, Oxford, 1978.
[22] O. Ouchterlony and L.-A. Nilsson, *in* "Handbook of Experimental Immunology" (D. M. Weir, ed.), p. 19.10. Blackwell, Oxford, 1978.
[23] J.-P. Vaerman, this series, Vol. 73, p. 291.

that still gives well-defined rings is selected for use. In our experience, this concentration will give rings of about 8.5 mm in diameter with 100 μg of IgG per milliliter.

To prepare the plates, an appropriate volume of antiserum (about 100 μl if high-titered) is mixed with 20 ml of 1% agar (Difco) in 0.01 M Tris-HCl pH 7.8, 0.14 M NaCl, 0.1% NaN$_3$ at 55°; 19 ml are spread evenly over a prewarmed, level, 4 × 5 inch glass plate and allowed to solidify. Plates may be stored at least 1 year in sealed containers (Tupperware) containing paper towels wetted with the same buffer. A 5 × 7 matrix of holes 3 mm in diameter is punched in the agar with a No. 1 cork borer attached to an aspirator, and 8-μl samples are applied, avoiding overflowing. Standards of mouse IgG or of the appropriate kappa chain allotype of rat IgG at 200, 100, 50, 20, and 10 $\mu g/ml$ are applied to each plate. Unknowns are applied at several different dilutions guessed to give concentrations of about 100 $\mu g/ml$. High concentrations can cause artifacts in neighboring wells.[23] Diffusion is allowed to occur for 2 days in a humid atmosphere at room temperature. Ring diameter is measured using a viewing stand. The stand is a 9 inch × 7½ inch piece of ¼ inch plywood with a 5½ inch hold cut out of its center, supported on 9 inch long dowel legs. The stand is placed in a darkroom on a light box with a piece of black paper directly beneath the stand to provide indirect illumination. Agar is cut from the corners of the plate that touch the stand, and the plate is inverted and placed on the stand. The agar remains adherent to the plate. Ring diameter is measured in two perpendicular directions through a measuring magnifier (Bausch and Lomb No. 81-34-38) placed on the back of the plate. A calibrating viewer, Model 2743, is also available from Transidyne General Co., Ann Arbor, Michigan. The standard curve is constructed by plotting the product of the two ring diameters against antigen concentration.

B. Radioimmunoassay for Quantitating Monoclonal Immunoglobulins[ii]

Monoclonal mouse antibodies to rat kappa chains (RG7/9) and rat IgG subclasses 1 (RG11/39), 2a (RG7/1), and 2b (RG7/11)[11] can be obtained from Boehringer-Mannheim (Indianapolis, Indiana) or by growth of the cell lines (American Type Culture Collection, Rockville, Maryland). Monoclonal rat antibodies to mouse kappa chains and IgG subclasses[20] are available from New England Nuclear (Boston, Massachusetts). Antisera to Fab fragments are prepared as described in Sections II,A,1 and I,A,1. Antibodies can be affinity purified and radioiodinated on Sepharose CL-4B conjugated to the appropriate antigen, as described in Section I,A,2. For mouse monoclonal antibodies, we have found it to be more convenient to purify milligram quantities by *Staphylococcus aureus* protein A affinity chromatography[24] and iodinate 10-μg quantities using chloroglycoluril.[25]

Monoclonal immunoglobulins in culture supernatants, ascites, or other form and normal IgG standards are serially diluted in 10% FCS–PBS in hard-well Microtiter plates. IgG, 10 μg/ml in PBS, is adsorbed to soft-well Microtiter plates. The competition assay is then carried out as described in Section I,A,4. Alternatively, an equilibrium inhibition assay may be carried out. The ^{125}I-MAb (5 μl) is mixed with antigen dilutions (50 μl) in hard-well plates, the plates are sealed with tape and allowed to stand overnight at 4°, and then 40-μl aliquots are transferred to the antigen-coated soft-well plates, and the assay is continued as described in Section I,A,4.

Acknowledgments

This work was supported by USPHS Grant CA 31798, Council for Tobacco Research Grant 1307, and a junior faculty award from the American Cancer Society.

[24] P. L. Ey, S. J. Prowse, and C. R. Jenkin, *Immunochemistry* **15**, 429 (1978).
[25] P. J. Fraker and J. C. Speck, *Biochem. Biophys. Res. Commun.* **80**, 849 (1978).

[12] Indirect ^{125}I-Labeled Protein A Assay for Monoclonal Antibodies to Cell-Surface Antigens

By JOSEPH P. BROWN, KARL ERIK HELLSTRÖM, and
INGEGERD HELLSTRÖM

The hybridoma technique of Köhler and Milstein[1,2] has greatly facilitated the identification and characterization of cell-surface antigens, for which monoclonal antibodies have many advantages over conventional serological reagents. However, some of the methods commonly used to test antisera are unsuitable for use with monoclonal antibodies, either because they fail to detect antibodies of certain immunoglobulin isotypes or because they are insensitive to antibodies bound to a single antigenic determinant. Mouse IgG1 antibodies, for example, are not detected by complement-dependent cytotoxicity,[3] and IgG2a antibodies to a single

[1] G. Köhler and C. Milstein, *Nature (London)* **256**, 495 (1975).
[2] G. Galfrè and C. Milstein, this series, Vol. 73, p. 3.
[3] M.-Y. Yeh, I. Hellström, J. P. Brown, G. A. Warner, J. A. Hansen, and K. E. Hellström, *Proc. Natl. Acad. Sci. U.S.A.* **76**, 2927 (1979).

epitope of a cell surface antigen are relatively inefficient in this assay.[4] Also, during the screening of hybridomas, many hundreds of samples of culture medium must be tested, and even more tests are required when cloning and recloning hybridomas. Consequently, there has been considerable emphasis on the development of rapid, sensitive screening assays capable of detecting monoclonal antibodies of any immunoglobulin isotype.

Antibodies bound to cell-surface antigens can be detected by using specific antibody purified from anti-immunoglobulin serum by immunoadsorption and labeled with a radioisotope, a fluorescent hapten, or an enzyme. An alternative method is based on the ability of protein A, a constituent of the cell wall of most strains of *Staphylococcus aureus,* to bind to the Fc region of IgG antibodies of many mammals.[5] Protein A has proved to be a useful serological reagent.[5–15] Unfortunately, protein A binds poorly or not at all to IgM, mouse IgG1, and to most rat IgG subclasses.[13] However, by using a rabbit anti-immunoglobulin serum in an intermediate incubation (e.g., a rabbit anti-mouse immunoglobulin serum when working with mouse antibodies), protein A can be used to detect any class of immunoglobulin. In this chapter, we describe an indirect ^{125}I-labeled protein A assay,[16] which we have used to screen mouse and rat hybridomas for antibodies to cell-surface antigens of human tumors.[3,17]

The indirect ^{125}I-labeled protein A assay has several attractive features as a method for screening hybridomas. First, the materials required are commercially available and inexpensive. Second, antibodies of any isotype can be detected. Third, multiple microtest plates with growing hybridomas can be screened rapidly, and a permanent visual record, in the form of an autoradiograph, is obtained.

Briefly, the assay is as follows. Target cells are plated out in flat-bottomed microtest wells to give an adherent monolayer of cells. The

[4] I. Hellström, J. P. Brown, and K. E. Hellström, *J. Immunol.* **127**, 157 (1981).
[5] G. Kronvall, U. S. Seal, J. Finstad, and R. C. Williams, Jr., *J. Immunol.* **104**, 140 (1970).
[6] K. I. Welsh, G. Dorval, and H. Wigzell, *Nature (London)* **254**, 67 (1975).
[7] G. Dorval, K. I. Welsh, and H. Wigzell, *J. Immunol. Methods* **7**, 237 (1975).
[8] J. P. Brown, J. M. Klitzman, and K. E. Hellström, *J. Immunol. Methods* **15**, 57 (1977).
[9] S. W. Kessler, *J. Immunol.* **115**, 1617 (1975).
[10] S. W. Kessler, *J. Immunol.* **117**, 1482 (1976).
[11] P. L. Ey, S. J. Prowse, and C. R. Jenkin, *Immunochemistry* **15**, 429 (1978).
[12] J. W. Goding, *J. Immunol. Methods* **20**, 241 (1978).
[13] J. J. Langone, *J. Immunol. Methods* **24**, 269 (1978).
[14] J. J. Langone, this series, Vol. 70, p. 356.
[15] L. Suter, J. Bruggen, and C. Sorg, *J. Immunol. Methods* **39**, 407 (1980).
[16] J. P. Brown, J. D. Tamerius, and I. Hellström, *J. Immunol. Methods* **31**, 201 (1979).
[17] R. G. Woodbury, J. P. Brown, M.-Y. Yeh, I. Hellström, and K. E. Hellström, *Proc. Natl. Acad. Sci. U.S.A.* **77**, 2183 (1980).

cells are incubated in turn with spent culture medium from the hybrido-mas, with a suitable dilution of rabbit anti-immunoglobulin serum, and finally with radiolabeled protein A. The plates are then autoradiographed, or the wells are counted individually.

The steps of the assay will now be discussed in detail.

Target Cells

For rapid screening of large numbers of hybrids, target cells are most conveniently tested as adherent monolayers in flat-bottomed 0.3 cm^2 microtest wells. The cells are plated out at 10,000 to 50,000 per well in 100–200 μl of culture medium and allowed to grow to confluence. Most human cells, such as fibroblasts, melanoma cells, and carcinoma cells, form a stable monolayer that lasts through the assay. On the other hand, mouse cells, for example, cells from chemically induced mouse sarcomas, have a tendency to detach from the wells during the assay. This can be prevented by fixing the cells with 10% buffered formalin at 0° for 10–30 min either before the assay or after the incubation with hybridoma culture medium. Fixation after incubation with antibody allows detection of anti-bodies to epitopes that might be destroyed by fixation. However, fixation at either stage of the assay tends to increase the level of nonspecific binding. Nonadherent cells can be tested either in suspension in V-bottomed microtest wells, the plates being centrifuged for 5 min at 1500 g after each incubation and wash (using commercially available plate carriers),[16] or bound to polylysine-treated flat-bottomed wells.[15] If the number of target cells available for the assay is limiting, 60-well microtest plates, with wells 1 mm in diameter, can be used (e.g., Falcon Plastics, Oxnard, California, catalog No. 3034). Other methods of binding cells to microtest wells have been described.[18] The same technique can also be applied to purified proteins, viruses, or cell membrane preparations adsorbed to flat-bottomed polystyrene wells.[19,20]

Unless the frequency of positive hybrids is expected to be extremely low, it is most efficient to screen simultaneously not only on the immunizing cell line, but also on a control cell line (such as autologous fibroblasts or B cells when screening for antibodies to tumor-associated antigens). In this way, one can identify antibodies recognizing antigens common to both cell types and also antibodies that bind nonspecifically, either to the target cells or to the microtest wells themselves.

[18] C. H. Heusser, J. W. Stocker, and R. H. Gisler, this series, Vol. 73, p. 406.
[19] R. C. Nowinski, M. D. Lostrom, M. R. Tam, M. R. Stone, and W. N. Burnette, *Virology* **93**, 111 (1979).
[20] D. Colcher, P. Horan Hand, Y. A. Teramoto, D. Wunderlich, and J. Schlom, *Cancer Res.* **41**, 1451 (1981).

Incubations and Washes

The three successive incubations with hybridoma medium, rabbit anti-serum, and protein A (see below) are essentially identical in terms of reagent volumes, incubation times, and subsequent washes. Volumes of 50 μl and incubation times of 30 min to 1 hr in a 37° incubator with 7% CO_2 in air are convenient. After each incubation, the contents of the wells are aspirated, and the cells are washed twice with 200 μl of either culture medium or phosphate-buffered saline pH 7.2 (PBS), containing 1% fetal calf serum. All additions to the wells are made with an 8-channel micropipette (Flow Laboratories, Inglewood, California), and an 8-channel aspirator is used to remove liquid from the wells. When 60-well plates are used, additions to the wells are made singly. However, for the washes the entire plate can be rinsed with 5 ml of culture medium.

Hybridoma Culture Medium

When the assay is used to screen hybridomas, samples of culture medium are taken 12–20 days after the cell fusion. At this time, the microtest wells are almost confluent with hybridoma cells, and the spent medium contains 1–10 μl of monoclonal antibody per milliliter. We generally transfer 150-μl samples of medium to sterile 96-well microtest plates using sterile pipette tips and add fresh culture medium to the hybridoma cultures. This operation is carried out in a laminar-flow hood to avoid contamination. The samples should be kept in a humidified incubator with CO_2 in air atmosphere until ready for use to maintain a physiological pH; otherwise the target cells may detach during the first incubation.

To start the assay, culture medium is aspirated from the target cells and replaced with 50 μl of spent medium from the hybridoma cultures. Normally, for screening purposes a single sample from each well is tested undiluted on two target cell types, one positive and one negative for the antigen of interest. When screening for antibodies to tumor-associated antigen, for example, we use tumor cells and autologous fibroblasts as targets.

Rabbit Anti-Mouse Immunoglobulin Serum

In the second incubation, rabbit antiserum to mouse immunoglobulin is used at a dilution of 1 : 500 to 1 : 2000. We have found it advisable to test several lots of antiserum, selecting for one with both a high titer of specific antibody and a low level of nonspecific binding to the target cells. It is important to use antisera from rabbits rather than goats, since goat antibodies bind protein A extremely poorly.[14] Pig antibodies are reported to bind protein A well,[13] and may also be useful in the assay. For determining immunoglobulin isotypes, we use subclass-specific antisera[5] (see below). When screening for rat, human, or other monoclonal antibodies, rabbit antisera of the appropriate specificity are used. The incubation with rabbit antiserum may be omitted if one wishes to select only hybridomas that produce antibodies binding protein A directly.

[125]I-Labeled Protein A

The final incubation is with ^{125}I-labeled protein A. Labeling methods have been discussed previously.[13,14] We use the chloramine-T technique.[21] Briefly, 30 μg of protein A are incubated with 1 mCi of Na[125]I in 0.5 ml of PBS, in the presence of 10 μg of chloramine-T for 10 min at 0°. The reaction is stopped by adding 10 μg of sodium metabisulfite, and the labeled protein is purified by gel filtration on a 5-ml column of Sephadex G-25 that has been pretreated with 1 ml of bovine serum albumin and equilibrated with PBS. The ^{125}I-labeled protein A is mixed with bovine serum albumin (1 mg/ml final concentration) and can be stored at $-80°$ for at least 2 months. Alternatively, ^{125}I-labeled protein A can be purchased (Amersham, Arlington Heights, Illinois). The ^{125}I-labeled protein A is diluted to 10^6 cpm/ml with culture medium before use (i.e., 50,000 cpm/well).

Autoradiography

For screening purposes, we find it more convenient to autoradiograph the microtest plates than to count wells individually.[5] A permanent record of the data is obtained with a minimum of manipulation, and the results can be related directly to the original microtest plates in which the hybridomas were grown.

After the incubation with ^{125}I-labeled protein A and subsequent washes, the plates are sealed with tape, and, if necessary, their edges are cut off with a band saw. For the autoradiographic exposure we use a lighttight, wooden cassette, designed to keep up to four 96-well plates in close contact with a 20 × 25 cm sheet of X-ray film. The cassette is constructed so that the plates are inverted on a foam rubber pad. Sensitivity is maximized by using an X-ray intensifying screen with preflashed film at $-70°$ as described.[22] The film is placed over the plates, and the intensifying screen is attached to the inside of the lid of the cassette. An

---

[21] P. J. McConahey and F. J. Dixon, *Int. Arch. Allergy Appl. Immunol.* **29,** 185 (1966).

[22] R. Swanstrom and P. R. Shank, *Anal. Biochem.* **86,** 184 (1978).

overnight exposure allows one to detect several hundred disintegrations per minute (dpm) in a 0.3 cm^2 well. An alternative, though less sensitive, method of autoradiographing microtest plates has been described.[23]

In some cases, one may prefer to obtain quantitative data. The cells are removed from the microtest wells in 200 μl of 2 M NaOH and transferred to tubes; the amount of bound ^{125}I is determined in a gamma counter. Alternatively, the plates are cut up with a band saw. However, a minimum of 82 cuts per plate is required, and the individual wells are difficult to fit into test tubes for counting. Therefore, we find the NaOH procedure to be more expeditious.

Evaluation of Results

Negative and positive controls are included in each plate to facilitate interpretation of the autoradiographs. For negative controls we use fresh culture medium or culture medium containing antibody to an antigen known to be absent from the target cells. For a positive control, either a known positive hybridoma supernatant or a suitable dilution of known positive antiserum in culture medium is used. In some experiments, particularly when the background (nonspecific) binding is low (see below), it can be difficult to align the plates with the autoradiograph. The positive controls are helpful in this regard. Working with Falcon No. 3040 96-well plates, we generally reserve wells E12, F12, G12, and H12 for controls, two negative and two positive. Different arrangements of positive and negative controls among these four wells enables us to distinguish up to four plates from one another and to align the plates with the autoradiograph.

Nonspecific Binding

The sensitivity of the assay depends to a large extent upon the background level of binding (i.e., in the negative controls). Several factors can contribute to a high background. First, if the target cells express cell-surface immunoglobulin, they will bind protein A either directly or via the rabbit antiserum. Second, rabbit antibodies can bind directly to antigens of the target cells. Third, antibodies can adsorb nonspecifically to either target cells or to the microtest wells. The last two factors can be minimized by screening rabbit antisera not only for a high-titer against mouse immunoglobulin (so that a high dilution can be used in the assay), but also for a low level of nonspecific binding to the target cells. Ideally the level of

[23] E. W. Weiler and M. H. Zenk, this series, Vol. 73, p. 394.

nonspecific binding should be at or below the threshold of detection by autoradiography, i.e., several hundred disintegrations per minute per well.

Determination of Immunoglobulin Isotype

By using isotype-specific rabbit antisera, the indirect protein A assay can be used to determine immunoglobulin isotypes.[3,16] Generally the specificity of commercial isotype-specific antisera is established by double immunodiffusion, a less sensitive method than the binding assay. Antisera that are monospecific by this criterion may, therefore, cross-react in the binding assay. Accordingly, one must determine for each antiserum a dilution at which it is operationally monospecific. We have generally used a dilution of 1 : 500.

Some mouse antibodies (IgG2 and IgG3) bind protein A directly. However, the protein A assay can be used to determine their isotype, since even with these antibodies there may be a considerable increase in binding with the appropriate rabbit antiserum.[3]

Discussion

The indirect ^{125}I-labeled protein A assay can be used to detect any class or subclass of immunoglobulin bound to cells (as described here), cell membranes, viruses, and proteins. It is rapid, simple, and sensitive and is, therefore, suitable for screening hybridomas. In addition, all the reagents are commercially available.

When using the assay for screening hybridomas, one commonly works with cultured cells and tests two cell lines in parallel, looking for antibodies that discriminate between the two targets. If this strategy is used, it is not desirable to have more than one hybrid per well, since antibodies of the desired specificity may be obscured by cross-reacting antibodies present in the same well. However, it is difficult to predict fusion efficiency in advance, and one frequently does obtain more than one hybrid per well. In this situation, one may omit the second step of the assay (incubation with rabbit anti-immunoglobulin serum) so that the only antibodies detected are IgG2 and IgG3, which bind protein A directly. Since hybridomas producing these antibodies comprise only about one-fifth of the total, few wells will contain more than one detectable hybridoma.

Another common problem is interpreting a weak reaction on the control cell line. This can be due either to a single antibody reacting with an epitope present in small amounts on the control line or to the presence of a second, cross-reacting antibody in the sample tested. These possibilities can be distinguished by cloning the hybridomas. One can, however, save time by determining whether the antibody binding to the control cell line is of the same isotype as that binding to the immunizing cell line; if not, it is clearly another, cross-reactive antibody.

Binding assays with cultured cells as targets have played an important role during the first few years of hybridoma research. They are rapid and simple and hence allow the screening of large numbers of hybridomas for antibodies of any isotype. However, it is becoming increasingly evident that screening should constitute only one part of the research endeavor. Careful and definitive study of individual antibodies is needed. Binding assays of cultured cells such as the one that we describe here should not, therefore, be used to the exclusion of other, more informative screening assays.

As a rule, we believe that the best approach to screening hybridomas is to use methods that allow detection of the type of antibodies or target antigens in which one is interested. If, for example, one wishes to obtain antibodies that can be used to immunoprecipitate protein antigens from cell lysates, immunoprecipitation is probably the best screening technique to use. If one needs antibodies that are cytotoxic in the presence of complement or function in lymphocyte-dependent cellular cytotoxicity, one should screen for antibodies with just those properties. If one wishes to obtain antibodies for immunoperoxidase staining of cryostat sections, then that technique may, in the final analysis, be the most efficient for screening. An efficient strategy is to use binding assays to select hybridomas that produce antibodies of the isotopes expected to have the desired activity, e.g., IgM antibodies for complement-dependent cytotoxicity and protein A-binding antibodies for use in immunoprecipitation, or to eliminate antibodies binding to, for example, fibroblasts, and then to investigate these antibodies further in a second screening step.

We would like to make two more comments. First, indirect binding assays such as that described here are not suitable for definitive determination of antigen distribution on a range of cell types, since they are designed for speed rather than sensitivity and quantitation. Direct binding assays with radiolabeled monoclonal antibody[24] or quantitative absorption studies are superior for this purpose.

Second, one can, instead of using ^{125}I-labeled protein A, employ alkaline phosphatase-conjugated protein A[15], in an enzyme-linked immuno-

---

[24] J. P. Brown, R. G. Woodbury, C. E. Hart, I. Hellström, and K. E. Hellström. *Proc. Natl. Acad. Sci. U.S.A.* **78**, 539 (1981).

assay (ELISA). The ELISA procedure may be a useful modification, particularly for laboratories not equipped for handling radioisotopes, since the two procedures appear to be comparable in terms of speed and sensitivity.

We conclude that the indirect ^{125}I-labeled protein A assay is a simple and rapid method for screening hybridomas for production of antibodies to cell-surface antigens, particularly of adherent cells. All the reagents are commercially available. The assay can detect antibodies of any immunoglobulin class or subclass, and it can be used to screen hundreds of samples per day.

Acknowledgments

This work was supported by grants CA27841 and CA19149 from the National Institutes of Health and by grant IM241A from the American Cancer Society.

[13] Enzyme-Linked Immunosorbent Assay for Screening Monoclonal Antibody Production Using Enzyme-Labeled Second Antibody

By J. Y. DOUILLARD and T. HOFFMAN

Production of monoclonal antibodies of predefined specificity by the technique of cell–cell hybridization[1] demands fast, sensitive, and reproducible screening assays to identify those with a desired reactivity from among the majority making irrelevant immunoglobulins. Although direct measurement of antibody activity (for example, complement-dependent lysis) may be employed, assays of antibody binding to fixed antigen are usually preferred, since they detect any and all classes or subclasses of immunoglobulins. Methods using fluorescein-labeled anti-antibodies to assay for supernatants with antigen-binding activity are available, but they are tedious, cumbersome, and time-consuming. Radioimmunoassays require hazardous, costly, and evanescent materials along with sophisticated equipment. These drawbacks led to consideration of the enzyme-linked immunosorbent assay (ELISA),[2] since its sensitivity and reproducibility have been shown to be comparable to other binding assays.[3] Furthermore, the possibilities for its automation and application in microsystems seemed particularly attractive.

[1] G. Köhler and C. Milstein, *Nature (London)* **256**, 495 (1975).
[2] J. Y. Douillard, T. Hoffman, and R. B. Herberman, *J. Immunol. Methods* **39**, 309 (1980).
[3] E. Engvall, this series, Vol. 70 [28].

Assay Method

Principle

The presence of antibody to a given antigen is determined by incubating hybridoma supernatant with immobilized antigen, followed by washing and the subsequent addition of anti-immunoglobulin antibody conjugated with enzyme. Further incubation with substrate results in a measurable reaction (usually a change in optical density) only in those wells in which antibody originally reacted with antigen.

Antigen Immobilization and Plate Preparation

Supernatants from hybridomas are usually screened initially against the immunogen itself and irrelevant antigen(s) as a control(s). Isolation and preparation of antigen for screening must therefore take into account preservation of its distinct properties. Soluble antigens, usually proteins, may be attached to a solid phase by passive adsorption to plastic without detectable influences on their antigenic structure. Cell-associated antigens require special manipulation to avoid disruption of their structure.

Materials

Solutions

Carbonate-bicarbonate coupling buffer, pH 9.6. Dissolve in 1000 ml of distilled H_2O: 1.6 g of Na_2CO_3, 2.9 g of $NaHCO_3$, and 0.9 g of NaN_3.

Phosphate-buffered saline-bovine (PBS), 0.1% bovine serum albumin (BSA), 0.1% sodium azide. PBS is available commercially or can be made as follows: to 127.5 g of NaCl, add 217.5 ml of 1 M Na_2HPO_4; add 97.5 ml of 1 M NaH_2PO_4; dilute to 15 liters with distilled H_2O and adjust the pH with 10 N NaOH.

Glutaraldehyde, 0.25% (can be stored at 4° for up to 8 weeks)

Washing solutions. Solutions of both types can be stored for several weeks at 4°.

Soluble antigen washing solution: PBS; $MgCl_2$, 1.5 mM; 2-mercaptoethanol, 2 mM; Tween 20, 0.05%; NaN_3, 0.1%.

Cellular antigen washing solution: PBS; $MgCl_2$, 1.5 mM; 2-mercaptoethanol, 2 mM; BSA, 0.1%; NaN_3, 0.1%.

Microplates: Ninety-six well Microtiter plates currently available from several manufacturers can be used. We routinely employ polystyrene Linbro Titertek plates (Linbro Scientific Company, Hamden, Connecticut) or Imulon I Cooke Microtiter plates (Dynatech, Alexandria,

Virginia), both with flat bottoms. Polyvinyl flexible plates have also been used successfully.

Spectrophotometers: Any of the multitude of instruments available today may be used to measure optical density. The applicability of this technique is in great part enhanced by the availability of automated "through the well" ELISA readers. These are sold as the Titertek Multiscan (Flow Laboratories, McLean, Virginia) and Dynatech (Alexandria, Virginia) Microelisa autoreader.

Procedure. Two hundred microliters of soluble antigen at a concentration of 1–10 μg/ml in coupling buffer are dispensed into the wells of a 96-well microplate and incubated at 4° for a period of at least 8 hr. Storage of these plates for up to 4 weeks has been feasible at either room temperature or −20°.

When intact cells are used as antigen, they may be used fresh or after fixation. For fresh cells, Microtiter plates are first incubated for 1 hr at room temperature with PBS–BSA–NaN₃. Plates are then washed with PBS–BSA. This may be accomplished by removing supernatant with a pipette or an automated washer apparatus. We have also successfully used "flicking" (i.e., removing supernatants with a short, rapid inversion motion using the opposite hand as a stop). Twenty to 30 μl of the cell suspension in serum-free medium are dispensed into the wells and allowed to settle for 1 hr prior to proceeding with the assay.

For glutaraldehyde fixation, 100 μl of the cell suspension in PBS are dispensed into the wells of untreated Microtiter plates then spun down for 10 min at 400 g. The medium is then carefully discarded by "flicking" and blotting the inverted plate on a tissue paper, without disturbing the cell layer. Fifty microliters of 0.25% glutaraldehyde are added to each well for 7 min at room temperature. Two hundred microliters of PBS–BSa–NaN₃ are dispensed into the well in order to wash away the glutaraldehyde. The plate is flicked, and 200 μl of PBS–BSA–NaN₃ are added once more. The plate can be used immediately in the assay or covered for extended storage at 4° for several weeks.

Enzyme-Labeled Second Antibody

Most monoclonal antibodies to date have been produced using mouse–mouse hybridomas and therefore require enzyme-labeled anti-mouse immunoglobulins. Such enzyme-labeled "second antibodies" or their F(ab')₂ fragments are readily available commercially. Coupling of enzyme to antibody can also be performed by simple techniques[4,5] and vary little

[4] P. K. Nakane and A. Kawaoi, *J. Histochem. Cytochem.* **22,** 1084 (1974).
[5] S. Avrameas, T. Ternynck, and J. L. Guésdon, *Scand. J. Immunol.* **8,** Suppl. **7,** 7 (1978).

Enzymes Commonly Used in Enzyme-Linked Immunosorbent Assay

Enzymes	Substrates	Wavelength (nm)
Horseradish peroxidase	o-Phenyldiamine	492
Calf mucosa alkaline phosphatase	p-Nitrophenyl phosphate	405
Escherichia coli β-galactosidase	O-Nitrophenyl β-D-galactopyranoside	414

for immunoglobulins directed against immunoglobulins from other species.

Several kinds of enzymes may be coupled to antibody including alkaline phosphatase from calf mucosa, horseradish peroxidase, and *Escherichia coli* β-galactosidase (see the table). Choice of the enzyme depends particularly on the idiosyncracies of the assay. For example, β-galactosidase was preferred by us since this enzyme is not generally present on the surface of eukaryotic cells. Sheep F(ab')$_2$ anti-mouse immunoglobulin (H and L chain) coupled to β-galactosidase (BRL No. 9502 SA, Lot No. 1211, Gaithersburg, Maryland) gave consistent results in our system. This antibody can be stored at 4° for several months, but dilution should be performed immediately before each assay.

Substrate

Similar conditions apply to the choice of substrate as for enzyme. For the β-galactosidase system, o-nitrophenyl β-D-galactopyranoside (Sigma, St. Louis, Missouri) is used. This substrate is prepared at a 0.1% concentration in the following buffer: PBS; MgCl$_2$, 1.5 mM; 2-mercaptoethanol, 100 mM. Substrate should be prepared fresh and cannot be stored. Cleavage of the o-nitrophenyl radical from the substrate produces a yellow color change that can be quantified at a wavelength of 414 nm.

Immunoassay

Incubation of Antigen with Supernatants. Prepared antigen-containing plates are flicked, filled with 0.2 ml of PBS–BSA 0.1%, and incubated at 4° for 1 hr in order to block the remaining electrostatic sites of the polystyrene plates. Buffer is removed, and the plate is allowed to dry by brisk inversion on tissue paper. Thirty to 50 μl of the test supernatant are dispensed into duplicate wells. A known antiserum serves as the positive control. Several negative controls are desirable including a set of unrelated antibodies in addition to PBS alone. The plate is incubated for 1 hr at room temperature.

Washing Procedure. After incubation, the plates are washed by dispensing 0.2 ml of the appropriate washing solution in each well. A 5-min incubation allows the buffer to act.

When using soluble adsorbed antigen or glutaraldehyde-fixed cells, the plates are emptied by flicking or automatic washing and filled again with 0.2 ml of washing solution. Four washes are usually performed. When freshly isolated cells are present in the wells, plates must be spun for 5 min at 400 *g* before flicking or washing in order not to lose cells. Automatic washing does not seem to be adequate for fresh cells, since the sucking action of the machine causes inordinate cell loss.

Incubation with Enzyme-Labeled Antibody. Optimally, two or three dilutions of enzyme-labeled antibody should be used to detect remaining hybridoma antibody. Practically, constraints of antigen availability and cost of reagents require one dose to be used in screening after preliminary titration experiments. In our system, 50 μl of enzyme-labeled second antibody, diluted 1:200 in washing solution, are dispensed into each well and incubated for 1 hr at room temperature. After four more washes, 50 μl of substrate are added to each well and allowed to react for 1 hr at room temperature.

Automatic Reading of the Plate. After incubation, the 96 wells of the plate can be read in an automatic ELISA reader, at the appropriate wavelength for the enzyme. The enzyme–substrate reaction can either be allowed to go on for further reading or be stopped, as in the case of β-galactosidase, by adding 50 μl of 0.5 *M* sodium carbonate. The reference well against which the zero value is set (blank) is given by reading the wells with PBS–BSA 0.1% during the first incubation.

Data Interpretation. The reference well chosen as blank determines which wells are negative. Any value above this could be taken as positive. For better precision, several negative controls should be included in the assay to take into account discrepancies among duplicates, level of sensitivity of the reader, and vagaries of nonhomogeneous antigens. We use a statistical approach to define the optical density value above which values are considered positive. All the values from "negative" controls are combined to obtain the arithmetic mean ($\bar{X}$) and the standard deviation (SD). We then consider as positive all wells with results greater than 3 standard deviations from the mean.

Discussion

A wide variety of soluble antigens have been coupled to plastic 96-well microtiter plates in concentrations ranging from 1 to 50 μg/ml. Higher antigen concentrations are not useful and may be responsible for prozone effects.

When dealing with cells, the number used per well is important. We previously investigated this parameter and found that numbers ranging from 2×10^5 to 5×10^5 fresh cells give optimal results. Glutaraldehyde fixation allows decrease of the number of cells to 8×10^4 to 2×10^5, since many fewer cells are lost during the washes. Higher numbers are not recommended and may lead to high background values or, in some cases, to false negative results by formation of clumps that detach during washing. In addition to requiring fewer cells, glutaraldehyde fixation allows safe storage, an important advantage when a special cell type is not available daily. It is possible that antigenic alteration or exposure of new (i.e., intracellular) antigen by glutaraldehyde treatment can occur. However, in none of our experiments was this observed, as reported elsewhere.[6-9]

Pretreatment of the microplates with poly(L-lysine) 0.1%, as recommended by Kennett[10] in order to increase the electrostatic potential, does not seem to be necessary and eventually leads to higher optical density values in the negative controls, even when glycine was used after fixation to block the remaining electrostatic sites, probably by nonspecific adsorption of the second antibody to the plastic. However, such treatment might be useful for certain cell types, as mentioned by Cobbold.[6]

Incubation time at the different steps of the assay can vary. In our hands as in others, 1 hr of incubation at room temperature has given consistent and reproducible results. Longer periods of time have sometimes been recommended, but we frequently observed in this case higher background values, without gain in positive or negative absolute values in experimental wells. Agitation during the incubation does not seem to be required and, if performed, should be very mild.

Color changes may be observed in the positive wells as early as 15 min after adding substrate. We nevertheless let the reaction go for 1 hr in order to detect supernatants with low antibody concentration. Longer incubation of enzyme and substrate gives higher background values without changing the final result of the assay. When needed, several intermediate readings can be performed at 30 min, 1 hr, or longer without stopping the reaction.

[6] S. P. Cobbold and H. Waldmann, *J. Immunol. Methods* **44**, 125 (1981).
[7] J. W. Stocker and C. H. Heusser, *J. Immunol. Methods* **26**, 87 (1979).
[8] S. D. Rockoff, K. R. McIntire, A.-H. Kaung, G. L. Princler, R. B. Herberman, and J. N. Larson, *J. Immunol. Methods* **26**, 369 (1979).
[9] R. D. G. McKay, *FEBS Lett.* **118**, 219 (1980).
[10] R. H. Kennett, *in* "Monoclonal Antibody" (R. H. Kennett, T. J. McKearn, and K. B. Bechtol, eds.), p. 376. Plenum, New York, 1980.

Choice of the second (enzyme-labeled) antibody for the assay is dictated by the hybridoma species, considering the fact that the antibody produced by the hybrid is most often that of the immune cell, not of the myeloma. Usually, anti-immunoglobulins without class restriction are preferred in order to detect any type of monoclonal antibody. $F(ab')_2$ fragments should be used when screening hybridoma supernatants against cells with receptors for Fc in order to avoid nonspecific binding. The purity, specificity, and affinity for different monoclonals of this antibody are key determinants of which hybridomas will be detected. These considerations are particularly important when comparing results of ELISA to other binding assays employing different anti-immunoglobulins.

As in any type of assay, a value above which results should be considered positive has to be chosen. The statistical approach that we prefer consists of adding all the optical density values of the negative controls, calculating the mean $(\bar{X})$ and the standard deviation (SD). In order to ensure accuracy and reproducibility, several negative controls have to be performed in duplicate, including PBS–BSA, supernatant of the nonsecreting parental myeloma, culture medium, and unrelated antibodies. According to this method, any value above the mean plus three standard deviations has only a 1% chance of being false positive. If one wants to be less conservative, two or even one standard deviation above the mean can be taken, with 5% or 33%, respectively, chance of having false-positive results. The choice of three standard deviations may seem very strict and may miss some supernatants with interesting antibodies. On the other hand, this conservative approach avoids maintaining hybridomas without proven specificity.

In our experience, ELISA has compared favorably with other binding assays (radioimmunoassay, indirect rosette assays) when side-by-side comparisons were performed using cells as antigen. In some cases discrepancies have been observed, notably with fluorescence-activated cell sorter (FACS) analysis. These differences may be ascribed to antigenic variation in different cell populations, varying levels of sensitivity of the assays, or (we presume, most significantly) different affinity for a particular monoclonal of the anti-antibodies employed in conjugation of fluorescent label or enzyme. Above all, ELISA should be regarded as a screening assay, not as an absolute definition of antigen–antibody specificity. This test (or any other screening assay) mandates ultimate characterization of monoclonal antibody reactivity by a wide variety of different techniques including other binding assays, immunoprecipitation of molecular species, and functional tests.

[14] Rosette-Forming Cell Assay for Detection of Antibody-Synthesizing Hybridomas

By PIERRE LEGRAIN, DOMINIQUE JUY, and GÉRARD BUTTIN

Detection of antibody-synthesizing hybridomas often represents a difficult problem and is always tedious work. The assay required for that screening must be highly specific, fast, and sensitive enough to allow an extremely selective screening of several hundred clones growing simultaneously. The rosette-forming cell micromethod, already described by Juy et al.[1] is certainly the method of choice when the antigen can be coupled to erythrocytes. We successfully used this assay for the screening of many different hybridomas such as anti-DNP,[2] anti-dextran, anti-levan,[3] anti-idiotype,[4,5] and anti-enzyme clones. The rosette assay may also be useful for the precise analysis of the specificity of the monoclonal antibody secreted by hybridoma cells. Here, we present an example of the analysis of cross-reactive idiotypes by means of monoclonal antibodies and the rosette assay.

Coupling Procedure

The method utilized for the coupling of proteins to sheep red blood cells (SRBCs) is the one by Truffa-Bachi and Bordenave[6] except for the duration of the reaction, which can range from 1 to 6 min.

Reagents

Saline: 0.9% NaCl
Acetate buffer: 0.9% NaCl; sodium acetate, 0.01 M, pH 5.5
Chromium chloride: $CrCl_3$, 6 H_2O (Merck No. 2847)
SRBCs: usual conditions of preparation

---

[1] D. Juy, P. Legrain, P.-A. Cazenave, and G. Buttin, *J. Immunol. Methods* **30**, 269 (1979).
[2] D. Primi, D. Juy, and P.-A. Cazenave, *in* "Oscillatory Dynamics in the Immune Response" (P. de Lisi and J. Hierniaux, eds.), CRC Press, Boca Raton, Florida 1982 (in press).
[3] B. Goud, P. Legrain, J. C. Antoine, S. Avrameas, and G. Buttin, *J. Receptor Res.* **2**, 63 (1981).
[4] P. Legrain, D. Voegtlé, G. Buttin, and P.-A. Cazenave, *Eur. J. Immunol.* **11**, 678 (1981).
[5] P. Legrain, G. Buttin, and P.-A. Cazenave, *in* "Monoclonal Antibodies and T-Cell Hybridomas" (G. J. Hämmerling, U. Hämmerling, and J. K. Kearney, eds.), p. 416. Elsevier-North/Holland, Amsterdam, 1981.
[6] P. Truffa-Bachi and G. R. Bordenave, *Cell. Immunol.* **50**, 261 (1980).

METHODS IN ENZYMOLOGY, VOL. 92

Procedure. A 1-ml pellet of SRBCs is washed three times in 25 ml of saline. One volume of the pellet of SRBCs is then mixed with 1 volume of the protein solution at a concentration of 1 mg/ml in saline. We must emphasize the importance of using a protein preparation that has been well dialyzed and, in particular, completely free of phosphate ions. One volume of a solution of $CrCl_3$ at the concentration of 1 mg/ml is dissolved *just before use* (30 sec) in acetate buffer and added to the reactants. The mixture is maintained at room temperature and gently shaken for several minutes. The reaction time depends on the protein and is normally in the range of 1–6 min (usually 5 min). The reaction is stopped by addition of a large excess of cold saline, and the SRBCs are centrifuged (400 g, 10 min) and washed twice. Protein-coupled SRBCs have been stored at 4° in saline containing fetal calf serum (3%) for several weeks. The efficiency of the coupling procedure may be tested by hemagglutination assays with any immune serum obtained against the coupled protein.

Remarks. The relatively large amount of antigen needed for the coupling procedure is one limiting parameter of the screening method that we propose since antigen must be available in milligram amounts. The purity of the protein preparation is also a limiting factor. We have obtained good results with protein preparations containing only 30% specific antigen. However, rosette formation with impure antigen preparation may be limited to the detection of high-affinity monoclonal antibodies only (see below the section "Specificity Studies by Rosette Assay).

Rosette Assay[1]

Material and Reagents

Rosette medium: Eagle's medium supplemented with 5% fetal calf serum and 0.01 M N-2-hydroxyethylpiperazine-N'-2-ethane-sulfonic acid (HEPES), pH 7.3
U-bottom or V-bottom Microtiter plates (96 wells, Greiner)
Flat-bottom Microtiter plates (96 wells, Greiner)
Centrifuge for Microtiter plates (Beckman TJ 6, for example)

Procedure. One-third of each hybridoma culture is harvested 2 or 3 weeks after the fusion event, when the cell density ranges from 2×10^4 to 2×10^5 cells/ml. Since only 1000 cells are required for one rosette assay, several tests can be performed from 300 μl of culture. Cells are centrifuged and resuspended in rosette medium at a concentration of about 5×10^4 cells/ml. An equal volume of antigen-coated SRBCs (4×10^7 cells/ml) diluted in rosette medium is added to 50 μl of hybridoma cell suspension in Microtiter-plate wells (U or V bottom).

TABLE I
HYBRIDOMA SCREENING BY ROSETTE ASSAYS

Antigen	Number of culture wells	Number of wells with growing hybridomas	Number of cultures tested by rosette assay	Number of positive cultures[a]
TNP-Ficoll	42	20	20	7
Dextran	144	100	100	1
Levan (polyfructosan)	336	122	100	4
ABPC 48 (BALB/c IgA)	960	387	352	23
UPC 10 (BALB/c IgG2a)	336	169	146	13
IDA10 (BALB/c IgG1)	864	644	644	202
IDA23 (BALB/c IgG2b)	720	240	227	12

[a] A positive culture contains more than 10% of rosetting cells.

Plates are centrifuged for 15 min at 400 g. Thirty minutes after centrifugation (cells may rest as a pellet overnight at 4°), cells are gently resuspended and transferred into flat-bottom wells. The numeration of the rosette-forming hybridoma cells is determined using an inverted microscope. A cell is considered as positive if a minimum of five SRBCs are attached to it.

Screening of Hybridomas

Rosette assays have been used in our laboratory for many screenings (Table I). Controls of the specificity of rosette assay included myeloma lines X63Ag8,[7] NS1,[8] and SP2-0Ag.[9] No rosetting cells were observed in any case (Ig-SRBCs, DNP-SRBCs, Dextran-SRBCs) with the exception of Levan SRBCs. This is attributed to the use of a stearoyl derivative of levan for the coupling to SRBCs.[10] A low frequency of rosetting cells may correspond to dead cells, which have been shown to rosette with unsubstituted SRBCs. Therefore, the viability of hybridomas should be assayed where nonspecific rosettes (i.e., low frequency) are observed. The fraction of rosetting cells within a positive culture depends on several parameters. The most important one is the presence in the same well of both positive and negative hybrids. This situation has to be suspected when the

[7] G. Köhler and C. Milstein, Nature (London) 256, 495 (1975).
[8] M. Shulman, C. D. Wilde, and G. Köhler, Nature (London) 276, 269 (1978).
[9] G. Köhler, S. C. Howe, and C. Milstein, Eur. J. Immunol. 6, 292 (1976).
[10] U. Hämmerling and O. Westphal, Eur. J. Biochem. 1, 46 (1977).

number of wells with growing cells is comparable to the number of wells in which cells were distributed after the fusion. For that reason, we prefer to distribute cells in many culture wells: under these conditions of "limiting dilution" (Table I), each culture is usually clonal. The simplicity of the rosette assay permits one to screen within a few days several hundred cultures for an antibody specificity directed against one antigen or 100 cultures for antibody specificities directed against a set of different antigens. Another source of heterogeneity in the cultures that must be pointed out is the genetic stability of intraspecific hybrid cells. This is usually good (see below the section Clone Stability); however, loss of chromosomes during the first divisions after the fusion is well documented and may lead to nonsecreting clones. For this reason it is particularly worthwhile to be able to screen the hybridoma cells as soon as possible.

In the results given in Table I, cultures were scored as positive when they contained at least 10% of rosetting cells. Some cultures were positive with 95% rosetting cells. The rosette assay is routinely repeated several times before cloning to control the number of positive cells within the cultures. Occasionally, the percentage of positive cells quickly decreases. In these cases and when the positive cultures contain few positive cells, cloning by micromanipulation of the rosetting cells is strongly recommended (see below).

Antibody Secretion by Hybridoma Cells

In principle, the rosette assay could detect nonsecreting cells bearing surface immunoglobulins. Since we started to screen hybridoma cells by rosette assays, more than 100 cloned hybridomas have been characterized. All these cloned hybridomas did secrete antibody *in vitro*. The concentration of antibody in culture supernatants ranges from 5 to 25 μg/ml. It must be pointed out that the rosette assay does not allow the quantification of the concentration of antibody secreted by the cells. With several hybridoma cultures we observed agglutination of antigen-coated SRBCs but no rosetting cells. This agglutination is specific and does not occur with non-antigen-coated SRBCs. The same pattern is still observed with clones derived from these cultures. All these particular hybridomas, which had different binding activities (anti-DNP, several anti-idiotypic activities) have been shown to secrete IgM molecules, and all IgM-secreting hybridomas exhibited this agglutination pattern. Repeated washings of the cells to eliminate soluble immunoglobulins did not restore rosette formation. We consider the agglutination pattern as a nonambiguous positive pattern for IgM-secreting hybridomas, despite our inability to demonstrate surface immunoglobulins.

TABLE II
CLONING BY MANIPULATION OF ROSETTING CELLS

Fusion	Hybrid	Percentage of rosetting cells before cloning	Number of micro-manipulated cells	Number of clones obtained	Percentage of rosetting cells in the clones obtained
AntiIDA10	Hybrid 54	48	5	2	80–98
	Hybrid 304	18	3	2	97–99
AntiIDA23	Hybrid 50–10	29	4	2	100–98
	Hybrid 120	4	6	4	100–100–100–99

Cloning by Micromanipulation of Rosetting Cells

This technique is very easy to perform[11] and requires no special equipment. Parallel screenings by rosette assay and binding assays (RIA or ELISA) for the detection of antibody in the culture supernatants generally show good correlation either for positive or negative cultures. However, only very sensitive assays may detect antibody at a concentration of 0.1 μg/ml often corresponding to a very low percentage of positive cells within the culture. Cloning by limiting dilution, in these cases, is usually a hopeless procedure. On the other hand, the availability of a rosette assay permits, as shown in Table II, the isolation of a rosetting cell among hundreds of negative ones and consequently the expansion of a positive clone.

For the cloning of rosetting cells antigen-coated SRBCs are prepared as described above under semisterile conditions (it is not necessary to sterilize the antigen preparation), and we do all previous manipulations in the hood with sterile buffer or saline.

Rosette formation is performed in sterile plate wells; 100–200 hybridoma cells are taken from these wells and laid in 1.5 ml of rosette medium into a 100-mm Petri dish. Rosetting cells are observed using an inverted microscope at low magnification ($\times$40). With a simple Pasteur pipette previously drawn out in the flame, it is possible to aspirate one single rosetting cell and to transfer it into a well prefilled with 0.2 ml of culture medium and feeder cells (10^5 thymocytes). As shown in Table II, cloning efficiency is good, and every clone formed rosettes with antigen-coated SRBCs.

[11] D. Zagury, L. Phalente, J. Bernard, E. Hollande, and G. Buttin, *Eur. J. Immunol.* **9,** 1 (1979).

Clone Stability

After cloning by limiting dilution or micromanipulation, the growing cell populations can be tested by the rosette assay. In all these assays, most positive clones formed 95–100% rosettes. Those clones forming 100% rosettes were selected for propagation and passage in mice for the production of ascitic fluid. Since we started to use the rosette assay, which is specific for producing cells, not for the antibody concentration in the supernatant, we have never lost a secreting hybridoma after cloning. After several months of *in vitro* culture, clones consisted of a minimum of 70% rosetting cells. Only 1 hybridoma among 50 had to be recloned because it was losing its binding activity.

Specificity Studies by Rosette Assay

The studies of cross-reactivities between related idiotypes by detection of rosette formation will be taken as an example of the advantages of this assay. Controls for the specificity of the rosette assay may be performed by inhibition experiments and have been presented and discussed elsewhere.[1] Here we shall present an analysis of cross-reactivity between idiotypes by means of the rosette assay and monoclonal anti-idiotypic antibodies, and we shall also discuss the problem of the affinity of the monoclonal antibody required for the formation of rosettes.

ABPC 48 and UPC 10 are BALB/c myeloma proteins with levan binding activity. We have obtained several monoclonal anti-ABPC 48 anti-idiotypic antibodies (IDAs) and monoclonal anti-UPC10 anti-idiotypic antibodies (IDUs). We have raised monoclonal antibodies designated AIDA10's and AIDA23's, respectively, against two IDAs, namely, IDA10 and IDA23. Analysis of cross-reactions between monoclonal antibodies and several idiotypes are tedious by classical binding assays. Inhibition of binding assays (in RIA or ELISA[4]) are less sensitive than direct binding assays. Contrary to this, it is very easy to test the capability of hybridoma cells to form rosettes with several idiotype-coated SRBCs. Table III summarizes the results obtained in different systems using this assay. Idiotypic cross-reactivity is observed between ABPC 48 and UPC 10 and already has been described.[4] Among the different IDAs, idiotypic similarities and differences have been defined by the analysis of their interactions with numerous monoclonal anti-idiotypic antibodies raised against IDA10[5] or IDA23 (Table IV). In most cases the result of the rosette assay gives a straightforward answer; i.e., we observe either 0% or more than 90% of rosetting cells within a cloned culture. Table IV gives examples of such a case. All AIDA23 hybridomas did rosette with IDA23-coated SRBCs, but not with IDA3 or IDA16. The hybridomas AIDA23/1, AIDA23/3, and AIDA23/8 rosetted with IDA4 and IDA10. In some cases, however, the rosette assay using cross-reactive idiotype-coated SRBCs

TABLE III
IDIOTYPIC SPECIFICITIES ANALYZED BY DIFFERENT ROSETTE ASSAYS

Hybridoma cells	Antigen	SRBCs coated with[a]	
		ABPC 48	UPC 10
IDA4, 5, 9, 10, 14, 15, 16, 19, 20, and 21	ABPC 48	+	+
IDA3, 6, 7, 8, 13, 17, and 23	ABPC 48	+	−
IDU1, 3, 5, 7, 8, 9, and 11	UPC 10	−	+
IDU8 and 10	UPC 10	+	+

Hybridoma cells	Antigen	IDA10	IDA14	IDA19	IDA20	IDA23
AIDA10/2, 3, 5, 6, 7, 8, 9, 10, 12, 14, 15, 16, 17, 20, 21, and 22	IDA10	+	+	+	+	+
AIDA10/1, 11, and 18	IDA10	+	+	+	+	−
AIDA10/4	IDA10	+	+	−	−	−
AIDA10/13 (IgM)	IDA10	A	A	A	A	A
AIDA10/19 and 23 (IgMs)	IDA10	A	A	A	A	−

[a] +: more than 70% of rosetting cells within a cloned hybridoma culture; −: no rosetting cell among 100 cells scored; A: no rosetting cell observed. Hemagglutination of antigen-coated SRBCs.

TABLE IV
ROSETTE ASSAYS WITH DIFFERENT IDIOTYPE-COATED SRBCS AND
ANTI-IDIOTYPIC HYBRIDOMA CELLS

Hybridoma cells	SRBCs coated with					
	IDA3	IDA4	IDA8	IDA10	IDA16	IDA23
AIDA23/1	0[a]	165	188	177	0	177
AIDA23/2	0	0	194	0	0	180
AIDA23/3	0	168	177	171	0	140
AIDA23/4	0	0	55	0	0	187
AIDA23/5	0	0	133	0	0	196
AIDA23/6	0	0	131	0	0	170
AIDA23/7	0	0	183	0	0	188
AIDA23/8	0	183	185	180	0	181
AIDA23/9	0	0	195	0	0	192
AIDA23/10	0	0	80	0	0	199
ASO1 (12) 7[b]	185	192	188	190	175	174

[a] Number of rosetting cells (at least 5 SRBCs attached to the cell). Two assays were performed, and 100 cells were counted each time.
[b] A hybridoma line secreting ABPC 48 myeloma protein.

were less clear cut. This is illustrated by the results of the rosette assay with IDA8-coated SRBCs. All AIDA23 hybrids were scored as positive, but 4 of them formed an abnormally low percentage of rosettes, especially AIDA23/4 and AIDA23/10. These rosettes were not bushy, having less than 10 IDA8-coated SRBCs around a cell (when less than 5 SRBCs were attached to a hybridoma cell, this cell was scored as negative). Therefore, a low percentage of rosetting cells within a cloned culture does not always mean that the cell population is heterogeneous for the synthesis of the monoclonal antibody. It may reflect the limit of the sensitivity of the rosette assay for low-affinity binding.

We have been able, by direct rosette assays and also by inhibition of the rosette assays (results not shown; see Legrain et al.[5]), to characterize the idiotypic pattern of various immunoglobulins. This kind of analysis can be developed for the comparison of the antigenic determinants of any group of structurally related molecules.

Acknowledgments

This work was supported by the University Pierre et Marie Curie, the Ligue Nationale Française contre le Cancer, the Fondation pour la Recherche Médicale Française, and the Centre National de la Recherche Scientifique (ATP No. 093613/00).

[15] Rapid Screening and Replica Plating of Hybridomas for the Production and Characterization of Monoclonal Antibodies

By RICHARD B. BANKERT

In 1975 Köhler and Milstein reported that mouse myeloma cell lines could be fused with antibody-producing spleen cells and that the resulting cloned hybrid cell lines "hybridomas" continued to secrete monoclonal antibodies.[1] This discovery initiated a new era in immunological research. It has been estimated that at least half of the immunology laboratories in this country are using the hybridoma technology either to make monoclonal antibody reagents or to generate continuously cultured cell lines for studying the biochemistry of immunocompetent cells and their effector molecules.[2] The opportunities for future research and the clinical applications in both diagnosis and therapy of disease appear to be enormous.

[1] G. Köhler and C. Milstein, Nature (London) 256, 495 (1975).
[2] "New Initiatives in Immunology." U.S. Department of Health and Human Services, N.I.H. Publication No. 81-2215, p. 113 (1981).

Details of the experimental procedures for the production of hybridomas have been adequately covered in this series.[3] The cell fusion technique itself is relatively simple, and with the development of improved fusion protocols[4] and the availability of several new parental myeloma cell lines,[5-7] the cell fusion efficiency has greatly increased. At present the most difficult aspect of the hybridoma technique is the identification and selection of the hybridomas that are producing antibody of the desired specificity. There are at least two initial stages at which the hybridoma cultures must be screened for the production of antibodies. The first screening is performed 2–3 weeks after the fusion of the myeloma cells and immune spleen cells. At this time (for each fusion in which a single spleen is used), there may be as many as 400–500 hybridoma cultures to test for antibody production. Subsequently, each of the hybridoma cultures that are identified as producing the desired antibody (typically between 10 and 20%) must be cloned. This is generally accomplished via a limiting dilution fractionation in which serial twofold dilutions of cells are seeded into microculture plates. The second-stage screening is then performed on the supernatant fluids of these cultures once they approach confluence. The positive cultures containing the minimum number of seeded cells are either fractionated again or cloned in semisolid agar. The cloning and recloning often require assaying a minimum of 300 culture supernatant fluids for each of the hybridoma cultures initially identified as positive for antibody production. Understandably, the initial screening and the subsequent cloning and rescreening is the most tedious and labor-intensive portion of the entire hybridoma production scheme.

Many screening assays have been used to identify the antibody-producing cultures, including a variety of radioimmunoassays, enzyme immunoassays, immunofluorescence assays, and hemolytic assays. Selection of the optimal screening assay is critical to the success of the development of a hybridoma. This selection will depend heavily upon the physicochemical characteristics of the antigen of interest.

In general the ideal assay should (a) detect very low levels of antibody; (b) be specific for the selected ligand; (c) be relatively simple to perform; (d) identify antibodies of all different immunoglobulin isotypes; and (e) be able to screen large numbers of hybridoma microcultures rapidly.

The inherent limiting factor in the screening assays has been the re-

[3] G. Galfrè and C. Milstein, this series, Vol. 73, p. 3.
[4] M. L. Gefter, D. H. Margulies, and M. D. Scharff, *Somatic Cell Genet.* **3**, 231 (1977).
[5] M. Shulman, C. D. Wilde, and G. Köhler, *Nature (London)* **276**, 269 (1978).
[6] J. F. Kearney, A. Radbruch, B. Liesegang, and K. Rajewsky, *J. Immunol.* **123**, 1548 (1979).
[7] S. Fazekas de St. Groth and D. Scheidegger, *J. Immunol. Methods* **35**, 1 (1980).

quirement for individually sampling and manipulating culture supernatants and/or cells from each hybridoma culture. The manual sampling of thousands of individual cultures is extremely tedious and time-consuming. Three very simple and rapid screening techniques that eliminate manual sampling are described in this chapter.

Transfer Template Localized Hemolysis in Gel Assay to Detect Hapten, Protein, and Carbohydrate Specific Antibodies

Principle

The detection of antibody in this assay is based upon the lysis of antigen-coated sheep red blood cells (SRBC) by specific antibody and complement and is a modification of the assay reported earlier.[8] Lysis is visualized as a clear area (hemolytic spot) on a slide containing a "lawn" of target SRBC suspended in gelled agar on a slide. This localized hemolysis in gel assay, or so-called "hemolytic spot assay," will be presented here for the detection of anti-hapten antibody. It has also been successfully employed to identify hybridomas secreting monoclonal anti-protein and anti-carbohydrate antibodies.[9]

The key to the rapid screening and replica plating assay reported here is a special fenestrated transfer template (Fig. 1). It is a 96-well plate with a calibrated hole in the bottom center of each well (Fig. 1a). To assay for anti-hapten antibodies the transfer template is positioned over a 96-well microculture plate containing the growing hybridomas (Fig. 1b). After making contact with the tissue culture supernatant, each orifice of the transfer template retains approximately 2 μl of tissue culture supernatant. The transfer template is then placed onto an assay slide containing a thin layer of hapten-conjugated target erythrocytes incorporated into agarose. After incubation with an anti-immunoglobulin (specific for multiple light-chain and heavy-chain determinants) and complement, areas of localized hemolysis in the gel indicate hybridomas that are secreting anti-hapten antibodies (Fig. 2a). The assay detects as little as 10 pg of antibody. Since the transfer template can be used as a replica plate, one can repeatedly transfer samples to various slides that contain either different hapten target cells or different hapten analog inhibitors in the agarose layer (Fig. 3). Therefore, in addition to rapidly screening microcultures for positive hybridomas, this procedure permits the characterization of each monoclonal antibody's fine specificity. One can also employ the replica-plating potential of this assay to determine the heavy-chain isotype of each antibody. Here individual slides are prepared and facilitated with antisera specific for each immunoglobulin heavy-chain isotype.

[8] R. B. Bankert, D. DesSoye, and L. Powers, *J. Immunol. Methods* **35**, 23 (1980).
[9] R. B. Bankert, D. DesSoye, and L. Powers, *Transplant. Proc.* **12**, 443 (1980).

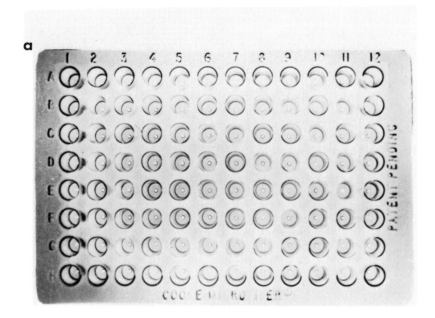

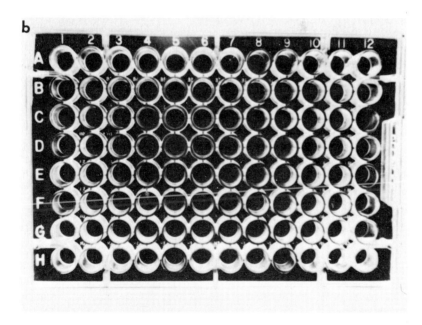

Fig. 1. (a) The transfer template consists of a flexible, round-bottom 96-well plate. Note the hole in the bottom center of each well. This plate can be prepared as described in text, or it is commercially available from Cooke Laboratory Products, Alexandria, Virginia. (b) This is the standard 96-well microculture plate in which the hybridoma cell lines are cultivated. The geometry of the transfer template is such that each well of the template meshes with a corresponding well of the microculture plate.

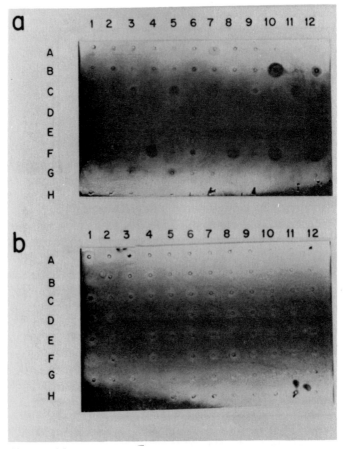

Fig. 2. (a) The hemolytic spot assay slide (Gel Bond film) contains phthalate-SRBCs as target cells; 96 individual 1-μl samples from a microculture plate of hybridomas were delivered to this plate with the transfer template. The hybridomas that were sampled were produced by the fusion of Sp2 cells with spleen cells from a BALB/c mouse which had received two immunizations with 4-azophthalate conjugated to the protein carrier, keyhole limpet hemocyanin. The photograph was taken with dark-field illumination, and the localized areas of hemolysis (dark circular spots) indicate supernatants containing anti-phthalate antibody. Examples of positive hybridoma supernatants are shown at positions B10, B12, C1, C5, C7, C11, C12, E7, E9, E11, F2, F4, F8, F10, and F12. The lysis occurred only after incubation of the slides with rabbit anti-mouse immunoglobulin and guinea pig serum as a source of complement. The very small rings in all the positions are artifacts created by the positioning of the transfer template onto the agar of the assay slide. The grid 1–12 and A–H coincides with the grid of the transfer template and microculture plate. (b) The hemolytic assay slide contains 5×10^{-3} M free phthalate incorporated into the agarose in addition to the phthalate-SRBCs. The same 96 microculture supernatant fluids were assayed on this slide. Note that all of the areas of lysis shown in (b) are now inhibited with the free hapten. This establishes the specificity of the antibody assay.

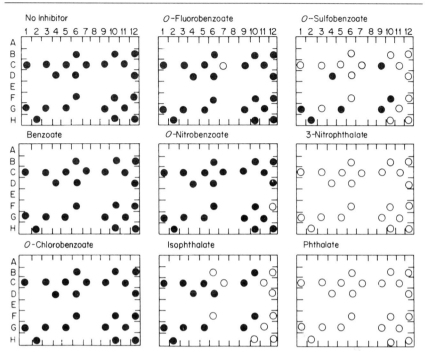

FIG. 3. A graphic display of the actual Gel Bond film lysis and lysis inhibition patterns produced by the supernatant fluids from microcultures of 23 cloned hybridomas secreting antibody specific for 5-azophthalate. Antibody was delivered to the Gel Bond slides from the microculture wells with a transfer template or replica plate. The transfer template deposits approximately 1 μl of tissue culture supernatant to the Gel Bond assay slide. The filled circles indicate areas of hemolysis on the assay slides, and the open circles indicate inhibition of hemolysis. Each assay slide contains phthalate-SRBC suspended in agarose and 5 $\times$ 10^{-3} M free hapten, where indicated. From Bankert et al.[9]

Materials

Agarose: Sea Plaque (Marine Colloids Division, FMC Corporation, Rockland, Maine)

Medium 199, 2$\times$ concentrate: Prepared from 10$\times$ concentrate (GIBCO Laboratories, Grand Island, New York).

Basal Medium Eagles (GIBCO Laboratories)

Guinea pig serum (GIBCO Laboratories)

Phosphate-buffered saline: 0.075 M KH$_2$PO$_4$–0.075 M NaCl, pH 7.2

Facilitators: rabbit anti-mouse Ig. This antiserum is prepared by injecting 1 mg of BALB/c γ-globulin plus Freund's complete adjuvant into rabbits in multiple intramuscular sites. The injection of 1 mg of γ-globulin is repeated three more times at 1-month intervals using

Freund's complete adjuvant. Rabbits are bled 1 week after the last immunization. Rabbit anti-mouse isotype-specific reagents were supplied by Miles Laboratory (Elkhart, Indiana). These were found by us to be specific for μ, γ_1, γ_{2a}, γ_{2b}, and γ_3 according to a radioimmunoassay.

Gel Bond film: 100 mm × 150 mm size, 7 mil thickness (Marine Colloids Division, FMC Corporation, Rockland, Maine). This is a clear flexible plastic film used to support the agar gel containing the target erythrocytes. The major advantage of this film is that the gel adheres very firmly to the hydrophilic side of the Gel Bond. The secure attachment of the gel to the slide is critical in this assay. The Gel Bond is also unbreakable and permits one to air-dry gels, which can then be fastened to a notebook for a permanent record of the screening assay results.

Transfer template: Cooke Microtiter (Cooke Engineering, Alexandria, Virginia) (Fig. 1). The transfer template can also be made by drilling holes 1.5 mm in diameter in the bottom of each well of the 96-well U Microtiter plates (Cooke Laboratory). By stacking several plates together and using a drill press many plates can be prepared in a short period of time. These plates can be sterilized by exposure to UV light. The commercially available transfer template is supplied as a sterile individually wrapped unit and was found to be ideal (albeit expensive) for this assay.

Microtiter plate: 12.7 × 9.5 cm, 96 round-bottom U wells (Dynatech Laboratories, Inc., Alexandria, Virginia) (see Fig. 1).

Target sheep red blood cells: 6% red cells (v : v) in phosphate-buffered saline, pH 7.2. The hapten 4-aminophthalate was attached to SRBC via a synthetic lipopolysaccharide[10] or by the coupling of 4-aminophthalate to cyanuric chloride.[11] These and other techniques for coupling protein, hapten, or carbohydrate ligands to erythrocytes are discussed in this volume[12] and in Vol. 74[13] of this series.

The hybridomas that were screened and reported here were produced and cloned according to a previous report.[14]

Equipment

Water bath, 46°. The level of water must be higher than the level of agarose in the test tubes.

[10] R. B. Bankert, G. L. Mayers, and D. Pressman, *J. Immunol.* **123**, 2466 (1979).
[11] P. K. Mazzaferro and G. L. Mayers, *J. Immunol. Methods* **46**, 327 (1981).
[12] Y.-H. Jou, P. K. Mazzaferro, G. L. Mayers, and R. B. Bankert, this volume [21].
[13] T. Borsos and J. J. Langone, this series, Vol. 74, p. 162.
[14] R. B. Bankert, D. Mazzaferro, and G. L. Mayers, *Hybridoma* **1**, 47 (1981).

Slide warmer, 37° (Fisher Scientific Company)
Cooling tray and leveling bubble (Gelman Instrument Company)
Ice bath for cooling tray, 36 cm × 43 cm × 6 cm
Gel spreader: made from a 9-inch Pasteur pipette or a bent glass rod
Glass plates, 100 mm × 165 mm × 2 mm. Supplied by local hardware store.
Humidified chamber, 10 cm × 20 cm × 30 cm; plastic container with an airtight lid
Diamond marking pencil (Fisher Scientific Company)
Pipettes: Pasteur 9-inch and bulb, 1 ml, 5 ml, and 10 ml; bulb-type safety pipette filler; pipetteman adjustable pipette (200 μl) and tips.
Incubation tray, Plexiglas, 19 cm × 41 cm
Erlenmeyer flask, 125 ml
Graduate cylinder, 25 ml
Test tube rack and test tubes, 13 mm × 100 mm
Incubator, 37°
Portable fan or hair dryer

Procedure

Preparation of 0.75% Agarose. Dissolve 0.1875 g of Sea-Plaque agarose in 12.5 ml of distilled water. Heat the water and agarose in a water bath until dissolved. The temperature of the agarose should reach approximately 70°. Add an equal volume (i.e., 12.5 ml) of 2× medium 199 (which has been warmed to 45°) to the agarose solution and place in an Erlenmeyer flask. Keep in the 45° water bath.

Preparation of Assay Slides. Glass culture tubes (13 × 100 mm) are placed in a test tube rack and put into a 45° water bath. For each slide to be prepared, 3.2 ml of the liquid agarose are added to a prewarmed test tube. All the tubes that will be required should be prepared ahead of time. Just prior to pouring onto the Gel Bond plastic film, 0.4 ml of the 6% suspension of hapten-SRBC is added to the agarose-containing tube, and the cells are thoroughly suspended in the agarose by covering the tube with Parafilm and inverting the tube several times. The Gel Bond films are numbered on the hydrophobic side using a diamond point marking pencil. A glass plate is placed on the slide warmer (at 37°). A few drops of 0.15 *M* NaCl are placed on the middle of the glass plate, and the Gel Bond film is placed onto the glass plate with the hydrophobic side of the Gel Bond facing down. The agarose hapten-SRBC suspension is poured onto the hydrophilic surface of the Gel Bond and spread evenly over the entire surface. This may require some practice. One may use either a glass rod bent at a 90° angle or a piece of the Gel Bond film (using an edge of the hydrophobic side of the film) to spread the agar.

The glass slide containing the Gel Bond, liquid agarose, and hapten-SRBC is transferred to the leveling tray that has been packed in ice. The chilled leveling tray ensures an even distribution of the gel plus hapten-SRBC, and promotes the gelling of the agarose.

Sampling the Hybridoma Microcultures with Transfer Template. After the agarose containing the hapten-SRBC has gelled, the Gel Bond (still on the glass plate) is placed in the laminar flow hood. The hybridoma microcultures to be tested are removed from the incubator and placed in the laminar flow hood. The sterile transfer template (Fig. 1a) is rinsed once in sterile saline 0.15 *M*. This is accomplished by positioning the template over a 96-well microculture containing 50 μl of saline in each well. After insertion into the saline, the template is blotted once on a sterile gauze sponge. The transfer template is now ready to sample the hybridoma microtubules. The template is positioned directly over the 96-well tissue culture plate (Fig. 1b) and lowered onto it so that the bottom of each well of the transfer template gets inserted into the corresponding well of the tissue culture plate. Upon making contact with the surface of the tissue culture supernatant fluid, the transfer template is removed and placed upon the hemolytic assay slide, i.e., Gel Bond, containing the layer of gelled agarose and hapten-SRBC. After 30 sec, the transfer template is removed from the test slide. This step delivers 96 1-μl aliquots of tissue culture supernatant fluids to the assay slide. The 96 samples can easily be seen on the surface of the gel. When first using this technique it is advisable to inspect the gels to ensure that supernatant fluid from each well has been transferred to the test slide. If any wells are missed, one may at this time sample the missed well using a Drummond Microdispensor. However, with only a limited amount of practice, this should not be necessary.

The procedure can be repeated several times, using another sterile transfer plate for each new assay slide. It is customary to plate the samples onto Gel Bonds that contain control target erythrocytes (i.e., SRBC to which no antigen has been attached). Repeated plating onto a variety of different assay slides containing other target cells, i.e., replica plating, and various facilitators have been used effectively to characterize each hybridoma for fine specificity, light- and heavy-chain isotypes and idiotypes.[9,15] Repeated plating onto slides containing the same target cell, but with one of several different inhibitors is also very useful in establishing a preliminary fine specificity characterization of hybridomas. An example of this approach is shown in Fig. 3.

[15] A. G. Bloor, Y.-H. Jou, C. Hoeplinger, J. Gartner, G. L. Mayers, and R. B. Bankert, *J. Immunol.* **128**, 1443 (1982).

The assay slides are then placed in a humidified chamber and incubated for 30 min at 37° to allow the tissue culture supernatant fluids to diffuse into the agarose of the assay slides.

Treatment of Assay Slides (Gel Bonds) with Rabbit Anti-Mouse Immunoglobulin. After the 30-min incubation, the Gel Bond films are separated from the glass plates and rinsed once by immersing the films in a beaker of saline in 0.15 M NaCl. Three milliliters of a 1 : 100 dilution of rabbit anti-mouse Ig (the optimal dilution must be determined empirically) are placed on a flat plastic tray, and the Gel Bond (without the glass plate) is placed gel-side down on the pool of anti-mouse Ig. The Gel Bonds are incubated floating in the anit-Ig for 30 min at 37°.

Treatment of Assay Slides (Gel Bond) with Complement. After the second 30-min incubation, the Gel Bonds are removed from the incubator. Three milliliters of a 1 : 10 dilution of guinea pig serum (as a source of complement) are placed on a flat plastic tray, and the Gel Bonds are placed gel-side down in the pool of diluted guinea pig serum.

After a final 30-min incubation at 37°, the slides are removed from the incubator and the results are recorded. The presence of antibody in a supernatant fluid is indicated by a clear area of lysis (Fig. 2a).

Fixing of Assay Slides (Gel Bonds). The flexible Gel Bond films are air dried. The films must be hung on clips and dried rapidly by placing them in the path of forced air (i.e., from fan or hair dryer or blower without heat). After 3 hr (when films are thoroughly dried) the Gel Bond films are immersed in absolute methanol for 30 min and washed in distilled water. This "fixed" Gel Bond provides a permanent record of the screening assay results, which can be placed into a notebook.

Replica Plating to Characterize Fine Specificity of Hybridomas

One of the major advantages of this technique is its ability repeatedly to sample and assay the same microculture plate. Employing the principle of replica plating originally described by Lederberg and Lederberg for use in bacteriology,[16] we have been able to characterize hybridoma antibodies according to their fine specificity and thereby to select unique clonotypes from a large repertoire of hapten (phthalate) specific hybridomas. By incorporating a panel of haptens cross-reactive with phthalate into the assay slides and recording the ability of each hapten to inhibit the antibody-mediated lysis of the phthalate-SRBC, the hybridomas can be separated into identifiable fine specificity sets. Figure 3 demonstrates how this fine specificity analysis works. In the upper left-hand panel, the pattern of lysis on phthalate-SRBC is depicted for 23 hybridoma clones se-

[16] J. Lederberg and E. M. Lederberg, *J. Bacteriol.* **63**, 399 (1952).

creting anti-phthalate antibodies. The remaining eight panels depict the patterns of lysis of these same 23 hybridomas when assayed on slides containing phthalate-SRBC and one of seven related haptens, including phthalate. Note that phthalate inhibits the lysis produced by all 23 supernatant fluids (lower right panel). Only one of the clones (C7 reading from the grid on Fig. 3) was inhibited by *o*-fluorobenzoate and one other clone (F12) was inhibited by *o*-nitrobenzoate. Ten clones were inhibited by isophthalate, and an entire set of 16 clones were inhibited by *o*-sulfobenzoate. An additional dimension can be added to this analysis by varying the concentrations of the inhibitors in the gels.

This type of replica-plating analysis can be applied whenever a library of soluble cross-reactive haptens/antigens are available. Based upon this type of fine specificity analysis, it has been possible to establish 11 distinct clonal sets.[9,15]

Comments

In developing this hybridoma screening assay, several important criteria were considered: sensitivity, specificity, simplicity, reproducibility, speed, and adaptability of assay to screen for antibodies with a broad spectrum of different specificities.

With regard to sensitivity requirements, one must first have some knowledge of how much antibody is produced by hybridomas in culture. The concentration of antibody in tissue culture supernatant fluids varies among different hybridomas and with different culture conditions. We have observed concentrations of antibody ranging between 0.4 μg/ml and 50 μg/ml using various myeloma cell lines for fusion and with different experimental antigens. The assay described here detects as little as 10 pg of antibody, and its specificity was established by inhibition with free antigen incorporated into the agarose. The table illustrates both the sensitivity and specificity of the assay. Three monoclonal anti-phthalate antibodies were utilized here. The results were essentially the same for several other antigen systems we have tested. The antibodies were purified by affinity chromatography[9,15] and assayed on gels containing phthalate-SRBC with or without free phthalate as inhibitor. Here, as in other antigen systems that we have tested, the sensitivity of the assay is at the level of 10–50 pg of antibody, which is well within the level required to detect antibody production by hybridomas in culture. The presence of free phthalate (but not the non-cross-reactive benzoate) inhibits the hemolysis of phthalate-SRBC (Fig. 2b; see the table), thereby demonstrating the specificity of the hemolytic spot assay.

SENSITIVITY OF THE HEMOLYTIC ASSAY FOR ANTI-PHTHALATE
MONOCLONAL ANTIBODY[a]

Hybridoma clone	Inhibitor[b]	Antibody proteins[c] required for hemolysis of phthalate-SRBC (ng)
2A8	None	0.01
2A8	Phthalate	100
5C11	None	0.05
5C11	Phthalate	100
3A2	None	0.01
3A2	Phthalate	100

[a] From Bankert et al.[8]
[b] Inhibitors present at a concentration of 5×10^{-3} M.
[c] Antibodies purified by affinity chromatography using an immunoabsorbent Sepharose 4B to which aminophthalate is covalently coupled. Antibodies are eluted from the column with phthalate.

The assay is very simple to set up and perform. Our laboratory has trained well over 300 individuals with a variety of scientific backgrounds in the use of this technique. We find that an investigator with very limited experience can easily screen 1000–2000 microcultures in a 3-hr period. It is also adaptable to a wide range of antigens. With the development of many new approaches to attaching antigens to red blood cells,[12] the assay can be used to screen for antibodies with specificity for virtually any soluble antigen (protein, carbohydrate, nucleic acid) or hapten.

It is important to note that most of the hybridoma supernatant fluids assayed did not produce hemolysis directly. This is because most antibodies (except for those with the μ heavy-chain isotype) either do not fix complement or are not efficient enough to effect complete hemolysis. By using a rabbit anti-mouse immunoglobulin antiserum, it is possible to detect most if not all antibodies regardless of immunoglobulin light- or heavy-chain isotype. In the course of our work, we have identified antibody-producing hybrids that were later determined to be expressing μ, γ_1, γ_{2a}, γ_{2b}, and γ_3 heavy-chain isotypes and both κ and λ light-chain isotypes according to radioimmunoassay of the affinity-purified monoclonal antibodies.[15] Subsequently, it was shown that the heavy- or light-chain isotype could be accurately predicted by facilitating with a library of rabbit anti-mouse isotype-specific antibody reagents.

Transfer Template Radioimmunoassay to Detect Cell Surface-Specific
Antibodies

Principle

The previously described screening assay can be applied to test hybridomas secreting antibodies with specificity for virtually any soluble antigen. Another assay is described here for screening hybridomas that secrete antibodies specific for cell-surface determinants.[17,18] This assay is a solid-phase radioimmunoassay that employs target cell monolayers cultured on transfer templates as the antigen. The calibrated aperture in the bottom of each of the 96 wells is small enough to retain fluid contents by surface tension during monolayer growth, but it also permits fluid to enter the wells when the transfer templates are lowered into the hybridoma microculture plates.

The particular assay described here is for hybrid cell lines secreting antibody with specificity for avian neural surface-membrane antigens.[17,18] This assay can be adapted for other cell lines that attach and grow on plastic and for nonadherent cell lines that can be attached to poly-L-lysine-coated transfer templates.

Procedure

Preparation of Retina Cell Monolayers on Transfer Template. Poly(vinyl chloride) 96-well V-bottom Microtiter plates (Dynatech Corporation) are fenestrated by a single passage through the bottom of each well with a 25-gauge needle. The commercially available transfer plate (Cooke Engineering) is also suitable for use as the transfer template.

Trypsin-dissociated embryonic chick retina cells (1.2×10^6) were inoculated into each of the 96 wells of the transfer template. The cells are cultivated for 2 days in 150 μl of Eagle's medium containing 10% fetal bovine serum. During monolayer growth, the transfer templates are suspended in 96-well flat-bottomed polystyrene microculture plates and covered with a microculture plate lid. When commercial transfer plates are used, they are elevated by 5 mm × 42 mm strips of filter paper along each edge of the microculture plate to prevent contact between the bottom of the microculture plate well and the transfer template aperture.

Assaying Hybridoma Microcultures. Hybridomas specific for the avian retina were produced as reported earlier.[17,18] Just prior to assaying

[17] M. D. Schneider and G. S. Eisenbarth, *J. Immunol. Methods* **29**, 331 (1979).

[18] G. D. Trisler, M. D. Schneider, and M. Nirenberg, *Proc. Natl. Acad. Sci. U.S.A.* **78**, 2145 (1981).

the microcultures, the medium is discarded from the transfer template bearing the retina cell monolayers by sterilely decanting the transfer plate with brisk inversion. The transfer plate is then inserted into the corresponding 96 wells of the hybridoma microcultures permitting the inflow of microculture supernatant fluid through the aperture in each well of the transfer template. The plates are incubated with the tissue culture supernatant fluids for 30 min at 37°. The wells are washed three times with phosphate-buffered saline (PBS) containing 1 mg of pigskin gelatin per milliliter of PBS (PBS-gel) at 4°. The presence of cell-bound antibody is determined either by binding [17]I-labeled protein A[17] or by ^{125}I-labeled F(ab')$_2$ antibody fragment directed against mouse IgG heavy and light chains, i.e., rabbit anti-mouse Ig.[18] The latter is preferable, since this affinity-purified reagent often reacts with hybridoma antibodies not detected using ^{125}I-labeled protein A.[17] Each cell monolayer is incubated for 30 min at 37° with nanogram amounts (the appropriate amount must be determined empirically for each antibody preparation) of the affinity-purified ^{125}I-labeled rabbit anti-mouse Ig, and 500 μg of bovine serum albumin. The cells are washed three times in the PBS-gel. Individual wells are cut from the transfer template, and the radioactivity bound is determined with a gamma counter.

Immunoadsorption Assay of Hybridomas

This assay, which also employs the principle of replica-plating, is another example of a sensitive and rapid method for the detection of monoclonal antibodies secreted by hybridomas.[19] The immunoadsorption procedure is for screening hybridomas that have been cloned directly in soft agarose. Nitrocellulose filters that have been coated with a specific protein, with antigen-coupled erythrocyte ghosts or with other cells used as antigens, are placed on the agarose surface. After incubation to allow immunoadsorption of any secreted antibodies specific for the filter-bound antigen, the filter is removed and overlaid with a suspension of antigen-coupled erythrocytes that react with the adsorbed antibodies. The unbound erythrocytes are allowed to fall off the filter, and the remaining erythrocytes appear as red spots that delineate the sites at which antibody-forming hybrid clones are present in the agarose. Alternatively, the filter may be treated with ^{125}I-labeled antigen followed by autoradiography to reveal the position of positive hybrid clones.

[19] J. Sharon, S. L. Morrison, and E. A. Kabat, *Proc. Natl. Acad. Sci. U.S.A.* **76**, 1420 (1979).

[16] Use of High-Resolution Two-Dimensional Gel Electrophoresis for Analysis of Monoclonal Antibodies and Their Specific Antigens

By TERRY W. PEARSON and N. LEIGH ANDERSON

High-resolution two-dimensional polyacrylamide gel electrophoresis (2-D gel electrophoresis) as developed by O'Farrell[1] allows the analytical separation of large numbers of proteins. The technique, when properly performed, enables a trained individual to separate reproducibly (on the basis of isoelectric point and molecular weight) both acidic and basic proteins in a mixture. In addition, by using internal isoelectric point and molecular weight markers and by correct interpretation of gel patterns, a fair amount of biochemical information on individual proteins can be obtained. Clearly the technique is one of the most powerful for resolving and characterizing proteins in complex biological materials.[2]

Hybridoma technology[3,4] also will enable a more complete analysis of antibody repertoires[5] than was possible in the past and in addition allows the production of large amounts of chemically homogeneous monoclonal antibodies, which can be used as highly specific probes for antigens in biological samples. For a general discussion of the properties of monoclonal antibodies, the principles behind their derivation and their uses, the reader is referred to the First Wellcome Foundation lecture by Milstein.[6] Detailed specific applications of monoclonal antibodies are reviewed in Hämmerling, Hämmerling, and Kearney.[7]

Although the 2-D gel and monoclonal antibody technologies are unrelated, taken together they are powerful techniques for analysis of complex antigenic mixtures[8] and antibody repertoires. In this report, we will discuss (a) general considerations for running 2-D gels, including the use of internal isoelectric point and molecular weight markers; (b) interpretation of 2-D gel patterns; (c) the use of 2-D gels for characterization of mono-

---

[1] P. H. O'Farrell, *J. Biol. Chem.* **250,** 4007 (1975).

[2] N. G. Anderson and N. L. Anderson, *Behring Inst. Mitt.* **63,** 169 (1979).

[3] G. Köhler and C. Milstein, *Nature (London)* **265,** 495 (1975).

[4] G. Köhler and C. Milstein, *Eur. J. Immunol.* **6,** 511 (1976).

[5] M. Reth, G. J. Hämmerling, and K. Rajewsky, *Eur. J. Immunol.* **8,** 393 (1978).

[6] C. Milstein, *Proc. R. Soc. London Ser. B* **211,** 393 (1981).

[7] "Monoclonal Antibodies and T-Cell Hybridomas" (G. J. Hämmerling, U. Hämmerling, and J. F. Kearney, eds.). Elsevier/North-Holland, Amsterdam, 1981.

[8] T. W. Pearson and N. L. Anderson, *Anal. Biochem.* **101,** 377 (1980).

clonal antibodies; and (d) the use of 2-D gels for analysis of antigens bound by specific monoclonal antibodies.

Two-Dimensional Gel Electrophoresis

General Considerations

In the 2-D gel technique as described by O'Farrell,[1] proteins are separated in the first dimension by electrophoresis in a pH gradient established using Ampholines and in the second dimension by electrophoresis in sodium dodecyl sulfate (SDS) through cross-linked polyacrylamide. Thus the proteins are separated by two independent parameters, their net charge (first dimension) and SDS–molecular weight (second dimension). With most 2-D gel systems, proteins within an isoelectric point range of approximately pH 3.5–8.5 and within a molecular weight range of approximately 8000–100,000 can be resolved. Proteins with more acidic or basic isoelectric points can, however, be resolved by altering the pH gradient or by running nonequilibrium gels (see below). Proteins with higher or lower molecular weights can be resolved by lowering or raising the degree of cross-linking or acrylamide concentration. For example, high molecular weight proteins (greater than one million) can be resolved by using N,N'-diallyltartardiimide to cross-link the acrylamide.[9] Thus, by adopting different conditions into the 2-D gel system, it is possible to analyze proteins of widely varying molecular characteristics and, depending on the method used, to detect them in extremely small quantities (see below).

The ISO-DALT System

Many different 2-D gel systems and apparatus for running them exist. Many of them work extremely well and give good 2-D gel patterns that yield important information, depending on the experiment being performed. Generally it is true that most investigators can run a 2-D gel and obtain good spot separation. Running reproducible gels is not as easy, however, and this fact often leads to frustration, especially if complex gel patterns are being compared between gels run on different samples. For this reason, and so that many gels could be run by an individual researcher, the ISO-DALT system for multiple 2-D gel electrophoresis was developed by Anderson and Anderson.[10,11] This system allows 20 isoelectric focusing gels (first dimension) to be formed simultaneously from the

[9] A. Ziegler and H. Hengartner, *Eur. J. Immunol.* **7,** 690 (1977).
[10] N. G. Anderson and N. L. Anderson, *Anal. Biochem.* **85,** 331 (1978).
[11] N. L. Anderson and N. G. Anderson, *Anal. Biochem.* **85,** 341 (1978).

same batch of materials and, in addition, enables them to be run simultaneously in the same apparatus used to pour them. Similarly, 10 or 20 polyacrylamide slab gels (second dimension) can be poured at once and, after loading with the first dimension tube gels, run simultaneously. Replicate samples run using this system give 2-D gel patterns that are almost, if not exactly, identical. In addition, the effort and skill used to pour, run, and develop 10 2-D gels using the ISO-DALT apparatus are often less than that required for running one or two gels with most conventional apparatus. The name "ISO-DALT" is derived from the first-dimension separation, which depends on the ISOelectric points of the proteins being separated, and the second-dimension separation, which depends on their apparent molecular mass in DALTons. The apparatus is available from Electronucleonics, P.O. Box 451, Oak Ridge, Tennessee. The procedures described throughout this chapter were all developed using the ISO-DALT system but nevertheless can be applied with few modifications to other 2-D gel techniques. Two points must be kept in mind, however, when adapting the procedures to other 2-D gel systems.

1. Samples are prepared in a buffer containing sodium dodecyl sulfate (SDS) prior to isoelectric focusing in the ISO-DALT system. The first-dimension gels contain the nonionic detergent Nonidet P-40 (NP-40) and urea, which strips the SDS off the proteins during electrophoresis.
2. The second-dimension slab gels are gradient gels and thus allow "tighter" focusing of spots than do nongradient gels. Detailed methods for ISO-DALT 2-D electrophoresis are described elsewhere,[12,13] as they are too lengthy to be included here.

Separation of Acidic and Basic Proteins (Nonequilibrium Gels)

It is usual in most 2-D gel electrophoresis experiments to load the appropriately solubilized samples at the basic end (high pH) of the isoelectric focusing gels and to allow electrophoretic separation to proceed toward the acid (low pH) end of the gels. In this way (depending on the Ampholine mixture used), most of the proteins with low or middle pI values will be resolved in the first dimension. Unfortunately, at equilibrium, highly acidic proteins (pI less than 3.0) and many basic proteins (pI

[12] S. L. Tollaksen, N. L. Anderson, and N. G. Anderson, *Argonne Natl. Lab.* (*Rep.*) ANL-BIM-81-1 (1981).
[13] R. P. Tracy and N. L. Anderson *in* "Clinical Laboratory Annual" (J. Batsakis and H. Homburger, eds.) Vol. 2, in press. 1982.

greater than 8.5–9.0) are at the gel ends and thus not separated. To overcome this problem, nonequilibrium gels are run in the first dimension. When acidic proteins need resolving, the standard isoelectric focusing gels are run, but electrophoresis is not allowed to proceed to equilibrium (usually 3500–5000 volt hours are used instead of 9000–10,000). Additionally the amount of low pH Ampholines can be increased in the Ampholine mixture. For basic proteins, the procedure is not quite so simple. Samples are first solubilized in phosphatidylcholine mixture (dipalmitoyl L-α-phosphatidylcholine 0.5%, pH 3.5–10, ampholytes 2%, dithioerythritol 5%, urea 9 M), ultracentrifuged, and applied to the acid end of the first-dimension tube gels. The polarity of the electrodes is reversed and nonequilibrium electrophoresis is performed, again using 3500–4000 volt hours depending on the sample. In this way, basic proteins can be easily resolved. Detailed methods for solubilization of samples containing basic proteins and for their 2-D analysis (BASO gels) have been described by Willard et al.[14]

Visualization of Proteins in 2-D Gels

For many applications, staining of protein spots on 2-D gels can be performed using Coomassie Blue.[11] Visualization of monoclonal antibody heavy and light chains in ascites fluids is a good example.[8] When proteins are present in extremely low amounts or when protein overloading of gels is a problem, then autoradiography or fluorography[15] of radiolabeled samples or silver staining[16,17] is often necessary. Several new modifications of the silver staining procedures allow routine processing of batches of 2-D gels (10–20 at a time), a feat not thought possible even a year ago. The method we use is that described by Sammons et. al.[18] In addition to allowing the use of extremely small amounts of material for running 2-D gels, the new silver stain procedures enable the visualization of many more proteins in 2-D gel patterns when compared to the already sensitive Coomassie Blue stain.

In some experimental situations, it is necessary or convenient, or both, to use radiolabeled proteins or glycoproteins for visualization of 2-D gel spots or patterns. For example, in immunoprecipitation procedures, it

---

[14] K. E. Willard, C. S. Giometti, N. L. Anderson, T. E. O'Connor, and N. G. Anderson, *Anal. Biochem.* **100**, 289 (1979).
[15] M. Bonner and R. A. Laskey, *Eur. J. Biochem.* **46**, 83 (1974).
[16] B. R. Oakley, D. R. Kirsch, and N. R. Morris, *Anal. Biochem.* **105**, 361 (1980).
[17] R. C. Switzer, III, C. R. Merril, and S. Shifrin, *Anal. Biochem.* **98**, 231 (1978).
[18] D. L. Sammons, L. D. Adams, and E. E. Nishizawa, *Electrophoresis '81* **2**, B5 (1981).

is much easier to identify the precipitated antigens if they are radiolabeled, as the other proteins (e.g., antibodies) remain undetected after autoradiography–fluorography of the 2-D gels. An elegant technique that allows extremely accurate comparison of two different 2-D gel patterns is that of double-label autoradiography.[19] By combining highly reproducible 2-D gels with sensitive detection methods the analytical dissection of protein mixtures has attained a new level of sophistication that will almost surely revolutionize many aspects of fundamental biology and clinical medicine.[20]

Isoelectric Point and Molecular Weight Standards

It is often desirable, or even necessary, to identify protein spots on 2-D gels by a set of gel coordinates that accurately define the position of the spots, at least in a relative sense. To do this, internal markers for both isoelectric points and molecular weights have been derived. Internal markers that cover a wide range of pI values and molecular weights are necessary because of the complexity of many 2-D gel patterns and the fact that local distortions in individual gels often occur. By correctly assigning coordinates (based on internal standards) to protein spots in 2-D gels, it becomes possible accurately to compare 2-D spot patterns between gels or between laboratories, even those that use different 2-D gel systems.

Internal Isoelectric Point Standards: "Carbamylation Trains"

When proteins are heated in a solution of urea, their free amino groups are carbamylated through the production of cyanate. By heating a protein for various lengths of time, the extent of carbamylation can be varied. The carbamylation of each free amino group results in the loss of a positive charge on the protein with the effect that the isoelectric point of the protein is altered to a more acid pH. A unit shift in pI is obtained with each amino group carbamylated. By mixing a number of differentially carbamylated samples of the protein followed by isoelectric focusing in urea and SDS–electrophoresis, a "carbamylation train" can be established. This "carbamylation train" appears on the 2-D gel as a row of spots of approximately uniform molecular weight and thus can be used as a regular series of internal pI markers.[21] The number of spots obtained by carbamylation of a given protein is one plus the number of free amino

[19] E. H. McConkey, *Anal. Biochem.* **96,** 39 (1979).
[20] *Clin. Chem.* **28,** 737 (1982).
[21] N. L. Anderson and B. J. Hickman, *Anal. Biochem.* **93,** 312 (1979).

FIG. 1. Isoelectric point (pI) standard carbamylation train. Unfractionated human muscle were coelectrophoresed in two dimensions with rabbit muscle creatine phosphokinase (CPK) carbamylation train standards. A is albumin (pI = -10 CPK); B is muscle actin (pI = -17.5 CPK); C is an unknown protein at -12.1 CPK; D is human muscle CPK; and E is an unknown protein with pI too basic to measure using rabbit CPK standard. All gels are shown with the acid end to the left, resulting in a system of pI values that run according to the Cartesian convention. From Anderson and Hickman.[21]

groups in the protein. Thus by using a fairly basic protein (pI 7–8.5) a "carbamylation train" spanning most of the gel can be obtained. An example of such a protein (rabbit muscle creatine phosphokinase) used as a "carbamylation train" is shown in Fig. 1. The use of even more basic proteins as internal pI markers will be necessary for standards on BASO-gels.

Internal Molecular Weight Standards

Internal molecular weight standards for 1-D or 2-D gels have been used for many years. Usually only a few protein markers are used. These are expensive for routine use and do not adequately cover a wide range of molecular weights evenly. By using heart proteins as internal molecular weight standards, these problems have been overcome.[22] When whole rat heart homogenate is used at a concentration of 10 mg/ml, 80 lines can be seen on a 10 to 20% linear gradient of acrylamide. Since the exact molecular weight of many of the proteins in heart muscle are known, a calibration curve can be drawn for determination of SDS (apparent) molecular weight of proteins in the second dimension (Fig. 2).

Interpretation of 2-D Gel Patterns

Often, 2-D gels show an extremely complex pattern of protein spots that at first overwhelms the 2-D gel novice. However, much information can be gained from gel patterns despite their complexity, once a few factors important in interpreting the microheterogeneity of proteins are learned.

Most proteins appear as more than one spot when analyzed by 2-D gel electrophoresis. It is clear that most of the micropatterns exhibited by pure proteins are truly a reflection of protein-modifying processes occurring *in vivo,* since the patterns are reproducible and are found in gels of whole solubilized material (in the presence of proteolysis inhibitors), of immunoprecipitates, of immunoadsorbent eluates, and of classically purified proteins. There is no evidence that the 2-D gel technique itself generates any heterogeneity. Nevertheless, it is important that care be taken to ensure that samples are properly solubilized, that proteolysis is eliminated, and that the 2-D apparatus and materials are clean and of high quality. Since urea is often used for sample solubilization and is present in the first dimension gels of the ISO-DALT system, it is imperative that the samples are not heated and that the isoelectric focusing apparatus is kept cool in order to avoid charge-shift variants of proteins caused by carbamylation of free amino groups (see previous section on isoelectric point standards). One additional caution is that many commercially available proteins are not as pure as they are purported to be, and, since the 2-D gel technique gives tremendous resolution, care must be taken in using these proteins as standards for comparison.

Microheterogeneity: Charge Shifts

A pure protein can appear as a single spot or (more usually) as a series of two or more spots in a horizontal row across part of the 2-D gel. There

---

[22] C. S. Giometti, N. G. Anderson, S. L. Tollaksen, J. J. Edwards, and N. L. Anderson, *Anal. Biochem.* **102,** 47 (1980).

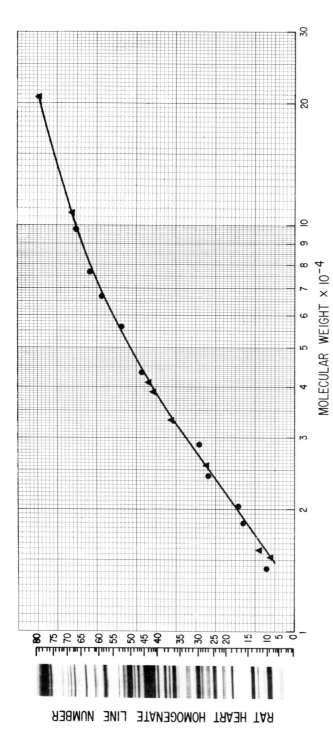

FIG. 2. Molecular weight standard curve. Rat heart homogenate lines were assigned an sodium dodecyl sulfate (SDS)-molecular weight value from a set of calibration proteins and used to establish this curve[22] for a gel composed of 10 to 20% acrylamide in a linear gradient. The gel lines of the rat heart homogenate pattern are shown in the photograph of the gel segment at the left of the figure. ●, Purified calibration proteins; ▲, rat or rabbit proteins. From Giometti et al.[22]

can be several different causes of this charge heterogeneity. First, the p*I* of some of the proteins may be altered by changes in the charges of individual amino acids, for example by carbamylation or other mechanisms that alter free amino groups. Phosphorylation or methylation of amino acids could cause similar charge modifications or, for that matter, any deamidation process. The change of a single charged amino acid can cause a unit shift in p*I*. It must always be kept in mind that allelic or polymorphic forms of the same functional protein may be present, and that these could differ by as little as a single charged amino acid.

Glycosylation of proteins with charged sugars can also cause shifts in isoelectric point. For example, the addition of sialic acids to a protein results in charge shifting toward the acid end of the gel. This shifting of isoelectric points to the left is usually accompanied by an upward sweeping arc of spots. This pattern can be explained by. increases in molecular weights of the most acidic spots as more sialic acids are added to the protein. One would predict therefore, that neuraminidase digestion should shift all species toward a single spot at the right if the charge shifts on such a protein are due solely to sialic acid residues, not to differences in polypeptide charge. As an example of charge shifting due to sialation of proteins, the 2-D gel patterns of native and neuraminidase-treated immunoadsorbent-purified human haptoglobin β chain are shown in Fig. 3. It can be seen that the native β chain consists of two arcs of spots that sweep upward toward the acid end of the gel (Fig. 3a). After neuraminidase digestion, all species (including those in the minor arc) are shifted toward a single isoelectric point at the right (Fig. 3b). This indicates that the charge shifts are all due to the addition of sialic acids to the protein and that the isoelectric point difference between major and minor arcs is due to different sialic acid contents rather than differences in polypeptide charge.

Microheterogeneity: Molecular Weight Variations

Many proteins, though pure, appear in 2-D gel patterns as spots of different molecular weights. Again, this can be due to changes in either polypeptide or carbohydrate portions of the protein or glycoprotein. If the spot pattern shows an arc curving up to the left (toward the acid end of the gel), the higher molecular weight spots are likely to contain more sialic acid residues than spots of lower molecular weight, as mentioned previously. If the spots are not in the form of an arc but still show different p*I*s, the higher molecular weight spots could be glycosylated with an oligosaccharide unit rather than with incremental additions of single sialic acid

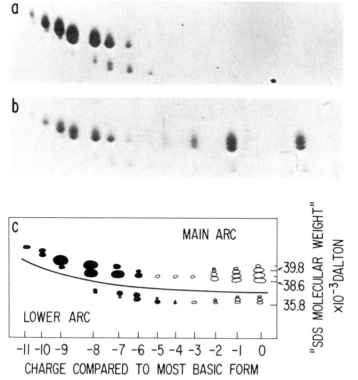

FIG. 3. Portions of two-dimensional gels showing (a) human haptoglobin β chain, (b) native β-chain plus products of neuraminidase digestion, (c) diagrammatic representation of major and minor arcs (filled ellipses) and their neuraminidase-digested forms (open ellipses). The minor arc reproduces the structure of the major arc but is shifted down (≈ 2800 SDS-daltons) and two charges to the right ($+2$). The neuraminidase-digested forms, however, are matched directly above one another as far as desialation proceeds. Micro-SDS-molecular weight heterogeneity in the form at position 0 exhibits increments of ≈ 600 SDS-daltons. From N. L. Anderson and N. G. Anderson [*Biochem. Biophys. Res. Commun.* **88,** 258 (1979)].

residues as in the previous example. It is important here to ascertain that glycosylation is responsible for the molecular weight differences, as pre-cursor–product forms or proteolyzed polypeptides can often give rise to different molecular weight forms of the proteins.

When the 2-D gel patterns appear as a row of vertically aligned spots, that is with the same p*I* values, it is likely that the variation in molecular weights is due to glycosylation with neutral sugars. Again, as an example, see Fig. 3, specifically the open ellipses at the basic end of the pattern

shown in Fig. 3c. Here the four spots differ by about 600-dalton increments. It is likely that these incremental increases are due to addition of
neutral sugars, as it is probable that removal or addition of short peptides
from the molecules would cause charge shifts.

There is a lot to learn regarding the interpretation of 2-D gel patterns.
Why do "carbamylation train" patterns arc up toward the left if only
charge is altered on the proteins?[21] By using endoglycosidases, lectins,[23]
and antibodies in gel-transfer techniques, much more understanding of
proteins and the molecular anatomy[24] of biological materials will be
gained. A more detailed discussion of the microheterogeneity of proteins
as reflected in 2-D gel patterns is presented by Anderson and Anderson.[25]

2-D Gel Analysis of Antibody Repertoires and Monoclonal Antibodies

The serum of an animal contains immunoglobulins of varying specificities, affinities, and classes. By using 2-D gel analysis, the repertoire of
these antibodies at any one time can be resolved.[26] Each antibody is
composed of two identical heavy chains (H chains) and two identical light
chains (L chains) and therefore under reducing conditions should appear
on a 2-D gel as a single high molecular weight spot (H chain) and a low
molecular weight spot (L chain). Usually this is not the case, however, as
charge (and sometimes molecular weight) variants of the H and L chains
give a slightly more complex 2-D pattern. Nevertheless, the individual H
or L chain species in a serum can be resolved (Fig. 4), especially the light
chains (Fig. 5), since they are usually not glycosylated and give much
more discrete spots than the H chains. Figures 4 and 5 illustrate two main
points important for the rest of this chapter. The wide variety of H- and L-
chain spots indicates, first, that the antibody repertoire at any one time is
finite and large, and, second, that individual H and L chains can be resolved from their neighbors. This is important, since the characteristics of
monoclonal antibodies as reflected in their component H and L chains can
thus be studied using 2-D gels.

Much information can be obtained from 2-D gel analysis of monoclonal antibodies. First of all, 2-D electrophoresis is the method of choice
for determination of the chain composition of antibodies secreted by
myeloma hybrids. This is important for establishing the monoclonality of
the hybrid cells and, in situations where immunoglobulin synthesizing
myeloma parent cells are used, for establishing which parental myeloma

[24] N. G. Anderson and N. L. Anderson, *Behring Inst. Mitt.* **63**, 169 (1979).
[25] N. L. Anderson and N. G. Anderson, *Biochem. Biophys. Res. Commun.* **88**, 258 (1979).
[26] N. L. Anderson, *Immunol. Lett.* **2**, 195 (1981).

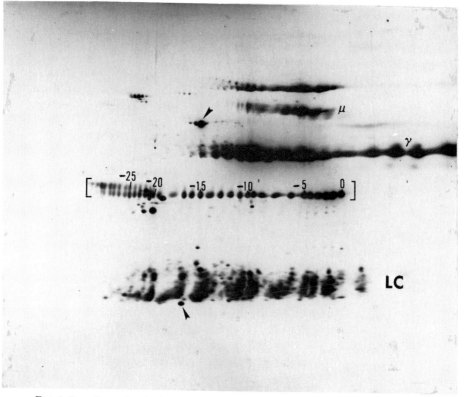

Fig. 4. Two-dimensional gel pattern of protein A-adsorbed[8] mouse serum immunoglobulin chains. The horizontal row of spots in brackets are rabbit muscle creatine phosphokinase (CPK) isoelectric point standards.[21] The upward arrowhead marks apoA-1 lipoprotein, and the downward arrowhead marks albumin. The immunoglobulin chains are the light chains (LC), the γ and μ are heavy chains, and an unknown higher molecular weight chain above μ, which is probably a γ-chain dimer. Remaining spots are probably complement components and various "sticky" proteins. From Anderson.[26]

immunoglobulin chains are secreted, if any, in addition to those of the specific antibody from the immune spleen cell partners used in fusion. Information on the immunoglobulin heavy- and light-chain isotypes (classes) can also be obtained from 2-D gel patterns, as can data on glycosylation and molecular weight and charge of the native and variant H and L chains. Although little information exists so far, it is possible that the above kinds of data collected from 2-D analysis of large numbers of monoclonal antibodies will aid in the understanding of antibody repertoires, for example, in response to disease.

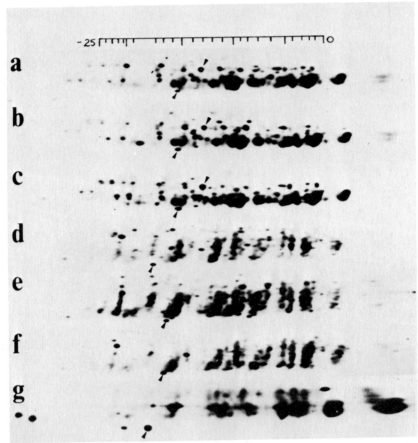

Fig. 5. Light-chain portions of several two-dimensional gels of protein A-adsorbed mouse sera. (a–c) From three individual BALB/c mice raised in the same cage; (d–f) from three outbred white-footed mice (*Peromyscus leucopus*). (g) From a healthy human (NLA). Upward pointing arrowheads mark the only known non-light-chain spot contaminating this region (apoA-1 lipoprotein), and downward pointing arrowheads mark the predominant BALB/c spot, which reacts with anti-mouse λ-chain antiserum (probably the λ_0 gene product). In both BALB/c and human, λ chains occur along the top of the light-chain band, with k chains below these. CPK standards are shown at the top of the figure. From Anderson.[26]

Preparation of Monoclonal Antibodies for 2-D Gel Analysis

Two-dimensional gel patterns of monoclonal immunoglobulins are usually obtained by analysis of tissue culture supernatants or of ascites fluids. Since the antibody concentration is relatively low in tissue culture supernatants, internal radiolabeling of the antibody chains is performed, followed by running of the gels and visualization of the labeled heavy and light chains by autoradiography or fluorography of the dried gels. Visualization of heavy and light chains on gels run on ascites fluids is performed by either Coomassie Blue or silver staining.

Internal Labeling of Monoclonal Antibodies. Approximately 1 to 2 × 10^6 doubly cloned hybridoma cells from a healthy culture (exponential growth phase) are centrifuged in a conical 10–15-ml centrifuge tube at 400 g for 5 min, resuspended in 5 ml of lysine-free RPMI-1640 medium (GIBCO-BIOCULT), and washed by a further centrifugation. The cells are gently resuspended by tapping the tube and mixed into 1.0 ml of lysine-free RPMI-1640 containing 10% dialyzed fetal calf serum and 2.5 μCi of L-[U-^{14}C]lysine monohydrochloride (Amersham, Catalog No. CFB.69). The tube is gassed with 5–10% CO_2 in air and incubated at 37°. After 12–16 hr of incubation, the cells are pelleted by centrifugation (400 g, 5 min), and the supernatant is aliquoted into 100-μl amounts and stored at −20°. No more than 2 hr before running on 2-D gels, an aliquot of monoclonal antibody-containing supernatant is thawed and centrifuged for 5 min on a microfuge to remove aggregates. Usually between 10 and 20 μl of supernatant are mixed with 30 μl of SDS-MIX (2% sodium dodecyl sulfate, 1% dithioerythritol, 10% glycerol, 0.05 M, pH 9.5, cyclohexyl-aminoethanesulfonic acid); the mixture is centrifuged on a microfuge for 10 min, and 10–20 μl are applied to the isoelectric focusing gel. Fluorography[15] of dried gels is usually performed, the incubation period depending on the extent of incorporation of [^{14}C]lysine into secreted antibody. Routinely, 7–10 days suffice with the above method.

Preparation of Monoclonal Antibody-Containing Ascites Fluids. Ascites fluids containing large amounts of monoclonal antibodies (1–20 mg/ml) are prepared by injection of hybridoma cells into the peritoneums of pristane-treated mice.[27] Ascites fluids harvested 7–14 days later often contain dead cells and pristane and are thus messy and often viscous owing to the presence of DNA, fibrin, and oil. Cells and large aggregates are removed from the ascites fluids by centrifugation at 800 g for 15 min. The supernatants (5–15 ml) are filtered through cotton wool (1 cm, packed loosely into a 5-ml syringe barrel) to remove DNA, fibrin, etc., then ultracentrifuged (100,000 g, 20 min) to remove aggregates, and finally filtered through 0.45 μm Millipore membranes to sterilize. They are used immediately for gel analysis or are stored at −20° in 100-μl aliquots after

[27] T. W. Pearson, M. Pinder, G. E. Roelants, S. K. Kar, L. B. Lundin, K. S. Mayor-Withey, and R. S. Hewett, *J. Immunol. Methods* **34**, 141 (1980).

addition of sodium azide to 0.1% final concentration. If the 2-D gels are to be silver stained, 1 μl of ascites is mixed with 30 μl of SDS-mix, and 15 μl of this mix are applied to the isoelectric focusing gel after centrifugation in a microfuge for 10 min to remove aggregates. For Coomassie Blue staining, 10 μl of ascites are mixed with 30 μl of SDS-mix, and (again after centrifugation) 15 μl are applied to the isoelectric focusing gels.

Solubilization of some monoclonal antibodies may require the use of phosphatidylcholine–urea mix (dipalmitoyl L-α-phosphatidylcholine 0.5%, pH 3.5–10, ampholytes 2%, dithioerythritol 5%, and urea 9 M[14]), as their H and/or L chains are too basic to be resolved on standard ISO gels. Nonequilibrium BASO-gels must therefore be run on these antibodies. Detailed methods for running BASO-gels are described elsewhere.[14]

Protein A Binding of Monoclonal Antibodies. Protein A from *Staphylococcus aureus* has the unique property of binding to the Fc portions of immunoglobulins of various subclasses.[28] Although it is generally true that certain subclasses (isotypes) bind better than others and that the immunoglobulins of some species bind better than those of other species, nevertheless the rules are not absolute. This is clearly true with monoclonal antibodies, which are primarily mouse or rat. In our Victoria laboratory, 8/12 monoclonal murine IgM antibodies produced to different antigens were found to bind strongly to protein A at pH 7.4 (unpublished observations). The dogma that in the mouse only IgG antibodies bind to protein A[29] breaks down when monoclonal immunoglobulins are examined. Thus we routinely examine most monoclonal reagents for protein A-binding activity, as this characteristic aids in antibody purification and in making immunoadsorbents for antigen and antibody analysis on 2-D gels. Disposable microimmunoadsorbent columns for protein A binding of monoclonal antibodies are made as follows: Four millimeters are cut off the small end of yellow Eppendorf pipette tips (1–100 μl, Brinkmann Instruments, Inc., Westbury, New York, catalog No. 2235 130-3) using a razor blade. The bottom 2 mm portion is then plugged with a wisp of surgical cotton using a 25-μl glass disposable microsampling pipette (Corning Glass Works, Corning, New York, catalog No. 7099-S) to press the cotton lightly into place. These plugged pipette tips (microcolumns) are then inserted into blue Eppendorf pipette tips (101–1000 μl, Brinkmann Instruments, Inc., Cat. No. 2235 090-1), which are cut off 1 cm from the large end, and the assembly is placed into a plastic test tube (Luckham LP3, Sussex, England) as shown in Fig. 6. The use of these

[28] A. Grov, P. Oeding, B. Myklestad, and J. Ausen, *Acta Pathol. Microbiol. Scand. Sect. B* **78B**, 106 (1970).
[29] G. Kronvall and R. C. Williams, *J. Immunol.* **103**, 828 (1969).

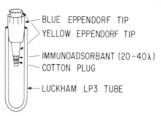

BLUE EPPENDORF TIP
YELLOW EPPENDORF TIP
IMMUNOADSORBANT (20-40λ)
COTTON PLUG
LUCKHAM LP3 TUBE

FIG. 6. Diagram of a microimmunoadsorbent column. A section of the large end of a blue Eppendorf tip serves as an adapter holding a yellow Eppendorf tip column. A wisp of cotton holds the immunoadsorbent in place. Antibodies coupled directly to a solid phase, such as Sepharose, or linked by binding to protein A–Sepharose can be used as immunoadsorbent. This arrangement can be centrifuged to dry the immunoadsorbent and, after rehydration with a dissociating elution buffer, to collect the eluate. Almost all the bound protein can be eluted in one column volume (20–40) μl). From Pearson and Anderson.[8]

particular Eppendorf tips and LP3 tubes is important as the size of the components and the flanges on the upper ends of the Eppendorf tips allow air to escape during sample application, washing, and elution procedures.

Columns are packed at room temperature by pipetting 400 μl of a 10% suspension of protein A–Sepharose CL-4B (Pharmacia, Uppsala, Sweden) in phosphate-buffered saline (PBS) into each pipette tip and allowing the PBS to run through. A diagram of a microimmunoadsorbent column is shown in Fig. 6. Fifty microliters of neat ascites fluid (containing the specific or control monoclonal antibodies) are applied to the column and gently blown into the protein A. After 30 min of incubation at room temperature, the column is filled with PBS and washed by gentle blowing until the PBS reaches the level of the protein A bed. The column is topped up with PBS once more, and the entire assembly is centrifuged at 800 g for 5 min, effectively drying the immunoadsorbent column. Finally, the column assembly is transferred to another LP3 tube, and 40 μl of a 1:2 dilution of sample elution buffer (2% sodium dodecyl sulfate–5% 2-mercaptoethanol–10% glycerol) is added. This rehydrates the dried immunoadsorbent. After incubation at room temperature for 30 min, the columns are eluted by centrifuging at 800 g for 10 min. The eluate (then in 40 μl of sample preparation buffer) is heated to 95° for 5 min and is either stored at $-20°$ or directly applied to isoelectric focusing gels for 2-D analysis. Ten to twenty different monoclonal antibodies can easily be analyzed for binding to protein A in this manner.

Monoclonal Antibody 2-D Gel Patterns

Representative 2-D patterns of monoclonal antibody-containing ascites fluids are shown in Figs. 7 and 8. Figure 7a shows the 2-D gel pattern

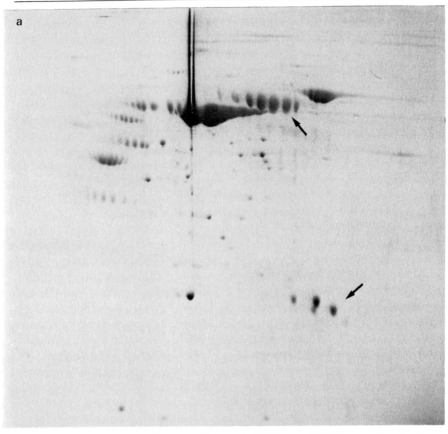

FIG. 7. Two-dimensional gel pattern of (a) ascites fluid from a BALB/c mouse bearing the myeloma-hybrid BCAT1/71.80.5 (the hybridoma secretes an IgM anti-human brain choline acetyltransferase). The heavy chain of the monoclonal antibody (μ) is designated by the upward-pointing arrow. The light chain is designated by the downward-pointing arrow. (b) Eluate from a protein A–microimmunoadsorbent column after passage of BCAT1/71.80.5 ascites fluid. Again the upward arrow designates the monoclonal H chain and the downward arrow the L chain.

of whole ascites fluid containing an IgM monoclonal antibody αBCAT1/71.80.5 (anti-human brain choline acetyltransferase). The IgM (μ) heavy chain (indicated by the upward-pointing arrow) is clearly visible as a row of charge-shifted spots. The light chain (downward-pointing arrow) appears as three major spots of different charge. Presumably, the charge heterogeneity of the μ chain is greater than that of the light chain owing to glycosylation with charged sugars. Figure 7b is the pattern obtained when

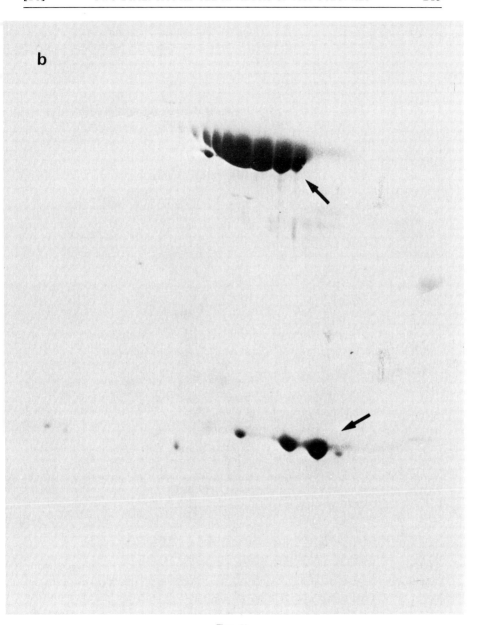

FIG. 7b.

FIG. 8. Two-dimensional gel pattern of (a) ascites fluid from a BALB/c mouse bearing the myeloma-hybrid HSA1/25.1.3 (the hybridoma secretes an IgG, anti-human serum albumin). (b) Eluate from a protein A–microimmunoadsorbent column after passage of HSA1/25.1.3 ascites fluid. The arrows designate H and L chains as in Fig. 7. The H chain in this case is γ_1.

the monoclonal antibody is eluted from a protein A–microimmunoadsorbent. Figure 8a shows the 2-D gel pattern of whole ascites fluid containing IgG monoclonal antibody αHSA1/25.1.3 (anti-human albumin). Charge-shifted IgG (γ_1) and light chains are clearly visible as major proteins in the ascites fluids. Again (Fig. 8b) the heavy and light chains are easily visualized after elution from protein A. Figure 9 shows a 2-D gel profile composed to indicate the *approximate* positions of immunoglobulin heavy-

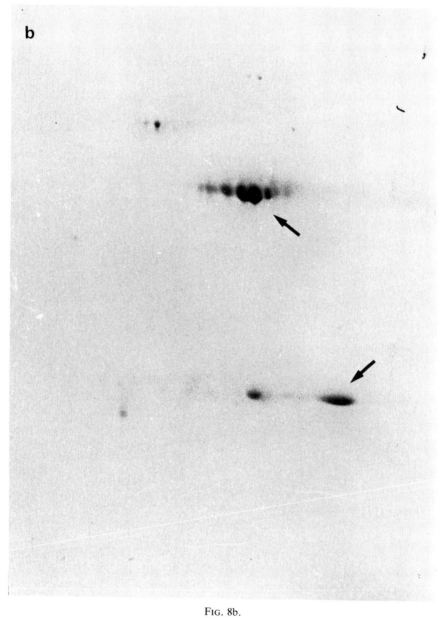

FIG. 8b.

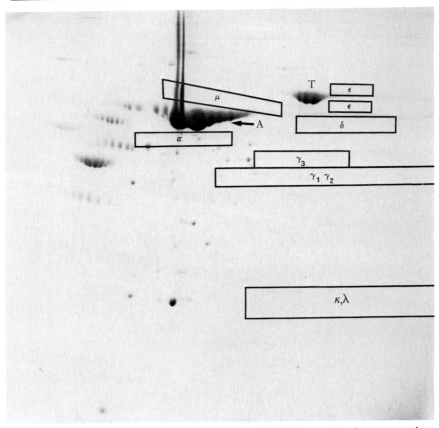

Fig. 9. Two-dimensional gel of ascites fluid from a BALB/c mouse bearing a nonsynthesizing myeloma cell line. The gel has been composed to show the approximate positions of the heavy-chain isotypes expected for most mouse and human immunoglobulins. T designates serum transferrin, and A serum albumin. κ and λ indicate the position of kappa and lambda light chains. Heavy and light chains of monoclonal antibodies can be more acidic or more basic than shown in the diagrammed rectangles and on occasion may exhibit slightly different molecular weight forms.

chain classes (isotypes). The assigned positions were obtained from 2-D gel data on mouse monoclonal antibodies (T. W. Pearson, Victoria) and from data on human myelomas (Egil Jellum and Anna Karina Thorsrud, Oslo). It is clear from our studies that many monoclonal heavy and light chains exhibit isoelectric points above 8.5 and thus do not appear on standard ISO gels. Extension of the pH range to encompass the higher pI values or, more usually, running of nonequilibrium BASO-gels is thus necessary for 2-D analysis of some immunoglobulins.

It must be emphasized that the positions assigned to isotypes on Fig. 9 are to be used only as a guide and that exceptions may occur. For interpretation of complex spot patterns of monoclonal antibodies, refer to the previous section on interpretation of 2-D gel patterns (this chapter). An excellent review of structural aspects of immunoglobulins may also aid here.[30]

2-D Gel Analysis of Antigens Bound by Monoclonal Antibodies

A detailed description of methods used for studying various antigens bound by specific antisera or monoclonal antibodies is beyond the scope of this chapter. However, several general procedures useful in combining monoclonal reagents and 2-D gels for analysis of antigens will be discussed. First of all, as a caution, it is important to realize that many cellular antigens or antigens present in biological fluids are not proteins or glycoproteins and thus are not suitable for gel analysis. This is useful to keep in mind, especially if difficulties are encountered in radiolabeling of certain antigens or if gels fail to show spots after extensive testing.

In general, 2-D gel analysis of specific antigens in complex mixtures involves solubilization of the antigen in question in conditions that minimize or eliminate proteolysis, aggregation, or complexing of antigens and denaturation of antigenic sites. In addition, the solubilizing agents must not interfere with antigen–antibody binding. Often the antigens are present in small quantities and must be labeled with radioisotopes in order that they may be detected by autoradiography or fluorography of dried 2-D gels. Methods for ^{125}I surface labeling of membrane proteins and internal labeling of proteins with ^{14}C-, ^{3}H-, or ^{35}S-labeled amino acids are described by Jones,[31] as are methods for 1-D and 2-D gel analysis of antigens.

Preparation of Antigen Mixtures for Immunoadsorption

We have found that solubilization of membrane proteins (and most other cellular proteins) is best done using the nonionic detergent Nonidet P-40 (0.1–0.5% in PBS[31]). In addition, for solubilization of most biological materials it is necessary to include the proteolysis inhibition cocktail: phenylmethylsulfonyl fluoride, 87 mg; pepstatin, 3 mg in 5.0 ml of 95% ethanol at a final concentration of 1 : 100 (J. V. Kilmartin, personal com-

[30] D. S. Secher, *Int. Rev. Biochem.* **23**, 1 (1979).
[31] P. P. Jones, *in* "Selected Methods in Cellular Immunology" (B. B. Michell and S. M. Shiigi, eds.), p. 398. Freeman, San Francisco, California, 1980.

munication). Solubilization conditions must be developed for each antigen if difficulties are encountered. All solubilized mixtures are centrifuged at 20,000–100,000 g to remove aggregates and other insoluble material prior to immunoadsorption procedures.

Monoclonal Antibody Immunoadsorbents

Monoclonal antibodies are coupled directly to Sepharose 4B[32,33] or bound to protein A–Sepharose as described above. Microimmunoadsorbent columns prepared with the Sepharose-coupled monoclonal antibodies are then used to bind specific antigens from mixtures by simply adding another step to the method used for testing the protein A-binding activity of monoclonal reagents: After the monoclonal antibody has been coupled to the protein A–Sepharose and the adsorbent has been washed once with PBS, 50 μl of antigen (or mixture containing the antigen) are added to the column, blown into the adsorbent, and incubated for 60 min at room temperature. After filling with PBS and washing as before, the column assembly is placed into a fresh LP3 tube and again topped up with PBS; the tubes are then centrifuged at 800 g for 5 min, effectively drying the immunoadsorbent. Elution and 2-D analysis of the bound antigens are carried out as before. An example of a monoclonal antibody immunoadsorbent-purified antigen is shown in Fig. 10. A monoclonal antibody specific for human serum albumin was used to bind albumin from human serum. Only the albumin (and light chains from the monoclonal antibody) are eluted from the column, showing the exquisite specificity of the method.

Immunoadsorption Using Staphylococcus aureus

If a monoclonal antibody binds to protein A from *Staphylococcus aureus,* then these organisms can be used as a solid-phase immunoadsorbent for isolation of specific antigens from mixtures. Usually the monoclonal antibody is incubated with an appropriately solubilized mixture of antigens internally labeled with [^{35}S]methionine or labeled with iodine-125, and the antibodies (and bound antigen) are isolated by adsorption to the *Staphylococcus* organisms. Bound material is solubilized, run on 2-D gels, and visualized by autoradiography or fluorography. Detailed methods for *Staphylococcus* A immunoadsorption have been developed by

[32] P. Cuatracasas, *J. Biol. Chem.* **245,** 3059 (1970).
[33] R. W. McMaster and A. F. Williams, *Immunol. Rev.* **47,** 117 (1979).

FIG. 10. Two-dimensional gel of the eluate from an HSA1/25.1.3 microimmunoadsorbent column over which human serum had been passed. The upper arrow designates the adsorbed human albumin; the lower arrow, the light chains from the monoclonal antibody coupled to the adsorbent. Presumably these L chains were not covalently linked to the solid-phase adsorbent.

Shapiro,[34] who has modified the basic procedure described by Kessler[35] so that background binding is minimized.

Discussion

Two-dimensional gel electrophoresis is an extremely powerful technique for analysis of individual proteins or mixtures of proteins. Its use-

[34] S. Z. Shapiro and J. R. Young, *J. Biol. Chem.* **256**, 1495 (1981).
[35] S. W. Kessler, *J. Immunol.* **115**, 1617 (1975).

fulness can be augmented by proper running of gels and correct interpretation of gel patterns. By combining the use of monoclonal antibodies with high-resolution 2-D gel analysis, a dissection of complex antigenic mixtures can be achieved to levels not thought possible even a few years ago. Indeed, by using monoclonal antibodies to identify proteins on gel transfers,[36] a "third dimension" may be added to the 2-D gels, potentially allowing a complete molecular anatomy of cells, tissues, and fluids.

[36] N. L. Anderson, S. L. Nance, T. W. Pearson, and N. G. Anderson, *Electrophoresis '82* 3, 135 (1982).

[17] Screening of Monoclonal Immunoglobulins by Immunofixation on Cellulose Acetate

By M. A. Pizzolato

The typing of monoclonal immunoglobulins (Ig) or "M" components has proved to be very important in the diagnosis of the so-called malignant monoclonal gammopathies such as multiple myeloma, Waldenström's macroglobulinemia, and the different types of heavy-chain diseases.

Usually the identification of the heavy-chain class and light-chain type is done by conventional immunoelectrophoresis. Nevertheless, in different circumstances immunoelectrophoretic analysis fails to identify a monoclonal protein and may produce ambiguous results. The use of immunofixation electrophoresis (IFE) described by Alper and Johnson[1] helps overcome this difficulty.

Principle of IFE

IFE involves an initial electrophoretic separation followed by an antigen (Ag)–antibody (Ab) reaction *in situ*. The principle of the method is simple and consists in the fact that the complexes of Ag and precipitating Ab are trapped in gels whereas all the soluble proteins are washed away. IFE is a procedure combining the high resolution of electrophoresis and the specificity and sensitivity of immunoprecipitation to give a method of considerable usefulness for the research laboratory as well as the clinical laboratory.

[1] C. A. Alper and A. M. Johnson, *Vox Sang.* 17, 445 (1969).

METHODS IN ENZYMOLOGY, VOL. 92

IFE reflects simultaneously the electrophoretic mobility as well as the antigenic specificity of a protein with better resolution than other techniques, such as one- or two-dimensional immunoelectrophoresis, diffusion being further minimized.

The principal uses of IFE are the detection of genetic polymorphism of different plasma proteins, the mobility conversion of proteins in the complement and coagulation systems, and the identification of multiple homogeneous bands and small monoclonal proteins.

Most workers have systematically used agarose gel or polyacrylamide gel and acrylamide and isoelectric focusing for high resolution.

Our technique is a modification of IFE, and cellulose acetate is used as a supporting medium.[2] It has several advantages over other methods and has proved to be suitable for rapid screening of monoclonal Ig.

Materials

Buffers: Veronal buffer, 0.05 ionic strength (μ), pH, 8.6. Alternatively Tris-Tricine buffer 0.05 μ, pH 9 or 0.4 M Tris-glycine buffer [tris(hydroxymethyl)aminomethane; 14.1 g; glycine, 22.6 g/liter] pH, 8.7, may be used.

Samples: Serum, plasma, urine, cerebrospinal fluid (CSF), isolated fractions, or other biological fluids.

Cellulose acetate strips: Different brands can be used: either dry, such as Beckman, Gelman, Helena, Sartorius; or gelatinized, such as Cellogel.

Antisera: Monospecific rabbit or preferably goat antiserum (Atlantic Antibodies, Scarborough 04074, Maine; Kallestad, Chaska 55318, Minnesota; Dako, Copenhagen, 2.000 Denmark; Behringwerke, Marburg-Lahn, Germany).

Volumetric distributor, DC/6 Chemetron (Milan)

Electrophoresis tank, Shandon Model U₇₇ after Kohn or similar for standard protein electrophoresis on cellulose acetate

Power source: A power supply with an output capacity of 300 V direct current and 60 mA.

Procedure

Sample Dilutions. Serum is diluted to give a monoclonal protein concentration between 0.3 and 1 g/liter. Depending on the antiserum titer used and the number and concentration of the bands present in the sam-

---

[2] M. A. Pizzolato, F. R. Goñi, and R. C. Salvarezza, *J. Immunol. Methods* **26**, 365 (1979).

ple, resolution may be improved by changing the concentration of Ag and/or Ab.

Sample Application. The strips of cellulose acetate (5.7 × 14 cm) are dipped in the buffer for 10 min. They are lightly blotted between two sheets of filter paper so that no excess moisture remains and are stretched across a removable bridge of Perspex (8.5 cm) and put into the electrophoresis tank.

The samples are applied on the strip by means of a multisampler applicator at 1 cm from the cathodic edge. Bromophenol blue may be used in the sample as a marker of the albumin front.

Electrophoresis. After sample deposition, electrophoresis is performed by applying 250 V (about 1 mA/cm width) for 15 min. In some cases, in order to improve resolution the migration should be prolonged. Electrophoresis is done at room temperature and in case a serum contains a cryoprotein the migration can be done at higher temperatures. Unlike the agarose gel method, no cooling system is needed.

Immunofixation. After electrophoresis, the immunofixation step is done by spreading undiluted monovalent antiserum of high titer (preferably of nephelometric grade) over the cellulose acetate strip from the cathodic end to the albumin front by means of a volumetric distributor that is moved backward and forward until the antiserum is completely absorbed.[3,4] Alternatively, a few drops of antiserum are placed on a glass plate, and the strip is pulled over it followed by incubation with the strip resting on glass as suggested by Kohn.[5] Other authors using agarose gel have proposed several methods for immunofixation, such as (*a*) layering of the antiserum over the surface of the gel[1]; (*b*) the use of filter paper or cellulose acetate premoistened with the antiserum overlay on the surface of the gel[6,7]; and (*c*) the use of a camel's hair brush in order "to paint" with the antiserum the surface of the gel[7]; however, the amount of antiserum required in all these cases is much higher.

Very weak precipitin bands may be accentuated by adding polyethylene glycol (M_r, 6000–20,000) to the antiserum at a final concentration of 2–4%.

Incubation, Washing, Staining, and Destaining. After the antiserum is spread, the strips are incubated in a moist chamber for 5 min at room temperature. Nonprecipitated proteins are removed from the strips by

[3] M. A. Pizzolato, *Clin. Chim. Acta* **45**, 207 (1973).
[4] M. C. B. Pizzolato, M. A. Pizzolato, and A. Agostoni, *Clin. Chem.* **18**, 237 (1972).
[5] J. Kohn, The Royal Marsden Hospital, London, England. Personal communication, 1980.
[6] R. F. Ritchie and R. Smith, *Clin. Chem.* **22**, 1982 (1976).
[7] L. P. Cawley, B. J. Minard, W. W. Tourtellote, B. I. Ma, and C. Chelle, *Clin. Chem.* **22**, 1262 (1976).

continuous shaking in saline for at least 15 min. The strips are then placed in a staining solution of Coomassie Brilliant Blue R 250 2.5 g/liter in methanol–water–acetic acid, 5:5:1 by volume, for about 10 min. Methanol–water–acetic acid, 5:5:1 by volume is used as destainer.

For a permanent record the strips should be dipped in ethanol–water (95:5, v/v) for 30 sec and in ethanol–acetic acid (72:25, v/v) for 1 min and then placed on glass plates and warmed at 50°–60° until they become completely transparent.

Results and Their Interpretation

Figure 1 demonstrates the effectiveness of IFE with various dilutions of an IgG κ monoclonal protein. The optimal Ag : Ab ratio is shown in Fig. 1B$_1$ (400 mg of monoclonal protein per liter), while no visible band is shown by conventional electrophoresis at the same concentration (Fig. 1A). We can also see other kinds of precipitation bands resulting from different incubation temperatures and times. Figures 1C and 1F show the pattern of IFE at 37°, where there is an accelerated dissolution of the complexes because of a faster diffusion. Excess Ag, high temperatures, or long periods of incubation cause dissolution resulting in fixed bands with unstained centers and staining at the margins (see Fig. 1C and, mainly, 1F).

Using different temperatures and times of incubation, we have determined that the best conditions with our system are 5 min at room temperature, between 20° and 25° (see Fig. 1B).

Correct dilution, temperature, and incubation times are important, particularly in the presence of oligoclonal proteins, which can make interpretation difficult because of their apparent restricted electrophoretic heterogeneity on IFE.

Regarding the antiserum, in addition to specificity, the titer and avidity are very important in obtaining sharply defined precipitation bands. In our experience, the use of goat antiserum is recommended in order to minimize Ag-Ab dissolution.

The optimal concentration for the individual proteins is about 100 ng/μl (20 ng in approximately 0.2 μl of sample; see Fig. 1B$_2$). This sensitivity agrees with the values reported by other authors using different techniques.[6–8]

The technique is particularly suitable for characterization of double or multiple components, mainly if they have similar electrophoretic mobilities or low concentrations, such as shown in Figs. 2 and 3, where a sharp

[8] C. H. Chang and N. R. Inglis, *Clin. Chim. Acta* **65**, 91 (1975).

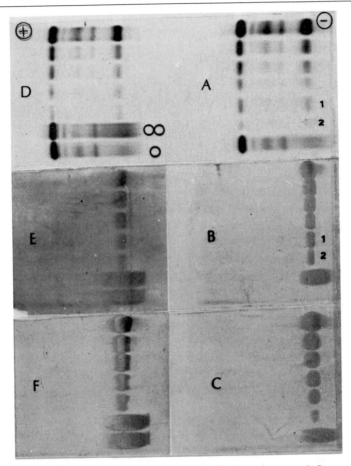

FIG. 1. (A and D) Electrophoretogram of several dilutions of a serum IgG κ monoclonal gammopathy. o, normal serum; oo, polyclonal gammopathy. (B,C and E,F) IFE on cellulose acetate of the same samples as A and D, respectively, with monovalent IgG antiserum. Incubation: (B) 5 min at room temperature; (C) 5 min at 37°; (E) 15 min at room temperature; (F) 15 min at 37°. From Pizzolato et al.[2]

resolution of the double and multiple band, respectively, can be seen. They cannot be satisfactorily resolved by conventional immunoelectrophoresis requiring the use of two-dimensional immunoelectrophoresis,[9] transfer immunoelectrophoresis,[10] or protein isolation. All these pro-

[9] C. B. Laurell, Anal. Biochem. 10, 358 (1965).
[10] J. Kohn, in "Methods in Immunology and Immunochemistry" (C. A. Williams and M. W. Chase, eds.), Vol. III, p. 275. Academic Press, New York, 1971.

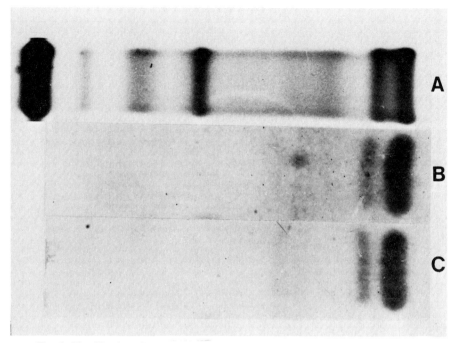

FIG. 2. Identification of a double IgG κ monoclonal component. (A) The electrophoretogram shows two "M" components of 12 g/liter and 21 g/liter, respectively. (B) IFE with anti-IgG; serum dilution 1:20. (C) IFE with anti-κ; serum dilution 1:20.

FIG. 3. Multiple monoclonal light chains (Bence-Jones protein type λ) in the urine of a myeloma patient. (A) Electrophoretogram staining with Amido Black 10B. (B) IFE with monovalent λ antiserum.

cedures are laborious and time consuming and require large amounts of material.

In comparison with conventional immunoelectrophoresis, IFE is simpler and faster and does not require a special degree of expertise. Since electrophoresis for immunofixation is done exactly like routine staining, bands made visible by both methods are easily compared.

Application. The most important uses of the technique are (*a*) characterizing monoclonal immunoglobulins when their concentration is below the sensitivity of immunoelectrophoresis, most of all when high concentrations of normal heterogeneous immunoglobulins occur; (*b*) localizing monoclonal proteins "hidden" by other serum proteins (monoclonal Ig masked by normal α_2 or β regions); (*c*) light-chain typing of monoclonal IgM or IgA, which is difficult to type by immunoelectrophoresis even after reduction with mercaptoethanol or dithiothreitol; (*d*) recognizing oligoclonal Ig; and (*e*) determining biclonal or multiple homogeneous bands.

In this laboratory IFE has also proved to be useful in identifying complexes between various proteins, such as IgM rheumatoid factor–IgG complexes and IgM–β-lipoprotein complexes (unpublished data).

Another practical extension of the method is the reverse immunofixation by which functional identification of Ab is possible by the use of appropriate antigens for fixation.

We have combined immunofixation with thin-layer gel filtration[11] in order to determine simultaneously the immunological identity plus the molecular weight of monoclonal Ig, particularly for Ig fragments, without their previous isolation.

Advantages. With reference to other IFE techniques using agarose gel, IFE on cellulose acetate gives similar resolution and has the following advantages: smaller quantities of reactants are used (2.5 μl of antiserum per sample); and the application of the antiserum by cellulose acetate or filter paper strips is avoided, thus reducing the risk of losing immunocomplexes. Furthermore, the time required for the different steps—electrophoresis, incubation, washing, staining, and destaining—is considerably reduced. The typing of a monoclonal Ig by this technique can be performed in only 60 min. Cellulose acetate is thinner than agarose gel, so smaller quantities of antiserum are required, and the Ag-Ab reaction is practically instantaneous because Ab diffusion is negligible. One possible disadvantage with this method is that less immune complexes can be washed in the soaking phase, especially when critical parameters ("M"

---

[11] M. A. Pizzolato and F. R. Goñi, *Int. Congr. Clin. Chem. 11th, 1981; J. Clin. Chem. Clin. Biochem. (Abstr.)* **19**, 802 (1981).

component concentration, temperature, and incubation time) are not in the correct range.

While the method is very sensitive, the sensitivity can be further amplified by radioisotope-labeled Ab (we have tried successfully the use of ^{125}I-labeled protein A as a second reactant in an indirect method) followed by autoradiography, by fluorescent-labeled Ab, or by conjugates of alkaline phosphatase or horseradish peroxidase.[1,7]

In total agreement with Ritchie,[6] we consider that the simplicity and economy of the method and the clarity of final results suggest that IFE should be considered as a possible successor to clinical immunoelectrophoresis for the study of monoclonal proteins. In our experience, IFE on cellulose acetate is the method of choice for routine screening of monoclonal Ig.

Acknowledgment

I am grateful to Elsevier/North-Holland Biomedical Press for permission to reproduce illustrations and quotations from previously published work.

[18] Solid-Phase Immunofluorescence Assay (SIFA) for Detection and Characterization of Monoclonal Antibodies against Soluble Antigens

By BURKHARD MICHEEL, HELMAR FIEBACH, and UWE KARSTEN

For the production of monoclonal antibodies, the most important step is the selection of sometimes only a few positive clones out of a vast majority of negative hybridomas. It is, therefore, necessary to have at one's disposal a serological assay that allows the screening of several hundred microtest plate wells per day. The general principle of such assays is to immobilize one of the reacting components (antigen or anti-mouse immunoglobulin) on a solid phase because this facilitates the washing steps to remove excess nonbound antibody and label. In most cases radioimmunoassays or enzyme immunoassays are used for screening antigen-specific hybridomas.[1]

Compared with radioisotopes and enzymes, fluorescein isothiocyanate (FITC), the most frequently used fluorescence marker, has several

---

[1] R. H. Kennett, T. J. McKean, and K. B. Bechtol, eds., "Monoclonal Antibodies, Hybridomas: A New Dimension in Biological Analysis." Plenum, New York, 1980.

advantages. It is of no potential hazard, very cheap, and very easy to conjugate to proteins. Fluoroimmunoassays are, therefore, used in different immunological laboratories as an alternative to the methods mentioned above.[2]

This chapter describes a simple solid-phase immunofluorescence assay (SIFA)[3] that is based on the adsorption of anti-immunoglobulins (anti-Ig) or antigens to cellulose nitrate membrane filters and is evaluated by using a microfluorometer.

Experimental Procedures

Hybridoma Technique

The production of monoclonal antibodies against soluble antigens is performed according to a previous publication.[3]

Materials for the Performance of SIFA

Requisites for solid-phase immunofluorescence assay (SIFA) are listed below.

Materials

Microtitration plates
Nylon gauze
Thin rubber sheet with 96 perforations
Plastic plate with 96 perforations
Clamps
Cellulose nitrate membrane filters (SM 113 07, Sartorius GmbH)
Punch for cutting 2-mm or 0.7-mm disks
Measuring plate with 96 spots
Thin plastic foil

Reagents

Phosphate-buffered saline (PBS, neutral pH) as diluent for immobilization of proteins on cellulose nitrate
Culture medium or PBS containing 10% fetal calf serum as diluent for subsequent steps
Anti-mouse Ig (IgG fraction or isolated antibodies)
Antigen

[2] E. Soini and E. Hemmilä, *Clin. Chem.* **25**, 353 (1979).
[3] B. Micheel, U. Karsten, and H. Fiebach, *J. Immunol. Methods* **46**, 41 (1981).

FITC-labeled antigen or FITC-labeled anti-mouse Ig (protein A, FITC-labeled protein A)

Equipment

Microfluorometer (fluorescence microscope with photometry device)

All incubations are done in Takatsy microtitration plates (LMM Laboratory Instruments Factory, Esztergom, Hungary) at room temperature under continuous shaking; 1 hr is chosen as general incubation time for each step. To avoid drying of the samples, the plates are covered with a plastic foil moistened on the margin.

Cellulose nitrate (SM 113 membrane filters, Sartorius GmbH, Göttingen, F.R.G.) is used as the solid phase. It shows no autofluorescence and passively adsorbs a constant quantity of protein per unit area. For most assays, disks 2 mm in diameter are cut from the filters by means of a revolving punch. In some of the later experiments disks 0.7 mm in diameter have been used, which had been cut out by a punch made from syringe needles.

For washing after each incubation, the microtitration plates are covered with nylon gauze, a thin sheet of rubber with 96 perforations aligned with the wells of the microtitration plates, and a plastic plate with 96 perforations. The plates are fixed together by clamps, and the cellulose nitrate disks are washed under a strong jet of tap water. Before adding new reagents, the water is aspirated with a water jet pump.

Microfluorometry

After finishing the assay and without removing the water after the last washing, the disks are transferred to a measuring plate. This consists of a plastic plate onto which a thin plastic foil with 96 corresponding perforations is glued. The transfer is done with a Labpipette fitted with a wide-aperture tip. The measuring plate is covered with a thin plastic foil. The other method of transferring is to put the measuring plate onto the incubation plate and turn both upside down so that the disks sediment into the corresponding spots of the measuring plate. A thin plastic foil is put between the plates, after which the incubation plate can be removed.[4]

The fluorescence intensity of the disks is then measured with a microfluorometer. This consists of a Fluoval 2 fluorescence microscope with incident illumination and an HBO 202 mercury lamp (VEB Carl Zeiss Jena, Jena, G.D.R.) combined with an MPV photometry device (Leitz GmbH, Wetzlar, F.R.G.). The following excitation filters are used:

[4] B. Micheel, U. Karsten, and H. Fiebach, *Acta Histochem.* **71**, 15 (1982).

W 301, B 229g (BG 23), B 223g (BG 12), 2× KP 490, and G 252g (GG 15), the emission filter is a G 247 (OG 4). A 16×/0.40 objective and a 3.2× eyepiece are used for measuring. The photomultiplier voltage is adjusted to 1.0 kV. Since the variation between different areas of one disk is minimal, only one square area of 6400 μm^2 per disk is measured. The microfluorometer is equipped with a handmade small set of pulleys that allow the simultaneous movement of the different shutters during the measuring process. In some experiments the disks have been measured directly in the wells by using a 3.2× objective.

Antigen–Antibody Combinations in SIFA (Table I)

In most experiments an antigen-labeled SIFA was performed, which required a very small quantity of antigen.[3] For this assay the IgG fraction

TABLE I

INCUBATION STEPS OF DIFFERENT SOLID-PHASE IMMUNOFLUORESCENCE ASSAY
(SIFA) COMBINATIONS

Step	Antigen-labeled SIFA	Volume	Anti-Ig-labeled SIFA	Volume	Time[a]
First incubation	Anti-mouse Ig (IgG fraction or isolated antibodies)	10 μl^b (1 mg/ml)	Antigen	1 μl^c (1 mg/ml)	1 hr
Washing					5 min
Saturation	Not necessary		Normal goat serum	10 μl (diluted 10^{-1})	30 min
Washing					5 min
Second incubation	Serum, ascites, or hybridoma culture fluid		As for antigen-labeled SIFA	5–10 μl	1 hr
Washing					5 min
Third incubation	FITC-labeled antigen	10 μl (2–10 $\mu g/ml$)	FITC-labeled anti-mouse Ig	5–10 μl (200 $\mu g/ml$)	1 hr
Washing					5 min
Transfer of membrane filter disks to measuring plate					20–30 min[a]
Microfluorometric evaluation					20–30 min

[a] For one plate.
[b] Tests with 2-mm membrane filter disks.
[c] Tests with 0.7-mm disks, antigen quantity sufficient for 10 disks.
[d] For simultaneous transfer, 5 min.

or the isolated antibodies of an anti-mouse Ig serum (10 μl containing 1 mg of protein per milliliter of phosphate-buffered saline) were passively adsorbed to the cellulose nitrate disks in the first incubation step. The next incubation was done with 10 μl of mouse serum or ascites fluid (diluted in medium containing 10% fetal calf serum) or the culture fluids to be tested for the presence of antibodies. In the last step the disks were incubated with 10 μl of the FITC-labeled antigen (2–10 μg of antigen per milliliter of medium containing 10% fetal calf serum; i.e., 10–100 ng per sample).

For an anti-Ig-labeled SIFA the disks were incubated first with the antigen solution (10 μl containing 1 mg of protein per milliliter) and then with 10^{-1} diluted normal goat serum to saturate remaining free sites on the cellulose nitrate surface. In the next step the disks were incubated with the samples to be tested and finally with FITC-labeled goat anti-mouse Ig (in most cases purified antibodies, 200 μg of labeled antibodies per milliliter).

It was found that the same results could be obtained if in the first step 1 μl of the antigen solution containing 1 μg of protein was added to each 2-mm disk, after drying, the next incubation steps were performed. By using 0.7-mm disks the same quantity of antigen was sufficient for 10 disks (i.e., 100 ng of antigen per sample). For the following incubations of the 0.7-mm disks, generally only 5 μl of solutions were used. In both SIFA combinations protein A has also been used. With protein A-reactive mouse antibodies, such assays showed similar results as the tests with anti-mouse immunoglobulins.

In the first experiments FITC-conjugation to proteins was carried out according to the general method using carbonate–bicarbonate buffer.[5] For most of the proteins the method of Gani et al.[6] using Na_3PO_4 to stabilize the alkaline pH resulted in our hands in a more effective labeling. This method was therefore generally used for our later experiments after adapting it to be applicable for protein concentrations lower than 1 mg/ml.

Solid-Phase Radioimmunoassay (SRIA)

For comparison a solid-phase radioimmunoassay was performed using poly(vinyl chloride) plates (pill blisters) as solid phase.[3] The same combinations were set up as for SIFA. The incubation volumes were 50–100 μl. In the antigen-labeled SRIA, 1–10 ng of ^{125}I-labeled antigen were used; in the anti-Ig-labeled SRIA, 250 ng of antigen were fixed to the solid phase by air-drying.

---

[5] L. Hudson and F. C. Hay, "Practical Immunology." Blackwell, Oxford, 1976.
[6] M. M. Gani, T. Hunt, and J. M. Summerell, *J. Immunol. Methods* **34,** 133 (1980).

Results and Discussion

For an efficient SIFA there must be a relatively high concentration of fluorescing molecules at the area that is measured with the microfluorometer. This is important because fluorescence signals have to be measured without the possibility of enhancing the values by a prolonged measuring time as in radioimmunoassays or by a secondary amplifying reaction as in enzyme immunoassays.

Cellulose nitrate membrane filters are well suited for this purpose. In a comparative study, the binding capacities of different cellulose nitrate filters were tested in SIFA (Table II). The greatest quantity of protein molecules per unit area is accumulated in SM 113 07 filters. The pore width of 0.2 μm allows probably all the proteins, including immunoglobulins, to penetrate and to be fixed onto the relatively large inner surface of this filter type. This conclusion could also be drawn from the results of radioimmunoassays. The values obtained with 2 mm-cellulose nitrate disks were comparable to those of assays with poly(vinyl chloride) as solid phase (the wells were about 8 mm in diameter). In addition the quantity of protein bound per unit area of these plates was not detectable with a microfluorometer. The adsorption of proteins to the cellulose nitrate is strongly dependent on concentration. Protein concentrations lower than 1 mg/ml (anti-Ig, IgG fraction, or isolated antibodies, or antigen) resulted in very low values in SIFA, especially in tests with monoclonal antibodies.

SIFA was used in our laboratory for the selection of monoclonal antibodies to bovine serum albumin (BSA)[7] and human serum albumin (HSA)[3] and in combination with radioimmunoassays for the selection and characterization of monoclonal antibodies to human α-fetoprotein (AFP; experiments carried out in cooperation with G. I. Abelev's group, All-Union Scientific Oncological Center, Moscow, USSR).[8]

In all tests positive and negative samples were assayed as controls. A diluted mouse immune serum or a monoclonal antibody of known specificity and titer served as positive control; diluted normal mouse sera and medium containing 10% fetal calf serum served as negative controls. Since for the same test performed on different days a variation in the fluorescence values between 15 and 20% was evident, the threshold for regarding a value as positive was calculated for each experiment according to the results of the controls.

[7] B. Micheel, U. Karsten, and A. Kössler, unpublished results, 1980.
[8] B. Micheel, H. Fiebach, U. Karsten, A. I. Gussev, A. K. Jasova, and F. Kopp, *Eur. J. Cancer Clin. Oncol.* (submitted).

TABLE II
EFFECTIVITY OF DIFFERENT CELLULOSE NITRATE MEMBRANE FILTER TYPES FOR USE IN
SOLID-PHASE IMMUNOFLUORESCENCE ASSAY (SIFA)

Filter type	SM 113 03	SM 113 05	SM 113 06	SM 113 07	SM 113 09	SM 113 10	SM 113 11
Pore width (μm)	1.2	0.65	0.45	0.2	0.1	0.05	0.01
Fluorescence intensity[a]	700	400	1300	2100	300	0	0

[a] Results are in arbitrary fluorescence units (FU); antigen-labeled SIFA with goat anti-mouse Ig (IgG fraction) immobilized on the solid phase, mouse anti-AFP serum (diluted 10^{-1}), and FITC-labeled AFP. The results indicate the differences between the values of the samples with the antiserum and those of the samples with normal mouse serum (diluted 10^{-1}).

For all systems both the antigen- and the anti-Ig-labeled SIFA variations showed good agreement and had a sensitivity that allowed the detection of monoclonal antibodies in the hybridoma culture fluid 2 weeks after fusion. The same culture fluids were in general found to be positive when a direct comparison was performed between antigen- or anti-Ig-labeled SIFA. An important advantage of the antigen-labeled SIFA was that no nonspecific staining was observed even when a high concentration of unrelated mouse Ig was used for the second incubation step (e.g., normal mouse serum diluted 10^{-1}).[3] The anti-Ig-labeled SIFA showed, however, a considerable nonspecific reaction with normal mouse serum (up to dilution 10^{-2}) provided no saturation with a nonrelated serum was performed after the antigen immobilization. After such a saturation step, the normal mouse serum up to dilution 10^{-1} showed only a low nonspecific staining. This was of no importance, since the concentration of mouse Ig in the hybridoma culture fluid was much lower. A comparison of the reactivity of culture fluids with anti-HSA and anti-AFP activity showed a reaction only with the corresponding antigen.

The sensitivity (derived from titration of antisera and hybridoma ascites fluids) was comparable for the antigen- and anti-Ig-labeled SIFA (Figs. 1 and 2). It is evident that in the antigen-labeled SIFA the fluorescence values for the samples with the monoclonal antibodies were as high as those with the conventional polyvalent antiserum. In the anti-Ig-labeled SIFA, the values for the polyvalent antiserum were in general higher than for the monoclonal antibodies. Since similar results were obtained in SRIA, this seems to be a general phenomenon (Figs. 1 and 2).

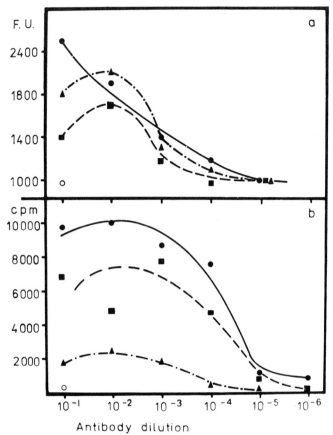

Fig. 1. Reaction of BALB/c anti-AFP serum and monoclonal anti-AFP ascites fluid (Mo anti-AFP) in antigen-labeled solid-phase immunofluorescence assay (a) and radioimmunoassay (b). Results are given in arbitrary fluorescence units (FU) and counts per minute (cpm). ●——●, BALB/c anti-AFP serum; ■———■, Mo anti-AFP Hy 34/38; ▲—·—▲, Mo anti-AFP Hy 34/41; ○, fetal calf serum.

When the sensitivities of SIFA and SRIA were compared for the HSA system, the SRIA was found to be only slightly more sensitive.[3] For the AFP system the antigen-labeled SIFA was at least one decimal exponent, for some antibodies even two decimal exponents, less sensitive than the antigen-labeled SRIA (Figs. 1 and 2). Hybridoma culture fluids could be diluted only 10^{-1} to 10^{-2} to give a positive reaction in SIFA, whereas in SRIA some culture fluids still reacted when diluted 10^{-3} to 10^{-4}. Since this effect is more pronounced for the AFP system than for the HSA system,

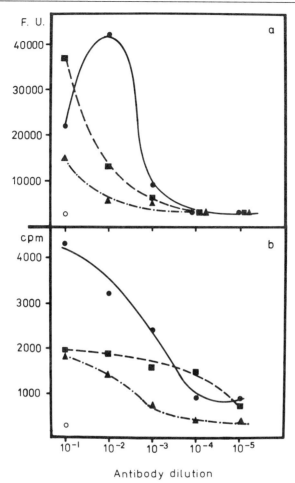

Antibody dilution

FIG. 2. Reaction of BALB/c anti-AFP serum and monoclonal anti-AFP ascites fluid in anti-Ig-labeled solid-phase immunofluorescence assay (a) and radioimmunoassay (b). For symbols see Fig. 1.

AFP seems to be better for labeling with iodine than HSA, whereas HSA has a higher labeling efficiency with FITC than does AFP.

The marker that is conjugated to the antigen is obviously also of importance for the detection of monoclonal antibodies in an antigen-labeled assay. This was concluded from comparative studies with antigen-labeled SIFA and SRIA. As shown in Fig. 1, anti-AFP hybridoma ascites fluid Hy 34/41 showed only a weak reaction in SRIA but a strong reaction in SIFA (the same effect was found with culture fluid). Since another anit-AFP clone showed the same reaction pattern and a similar clone was found for the HSA system,[3,9] it was concluded that these antibodies detected an epitope that contains tyrosine as an integral part. Further experiments must be carried out to prove this assumption.

When protein A was used in our assays, our monoclonal antibodies showed only a very weak or negative reaction. Reasonable results were obtained, however, by performing an additional incubation with diluted rabbit anti-mouse Ig serum.[4] It should, therefore, be possible to develop a general assay using FITC-labeled protein A or protein A immobilized on the cellulose nitrate. Experiments are in progress to standardize such an assay.

Conclusions

Solid-phase immunofluorescence assays (SIFA) using cellulose nitrate filters with a pore width of 0.2 μm are well suited to select and characterize monoclonal antibodies against soluble antigens. Different test combinations are possible with FITC-labeled antigens or FITC-labeled anti-mouse Ig. When performing assays with protein A it is advisable to include an additional incubation with rabbit anti-mouse Ig to avoid missing monoclonal antibodies of a class or subclass that does not react with protein A. SIFA can be performed with a relatively small quantity of antigen (20–100 ng per sample). On the other hand, especially for the anti-Ig-labeled SIFA, a protein concentration of 1 mg/ml is required for adsorption to the solid phase. Very small cellulose nitrate filter disks 0.7 mm in diameter can be used, since they still allow convenient handling during incubation and measurement.

Measurement with the microfluorometer offers several possibilities for automation.

When antibody titration is taken as a parameter for comparison, SIFA is in general at least one decimal exponent less sensitive than solid-phase radioimmunoassay.

Because of the high binding capacity of cellulose nitrate membrane filters, they could also be an excellent solid phase for radioimmunoassays and enzyme immunoassays. For some immunoassays this material has already proved its value.[10,11]

[9] B. Micheel, H. Fiebach, J. Kopp, and U. Karsten, *Acta Biol. Med. Ger.* **41,** 275 (1982).
[10] R. Wang, B. Merrill, and E. T. Maggio, *Clin. Chim Acta* **102,** 169 (1980).
[11] J. Sharon, S. L. Morrison, and E. A. Kabat, *Proc. Natl. Acad. Sci. U.S.A.* **76,** 1420 (1979).

Acknowledgments

The authors would like to thank Ms. G. Roloff and Ms. S. Käppler for excellent technical assistance, Drs. J. Mohr, M. Wagner, and T. Porstmann for providing goat anti-mouse Ig (IgG fraction), protein A, and AFP; and Dr. J. Denner for help in iodinating the proteins.

[19] Identification and Characterization of Lymphocyte Hybridomas by Electrophoresis of Glucose-6-phosphate Isomerase Isozymes

By THOMAS J. ROGERS and KATHLEEN O'DAY

A great deal of interest has been generated in the use of lymphocyte hybridomas for the production of various lymphocyte products. This technology is quite useful for the production of monoclonal antibody from the fusion of a B lymphocyte and a plasmacytoma cell line. The products of specific T-lymphocyte subsets may also be generated by fusion of the primary lymphocyte with a T-lymphocyte cell line. Care must be taken, in any case, that the lymphocyte products obtained are normally expressed by the primary cell type and are not an exclusive tumor cell product or an artifact of the hybridized cell state.

Most tumor cell lines employed for the generation of hybridomas are hypoxanthine–guanine–phosphoribosyltransferase defective. After fusion with the primary lymphoid cell, tissue culture medium containing hypoxanthine, aminopterin, and thymidine is used to select against tumor cell–tumor cell hybrids. Primary cell–primary cell hybrids tend not to proliferate after the initial 4–5 days of culture. The highest frequencies of hybridization have been achieved in our laboratory with primed lymphocytes, but experimentally "naive" lymphocytes can also be used as a source of primary cells.

We have maintained hybridomas generated with primary splenic lymphocytes for better than 18 months, although the average lifespan is between 4 and 6 months. The accumulation of cells bearing only a partial complement of both parental cells is of particular importance. Several methods are currently in use for the analysis and detection of hybridomas. The detection of surface markers expressed by both genotypes (e.g., Thy-1 and H-2) is very popular and represents a functional characteristic of a measurable percentage of the hybridoma culture. We commonly employ, and report here, the analysis of hybridomas by identification of both parental isotypes of glucosephosphate isomerase (GPI; EC 5.3.1.9). The technique is useful only when the tumor and lymphocyte parents express different isozyme types. Another limitation to any of these techniques is that they typically depend on the gene function expressed on only a single chromosome. It is possible, however, to employ this simple, inexpensive, and rapid technique as only a first step in a longer and more elaborate analysis for the detection of "complete" hybridomas.

This method of GPI analysis employs nondenaturing polyacrylamide gel electrophoresis followed by *in situ* enzyme detection with an indirect stain. The technique takes a total of 4–5 hr and requires very little enzyme, and consequently few cells.

Procedures

Extraction of the Hybridoma Culture. Cells must be in log phase at the time of analysis. Tissue culture media containing serum may be used, since all growth medium is removed before extraction. Cells are spun at 750–2000 g for 5–10 min. The cellular pellet is transferred to an Eppendorf microcentrifuge tube with 0.2–0.5 ml of saline. The cells are spun in a Beckman Microfuge B (8740 g) for 1 min with 0.8 ml of saline. The saline is removed, and the pelleted cells are vortexed in 1 ml of distilled water for 15–30 sec. The cells are further treated by repeated rapid freezing (in a liquid nitrogen bath) and thawing (in a 37° water bath). Cellular debris is pelleted by centrifugation (8740 g), and the cell lysate is transferred to a second centrifuge tube for further analysis. The cellular extract may be used immediately or stored at −20° for use at a later time (enzyme activity is stable up to about 1 week at −20°).

Electrophoresis of Glucosephosphate Isomerase Isozymes. The gel composition is 5.5% acrylamide (30:1 acryl-bis); 0.01 M phosphate, pH 7.0; 0.04% tetramethylethylenediamine (TEMED); and 0.04% ammonium persufate. The gel is cast by first mixing 0.920 ml of 30% acrylamide (Bio-Rad 161-0100), 0.920 ml of 1% bisacrylamide (Bio-Rad 161-6200), 0.250 ml of 0.2 M phosphate buffer (pH 7.0), and 2.91 ml of H_2O. At this point the solution is briefly degassed, and then 2 μl of TEMED (Bio-Rad 161-0800) and 20 μl of 10% ammonium persulfate (Bio-Rad 161-0700) are added. The gel should polymerize in 40 min to 1 hr. We commonly employ a "minislab" electrophoresis apparatus (Idea Scientific Company, Corvallis, Oregon), and the polymerized gel typically measures 10 × 8 × 0.05 cm. The small, thin gel size has several advantages over more conventional gels. It provides excellent resolution in a short time period; it allows very small quantities of a stained substance to be detected; and the demands for time and expensive materials are greatly reduced.

In order to perform the electrophoresis, the thawed (or fresh) cell extract is diluted in water to obtain a sample strength of at least 8×10^5 cells per milliliter, or 800 cells per microliter. Cell cultures weakly expressing isomerase activity, and cultures taken prior to or after log phase growth, may require a greater concentration. The mini-gel apparatus requires a sample size of 2–5 μl. Strong tumor cell cultures typically yield 0.1–0.2 unit of isomerase activity (as measured by the method of Roe[1]) per 10^6 cells, and hybridoma cultures typically yield 0.07–0.1 unit per 10^6 cells. A unit of enzyme is that amount which catalyzes the formation of 1 μmol of fructose 6-phosphate per minute.[2]

After dilution of the cellular extract in water, the extract is further diluted (1 : 2) in sample buffer. Sample buffer is composed of 0.1 M phosphate (pH 7.0), 50% glycerol, and 15 μM bromophenol blue. The bromophenol blue is not useful as a tracking dye in this application, since it travels much faster than the isomerase. We employ the dye to visualize the sample as it is loaded into the gel wells and also as a check that the electrodes are properly placed during the electrophoresis (the electrophoresis is performed "negative-to-positive").

The ideal sample concentration is 200–250 microunits in a volume of 2–5 μl. In order to obtain 200 microunits in 5 μl, there must be 0.04 unit per milliliter of sample buffer (or 0.08 unit per milliliter of cellular extract). If the cell culture yields 0.1 unit per 10^6 cells, the cell extract must be prepared from at least 8×10^5 cells per milliliter. We typically perform the extraction in small volumes (10–100 μl), and this reduces the demand for cells to very low levels.

The electrophoresis running buffer is 0.2 M phosphate, pH 7.1. The electrophoresis is performed at 4°, since the apparatus generates heat very rapidly. The rapid form of the isomerase migrates about 1.5–2.0 cm in 4 hr at 100 V and 35 mA, or in 3 hr at 160 V and 35 mA.

In Situ Stain of Glucosephosphate Isomerase Isozyme Activity. The isomerase activity is detected by providing the necessary components for Reactions (1)–(3).

$$\text{Fructose-6-P} \xrightarrow{\text{glucose-P-isomerase}} \text{glucose-6-P} \qquad (1)$$

$$\text{Glucose-6-P} \xrightarrow[\text{NADP}^+ \quad \text{NADPH + H}^+]{\text{Glu-6-P-dehydrogenase}} \text{glucose-P-}\delta\text{-lactone} \qquad (2)$$

$$\text{NADPH}^+ + \text{MTT} \xrightarrow{\text{phenazine methosulfate}} \text{NADP}^+ + \text{MTT formazan} \downarrow$$
$$\text{(colored)} \qquad (3)$$

[1] J. Roe, *J. Biol. Chem.* **107**, 15 (1934).
[2] T. J. Rogers and K. O'Day, *Anal. Biochem.* **116**, 389 (1981).

Our procedure for detecting isomerase activity is a modification of a technique originally described by DeLorenzo and Ruddle.[3] Five milligrams of NADP⁺ (Sigma N-0505), 40 mg of MgCl₂, 0.5 mg of phenazine methosulfate (PMS; Sigma P-9625), 5 mg of 3-(4,5-dimethylthiazol-2-yl)-2,5-diphenyltetrazolium bromide (MTT; Sigma M-2128), 5 μl of glucose-6-phosphate dehydrogenase (320–400 units ml⁻¹; Sigma G-6378), and 80 mg of fructose 6-phosphate (Sigma F-1502) are dissolved in 50 ml of 0.1 M Tris-HCl containing 1–1.5% agarose (Bethesda Research Laboratories, Gaithersburg, Maryland) at 45°. The PMS and MTT must be handled in the dark (or quickly in a room with reduced light). We find it most convenient to prepare stock solutions of these and all other solutions in Eppendorf microcentrifuge tubes and store these at −20° in a dark container. It is critical to add the reagents to the 45° agarose solution *immediately* before use, so the heat does not destroy activity. Once the reagents are added, and mixed, the solution is poured into two Microtiter plate lids (Falcon 3041) and allowed to solidify at 4° in a dark chamber. This is referred to as the "stain gel." Gloves are used to remove the stain gel from the Microtiter plate lid when necessary. If only one of the two stain gels is required, it is possible to store the other gel at 4°, in the lid, inside an airtight plastic bag, and in a dark chamber, for at least 24 hr.

The glass plates in the polyacrylamide gel apparatus (after the electrophoresis) are separated, and the polyacrylamide gel typically sticks to one of the plates. This polyacrylamide gel–glass plate sandwich is placed in a covered pan, and the stain gel is layered over the polyacrylamide gel. After incubation at 37° (for at least 1 hr), the isozyme diffuses into the agarose, where the intense blue color develops [from the MTT formazan; Reaction (3)]. Some color development occurs in the polyacrylamide also, but much less, and the resolution of isozyme forms is not significantly better. Both the stain gel and the polyacrylamide gel may be dried over white electrophoresis drying paper (Bio-Rad) and stored indefinitely.

Results and Discussion

Analysis of parental cell (a T lymphoma and C57B1/6J splenocytes) extracts and an extract of hybridoma Hyb-9-SC2 is shown in Fig. 1. The figure shows that about 180 microunits of total enzyme activity are necessary for visualization of the fast, intermediate, and slow forms of isomerase from this hybridoma. Complete resolution of the isozyme forms requires as little as 3 hr of electrophoresis. As little as 83 microunits of T-lymphoma BW 5147 could be detected, and this represented only 6900

[3] R. DeLorenzo and F. Ruddle, *Biochem. Genet.* **3,** 151 (1969).

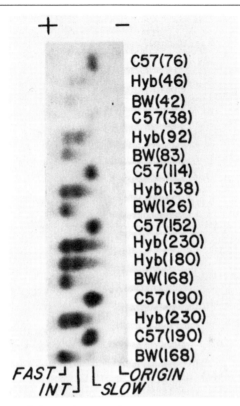

FIG. 1. Photograph of glucosephosphate isomerase zymogram showing isozymes from C57B1/6J, Hyb-9-SC2, and BW 5147 cells. The extracts were diluted so that the amount of enzyme shown in parentheses was present in each channel of the polyacrylamide gel. About 180 microunits of the Hyb-9-SC2 extract is necessary for visualization of the three isozyme types.

cells of this particular culture. The phenotypes of several mouse strains are shown in the table.

We have attempted to adjust the conditions of electrophoresis so that more complete resolution of the isozyme forms is attained. We have found, for example, that 5.5% acrylamide works quite well, but 5%, 6%, and 7.5% do not. Furthermore, a very critical aspect of the electrophoresis is the pH of the running buffer. A pH of 7.1 works well, but buffers at higher and lower pH values do not allow adequate isozyme resolution.

Analysis of numerous putative hybridoma clones can be carried out, from cellular extraction to final stained gel, in less than a day. The current cost of all materials, including the gel apparatus (but not the power supply), is less than $250. We find that this technique works well when

GLUCOSE-6-PHOSPHATE ISOMERASE
ISOZYME PHENOTYPES

Mouse strain	
Slow	Fast
C57[a]	BALB/c[a]
C56[b]	A/J[a]
CBA[b]	AKR[a]
	SJL[c]

[a] Rogers and O'Day.[1]
[b] N. Ruddle, *Curr. Top. Microbiol. Immunol.*
81, 203 (1978).
[c] DeLorenzo and Ruddle.[3]

cloning and subcloning hybridoma lines. The cloning may be carried out in microtiter wells (0.25-ml volumes), and the cell numbers in these wells (10^3 to 10^5) is sufficient for analysis by this technique. This technique should be easily adapted to any polyacrylamide gel apparatus, including the larger slab-gel systems, which are very popular.

Acknowledgments

This work was supported by Grant PCM-8003635 from the National Science Foundation and by PHS Biomedical Research Grant PRO7079. The authors wish to thank Gregory E. Harvey for excellent editorial assistance.

[20] Distinction of Epitopes by Monoclonal Antibodies

By C. STÄHLI, V. MIGGIANO, J. STOCKER, TH. STAEHELIN,
P. HÄRING, and B. TAKÁCS

Hybridoma technology has become very efficient in the production of large numbers of specific monoclonal antibodies (mAbs)[1] (compare this volume [3]). For instance, from five hybridization experiments (H-9 to H-18; this volume [3]) with the protein antigen human chorionic gonadotropin (hCG, M_r 50,000), over 100 stable hybridomas have been obtained.

[1] Abbreviations: mAb, monoclonal antibody; BSA, bovine serum albumin; hCG, human chorionic gonadotropin; hLH, human luteinizing hormone; PBS, phosphate-buffered saline without calcium or magnesium; RT, room temperature; POD, horseradish peroxidase; SABA, solid-phase antibody-binding assay.

Their characterization and the selection of the most useful ones has taken much more work than the effort spent obtaining them. One efficient strategy is to divide them into groups specific for different epitopes (=antigenic determinants) and then to select the optimal one from each group. "Optimal" can mean highest affinity, but it may also mean highest forward reaction rate constant (k_{+1}, e.g., for certain agglutination tests) or a particular antibody isotype, if immunological effector functions (e.g., complement fixation) are desired. To proceed in the opposite order, i.e., selecting first for high affinity and analyzing later for specificity, is less rewarding, as affinity frequently depends on the epitope. Thus, while the best antibodies against the β_4-epitope of hCG (see below) have affinities around $10^{11}\ M^{-1}$, the best against another isotope, β_1, are around 10^9. If the initial selection had been carried out on the basis of affinity, only anti-β_4 antibodies, which also react with hLH (see below), might have been selected for further study.

We describe in detail three main methods for the distinction of specificity of different mAb: competition SABA, binding to epitopes on different molecules, and epitope destruction by iodination.

Competition SABA is a generally applicable method based on the assumption that two mAb can bind to one antigen molecule simultaneously if they bind to different epitopes. Obviously, epitopes are recognized as different by this assay only if they are far enough apart to allow simultaneous mAb binding. Competition SABA is based on the solid-phase antibody binding assay (SABA) described in this volume [3]. Antigen is adsorbed to a surface (a well of a Microtiter plate). Saturating amounts of a first (inhibitory) mAb (mAb-1) are allowed to bind to this antigen. In a second binding reaction, an mAb-2 is allowed to bind to the same antigen. If the two mAbs recognize the same epitope, the bound mAb-1 will prevent mAb-2 from binding. Different methods are discussed by which mAb-2 is measured selectively in the presence of mAb-1. As part of one mAb-2 detection method, a simple technique to determine mAb isotypes is also described. With large numbers of mAbs, competition assays become laborious, as the number of individual competition experiments increases quadratically with the number (n) of mAbs [$n(n - 1)/2$].

Binding to epitopes on different molecules depends on the availability of natural or artificial alternative antigens (e.g., cross-reactive antigens). In SABA against these different antigens, antibodies display patterns of binding that may be characteristic for different epitopes (see below). Differences in maximum binding of ^{125}I-labeled antigen by different mAb can also be epitope-specific characteristics (see below). The choice of method depends on several factors including convenient availability of different reagents.

Competition SABA

Competition SABA consists of the following steps: preparation of antigen-coated Microtiter plates, binding of mAb-1 to antigen, binding of mAb-2, and detection of bound mAb-2.

Preparation of Plates

Microtiter plates (96-well) are coated with protein antigen (50 μl of a 1–10 μg/ml solution) as described in this volume [3]. Cells may also be attached to wells.[2,2a] Since many types of cells contain some peroxidase activity on their surfaces, mAb-2 detection systems using POD (see below), however, may not be used with cells.

Binding of mAb-1 and mAb-2 (Competition)

Wells of Microtiter plates coated with antigen or with cells attached are incubated $\geq$2 hr (or overnight for convenience) at RT with 50 μl of unlabeled mAb-1 (inhibitory mAb) at a concentration of at least 20 μg/ml [culture supernatant can be concentrated 10 times by use of a B15 Minicon concentrator (Amicon Corp., Lexington, Massachusetts) or purified mAb is diluted in 0.5% BSA]. Without removing these 50 μl, 10 μl of mAb-2 are added for an additional 3–5 hr of incubation at RT at a concentration of mAb-2 that is lower than that of mAb-1. Positive and negative controls are included by the use of an irrelevant mAb or 0.5% BSA instead of the antigen-specific mAb-1 (negative, uninhibited control) and by including unlabeled mAb-2 in the first incubation step (positive, inhibited control). After the second incubation, contents of wells are removed by flicking the plate or by aspiration if the contents are radioactive. Plates are then washed 4 times by filling wells with PBS (poured onto the plate directly) and removing it, as described. Different ways of measuring bound mAb-2 are now described.

Radiolabeled mAb-2

Plates are cut apart and individual wells are counted in a γ-counter (^{125}I), or submersed in scintillation fluid in a β-counter (^{3}H).

Purification of mAb-2. In order to label mAb with ^{125}I, the immunoglobulin must first be obtained in pure form. This is conveniently performed for immunoglobulins with γ_{2a} or γ_{2b} heavy chains and for many

[2] J. W. Stocker and C. H. Heusser, *J. Immunol. Methods* **26,** 87 (1979).

[2a] J. W. Stocker, F. Malavasi, and M. Trucco, *in* "Immunological Methods," (I. Lefkovits and B. Pernis, eds.), Vol. 2, pp. 299–308. Academic Press, New York, 1981.

with γ_1 heavy chains by the use of protein A–Sepharose (Pharmacia Fine Chemicals, Uppsala, Sweden). A 1-ml protein A–Sepharose 4B column is equilibrated with PBS. Then about 50 ml of tissue culture supernatant are passed through, followed by 5 ml of PBS (washing), and finally 0.1 M sodium acetate (pH ~ 2.9) for elution. To 500-μl column fractions, 100 μl of 1 M Tris-HCl, pH 9, are added with mixing. At pH 2.9 all mAbs that we have purified have been released. In some cases mAbs have already been released at pH 5.0. For antibodies that do not bind to protein A, immunoglobulin purification may be performed by a similar procedure from tissue culture supernatant using a column of anti-mouse Ig bound to Sepharose. This column is prepared by binding an Ig preparation (50% ammonium sulfate precipitate of antiserum) to CNBr-activated Sepharose 4B (Pharmacia) according to the manufacturer's instructions. After elution the mAb is then dialyzed against PBS.

Labeling with ^{125}I. Twenty micrograms of purified antibody can now be labeled with 0.5 mCi of carrier-free ^{125}I by a number of methods described elsewhere in this series.[3] The simplest one is the chloramine-T method. Under a hood, the following reagents are mixed: 200 μl of mAb + 200 μl of 0.1 M KPO$_4$ buffer (pH 7.0) + 0.5 mCi of carrier-free ^{125}I in 20–40 μl + 100 μl of chloramine-T (2 mg/ml in 0.1 M KPO$_4$ buffer pH 7.0). After 1–2 min of incubation at RT, 400 μl of sodium metabisulfite (2.5 mg/ml in 0.1 M KPO$_4$ buffer, pH 7.0) are added followed by 0.1 ml of 0.1 M KI. Labeled mAb is now separated from free ^{125}I by passage over a Sephadex G-25 column (1 cm × 20 cm) equilibrated with PBS containing 1 mg of BSA per milliliter. At this low ratio of ^{125}I to protein, each antibody molecule will contain at most 1 atom of ^{125}I. Higher ratios may also be tried, but at the risk of destruction of immunological activity. A survey in our laboratories of the frequency of destruction of mAb by iodination by the chloramine-T method has yielded widely disparate results. Some investigators have reported no inactivation problems in many iodinations of mAb, whereas others have had almost all of their antibodies inactivated. These latter mAbs, however, remained active after labeling with biotin (see below).

Biosynthetic labeling is a very simple way to label large numbers of mAbs. Hybridoma cells (~10^6/ml) are cultivated for 24 hr at 37° in leucine-free medium (RPMI-1640 minus leucine) supplemented with 10% fetal calf serum dialyzed against PBS and 20 μCi of [^{3}H]leucine per milliliter. Culture supernatants (10 μl per assay) are directly employed as mAb-2 without purification.

---

[3] This series, Vol. 70.

TABLE I
COMPETITION BINDING ASSAYS WITH β-THROMBOGLOBULIN (βTg) ANTIBODIES

mAb-1 ∇	Binding of labeled mAb-2 (cpm)							
	³H						¹²⁵I	
	βT-1	βT-2	βT-4	βT-6	βT-5	βT-9	βT-1	βT-5
βT-1	90	240	320	480	640	520	120	1390
βT-2	90	200	340	530	700	470	150	2080
βT-4	690	660	150	150	550	410	1010	2710
βT-6	650	650	120	120	680	440	860	2700
βT-5	690	880	380	580	210	160	1020	170
βT-9	760	850	390	540	250	150	1470	230
c-19	1350	1240	850	1280	1250	790	1440	1820

Example. Table I shows the results of such an experiment using both types of radiolabel with six mAbs against the platelet secretory protein β-thromboglobulin (βTg) and with an unrelated mAb (c-19: anti-carcinoembryonic antigen). Numbers are counts per minute of radiolabeled mAb-2 (listed in headings of columns) bound to antigen in the presence of 10× concentrated unlabeled culture supernatants (mAb-1) listed in column 1. βT-1 and βT-2 recognize the same epitope, as they are inhibited equally by each other and much less by other antibodies. Similarly βT-4 and βT-6 form one group, and βT-5 and βT-9 form another group recognizing the same epitopes, respectively. Similarly ¹²⁵I-labeled βT-1 and βT-5 are inhibited equally by βT-1 and βT-2, or βT-5 and βT-9, respectively, and much less by antibodies recognizing other epitopes (Table I).

Labeling mAb-2 via Isotype

If two mAb differ in their isotype, γ-chain subclass, or light chain, they can be detected individually in a competition SABA by the use of anti-isotype sera. This obviates the need for Ig purification and labeling. The assay comprises the two incubations for binding of mAb-1 and mAb-2 (competition) as described. After washing, wells are then incubated for 2 hr with 50 μl of appropriately diluted isotype-specific rabbit-anti-mouse-Ig sera (Nordic Immunological Laboratories, Tilburg, Holland). After washing 4 times, wells are incubated another 2 hr with 50 μl of goat-anti-rabbit-IgG antibodies conjugated with POD (Nordic). After washing 4 times, wells are finally incubated with substrate (6 mM H_2O_2 + 40 mM 2,2'-azinodi(3-ethylbenzothioazoline sulfonate), diammonium salt, or 6 mM H_2O_2 + 4 mM o-phenylenediamine in 0.1 M potassium citrate buffer, pH 5.5; a buffer containing 6 mM H_2O_2 can be stored at 4° wrapped in aluminium foil; the color reagents are prepared freshly before use and read by eye as the color develops. Obviously, by omitting the first incubation step with mAb-1, an unknown isotype of "mAb-2" can be determined by this method.

Biotin-Avidin Label

Labeling of purified mAb with ^{125}I or POD in many cases destroys their immunological activity (see above). The low molecular weight vitamin biotin in most of these cases can be covalently coupled to the mAb without destruction of the antibody binding site.

Labeling of mAb-2 with biotin. Purified mAb, 2 mg/ml, in 0.1 M NaHCO$_3$ + 0.15 M NaCl is mixed with one-tenth volume of a fresh solution in dimethyl sulfoxide of 2.5 mg of biotinyl-N-hydroxysuccinimide per milliliter (obtained as a powder from Biosearch, San Rafael, California), allowed to react for 2 hr at RT, dialyzed against PBS, and frozen at −70° in aliquots.

Labeling of Avidin with POD. The protein avidin (M_r 68,000), which binds to biotin with very high affinity, is conjugated with POD by a modification of the method of Nakane and Kawaoi.[4] Five milligrams of avidin are dissolved in 2.5 ml of 0.1 M Na$_2$CO$_3$ buffer, pH 9.5. Ten milligrams of POD (Boehringer, Mannheim, Germany) are dissolved in 3 ml of distilled water and allowed to stand for 1 hr at RT before adding 0.5 ml of freshly prepared 0.1 M sodium periodate solution. After mixing, the preparation is left at RT for 20 min, and then excess periodate is removed by dilution in 10 ml of 1 mM sodium acetate buffer, pH 4.5, and ultrafiltration in a stirred cell (Amicon) to a volume of 1–2 ml. This cycle of dilution and ultrafiltration is repeated twice, and the activated POD is finally concentrated in 2.5 ml and added to the avidin solution. After mixing, the reaction is allowed to proceed for 2 hr at RT. Then 0.5 ml of a fresh 0.1 M solution of sodium borohydride is added and mixed; the solution is left for 16 hr at 4°. Finally, the avidin–POD is dialyzed against PBS. For use, avidin–POD is diluted in 1% BSA in PBS.

Example. Table II shows a competition experiment with 2 mAbs specific for hepatitis B surface antigen. The unlabeled mAb-1 was HBS-1 or HBS-2 diluted to 50 μg/ml in 1% BSA after purification. Wells coated with

---

[4] P. K. Nakane and A. Kawaoi, *J. Histochem. Cytochem.* **22**, 1084 (1974).

TABLE II
COMPETITION SABA WITH 2 mAb TO HEPATITIS
B SURFACE ANTIGEN[a]

mAb-1 ∇	mAb-2-biotin	
	HBS-1	HBS-2
HBS-1	0.19	1.38
HBS-2	1.34	0.20
None	1.41	1.38

[a] Numbers are A_{492} of reacted substrate.

antigen were preincubated with 50 μl of one or the other unlabeled mAb or with 1% BSA for 2 hr at RT before adding 25 μl of 1 μg of HBS-1 or HBS-2 per milliliter labeled with biotin. After an additional 2 hr of incubation, plates were washed 3 times with PBS. Then 50 μl of avidin–POD (1 : 2000) in PBS + 1% BSA were added for 30 min. After washing another 3 times, 200 μl of substrate (6 mM H_2O_2 + 4 mM o-phenylenediamine in 100 mM citrate buffer, pH 5.5) were added to each well. After 15 min, the reactions were stopped by the addition of 50 μl of 2 M HCl, and the absorbance at 492 nm was read spectrophotometrically. Both HBS-1 and HBS-2 block only their own binding, not that of the other mAb, and are thus specific for different epitopes (Table II).

Binding to Epitopes on Different Molecules

If different molecules having one or more particular epitopes in common are adsorbed to a surface, the binding pattern of individual monoclonal antibodies may be distinctive and allow assignment of specificity. An example of such a situation is provided by our experiments with monoclonal antibodies against hCG.

Example. Plates were coated with hCG, hCG-α (α-subunit of hCG), hCG-β (β-subunit of hCG), a fragment consisting of the 28 C-terminal amino acids of hCG-β, and hLH (human leuteinizing hormone). hLH has the same α-subunit as hCG. Its β-subunit is shorter at the C terminus by 28 amino acids and has approximately 80% sequence homology with the equivalent piece of hCG-β. Microtiter plates were coated with approximately equimolar amounts of these molecules (1 μg/ml for hCG and hLH, 0.4 μg/ml for hCG-α, 0.6 μg/ml for hCG-β, and 4 μg/ml for the fragment), and SABA was carried out with a collection of 35 hCG–mAb (undiluted culture supernatants), as described in this volume [3]. The results are shown in Table III. Of the 35 mAbs, 30 bound not only to intact hCG, but also to one of the isolated subunits, hCG-α (3 mAbs) or hCG-β (27 mAbs). The remaining 5 bound only to intact hCG and thus apparently bind to an epitope that is destroyed upon chain separation (h epitope, Table III). In an attempt further to subdivide the large number of hCG-β specific mAbs into groups with different specificities, the ratios (r) of cpm of ^{125}I bound to wells coated with hCG-β and with hCG, which reflects the binding to the two different antigen molecules, were analyzed [r(hCG-β:hCG)]. On the basis of these ratios, four groups of hCG-β-specific mAbs were formed (Table III). The ratios of the members of each group varied within ± 10–15% of the average ratio for each group. The corresponding 4 epitopes were named β_1, β_2, β_3, β_4 (Table III). Although mAbs against β_1 and β_4 differ very little in their binding ratios r(hCG-β:hCG) (1.0 vs 1.2), this difference is nevertheless significant, as the distinction of β_1 and β_4 is clearly confirmed by the pattern of binding to the C-terminal segment of hCG-β and to hLH.

The differences in binding to different antigen molecules, e.g., the low binding of anti-β_2 mAb to hCG relative to hCG-β, probably reflect preferential orientations of attachment of the antigen molecules to the plastic surface. Thus, hCG may attach preferentially in an orientation in which the β_2 epitope is hidden. We may conclude from differences in the binding patterns of 2 mAbs that they are specific for different epitopes (e.g., anti-β_1 vs anti-β_2 mAb). Similar or identical binding patterns, on the other hand, do not permit the conclusion that the mAbs are directed against the

TABLE III

BINDING CHARACTERISTICS IN SABA OF GROUPS OF hCG
ANTIBODIES SPECIFIC FOR DIFFERENT EPITOPES

Antibody	Epitope					
	α	β_1	β_2	β_3	β_4	h
Number of mAbs	3	21	1	1	4	5
hCG	+	+	+	+	+	+
hCG-α	+	−	−	−	−	−
hCG-β	−	+	+	+	+	−
r(hCG-β:hCG)	0	1.0	2.2	1.7	1.2	0
C-terminal segment	−	+	−	−	−	−
hLH	+	−	−	+	+	+
r(hLH:hCG)	2.0	0	0	0.7	1.1	0.6

same epitope. Thus, the mAbs of the group specific for β_1 may be specific for a collection of epitopes β_1, β_1', etc. (Table III). To test this possibility a competition SABA experiment was carried out with the anti-β_1 specific mAb. Since just one of the 21 mAbs had a γ_{2a} heavy chain, all of the other 20 were employed as inhibitory mAb-1. mAb-2 binding was measured by detecting bound γ_{2a} chain by the labeling-via-isotype method described above. All 20 mAbs employed as mAb-1 inhibited the binding of the γ_{2a} mAb. Thus all 21 mAbs bind to a single epitope β_1. β_4 and h were shown to be single epitopes as well by similar experiments carried out with the β_4 and h antibodies.

"Iodination Sensitivity" of Epitopes

Iodination has been observed to inactivate some epitopes ("iodination sensitive" epitopes). The inactivation of a particular epitope may be the result of the iodination of a tyrosine within that epitope or of the iodination procedure (partial denaturation). Frequently, a particular "iodination sensitive" epitope e_1 is inactivated only on a fraction of antigen molecules. Thus, at high concentration a mAb against e_1 binds that percentage (f_1) of a particular ^{125}I-labeled antigen preparation that has an intact e_1 epitope. Another mAb-2 directed against a second iodination-sensitive epitope e_2 may bind a percentage f_2. If f_1 and f_2 differ, mAb-1 and mAb-2 can be inferred to bind to different epitopes.

Procedure. The assay is carried out with both ^{125}I-labeled antigen and antibody in solution. Incubation should be allowed to proceed to completion (conveniently overnight). With ^{125}I-labeled antigen held constant (at the lowest convenient concentration) the mAb concentration is increased to reach a plateau of ^{125}I-labeled antigen binding.

Bound and free ^{125}I-labeled antigen are separated according to any of the methods used in radioimmunoassays. However, the most suitable of these procedures differs from one antigen to another. Alternatively, one of the following antigen-independent methods may be employed. After completion of the antigen–antibody binding reaction, the incubation mixture is supplemented with one of the following particulate suspensions to approximately 1% final concentration: (*a*) *Staphylococcus aureus,* bearing protein A on its surface ("Pansorbin", Calbiochem AG, Luzern, Switzerland) if the mAb can be bound by protein A (see above); or (*b*) Pansorbin preincubated for 4 hr with a rabbit anti-mouse-Ig serum (see above), and then washed; or (*c*) protein A–Sepharose (see above); or (*d*) anti-mouse-Ig–Sepharose (see above). After 30 min of incubation with one of these four precipitating agents, suspensions are centrifuged, supernatants are removed, and pellets are counted.

Additional Methods

Competition Assay with Labeled Antigen

Wells of Microtiter plates are coated with 1 mAb (mAb-1) by leaving 50 μl of purified mAb (5–10 μl/ml) in each well for 4 hr, while in a separate vial labeled antigen is incubated with a large excess of mAb-2. Then mAb-1 is removed from the well and replaced by the mAb-2–antigen mixture. Labeled antigen is then bound by mAb-1 adsorbed to the well only if it does not recognize the same epitope as mAb-2. This assay may be more convenient than the standard version of competition SABA described earlier because it does not require labeling of mAb. However, this competition assay has the same limitations as the "sandwich assay" (see below) and the additional limitation that it will not work with epitopes that are completely inactivated by iodination.

Sandwich Assay

Wells of Microtiter plates are coated with 50 μl of 5–10 μl/mg of one mAb (purified mAb-1) and then incubated with 50 μl of a mixture of antigen (100 ng/ml or less) and a second, labeled mAb (^{125}I-mAb-2). If mAb-1 and mAb-2 can bind to antigen simultaneously, ^{125}I-mAb-2 is bound to the well via a "bridge" consisting of antigen and mAb-1 ("sandwich"). mAb-1 and mAb-2 can bind antigen simultaneously if they are directed against different epitopes. Two mAbs directed against the same epitope may also bind simultaneously if this epitope occurs more than once on the antigen (repeated epitope). This would occur, for example, in dimeric proteins with equal subunits, provided the ^{125}I-mAb-2 concentration is not so high as to bind to all available epitopes and prevent them from binding to mAb-1 on the plate. If the aim of the experiment is to show whether the epitope specificity of 2 mAb is the same or different, high concentrations of unlabeled mAb-2 should be added to ^{125}I-mAb-2 to exclude this multipoint binding effect. On the other hand, by the use of the same mAb as mAb-1 and ^{125}I-mAb-2 (at low concentration), the assay can be used to establish whether an antigen bears repeating epitopes. Although the sandwich assay has worked well in a number of experiments, we generally prefer competition SABA, because it functions equally with repeated and nonrepeated epitopes and because it is much less affected by low affinities of the mAb than the sandwich assay (and than most other assays described herein). This is because, in the sandwich assay, antigen and mAb are held together by single interactions of one epitope with one mAb combining site, whereas in SABA a mAb is bound to the surface by at least two such interactions.

Additive Effect in SABA

A standard SABA using Microtiter plates coated with antigen is carried out in which a single mAb or mixtures of two mAbs are allowed to bind to the antigen adsorbed to the wells. If the two mAbs are specific for two different epitopes, the total amount of mAb-1 + mAb-2 bound can be expected to be higher than if they bind to the same epitope or than the amount bound when either mAb is incubated alone (additive effect), because the total number of binding sites is higher. To reveal this effect the ^{125}I-labeled second antibody (anti-mouse-Ig) is diluted with unlabeled second antibody to a concentration of 5–10 μg/ml. Additive effects have been observed in some experiments with mAb to hCG and with mAb to leukocyte interferon. They could not, however, always be reproduced.

Double Diffusion in Agar

mAb specific for epitopes occurring only once on a single protein cannot precipitate with antigen. Two mAbs against two different epitopes theoretically can form a linear polymer. However, in experiments with many different pairs of mAbs specific for hCG, only a small proportion precipitated with hCG even though many of these pairs were known from other experiments to bind to antigen simultaneously. Thus, we do not routinely use this method.

Choice of Method

Comparison of the Methods Described

The question of how many different epitopes are recognized by a collection of mAbs and which mAbs recognize the same epitopes can in principle always be answered by a competition-SABA experiment. In this assay two epitopes are recognized as different if they are not too close to each other so that the respective two mAbs can bind simultaneously. The assay is conducted easily and quickly and can thus be performed on a large scale. However, the isolation and labeling of mAb required by this method is laborious, and thus, where possible, other methods are used first. This approach may greatly reduce the number of pair combinations of mAbs that finally need to be tested by competition SABA. Depending on the availability of the required reagents (different antigen molecules bearing the same epitopes or ^{125}I-labeled antigen), these screening methods may comprise (a) binding to epitopes on different molecules; (b) distinction of epitopes on the basis of iodination sensitivity; and (c) competition assay with labeled antigen.

Proposal for a Strategy

A collection of mAbs, especially if large, may be analyzed in the following way.

1. Isotypes of all mAb are determined as described.

2. If different isotypes are found, competition-SABA is carried out by use of the labeling-via-isotype method. With mAbs from hybridizations following several immunizations with protein antigens, the majority of mAbs usually contain γ_1 and κ chains. Other isotypes are rare. As mAb-2, one thus employs every mAb with a rare isotype. All mAbs that block binding of such an mAb-2 are assumed to recognize the same epitope. Obviously in subsequent experiments only one mAb representing such a group need be included.

3. If different antigen molecules are available that carry relevant epitopes, a SABA is carried out with each and binding patterns and ratios are analyzed. This obviously not only may lead to further group formations, but also information about the location of the epitopes may be gained, e.g., on which subunit, α or β of a dimer, the epitope is located. Additionally or alternatively, iodination sensitivity of epitopes recognized by different mAb is analyzed, if ^{125}I-labeled antigen and a suitable precipitation method are conveniently available.

4. Competition SABA is now carried out with directly labeled mAb. If many mAb still remain to be tested, one may label the mAb internally by culturing them in ^{3}H-labeled amino acids. If high backgrounds are encountered in competition SABA, where complete inhibition should occur (see Table I), these are probably the result of unspecific binding of proteins other than Ig, which are also secreted by the hybridomas. This may occasionally prevent unequivocal interpretation (inhibition vs no inhibition). Thus, this experiment may allow only further elimination of mAbs that inhibit each other, leaving one with a (hopefully) small residual number of mAbs that must be purified and labeled with ^{125}I or preferably with biotin (as iodination may lead to inactivation) for a final competition-SABA experiment.

5. After the complete resolution of the epitope specificity profile, the optimal group(s) of mAbs (i.e., the optimal epitopes) and the optimal mAbs within each group can now be selected for a particular task.

Section II

Immunoassay of Antigens and Antibodies

A. Labeling of Antigens and Antibodies
Articles 21 through 24

B. Separation Methods in Immunoassay
Articles 25 through 28

C. Immunoassay Methods
Articles 29 through 40

D. Data Analysis
Articles 41 through 44

[21] Methods for the Attachment of Haptens and Proteins to Erythrocytes

By Yi-Her Jou, Paul K. Mazzaferro,
George L. Mayers, and Richard B. Bankert

Erythrocytes to which antigens or antibodies are adsorbed or conjugated have been used extensively as target cells for the detection of antibodies and antigenic substances. The reaction of antigen-bearing erythrocytes with the complementary antibody results in agglutination or initiates a complement-dependent hemolysis.[1-28] The antigen-bearing erythrocytes have been used in the quantitation of antigen by inhibition of passive immune hemolysis.[28] Such indicator erythrocytes have also been used to

[1] J. Hamburger, *Transplant. Proc.* **4**, 335 (1972).

[2] G. A. Molinaro, B. A. Bessinger, A. Gilman-Sachs, and S. Dray, *J. Immunol.* **114**, 908 (1975).

[3] R. B. Bankert, D. DesSoye, and L. Powers, *J. Immunol. Methods* **35**, 23 (1980).

[4] R. B. Bankert, this volume [15].

[5] G. Middlebrook and R. J. Dubas, *J. Exp. Med.* **88**, 521 (1948).

[6] E. Neter, *Bacteriol. Rev.* **20**, 166 (1956).

[7] M. B. Rittenberg and K. L. Pratt, *Proc. Soc. Exp. Biol. Med.* **132**, 575, (1969).

[8] J. K. Inman, B. Merchant, L. Clafton, and S. E. Tacey, *Immunochemistry* **10**, 165 (1973).

[9] E. R. Gold and H. H. Fudenberg, *J. Immunol.* **99**, 859, 1967.

[10] S. V. Boyden, *J. Exp. Med.* **93**, 107 (1951).

[11] F. I. Alder and C.-T. Liu, *J. Immunol.* **106**, 1684 (1971).

[12] E. S. Golub, R. K. Mishell, W. O. Weigle, and R. W. Dutton, *J. Immunol.* **100**, 133 (1968).

[13] H. M. Johnson, K. Brenner, and H. E. Hall, *J. Immunol.* **97**, 791 (1966).

[14] D. Levin, J. Jonak, and T. N. Harris, *J. Immunol. Methods* **17**, 101 (1977).

[15] N. R. Ling, *Immunology* **4**, 49 (1961).

[16] T. Ternyck and S. Avrameas, *Ann. Immunol.* (*Paris*) **127C**, 197 (1976).

[17] D. E. Mahan and R. L. Copeland, Jr., *J. Immunol. Methods* **19**, 217 (1978).

[18] L. Gyenes and A. H. Sehon, *Immunochemistry* **1**, 43 (1964).

[19] S. Lemieux, S. Avrameas, and A. E. Bussard, *Immunochemistry* **11**, 261 (1974).

[20] D. Pressman, D. H. Campbell, and L. Pauling, *J. Immunol.* **44**, 101 (1942).

[21] F. L. Adler and L. T. Adler, this series, Vol. 70, p. 455.

[22] G. A. Molinaro, W. C. Eby, and C. A. Molinaro, this series, Vol. 73, p. 319.

[23] G. A. Molinaro, W. C. Eby, and C. A. Molinaro, this series, Vol. 73, p. 326.

[24] G. Galfrè and C. Milstein, this series, Vol. 73, p. 3.

[25] N. K. Jerne, C. Henry, A. A. Nordin, H. Fuji, A. M. C. Kores, and I. Lefkovits, *Transplant. Rev.* **18**, 130 (1974).

[26] W. J. Herbert, *in* "Handbook of Experimental Immunology" (D. M. Weir, ed.), p. 20. Blackwell, Oxford, 1978.

[27] J. S. Garvey, N. E. Cremer, and D. H. Sussdorf *in* "Methods in Immunology," 3rd ed., p. 347. Benjamin, Reading, Massachusetts, 1979.

[28] T. Borsos and J. J. Langone, this series, Vol. 74, p. 161.

METHODS IN ENZYMOLOGY, VOL. 92

quantitate antigen-binding cells by rosette assay[1] and antibody-forming cells by plaque-forming cell assays.[21,25] When antibody is attached to erythrocytes, one is able to quantitate antigen either by a reverse hemagglutination or by a radial reverse hemolytic assay.[21,22] Moreover, these indicator cells can be used for enumerating antigen-bearing cells by a reverse rosette assay[2,21] and for counting antigen-secreting cells by a reverse plaque-forming cell assay.[21,23] Assays using antigen-bearing erythrocytes to detect hybridomas secreting the desired antibody can be found elsewhere.[3,4,24]

To maximize the usefulness of the above-mentioned assays, a simple and versatile method for the preparation of the required erythrocyte targets is indispensable. Ideally, the procedure should use a minimum quantity of the often precious antigen; result in a firm attachment of antigen to the erythrocyte membrane; affect rapid and efficient coupling under mild conditions so that the structural integrity of antigen and erythrocytes is maintained; produce target cells with long-term stability; yield specific and reproducible results; and be as simple as possible.

Some antigens, such as many polysaccharides of microbial or protozoan origins, adhere to erythrocytes with sufficient tenacity and in adequate amounts to obviate the need for anything more elaborate than incubation of cells with the antigen.[5,6] Some haptens, such as 2,4,6-trinitrobenzenesulfonate, can be conjugated directly to erythrocytes by covalent bond formation.[7] Most of the other antigens require pretreatment of the erythrocytes or the antigen, or both, by chemical means in order to facilitate the adsorption or conjugation. Several methods for coupling antigens, primarily haptens, to erythrocytes by this approach can be found in the literature.[1–28] These include the covalent coupling of haptens to a tripeptide carrier that permits covalent linkage to the cell surface[8]; the treatment of red cells and antigens with chromic chloride[9,24,28]; the pretreatment of red cells with tannic acid to affect the adsorption of antigens[10,11,21]; the covalent linkage of protein antigens to red cells via a condensation reagent, such as carbodiimide,[12–14,28] or via homobifunctional cross-linking reagents, such as 1,3-difluoro-4,6-dinitrobenzene,[15] p-benzoquinone,[16] toluene-2,4-diisocyanate,[17,18] glutaraldehyde,[19] bisdiazotized benzidine,[20,21] and others.[23,25] These techniques have been reviewed elsewhere,[21–28] none of these procedures fully meet the criteria set forth above.

Several methods for coupling proteins and haptens to erythrocytes have been developed.[29–34] One of the methods involves the use of a het-

[29] Y.-H. Jou and R. B. Bankert, Proc. Natl. Acad. Sci. U.S.A. 78, 2493 (1981).
[30] Y.-H. Jou, G. Johnson, and D. Pressman, J. Immunol. Methods 42, 79 (1981).

erobifunctional reagent, N-succinimidyl 3-(2-pyridyldithio)propionate (SPDP), to link proteins to sheep red blood cells (SRBC) through disulfide bond formation.[29] The second method takes advantage of succinylation of hapten–protein conjugates to facilitate the coupling to the surface of SRBC by carbodiimide.[30] Two methods are designed to couple haptens to the surface of SRBC by using another heterobifunctional reagent, methyl-p-hydroxybenzimidate,[31,32] or a multifunctional reagent, 1,3,5-trichloro-triazine.[32,33] The last method employs the noncovalent attachment of proteins and aminohaptens to the surface of SRBC via a synthetic lipopolysaccharide reagent.[34] These new coupling methods are outlined and discussed in this chapter.

Use of a Heterobifunctional Reagent N-Succinimidyl 3-(2-Pyridyldithio)propionate (SPDP) to Couple Proteins to the Surface of Erythrocytes

Principle

Condensation reagents, such as carbodiimide or carbonyldiimidazole, and homobifunctional reagents where the two reactive groups are identical (e.g., bisdiazotized benzidine or glutaraldehyde) have been the most prevalent cross-linking reagents for coupling proteins to the SRBC membrane. However, it is very difficult to achieve control of intramolecular versus intermolecular cross-linking with these reagents. The formation of homoconjugates of proteins (the result of homopolymerization) is another disadvantage of using condensation or homobifunctional reagents. Such side reactions obviously decrease the efficiency of protein–SRBC coupling. Furthermore, intramolecular cross-linking of proteins may alter the functional and antigenic integrity of the protein molecules by introducing additional tertiary structure into the proteins. With a heterobifunctional reagent, where the two reactive groups are directed toward different functional groups, we have shown that the protein–SRBC conjugation can be easily achieved in an efficient and controllable manner without unwanted intramolecular cross-linking and homoconjugation of proteins.[29]

The heterobifunctional reagent, SPDP, developed by Carlsson et al.,[35] contains one N-hydroxysuccinimide ester moiety and one 2-pyridyldi-sulfide moiety (Fig. 1a). The hydroxysuccinimide ester reacts with amino

[31] P. C. Isakson, J. L. Honegger, and S. C. Kinsky, J. Immunol. Methods **25**, 89 (1979).
[32] P. Mazzaferro and G. L. Mayers, J. Immunol. Methods **46**, 327 (1981).
[33] D. Blakeslee and M. G. Baines, J. Immunol. Methods **23**, 375 (1978).
[34] R. B. Bankert, G. L. Mayers, and D. Pressman, J. Immunol. **123**, 2466 (1979).
[35] J. Carlsson, H. Drevin, and R. Axén, Biochem. J. **173**, 723 (1978).

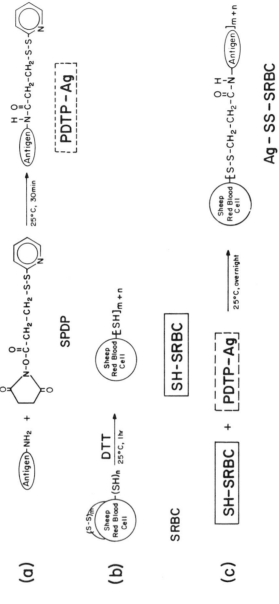

Fig. 1. Coupling of protein antigen to sheep red blood cells (SRBC). (a) Introduction of 3-(2-pyridyldithio)propionyl (PDTP) groups into a protein antigen by aminolysis. (b) Reduction of the disulfide bonds of the SRBC by dithiothreitol (DTT). (c) Reaction between the modified protein antigen containing PDTP groups and the reduced SRBC (SH-SRBC) through thiol-disulfide exchange to form the disulfide-linked protein-SRBC conjugate. Taken in part from Jou and Bankert.[29]

groups to form stable amide bonds, and the 2-pyridyldisulfide group reacts with aliphatic thiols to form aliphatic disulfide bonds (Fig. 1). Both of these reactions proceed rapidly under very mild conditions in aqueous media.[35] A general method for the coupling of proteins to the surface of SRBC by using SPDP is summarized in Fig. 1a–c. The procedure involves three steps. First, the 3-(2-pyridyldithio)propionyl (PDTP) groups are introduced into the protein molecules by the reaction of SPDP with a portion of the amino groups on the protein (Fig. 1a). Second, the disulfide bonds of the SRBC are reduced by dithiothreitol (Fig. 1b). Finally, the PDTP–protein conjugates are covalently coupled to the reduced SRBC (SH-SRBC) through thiol disulfide exchange to form the disulfide-linked protein-SRBC conjugate (Ag-SS-SRBC) (Fig. 1c). The procedure generally requires only 10–500 μg of protein for the preparation of 50 μl of packed protein-coupled SRBC, which are stable for at least 3 months at 4° in phosphate-buffered saline, pH 7.2, and are capable of detecting as little as 5 pg of anti-protein antibody in a hemolytic assay without noticeable nonspecific lysis.

The content of PDTP groups in the modified protein (PDTP-Ag) can be varied by using different amounts of reagent (Table I). The degree of

TABLE I

INTRODUCTION OF 3-(2-PYRIDYLDITHIO)PROPIONYL (PDTP)
GROUPS INTO PROTEINS BY
N-SUCCINIMIDYL-3-(2-PYRIDYLDITHIO)PROPIONATE (SPDP)[a]

Protein[b] type	Protein (mg/ml)	Protein (ml)	SPDP[c] (ml)	Substitution degree, [PDTP]/[protein][b]
hGG	40	2	0.25	1.9
bGG	40	2	0.3	4.0
BSA	30	2	0.25	3.2
Hy	1	2	0.3	8.2
PAPase	12	0.5	0.05	5.6
MIgG	12	0.5	0.05	11

[a] From Y.-H. Jou and R. B. Bankert.[29]

[b] The molecular weights used for the calculation of molar concentration of proteins are as follows: human γ-globulin (hGG), 160,000; bovine γ-globulin (bGG), 160,000; bovine serum albumin (BSA), 67,000; sea urchin egg hyalin (Hy), 300,000; human prostatic acid phosphatase (PAPase), 100,000; monoclonal mouse IgG (MIgG), 160,000.

[c] The 20 mM SPDP solution was freshly prepared in absolute ethanol.

modification, i.e., moles of PDTP groups per mole of protein, can be determined by treating the PDTP-Ag with an excess amount of dithiothreitol. This treatment effects the release of pyridine-2-thione, which has a molar absorptivity of 8.08×10^3 at 343 nm.[35] The amount of pyridine-2-thione released is equivalent to the content of PDTP groups in PDTP-Ag. Table I illustrates the results of the reaction of some proteins with SPDP. It has been shown that extensive modification of proteins often results in precipitation due to the decrease in solubility of these modified proteins.[35,36] The extent of modification before such a change in solubility can be observed, varies from one protein to another.

It is worthwhile to note that the procedure described here is applicable for coupling any molecule that contains amino group(s) to the surface of SRBC. However, it is important to point out that thiol-containing proteins or other molecules containing both amino and thiol groups cannot be coupled to the surface of SRBC using this protocol. It is obvious that the reaction of SPDP with molecules containing both amino and thiol groups will result in the intramolecular cross-linking and homoconjugation of the molecules. Alternative approaches for coupling thiol-containing proteins to the surface of SRBC through disulfide bond formation have been proposed and are being developed by us.[37]

The attachment of proteins (antibodies) has been extended to nucleated cell plasma membranes[37,38] and liposomes[39,40] using a modification of the coupling technique described here.

Materials and Reagents

The heterobifunctional reagent SPDP can be synthesized according to the procedures developed by Carlsson et al.[35] It is also commercially available from Pharmacia Fine Chemicals (Piscataway, New Jersey) and Pierce Chemical Company (Rockford, Illinois). It is important to store SPDP dry, below 8°. The dissolution of SPDP in absolute alcohol for the preparation of a 20 mM solution can be facilitated by brief sonication. An ethanol solution of SPDP (20 mM) kept at 23° for 20 days retains 80% of the initial ester content and retains 100% of the initial content of the 2-pyridyldisulfide structure.[35] However, we recommend preparation of the SPDP solution in ethanol immediately before use. Phosphate-buffered

[36] A. F. S. A. Habeeb, *Biochim. Biophys. Acta* **673**, 527 (1981).
[37] Y.-H. Jou and R. B. Bankert, U.S. Patent entitled, "Coated Cells and Their Use." Application filed August, 1981.
[38] Y.-H. Jou, B. Schepart, and R. B. Bankert, *Immunol. Commun.* (in press, 1982).
[39] L. D. Leserman, J. Barbet, F. Kourilsky, and J. N. Weinstein, *Nature (London)* **288**, 602 (1980).
[40] F. J. Martin, W. L. Hubbell, and D. Papahadjopoulos, *Biochemistry* **20**, 4229 (1981).

saline (PBS) consisting of equal volumes of 0.15 M NaCl and 0.15 M phosphate buffer, pH 7.2 is used in the procedure. Dithiothreitol is prepared as a 1 M solution in PBS.

Procedure

Defibrinating Freshly Drawn Sheep Blood with Glass Beads[41]

1. Draw 160 ml of whole sheep blood into an evacuated 500-ml sterile bottle containing approximately 300 glass beads of 3 mm diameter.
2. Rotate the blood with the beads for 30 min at room temperature and thereafter cool it to 4°.
3. Decant the blood.
4. Centrifuge the blood at 4° for 11 min at 500 g.
5. Remove the serum by decantation and resuspend the packed SRBC in Basal Medium Eagles (BME).
6. Repeat steps 4 and 5 four times.
7. Resuspend SRBC in 500 ml of BME and store the cell suspension at 4°.

Preparation of PDTP-Proteins

1. Estimate the quantity of SPDP for reaction with protein. Protein concentrations in the range of 0.1–40 mg/ml in PBS have been successfully used in this experiment. A volume of protein solution in the range of 0.2–2 ml can be used.

$$\text{Volume (ml) of 20 m}M \text{ SDPD to be added} = \frac{Q/\text{MW} \times n \times 5}{20 \times 10^{-3}}$$

where Q = quantity of protein (mg) = protein concentration (mg/ml) times volume of protein solution (ml); MW = molecular weight of protein; n = number of amino groups per protein molecule to be modified (i.e., to be converted into thiol reactive groups). For the first trial, calculation based on n = 5 is a good start.

2. Prepare 20 mM SPDP in *absolute* ethanol.
3. Add the appropriate volume of 20 mM SPDP to the protein solution. The resultant mixture is incubated at room temperature for 30 min.
4. Dialyze the reaction mixture against PBS overnight at 4°. The volume of PBS used is 1000 times the volume of the protein solution.
5. The modified protein (PDTP-protein) can be used to couple to red cells immediately or kept frozen at −20° until needed.

[41] R. B. Bankert, G. L. Mayers, and D. Pressman, *J. Immunol.* **118**, 1265 (1977).

Determination of the [PDTP]/[Protein] Ratio. A portion of the PDTP-protein may be used to determine the content of the PDTP groups in the molecule. This step can be omitted and one may proceed directly to the procedures for coupling PDTP–protein to red cells described below under Reduction of SRBC in Dithiothreitol and under Coupling of PDTP-protein to HS-SRBC.

1. Prepare 1 ml of a properly diluted PDTP-protein solution (2.5 to 15×10^{-6} M) in PBS.
2. Read and record the absorbance of PDTP-protein at 280 nm (A_{280}) and 343 nm (A_{343}°).
3. Add 0.1 ml of 1 M DTT to the PDTP-protein solution. The reaction mixture is left to stand at room temperature for 40 mins.
4. Read the absorbance of the DTT-treated solution at 343 nm (A_{343}).
5. Calculate the molar concentration of PDTP groups as follows.

$$[PDTP] = \frac{(A_{343} \times 1.1) - A_{343}^{\circ}}{8.08 \times 10^3}$$

6. Determine protein concentration from A_{280}, if the molar absorptivity of protein is known. Since 2-pyridyldisulfide groups contribute to the adsorbance at 280 nm, a correction should be applied in the calculation of protein concentration.

$$[Protein] = \frac{A_{280} - [PDTP] \times 5100}{molar\ absorptivity\ of\ protein}$$

7. Calculate the ratio of [PDTP]/[protein].

Reduction of SRBC by Dithiothreitol (DTT). Because the thiol group is very reactive and can take part in unwanted reactions, the DTT-reduced SRBC (HS-SRBC) should be prepared just prior to use for coupling to PDTP-protein.

1. Wash SRBC three times with PBS. The volume of the buffer used for each wash is about 15 times the packed volume of SRBC.
2. Add the appropriate amount of PBS to the packed SRBC to make a 2% (v/v) suspension.
3. Mix 0.5 ml of freshly prepared 1 M DTT and 12.5 ml of a 2% suspension of SRBC in PBS at room temperature on a rotating mixer for 1 hr.
4. Centrifuge (300 g) the cells and remove the supernatant.
5. Wash the cells four times with 15-ml portions of PBS.

Coupling of PDTP-Protein to HS-SRBC. Between 200 and 500 μg of PDTP-protein are allowed to react with 50 μl (packed volume) of HS-

SRBC. The optimal ratio of PDTP-protein to HS-SRBC varies for different proteins; therefore, this optimal ratio has to be determined individually for each protein.

1. Add 200–500 μg of PDTP-protein in PBS to 50 μl (packed volume) of HS-SRBC. Mix the contents on a rotating mixer at room temperature overnight.
2. Add PBS to a final volume of 15 ml. Centrifuge and remove supernatant.
3. Wash the cells two more times with 15-ml portions of PBS.
4. Adjust the volume of the cell suspension to the appropriate percentage for assay by adding PBS.

Succinylation of Hapten–Protein Conjugates to Facilitate Coupling to the Surface of SRBC

Principle

The introduction of a large number of haptenic residues on the surface of SRBC has been achieved by coupling haptens to an antigenically irrelevant large molecular weight carrier, such as a polysaccharide or a protein.[14,34] The large molecule itself may function as a spacer to separate the antigen from the conjugated cell surface. Previous studies have suggested that such spacer molecules enhance the sensitivity of the antigen-coupled erythrocytes used in immunoassays.[8] As indicated above, a controllable and efficient coupling of hapten–protein conjugates to the surface of SRBC is possible only when the procedure used results in limited intramolecular cross-linking and limited homopolymerization of the hapten–protein conjugates. When carbodiimide is used as the condensation reagent to couple hapten–protein conjugates to the surface of SRBC, both of these side reactions can occur because each protein molecule contains both free amino and free carboxylate groups, which can be condensed by carbodiimide to form amide bonds intramolecularly or intermolecularly.

It has been demonstrated by us that the conversion of all the free amino groups on the hapten–rabbit serum albumin (hapten-RSA) conjugates to carboxylate groups by succinylation facilitates carbodiimide-induced coupling to the surface of SRBC.[30] The rationale for the succinic anhydride treatment is as follows: First, all the free amino groups on the RSA molecule are blocked, thus preventing a carbodiimide-induced irreversible polymerization and intramolecular cross-linking of the hapten–protein conjugates, which would otherwise decrease the extent of its coupling to the surface of SRBC. Second, the increased number of carboxylate groups on the hapten–succinyl-RSA conjugates facilitates its coupling to the free amino groups of SRBC membrane proteins by increasing the concentration of the substrate that is initially reactive toward carbodiimide to form a carbodiimide–carboxylate reactive complex.

The hapten–succinyl-RSA conjugates can be prepared by several simple methods. The choice of the method is dependent upon the kind of hapten to be conjugated to RSA and the type of chemical linkage to be formed between them. The procedures described below are those that involve the most frequently used chemical reactions for conjugating haptens to proteins. Aminohaptens can be coupled to extensively succinylated RSA through amide bond formation by using carbodiimide. Aromatic aminohaptens can also be coupled to extensively succinylated RSA by diazotization of the haptens. Haptens that are reactive toward amino groups (e.g., 2,4,6-trinitrobenzenesulfonate, acid chlorides, and anhydrides of carboxylic acids) can be coupled to the uñsuccinylated amino groups in a partially succinylated RSA or coupled to a fraction of the amino groups of RSA followed by extensive succinylation of the conjugates.

The haptenated SRBC prepared by this succinylation–carbodiimide method are stable for at least 3 months and are capable of reproducibly detecting picograms of anti-hapten antibodies in hemolytic spot assays. Furthermore, a hapten–succinyl-RSA conjugate can be prepared in a large batch and kept frozen so that target SRBC can be easily prepared in a few hours using a small amount of hapten–succinyl-RSA conjugate (0.5 mg per 50 μl of packed SRBC).

Materials and Reagents

SRBC can be prepared as described above for the SPDP coupling procedure. The coupling saline used in the procedure was prepared by diluting borate buffer, pH 8 (0.167 M H$_3$BO$_3$, 0.134 M NaCl, 0.022 M NaOH) 12-fold in 0.147 M NaCl and then adjusting the pH to 6 or 7 by addition of 1 M HCl. 1-Ethyl-3-(3-dimethylaminopropyl)carbodiimide (EDCI) and succinic anhydride may be purchased from Aldrich Chemical Co. (Milwaukee, Wisconsin). p-Dioxane obtained from Burdick and Jackson Laboratories (Muskegon, Michigan) is kept dry over molecular sieve type 4A (Fisher Scientific Co., Fairlawn, New Jersey). 2,4,6-Trinitrobenzenesulfonic acid · 3H$_2$O can be obtained from Eastman Kodak Co. (Rochester, New York). All other reagents and solvents are commercially available reagent grade and are used without further purification.

TABLE II
NUMBER OF SUCCINYLATED AND UNSUCCINYLATED AMINO GROUPS PER
RABBIT SERUM ALBUMIN (RSA) MOLECULE[a]

Succinic anhydride (mg) used to react with 1 g of RSA	Number of succinylated amino groups per RSA molecule[b]	Number of unsuccinylated amino groups per RSA molecule[b]	Abbreviation of the succinyl-RSA
0	0	67	RSA
46	30	37	Suc_{30}-RSA
93	46	21	Suc_{46}-RSA
186	61	6	Suc_{61}-RSA
372	65	2	Suc_{65}-RSA

[a] From Y.-H. Jou et al.[30]

[b] Determined by the 2,4,6-trinitrobenzenesulfonic acid method.[42] A molecular weight of 70,000 was taken for RSA.

Procedure

Succinylation of Rabbit Serum Albumin. Rabbit serum albumin (RSA) is succinylated according to the method of Habeeb[42] except that borate-buffered saline (BBS), pH 8, rather than 0.1 M sodium phosphate buffer, pH 8, is used to prepare a 2% (w/v) solution of RSA.

1. Prepare succinic anhydride solution in dioxane at a concentration of 50 mg/ml.
2. Add an appropriate quantity (see Table II) of succinic anhydride to 50 ml of 2% (w/v) RSA at room temperature with constant stirring. The pH is maintained at 8 by the addition of 1 N NaOH. Continue the stirring for 30 min.
3. Dialyze the reaction mixture against two changes of 4 liters of water and then 4 liters of BBS, after which the mixture is kept frozen until needed.
4. Determine the residual free amino groups of succinyl-RSA by the 2,4,6-trinitrobenzenesulfonic acid method as described by Habeeb.[42]

Coupling of Diazotized Aromatic Aminohaptens to Suc_{65}-RSA

1. Prepare the diazonium salt of 0.15 mmol of aromatic aminohapten. A solution or suspension of 0.15 mmol of aromatic aminohapten in

---

[42] A. F. S. A. Habeeb, *Arch. Biochem. Biophys.* **121,** 652 (1967).

0.8 ml of 1 N HCl is cooled in an ice bath and treated with 0.15 ml of 1 N NaNO$_2$ for 20 min with rapid stirring. If the reaction mixture does not give positive results on KI-starch paper, additional 1 N NaNO$_2$ is added dropwise until a positive result is observed.

2. Add the diazonium salt solution prepared above to an ice-cooled solution of 70 mg of Suc$_{65}$-RSA in 12.5 ml of BBS with rapid stirring. The pH of the reaction mixture is maintained at 9 by the addition of 1 N NaOH. The stirring is continued in an ice bath for 3 hr.
3. Dialyze the reaction mixture against two changes of 4 liters of water followed by 4 liters of pH 6 coupling saline. Keep the mixture frozen until needed.

Coupling of Aminohaptens to Suc$_{65}$-RSA by Carbodiimide

1. Prepare a solution containing 35 mg of Suc$_{65}$-RSA and 3.5 mmol of the aminohapten in 15 ml of water. The pH of the solution is brought to 5 with 1 N HCl.
2. Add dropwise a freshly prepared solution of 60 mg of EDCI in 1 ml of ice-cold water with rapid stirring at room temperature. Approximately 10 min after completing the addition of the EDCI solution, a white precipitate will form.
3. Add 5 M NaOH to adjust the pH of the reaction mixture to 10, and continue stirring overnight at room temperature. The suspension becomes a clear solution at this point.
4. Dialyze the reaction mixture against two changes of 4 liters of water followed by 4 liters of pH 6 coupling saline. Keep the mixture frozen until needed.

Coupling of 2,4,6-Trinitrobenzenesulfonic Acid (TNBS) to Suc$_{30}$-RSA or Suc$_{46}$-RSA.

1. Prepare a solution of 35 mg of partially succinylated RSA (Suc$_{30}$-RSA or Suc$_{46}$-RSA) in 6.3 ml of BBS.
2. Add 3.5 ml of a 1% (w/v) aqueous TNBS solution.
3. Incubate the reaction mixture at 37° for 5 hr.
4. Dialyze the reaction mixture against two changes of 4 liters of water followed by 4 liters of pH coupling saline. Keep the mixture frozen until needed.

Coupling of Haptenated Succinyl-RSA to SRBC

1. Wash SRBC three times with coupling saline, pH 7. The volume of saline used for each wash is about 15 times the packed volume of SRBC.

2. Add 50 μl of packed SRBC to a solution of haptenated succinyl-RSA in 2.5 ml of coupling saline, pH 6, and mix the suspension thoroughly.
3. Add to the SRBC suspension a freshly prepared solution of 25 mg of EDCI in 0.5 ml of coupling saline, pH 6, and mix the reagents thoroughly.
4. Leave the reaction mixture to stand at room temperature for 30 min with occasional swirling to keep the red cells suspended.
5. Wash the cells three times with 10-ml portions of PBS (0.075 M KH_2PO_4, 0.075 M NaCl, pH 7.2) and store at 4° until used.

Coupling of Aromatic Aminohaptens to the Surface of SRBC via Methyl-p-hydroxybenzimidate (HB)

Principle

Methyl-p-hydroxybenzimidate (HB) is a heterobifunctional reagent with an amino-reactive imidoester and a phenolic group that is reactive toward diazonium salts, allowing addition of a hapten ortho to the hydroxyl in the ring structure. The imidoester reacts with amino groups of proteins to form an amidine linkage.[43] Thus, a protein or cell surface may be haptenated via this reagent.[31,33,43–47] A number of papers have also been published by Leon Wofsy and colleagues, who use this reagent in a hapten-sandwich labeling technique.[44–47]

Procedures have been developed that allow methyl-p-hydroxybenzimidate to be used as a coupling reagent for haptenating the surface of both erythrocytes and nucleated cells without extensive damage to the cells.[31,32] The modified erythrocytes have been shown to be both sensitive and stable for use as target cells in immunoassays.[31,32] Thus, this phenolic imidoester has proved to be a useful reagent for the coupling of aromatic aminohaptens to the surface of red blood cells.

The procedures given in this chapter for the coupling of haptens to the surface of SRBC via methyl-p-hydroxybenzimidate are those worked out by us[32]; however, an equally effective procedure has been worked out by Isakson et al.[31]

The procedure for coupling of haptens to the surface of SRBC involves preparation of a diazonium salt followed by addition of HB to the

[43] F. T. Wood, M. M. Wu, and J. C. Gerhart, *Anal. Biochem.* **69**, 339 (1975).
[44] S. Cammisuli and L. Wofsy, *J. Immunol.* **117**, 1695 (1976).
[45] M. K. Nemanic, D. P. Carter, D. R. Pitelka, and L. Wofsy, *J. Cell Biol.* **64**, 311 (1975).
[46] L. Wofsy and C. Henry, *Curr. Top. Mol. Immunol.* **7**, 215 (1978).
[47] E. F. Wallace and L. Wofsy, *J. Immunol. Methods* **25**, 283 (1979).

diazonium salt solution, which results in formation of a hapten–HB conjugate. This substituted imidoester is then mixed with SRBC where an amidine linkage forms between the hapten–HB conjugate and the SRBC membrane.

Haptenated red cells produced by this method have been shown to be both stable and sensitive for use in a hemolytic plaque assay.[31] Similarly, in the hemolytic spot assay these cells have been shown able to detect as little as 17 pg of specific anti-hapten antibody.[32]

Materials and Reagents

Methyl-*p*-hydroxybenzimidate can be purchased commercially from Pierce Chemical Co. (Rockford, Illinois) or can be synthesized easily in the laboratory according to Wood *et al.*[43]

Sheep red blood cells (SRBC) prepared as described above for the SPDP coupling procedure are washed and stored in 0.15 *M* phosphate-buffered saline (PBS) pH 7.2.

Procedure

Preparation of the Hapten–HB Conjugate. The methods given below for preparing the hapten–HB conjugate solution are such that a nearly isotonic solution will result. This is important, since the solution will be added directly to the SRBC for further coupling and the red blood cells will not tolerate great deviations from isotonicity without lysis or the production of fragile, unstable target cells.

The HB should not be added to the water in the procedure given below until immediately before use, since the imidoester will slowly hydrolyze in an aqueous medium.

1. Add 0.75 mmol of the aromatic aminohapten to 8 ml of 0.1 *M* HCl with stirring. If solution does not occur, the mixture may be briefly sonicated to form a cloudy suspension.
2. Prepare the diazonium salt of the aromatic aminohapten at 4° by dropwise addition of a cold 1 *M* sodium nitrite solution until a slightly positive color (gray) is observed on KI-starch paper.
3. Neutralize excess sodium nitrite by the addition of a few crystals of sulfamic acid, if the test paper indicates too much sodium nitrite (black color) has been added.
4. Add 1 *M* sodium nitrite periodically to the mixture in order to maintain a slightly positive test. The reaction should stabilize in 2–3 hr at 4°.
5. Dissolve 0.50 mmol of HB in 8 ml of distilled water and slowly add it to the diazonium salt solution at room temperature while raising

and maintaining the pH at 8.5 with 1 M NaOH. Within 2–3 hr the pH of the solution should stabilize at 8.5. The final volume of the solution should be about 20 ml. This yields a solution that is slightly hypertonic (0.24 M) but has little effect on the SRBC during the final coupling. If cellular fragility is encountered, the volume of 0.1 M HCl or water used in the procedure may be adjusted as needed.

6. Freeze aliquots of the hapten–HB conjugate solution at $-20°$ for later coupling to SRBC. This solution will remain usable for up to 2 weeks when frozen.

Coupling the Hapten–HB Conjugate to SRBC

1. Wash the SRBC twice in 15 volumes of PBS, pH 7.2.
2. Wash the SRBC once in 15 volumes of PBS brought to pH 9.0 with 1 M NaOH.
3. Add 0.5 ml of freshly thawed hapten–HB conjugate solution (pH to 9.0 with 1 M NaOH) to an equal volume of packed SRBC.
4. Incubate the cells at 37° for 5 hr with gentle tumbling.
5. Wash the cells 4 times in 30 volumes of PBS, pH 7.2.
6. Store the cells at 4° in PBS, pH 7.2. They should remain stable for at least 1 month.

Use of 1,3,5-Trichlorotriazine to Couple Haptens or Proteins to the Surface of Erythrocytes

Principle

1,3,5-Trichlorotriazine (TCT) is a multifunctional reagent possessing three reactive chloro groups. The chloro groups are reactive toward both hydroxyl groups, as found in carbohydrates, and primary amines, as found in many haptens and proteins. The nucleophilic displacement reaction may be selectively controlled by varying the temperature of the reaction mixture or the order in which the substituents are added to the triazine ring.[33,48,49] Thus, one, two, or three of the chloro groups may be selectively modified.

The procedure for coupling an aminohapten or protein involves the formation of a monosubstituted triazine at 4°. Since the reaction with triazine produces HCl, a good buffer is needed in order to serve as a proton sponge. A carbonate or borate buffer was found to be adequate for these purposes. Alternatively, the reaction between triazine and hapten

[48] T. H. Finlay, V. Troll, M. Levy, A. J. Johnson, and L. T. Hodgins, *Anal. Biochem.* **87**, 77 (1978).

[49] L. T. Hodgins and M. Levy, *J. Chromatogr.* **202**, 381 (1980).

may be carried out in an organic solvent. This method prevents hydrolysis of the triazine, which can occur in aqueous solution. p-Dioxane has been found to be an excellent solvent for the organic-phase reactions because it is miscible with water and because it will dissolve high concentrations of the triazine.[49] Organic solvents that are reactive toward acyl halides (such as N,N-dimethylformamide and dimethyl sulfoxide) should be avoided, since an explosive compound may be formed.

The monosubstituted triazine is allowed to react with erythrocytes at 37°; this results in coupling between the dichlorotriazine derivative and amino groups on the membrane surface via a second reactive chloro group. The last chloro group on the triazine ring is reported to be inactive toward primary amines below 100°, thus obviating the need to block this group.[33]

The procedure to be described is extremely easy to perform, requires a minimum of time, and produces target cells that are both stable and sensitive for use in hemolytic assays.[32,33] In the hemolytic spot assay these target cells were shown to detect less than 17 pg of anti-hapten antibody and remained stable for longer than 1 month.

The triazine method has also been used to couple a hapten to the surface of nucleated cells. These cells were shown to retain both their viability and growth properties after coupling. Further, these cells were shown to lose the surface-coupled hapten within 48 hr of culture.[32]

An extensive literature is also developing describing the use of trichlorotriazine as a method of activating affinity columns.[48-51]

Materials and Reagents

1,3,5-Trichlorotriazine can be purchased from Aldrich Chemical Co. (Milwaukee, Wisconsin). This reagent should be stored in the cold, preferably in a desiccator, since triazine will slowly hydrolyze on exposure to water, forming an inactive product. p-Dioxane obtained from Burdick and Jackson Laboratories (Muskegon, Michigan) is kept dry over molecular sieve type 4A (Fisher Scientific Co., Fairlawn, New Jersey). Sodium carbonate buffer is prepared by adjusting the pH of a 0.1 M Na_2CO_3 solution with KH_2PO_4 to 10. Phosphate-buffered saline (PBS) consisting of equal volumes of 0.15 M NaCl and 0.15 M phosphate buffer, pH 7.2, is used in the procedure.

[50] T. Lang, C. J. Suckling, and H. C. S. Wood, J. Chem. Soc. Perkin Trans. 1, p. 2189 (1977).

[51] A. Kessner, L. T. Hodgins, and W. Troll, Anal. Biochem. 92, 383 (1979).

Procedure

Preparation of a Hapten–TCT Conjugate. As mentioned previously, the conjugate may be formed in either an aqueous or an organic solution. In order to minimize the contact with water and problems of hydrolysis, the organic-phase method will be given. However, the reaction performed in an aqueous solution will work and can be adopted from a paper by Blakeslee and Baines that describes the formation of a lysine–TCT conjugate.[33] The procedure outlined below is a modified version of a procedure given by Finlay *et al.* for coupling enzymes to Sepharose via TCT.[48]

If significant hydrolysis of the triazine has occurred during storage, which can be determined by melting point or thin-layer chromatography, recrystallization from chloroform may be necessary before proceeding further (mp 147–148° after recrystallization twice from chloroform).

1. Dissolve 5 mmol of an aminohapten in 30 ml of *p*-dioxane.
2. Add 10 ml of 2 *M* *N,N*-diisopropylethylamine in *p*-dioxane to the solution at 4°.
3. Add a solution containing 5.5 mmol of TCT in 10 ml of *p*-dioxane to the hapten mixture to start the reaction, and allow the reaction to proceed for 2 hr at 4°.
4. Precipitate the product (hapten–TCT) with 50 ml of 1 *M* HCl.
5. Filter the precipitate and wash with 20 ml of cold 1 *M* HCl.
6. Dry the precipitate *in vacuo* and store at −20° over anhydrous CaSO$_4$. The hapten–TCT product is stable under these conditions of storage for at least 1 year.

Coupling of the Hapten–TCT Conjugate to the Surface of SRBC. The procedure as published by Blakeslee and Baines is followed for this coupling.[33]

1. Wash the SRBC twice in 15 volumes of PBS, pH 7.2.
2. Dissolve 5–10 mg of the hapten–TCT reagent in 8 ml of borate buffer, pH 8.0 (sonication may be needed). Adjust the pH to 8 if necessary.
3. Add the solution (suspension) to 0.5 ml packed volume of SRBC.
4. Tumble the cells gently at room temperature for 1 hr.
5. Wash the SRBC 3 times in PBS, pH 7.2, and store at 4°.

Coupling Proteins to the Surface of SRBC[32]

1. Dissolve 10 mg of protein in 1 ml of sodium carbonate buffer, pH 10.0, at 4°.

2. Add 20 μl of a *p*-dioxane solution containing 20 mg of TCT per milliliter to the protein solution, and allow to react overnight at 4°.
3. Dialyze the protein solution for 2–4 hr against 4 liters of borate-buffered saline, pH 8.
4. Add 100 μl of packed SRBC (washed in PBS, pH 7.2) to the dialyzed protein–TCT conjugate at 37° and tumble the mixture on a rotating mixer overnight. Adjust the pH to 8.0 with 0.1 *M* NaOH as necessary.
5. Wash the SRBC 3 times in PBS, pH 7.2, and store at 4°.

Coupling of Haptens and Proteins to the Surface of SRBC via a Synthetic Lipopolysaccharide

Principle

The previous methods involved the coupling of a hapten or a protein to the surface of erythrocytes by means of a covalent bond. This last method attaches the antigen to the surface of a cell without a covalent bond to the membrane molecules. A lipopolysaccharide (myristoyl-oxidized dextran, MOD) has been designed to which haptens and proteins can be covalently coupled. The antigen–MOD directly attaches to the surface of cells via a stable hydrophobic interaction with the plasma membrane.[34] It is believed that the antigen–MOD is attached to the surface of red cells via the intercalation of the lipid moiety of the polysaccharide into the hydrophobic portion of the plasma membrane. The covalent attachment of haptens and proteins to the MOD occurs between free amino groups present on the haptens or proteins and reactive aldehyde groups on the MOD. Thus, the synthetic lipopolysaccharide can be used as a general method for coupling haptens or proteins to red blood cells. This same approach has been used to attach various polysaccharides to erythrocytes in order to assay for anti-carbohydrate antibodies.[52–60] In the preparation of MOD it was found that a high molecular weight polysaccharide, such as dextran B-1355 (M_r, 40×10^6; Allene Jeanes, personal communication), gave the most sensitive target cells. The large dextran molecule may increase the sensitivity of the target cells by functioning as a spacer molecule to separate the

[52] W. J. Halliday and M. Webb, *Aust. J. Exp. Biol. Med. Sci.* **43**, 163 (1965).
[53] M. Landy, R. P. Sanderson, and A. L. Jackson, *J. Exp. Med.* **122**, 483 (1965).
[54] G. Möller, *Nature (London)* **207**, 1166 (1965).
[55] V. Hämmerling and O. Westphal, *Eur. J. Biochem.* **1**, 46 (1967).
[56] P. J. Baker and P. W. Stashak, *J. Immunol.* **103**, 1342 (1969).
[57] O. Pavlovskis and H. D. Salde, *J. Bacteriol.* **100**, 641 (1969).
[58] R. Kearney and W. J. Halliday, *Aust. J. Exp. Biol. Med. Sci.* **48**, 227 (1970).
[59] C. J. Sanderson, *Immunology.* **18**, 353 (1970).
[60] W. K. Cheung, M. E. Dorf, and B. Benacerraf, *J. Immunol.* **119**, 901 (1977).

antigen from the target cell surface.[8] Saturated fatty acids of C_{14} to C_{18} chain length appear to be suitable for use as the lipid moiety, although we have used primarily myristoyl dextran. It should be noted that target cells for some proteins (i.e., those proteins that will bind to the lipid groups such as serum albumin) cannot be prepared by this method.

The hapten or protein myristoyl dextran compounds attach directly to the cell membrane when incubated with erythrocytes. This occurs rapidly and under physiological conditions. The conjugating reagent, hapten–myristoyl dextran or protein–myristoyl dextran, is chemically stable and can be stored at 4° for extended periods of time. The target cells, which can be prepared in about 2 hr, can detect picogram levels of antibody in hemolytic assays. The target cells can also be used in hemagglutination assays or immunocytoadherence assays. In addition, only small amounts of antigen–myristoyl dextran reagent are required to produce the target cells.

A variety of haptens, N-ε-DNP-lysine, 3-aminopyridine, 4-aminopyridine, 4-aminophthalate, and 5-aminoisophthalate, have been successfully conjugated to cell membranes using MOD as the linking reagent. Proteins that have been successfully coupled by this method include Bence-Jones protein, human γ-globulin, bovine γ-globulin, rabbit anti-human Fab antibody, mouse anti-phthalate antibodies, and monoclonal anti-phthalate antibody.

Materials and Reagents

Myristoyl chloride may be purchased commercially from Pfaltz and Bauer (Fairfield, Connecticut). Dextran fraction S from *Leuconostoc mesenteroides,* NRRL B-1355, can be obtained by writing Dr. Slodki, c/o Biochemistry Research Fermentation Laboratory, 1815 North University Street, Peoria, Illinois 61604. SRBC can be prepared as described above for the SPDP coupling procedure. All other chemicals are reagent grade and used as commercially supplied.

Procedure

Preparation of Myristoyl Oxidized Dextran. Myristoyl dextran is prepared as described by Tsumita and Ohashi[61] and later modified by Leon *et al.*[62]

1. Suspend 40 mg of dextran B-1355 in 4 ml of anhydrous pyridine.
2. Add 25 μl of myristoyl chloride and stir the reaction mixture at room temperature for 48 hr.

[61] R. Tsumita and M. Ohashi, *J. Exp. Med.* **119,** 1017 (1964).
[62] M. H. Leon, N. M. Young, and K. R. McIntire, *Biochemistry* **9,** 1023 (1970).

3. Add 20 ml of isopropyl alcohol to precipitate the myristoyl dextran, and wash the precipitate three times with isopropyl alcohol to remove excess fatty acid.
4. Wash the precipitate with ethyl ether and air dry.
5. Dissolve myristoyl dextran in 0.01 M acetate buffer at pH 6 (concentration of 10 mg of dextran per milliliter of buffer).
6. Oxidize myristoyl dextran by addition of sodium periodate to make a final concentration of $1 \times 10^{-2} M$ periodate and allow to react at room temperature for 1 hr.
7. Dialyze against 13 liters of borate-buffered saline, pH 8.
8. Store the solution at 4°.

Preparation of Hapten–Myristoyl Dextran

1. Add 2 ml of a $2.5 \times 10^{-2} M$ aminohapten solution in borate buffer, pH 8.6, to 2 ml of $2.5 \times 10^{-7} M$ myristoyl dextran solution as prepared above (assuming the dextran has a molecular weight of 40×10^6).
2. Adjust the pH to 8.6 and stir at room temperature for 18–24 hr.
3. The resulting product can be used without further modification; however, if storage for an extended period of time is expected, then the Schiff base product should be reduced with excess sodium borohydride followed by dialysis to stabilize the linkage between the hapten and the myristoyl dextran.

Preparation of Protein–Myristoyl Dextran

1. Add 2 ml of a protein solution (1–10 mg of protein per milliliter of borate buffer, pH 8.6) to 2 ml of myristoyl-oxidized dextran (10 mg/ml).
2. Stir the reaction mixture at room temperature for 18–24 hr and use the reaction mixture directly for preparing target cells.

Attachment of Haptens and Proteins to SRBC via the Myristoyl Dextran Conjugating Reagent

1. Wash the SRBC twice at room temperature with phosphate-buffered saline (PBS), pH 7.2, immediately before sensitizing.
2. Incubate 1 volume of a 2.5 suspension (v/v) of SRBC in PBS with an equal volume of 25 μg/ml of the hapten–myristoyl dextran or protein–myristoyl dextran on a rotating mixer for 1 hr at room temperature.
3. Wash the cells three times with PBS.
4. Store at 4° until used for assays.

[22] Iodine Monochloride (ICl) Iodination Techniques

By M. Angeles Contreras, William F. Bale,* and Irving L. Spar

For labeling proteins and related substances in solution with ^{131}I or ^{125}I, the use of the ICl isotopic exchange procedure may often be the method of choice. The actual procedure is simple, since the reagents used are stable over long periods of time. In this technique the total amount of iodine incorporated into the iodinated material is known and can be controlled, and it avoids the use of oxidizing agents that may damage protein. These advantages were pointed out by McFarlane,[1] who developed an early version of this labeling procedure.

The basic method, as it is most widely used today,[2] is as follows: [131]I or ^{125}I in the form of iodide ions (often described by commercial sources as NaI) are mixed with the protein solution to be iodinated, buffered to a pH of approximately 8. An appropriate amount of ICl, in a weakly acidic NaCl solution, is then stirred rapidly with the protein–NaI mixture. The total buffering capacity must maintain the entire mixture at a pH of approximately 8.

As a result of this addition of ICl, two reactions occur. Radioactive iodine, e.g., ^{125}I, present as iodide, is converted to [125]ICl by very rapid chemical exchange with nonradioactive ICl. Iodination of protein takes place; it can be considered to occur as a result of the following reactions.

$$ICl + NaOH \rightarrow HOI = NaCl$$
$$HOI + \text{tyrosine residue} \rightarrow \text{iodinated tyrosine residue} + H_2O$$

This latter reaction is also rapid, so in 1 min a protective protein may be added and the labeled preparation be removed from the reaction vial.

Scope of Usefulness

Since the only oxidizing agent present during the iodination procedure is the ICl itself, no harm can come to protein that would be produced by other oxidizing agents. This contrasts with the oxidizing effects of chloramine-T used to oxidize iodide to iodine in the labeling procedure of that name. This advantage may sometimes be important when working with labile proteins. Since the amount of total iodine incorporated in the la-

---

* Deceased, June 28, 1982.

[1] A. S. McFarlane, *Nature (London)* **182**, 53 (1958).

[2] W. F. Bale, R. W. Helmkamp, T. P. Davis, M. J. Izzo, R. L. Goodland, M. A. Contreras, and I. L. Spar, *Proc. Soc. Exp. Biol. Med.* **122**, 407 (1966).

METHODS IN ENZYMOLOGY, VOL. 92

beled protein can never exceed the amount added as ICl, this sets an upper limit on the average number of iodine atoms attached to each molecule of labeled protein. This level of attached iodine is sometimes uncertain by other labeling procedures, since it cannot be readily controlled.

Three variants of the basic ion-exchange ICl iodination method will be considered. First to be described is a macro method in which the amount of protein to be labeled amounts to several milligrams, and the volume of solution in which it is contained is in the range of a few milliliters. This procedure is easily adapted to remote handling techniques behind γ-ray shields and has been used for attaching to 4 mg of immunoglobulin (IgG) 100 mCi or more of [131]I. Second to be described is a micro method for labeling amounts of protein ranging from less than 100 μg to several milligrams. The volume of the product after completing the labeling operation can be held to less than 1 ml. It is a good general procedure, but has not yet been adopted for remote shielded handling of large amounts of radioactivity. The third procedure to be described is a variant of the second in which the labeling procedure is carried out directly in suitable ^{125}I or ^{131}I shipping vials. Finally there will be a short discussion of techniques in which radioactive ICl is produced by oxidation of total iodide in radioactive NaI preparations.

Investigators proposing to use the ICl method for labeling should be aware of two limitations of this method. Since in this procedure no oxidation of iodide ion is carried out in the presence of protein to be labeled, any reducing substances present can compete for positive iodine with this protein. Therefore the presence of such substances should be kept low. Also, the ICl method has not been found to be a useful procedure for directly labeling cells or cell membranes; it is useful in preparing reagents for other types of ^{125}I or ^{131}I labeling—e.g., the preparation of diazotized diiodosulfonilic acid for labeling of cell membranes.[3]

Reagents

Iodine Monochloride Reagent. A convenient stock solution of ICl is 0.02 M in ICl, 2.0 M in NaCl, 0.02 M in KCl, and 1.0 M in HCl. It should be noted that iodine monochloride has a long history as a convenient term for referring to what is more strictly the hydrochloric acid solution of the trihalide ion (ICl_2^-). As described more fully elsewhere,[2] it can be prepared as follows: To a solution of 0.5550 g of KI, 0.3567 g of KIO_3, and 29.23 g of NaCl is added 21 ml of concentrated HCl (sp. gr. 1.18) and water to make the volume to 250 ml. The slight amount of free iodine is

[3] R. W. Helmkamp and D. A. Sears, *Int. J. Appl. Radiat. Isot.* **21**, 683 (1970).

then removed by passing through the solution a current of air saturated with water vapor for a few hours. It is worthwhile to ascertain that the air for aeration is truly saturated and that no reduction in volume occurs during this aeration procedure. Completeness of free iodine removal is tested by extracting a few milliliters of the ICl preparation with CCl_4. The separated CCl_4 fraction should be colorless with only a faint pink blush when the extraction is complete. If care has been taken that the air being used for iodine extraction is fully water saturated, the final ICl solution should be within 1% of the calculated 0.02 M value. It is indefinitely stable at room temperature. The exact molarity of the solution can be determined by adding an excess of KI to an aliquot and titrating the liberated iodine with a standardized thiosulfate solution.

^{125}I *Solutions.* Several sources of ^{125}I-labeled iodide solutions are available suitable for use by procedures described here. High specific activity preparations prepared without reducing agent are preferable. For use with the micro method, some necessary restrictions on volume and alkali content will be given later.

Some quantitative knowledge of the characteristics of ^{125}I are needed in setting up procedures that give a satisfactory degree of ^{125}I coupling to protein and little or no damage to the labeled protein by attaching too much iodine to it. Pure ^{125}I contains 4.596×10^{-7} mole or 57.44×10^{-6} g of iodine per Curie, which by definition is undergoing 3.7×10^{10} disintegrations per second. Let it be assumed that 10 mCi of ^{125}I, 4.596×10^{-9} mol, without additional iodine present as impurity, is used to label 4 mg of IgG, and that 4 mol of ICl are used per mole of protein (that is, 4 molecules of ICl per protein molecule). Four milligrams of this protein equals 2.5×10^{-8} mol, and the ICl to be used is $4 \times 2.5 \times 10^{-8} = 1.0 \times 10^{-7}$ mol. Thus after complete isotopic equilibrium has been reached in the exchange reaction between iodide and ICl, the fraction of ^{125}I available for labeling will be $(1.0 \times 10^{-7})/(1.0 \times 10^{-7} + 4.596 \times 10^{-9}) = 0.9561 = 95.6\%$.

Even when the amount of IgG to be labeled is reduced to 1 mg, and the amount of ICl is reduced accordingly, still most of the ^{125}I, in this case 84.5% is available for labeling. The situation changes significantly when 0.1 mg of IgG is to be labeled with 10 mCi ^{125}I, at the previously considered ratio of 4 mol of ICl to 1 mol of IgG (1 ICl molecule per 40,000 daltons protein). The protein represents 6.25×10^{-10} mol of IgG, and the corresponding ICl equals 2.5×10^{-9} mol. Thus, after the exchange reaction is completed, the fraction of ^{125}I available for labeling will be $(2.5 \times 10^{-9})/(4.596 \times 10^{-9} + 2.5 \times 10^{-9}) = 0.352$, or 35.2%.

The fraction of ^{125}I available for iodination can be increased by using larger amounts of ICl, but at the same time increasing the possibility of protein damage by overiodination. However, one may note that the labeling of hormones with ^{125}I by the chloramine-T method for radioimmunoassay is often carried out at levels of 50–100 mCi of ^{125}I per milligram of protein.[4]

^{131}I *Solutions.* Stable ^{127}I and ^{129}I (1.6 × 10^7 year half-life) are produced in substantial amounts,[2,5] when ^{131}I is produced by neutron activation of normal tellurium, probably the method most used at the present time. If the neutron radiation time is 24 hr, and the resulting iodine is processed and shipped during the next 48 hr, the iodine at shipping time would be 44% ^{127}I, 23% ^{129}I, and 33% ^{131}I. Thus, in practice, it is difficult to obtain and use in the laboratory ^{131}I in which the ^{131}I exceeds 20% of the total iodine content.

Increases in the time of irradiation and delays in processing, shipping, and use can lead to large increases in non-^{131}I iodine. For example, if the neutron irradiation time is 10 days, and the time before use for iodination another 8 days, ^{131}I will amount to only 12% of the total iodine. Occasionally iodine obtained from suppliers contains substantial amounts of nonradioactive iodine which have been added to minimize absorption effects. Such iodine is entirely unsuited for labeling purposes, and the labeling efficiencies achieved are close to zero.

Iodine-131 less diluted by other iodine isotopes can be obtained as iodine produced by uranium fission. Such iodine has been used in experimental and clinical cancer therapy attempts with antibody carrying doses of ^{131}I in the therapeutic range.[6,7] Tellurium, highly enriched in ^{130}Te, which was produced at Oak Ridge National Laboratory, would be an ideal source for ^{131}I production. By this latter procedure, iodine could be produced that would be almost entirely ^{131}I.

Borate Buffer. This buffer, pH 8, is prepared by adjusting a distilled water solution of 0.16 M NaCl and 0.20 M H$_3$BO$_3$ with 1.6 M NaOH to pH 8, with a final NaOH concentration of approximately 0.04 M. Borate buffer with twice these amounts of NaCl and H$_3$BO$_3$, adjusted to pH 7.65 with 1.6 M NaOH, is termed 2× borate buffer. When diluted with an equal volume of distilled H$_2$O, the pH becomes 8.

Protein Solutions for Labeling. The protein to be labeled is dissolved or dialyzed against borate buffer. The glycine buffer suggested by Mc-Farlane,[1] is also satisfactory but gives slightly lower labeling yields. As

[4] D. N. Orth, this series, Vol. 37, p. 22.
[5] W. F. Bale, R. W. Helmkamp, T. P. Davis, M. J. Izzo, R. L. Goodland, and I. L. Spar, "High Specific Activity Labeling of Proteins with ^{131}I," Report UR-604, AEC TID-4500 (16th ed.), 1962.
[6] I. L. Spar, W. F. Bale, D. Marrack, W. C. Dewey, R. J. McCardle, and P. V. Harper, *Cancer* **20**, 865 (1967).
[7] W. C. Dewey, W. F. Bale, R. G. Rose, and D. Marrack, *Acta Unio Int. Cancrum* **19**, 185 (1963).

noted earlier, any reducing substances present in the iodination mixture can compete with tyrosine residues for positive iodine. One common source of such reducing substances is cellophane dialysis tubing, particularly if the substance to be labeled has been concentrated in such tubing under reduced pressure. Therefore, if dialysis tubing is used, it should be purchased or extracted so as to be as free of reducing substances as possible. Proteins concentrated by reduced-pressure dialysis in cellophane or other membranes containing reducing substances that may be solubilized, should be dialyzed further to remove such reducing substances.

Under some conditions to be mentioned later, dialysis against 2× borate buffer may be preferable because of its greater buffering power.

Gel Separation Columns. For separating bound and unbound radioactive iodine, Sephadex G-25 or G-50 may be used. The size of the column should be such that an adequate separation is achieved between the fractions containing the labeled protein and the unreacted iodine or iodate. It is best to first ascertain the void volume with Blue Dextran and the point of inorganic iodine breakthrough using a small test sample of unreacted radioactive iodide. When passing the final labeled preparation through the column, the first burst of radioactivity should appear just behind the void volume for proteins in the immunoglobulin size range.

Ion Exchange Column. Dowex 1-X4 resin (50–100 mesh) is allowed to stand in contact with 1 N HCl for several hours, then washed successively with 20% NaCl solution, 0.85% NaCl solution until the filtrate is neutral, and then stored under the latter solution. A column is formed by placing 2.5 ml of this resin in a 3-ml disposable syringe containing a small plug of glass wool. The syringe is filled with 0.85% NaCl and then drained just before use.

It was found that such a column removed all iodide ions from an iodinated protein solution, and that no significant iodide ion was eluted on subsequent washing of the column with 1.5 ml of normal rabbit serum or 0.5 ml of this serum followed by 1 ml of saline solution. Despite this rinse, however, 4–8% of the iodinated IgG was still retained by the resin.

Catalase Solution. Sterile preparations, without preservatives, assaying 30,000 units or more per milligram of protein are used to destroy H_2O_2 that is sometimes present in high-activity commercial ^{131}I preparations.[2]

Macro Method for ^{125}I or ^{131}I Labeling

Apparatus

The setup shown in Fig. 1 is widely used for low and high level labeling with ^{131}I and ^{125}I and is well adapted for remote handling and good radia-

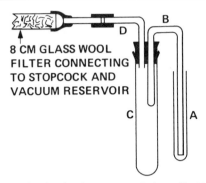

8 CM GLASS WOOL
FILTER CONNECTING
TO STOPCOCK AND
VACUUM RESERVOIR

FIG. 1. Apparatus for iodination by the macro technique. Test tubes used at location A can be of any appropriate size to hold reagents. C is a 2.2 cm internal diameter × 15 cm Pyrex test tube. B is a 0.1 cm internal diameter capillary tube, slightly constricted at the end in tube C. Its effective aperture is such that the application of strong suction through D will jet 5 ml of water from A to C in about 1 sec.

tion shielding for ^{131}I preparations in the 100–300 mCi range. A combination of lead bricks and transparent lead glass shielding, of thick-walled lead tube holders, and remote-handling grip tools that can work around corners make isotopic labeling followed by ion exchange or molecular sieve passage possible with negligible exposure to the operator.

This procedure will first be described for labeling 4 mg of IgG, assumed molecular weight 160,000, dissolved in or dialyzed against borate buffer, with 4 equivalents of ICl. This is a level of 1 ICl molecule per 40,000-dalton protein.

The necessary dilution of ICl from the 0.02 M stock solution is calculated, and a subdilution is made as follows: 4 mg of IgG represents 2.5×10^{-8} mol of protein; four equivalents of ICl is therefore 1×10^{-7} mol. To provide this amount of ICl in 0.2 ml of solution, a dilution containing 5×10^{-7} mol/ml is needed. One milliliter of the stock ICl solution will contain $0.02/1000 = 2 \times 10^{-5}$ mol. Therefore the dilution factor needed will be $(2 \times 10^{-5})/(5 \times 10^{-7}) = 40$.

Step 1. To make this dilution 0.2 ml of the stock ICl solution is mixed with 7.8 ml of 2 M NaCl solution already added to a clean disposable test tube. (In a labeling procedure where a different amount of ICl is required, the dilution is calculated and made in such a way that the required amount of ICl is present in 0.2 ml of solution close to 2 M in NaCl.)

Step 2. The protein to be iodinated, in a volume of 3 ml or less, is placed in tube C, and the iodination apparatus is assembled. Alternatively, it may be added by suction through tube B after the apparatus is assembled.

Step 3. The radioactive NaI solution, made to a volume of 2–3 ml with borate buffer, is added to the protein solution by suction.

Step 4. Immediately 0.2 ml of the subdilution of ICl prepared under step 1 is added to 1.8 ml of 0.85% NaCl solution and mixed. Then, by rapid suction, this preparation is added as a jet to the iodine–protein solution.

Step 5. Labeling occurs rapidly, so 1 min later, 1 ml of protective protein (20–60 mg) is added to protect against radiation effects and destruction of protein that may occur by such phenomena as surface denaturation in dilute solution. Typical protective proteins used are 6.25% solutions of human or animal albumin, citrated human or animal plasma, normal serum, or a solution of gelatin.

Unbound radioactive iodine can be removed from the labeled preparation by ion-exchange resin, dialysis, or passage through a suitable gel filtration medium, such as Sephadex G-25.

The ion-exchange resin is probably the easiest to use for moderate to high molecular weight proteins such as IgG. After passage of the labeled preparation through the resin at the rate of about 1 ml/min, the resin is rinsed with an additional 1–1.5 ml of protective protein. This procedure has a disadvantage in that the resin does not retain iodate, and with a rinse of 1 ml of normal serum it still retains a few percent of IgG. However, the presence of iodate is rarely, if ever, a practical problem, except in old ^{125}I-labeled preparations. Also, it can be tested for by a preliminary passage of a small portion of the radioactive preparation to be used for iodination through the resin. Iodide will be retained by the column, but iodate will come through in the first 10 ml of the effluent. If the iodinated material is one retained by the resin, such as insulin, glucagon, and many other substances of relatively low molecular weight, dialysis or gel filtration can be used. Dialysis has the disadvantage that dialyzable radioactive iodine is distributed through large volumes of fluid that may present a disposal problem.

Radioisotope calibrators, of the type used in nuclear medicine, are convenient for measuring the ^{131}I content of different fractions during iodination procedures, but almost any equipment capable of giving good relative γ-ray intensity measurements can be utilized.

This general procedure has proved to be well adapted for the attachment of ^{131}I to antibody for the purposes of human cancer therapy. Here it is desirable that relatively large amounts of ^{131}I be attached to protein, 100 mCi or more for individual patients, so that the amount of administered foreign protein can be kept small, not more than 4–5 mg if possible if the antibody is a protein of xenogenic or allogenic origin, to slow or avoid the possibility of allergic sensitization. Up to the time iodination is complete, the ^{131}I is transferred only by suction into closed containers, greatly reducing the possibility of radioactive aerosol formation. Equipment can easily be devised for carrying out the necessary procedures in a hood behind shields of lead or lead glass.

Use of Catalase

In earlier versions of the ICl iodination procedure,[8] in which radioactive ICl was formed first and then jetted into the substance to be labeled, it was found that hydrogen peroxide generated by the action of ^{131}I β-rays on water, first oxidized any reducing substances present and then produced free H_2O_2 that was able to destroy ICl. This reduced or prevented iodination. This H_2O_2 effect is reduced by the procedure in which ICl is added to the protein mixture, as described here, but this effect has not been studied at high ^{131}I levels. Catalase under these circumstances may still be useful in increasing ^{131}I iodination efficiency. As noted earlier, the catalase preparation should be of such purity that the amount used should be, in total protein content, a very small fraction of the material to be iodinated with ^{131}I.

To use catalase for H_2O_2 destruction before the labeling procedure is carried out, the Na^{131}I solution should first be adjusted to approximately pH 8; 600 units of catalase are added, the mixture is allowed to stand for 10 min, and then step 3, the addition of the Na^{131}I to protein, is carried out. Protein, added as catalase, is 0.5% or less of the 4 mg of protein to be coupled to ^{131}I.

Micro Method for Radioactive Iodine Labeling

The apparatus shown in Fig. 2 can in principle, and often in fact, be used for labeling protein in approximately the same amounts as in the macro method just described. However, procedures have not been developed for using it in a completely shielded operation. It has the advantage that, since the volumes used are small compared to those used in the macro method, protein concentrations can be kept higher during equivalent labeling procedures, and the effects of surface denaturation can be reduced.

The general procedure adopted to labeling 0.1-mg portions of IgG with 10 mCi amounts of ^{125}I, and some of the precautions to be taken, is as follows:

Step 1. A 1-inch, 21-gauge needle is inserted partially (0.5 inch) through the vial septum, and a 3.0-ml disposable syringe barrel, packed

---

[8] R. W. Helmkamp, R. L. Goodland, W. F. Bale, I. L. Spar, and L. E. Mutschler, *Cancer Res.* **20**, 1495 (1960).

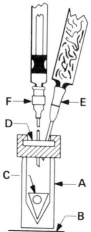

FIG. 2. Apparatus for iodination by the micro technique. A is the iodination V-vial.* B is the magnetic stirrer top. C is a triangular matrex Teflon-coated magnetic stirrer bar.* D is a self-sealing laminated disk.* E is a 21 gauge, 1-inch needle attached to a 3-ml disposable syringe barrel that has been packed, not too firmly, with glass wool. F represents a series of three hypodermic needles and syringes used in sequence during the iodination procedure. Stainless steel needles are recommended because of their greater strength. Lock-type syringes and stainless steel needles should be used when possible. Two-inch, 18 gauge needles are required for removing the labeled material from the iodination vial and often are necessary in transferring radioactive iodine from the shipping container to the iodination vial. * Purchased from Pierce Chemical Co., Rockford, Illinois.

with glass wool, is attached to the needle. (This step may be omitted if great care is taken that subsequent steps are conducted in such a way that positive pressure does not build up inside the vial).

Step 2. Preparation of Diluted ICl. The amount of protein to be labeled by this procedure, 0.1 mg, is 1/40th (2.5%) of the amount used with the macro procedure just described. Therefore at the same protein: ICl ratio used for the macro procedure, an additional 1:40 dilution will be required. This is obtained by making the first 1:40 dilution described under the macro procedure. It will be used for making a further 1:40 dilution by mixing 0.2 ml of this preparation with 0.85% NaCl solution as a part of step 5.

Step 3. The protein to be labeled, preferably not exceeding 0.2 ml in volume, is added with a tuberculin syringe that is equipped with a 20-gauge needle and has been prerinsed with borate buffer.

Step 4. The Na^{125}I solution, preferably not exceeding 0.2–0.3 ml in volume, diluted if necessary in borate buffer, is added with a syringe attached to an 18-gauge needle prerinsed with buffer. If feasible, it is desirable that the needle be locked to the syringe, since accidental disengagement at this union has been the site of at least one iodine spill.

Step 5. Iodination. The rotating vane is actuated briefly to mix protein and ^{125}I solutions. Then the final dilution of ICl is made by mixing 0.2 ml of the first 1 : 40 dilution to a tube containing 7.8 ml of 0.85% NaCl solution. Then 0.2 ml of this 1 : 1600 dilution of the stock ICl solution is drawn into a tuberculin syringe and, with the vane rotating briskly, expelled rapidly through the septum into the $Na^{125}I$–protein mixture. The vane is stopped after 2–3 sec.

Step 6. One minute later, protective protein is added and the labeled protein is withdrawn with a syringe with a lock-type attachment to a 2-inch, 18-gauge needle. The labeled protein is passed through the resin column or another equivalent method for removing unbound radioactive iodine. The reactive vial is then rinsed with an additional amount of protective protein that is passed through the inorganic iodide removal apparatus.

Labeling in Shipping Containers

Some shipping containers, for example, the Combi-V-Vials supplied by New England Nuclear, can be used directly as reaction vessels for iodination. The main requirements are (*a*) a self-sealing septum through which reactants can be added and removed; (*b*) a vial volume adequate to accommodate the necessary reagents and mixing procedures; and (*c*) a method for rapid mixing during the addition of ICl. Experience shows that a Genie-Vortex mixer can be used with the Combi-V-Vial to provide such mixing. However, it is desirable to carry out preliminary tests with a nonradioactive vial and visual control to assure that mixing is adequate with the volumes of reactants used during the iodination procedure.

This third method, a modification of the micro procedure for labeling 0.1 mg of protein with 10 mCi of ^{125}I, will not be described. ^{125}I should be shipped at high concentration (>350 mCi/ml) in 0.1 N NaOH. Protein to be labeled should be dissolved in or dialyzed against borate buffer at as high a concentration of protein as feasible and then diluted in borate buffer (made from water and reagents free of reducing substances) to a level of 0.5 mg of protein per milliliter.

Step 1. A 1 : 40 dilution of stock 0.02 M ICl is prepared as described under step 2 of the micro iodination procedure.

Step 2. Protein to be labeled, 0.1 mg in a volume of 0.2 ml, is injected with a syringe into the vial containing $Na^{125}I$, and the two components are mixed. This may be done by inversion of the vial after the protein addition, to wash down any ^{125}I adhering to top of vial, or by a preliminary centrifugation before adding the protein.

Step 3. Iodination. This step is the same, in principal, as the step 5 of the micro iodination procedure. The final dilution of ICl is made by mixing 0.2 ml of the 1 : 40 dilution with 7.8 ml of 0.85% saline. Then, while the ^{125}I-protein mixture in the shipping vial is swirled with the Vortex mixer, 0.2 ml of the final ICl dilution (1 : 1600) is added rapidly by syringe. The mixing is stopped in 3–4 sec.

Step 4. One minute later, the protective protein is added. The continuing procedure is the same as described under step 6 of the micro procedure.

Iodination Using Radioactive ICl Produced by Oxidation of Total Iodine in NaI Preparations

Helmkamp *et al.*[9] have described a method for converting ^{131}I of $Na^{131}I$ preparations quantitatively to ICl. No exchange reaction is involved. The original paper should be consulted for details. It has the advantage that all the radioactive iodine can be utilized for iodination, no matter how much there may be. This technique has the corresponding disadvantage that the actual amount of total iodine incorporated in the labeled material will be unknown unless one knows in advance the total iodine content of the $Na^{131}I$ preparation used. Helmkamp and co-workers[10] have also developed a technique for determining the total iodine content of such $Na^{131}I$ preparations.

In this labeling procedure initially designed for labeling 4 mg of γ-globulin with 50–200 mCi of ^{131}I, any H_2O_2 present in the $Na^{131}I$ preparation is destroyed by catalase. The solution is then made acidic (pH approximately 3) with 0.25 N HCl. Saturated NaCl solution and 1 N HCl are then added in amounts such that, on subsequent addition of KIO_3 containing 5 times the estimated total iodide content of the $Na^{131}I$ preparation used for labeling, the solution will be approximately 0.05 N in HCl and 1 M in NaCl. Under these conditions the iodide ion is converted to ICl with reasonable rapidity; the iodate ion undergoes a negligible amount of ion exchange. Thus at the end of 10 min that mixture can be adjusted rapidly by addition of a calculated amount of NaOH to a pH around neutrality and jetted into the solution to be iodinated buffered to a pH of 8. From then on, the procedure follows the ICl exchange procedure described above.

Doran and Spar[11] have described an adaptation of the oxidative ICl iodination technique adapted to labeling microgram amounts of protein

[9] R. W. Helmkamp, M. A. Contreras, and M. J. Izzo, *Int. J. Appl. Radiat. Isot.* **13,** 747 (1967).
[10] R. W. Helmkamp, M. A. Contreras, and W. F. Bale, *J. Nucl. Med.* **7,** 491 (1966).
[11] D. M. Doran and I. L. Spar, *J. Immunol. Methods* **39,** 155 (1980).

with ^{125}I to high levels of specific radioactivity. They include a review of the qualities of the iodinated proteins produced by the ICl and other iodination methods.

Additional Considerations

Although only two examples of labeling of proteins with radioactive iodine are discussed in detail, it is generally simple to extrapolate from them to suitable procedures for different amounts of protein to be labeled, different levels of radioactivity, alternative iodine isotopes, and different ratios of moles of iodine to moles of protein.

In general, ^{125}I can be substituted for ^{131}I, and vice versa, with no change in procedure. The amount of radioactive isotope used can vary from a small fraction of 1 μCi to many millicuries provided volumes are kept to the described range and proper safety precautions are taken. The molar ratio of ICl to protein can be varied, except that one must take into account the fact that decreasing the molar ratio of ICl to protein increases the effect of reducing substances in a protein solution, and the fraction of radioactive iodine incorporated into protein may be reduced.

It is important to calculate and carry out ICl dilutions so that a desirable molar ratio of ICl to protein is used. It is common and reasonable to assume, with larger proteins, that one ICl per 40,000 daltons protein is a reasonable level. The examples given here, labeling at a level of 4 molecules of ICl per IgG molecule of assumed 160,000 molecular weight, represents this level. There is, however, reported evidence that some proteins, e.g., fibrinogen, are damaged at this level of labeling. When labeling proteins with a molecular weight less than 40,000, a molecular ratio of 1 between protein and ICl represents a reasonable starting point. A final criterion is that the labeled protein fulfills the function for which the labeling operation was carried out.

Probably the macro method of labeling should not be routinely used for labeling less than 2 mg of protein. Low iodination yields may result, since reducing substances are not easy to eliminate from reagents, and in addition, surface denaturation may produce unacceptable protein damage. On the other hand, the micro method has become the routine procedure in our laboratory for labeling 4-mg amounts of protein with 10 mCi of ^{125}I or ^{131}I.

It is necessary that the buffering capacity of the protein solution, or of the buffer added with it, be sufficient to bring the pH of the ^{125}I or ^{131}I solution to approximately 8, possibly from a much more basic pH, and then keep the reaction mixture at approximately pH 8 after addition of the acidic ICl solution. The procedures as described do not cover all possibilities. As one extreme example, one investigator desired to iodinate, if possible, all tyrosine residues of a protein and used enough ICl to do so. This introduced so much acid with the ICl, that the yellow ICl color persisted and no iodination occurred. The remedy was to prepare and use an ICl of much higher molarity. Some investigators have attempted to use ^{131}I solutions that contained so much base that the pH remained very high after adding the protein solution, and the protein was denatured and precipitated.

Effect of Reducing Substances

Use of ICl iodination techniques, with a ratio of 1 ICl molecule per 40,000 daltons protein, never resulted in radioactive iodine attachment to proteins approaching 100%. There are probably multiple reasons for this, but the major one is reported to be the presence of reducing substances present as a part of the protein to be labeled or in the fluid where the reaction occurs. McFarlane[1] has suggested a preliminary oxidation of protein by iodine at pH 4.5 as a means of eliminating reducing properties of protein to be iodinated, but it has never been adopted as a working procedure. As noted earlier, one common source of reducing substances is dialysis tubing. In one series of experiments on the micro iodination technique, the use, as a diluent, of borate buffer that had been in contact with cellulose dialyzing tubing under various conditions reduced efficiency of ^{125}I coupling from values of about 50% to values ranging from 39 to 2.5%.

Importance of Rapid ICl Mixing after Addition

If complete mixing has not occurred during the ICl addition to the radioactive iodine–protein mixture before the ICl has reacted with protein or has otherwise been destroyed, then some radioactive iodine will remain as unexchanged iodide and have no chance to react with protein. This will result in lower labeling effectiveness. In some experiments it can also lead to misinterpretation of results unless taken into account. For example, let it be supposed that the labeled preparation was an enzyme, and that the experiment was intended to determine whether labeling with approximately one atom of iodine per enzyme molecule would lead to complete loss of enzymic activity. If mixing were so incomplete that the ICl reacted with only one-half of the enzyme molecules, then 50% would still be unlabeled and would retain full enzyme activity, irrespective of whether one iodine atom per enzyme molecule could destroy enzyme activity, or not.

Radiation Damage

Proteins are particularly susceptible to radiation damage when in dilute solutions, since the ratio of free radicals and peroxides produced from irradiation of water to protein molecules is high. At a level of 10 mCi/ml the self-irradiation from ^{131}I is at a level of about 4,000 rad/hr. In one study,[12] ^{131}I-labeled rabbit antibody to rat fibrin was purified and irradiated with [60]Co γ-radiation or lightly filtered 250 KV X-radiation. In dilute solutions (less than 0.1 mg of protein per milliliter), 10,000 rad of ^{60}Co radiation rendered the uptake of the purified labeled antibody by rat fibrin to only 17% and at 40,000 rad to about 47%. Dosage of 200,000 rad of [60]Co γ-rays or 100,000 rad of X-ray reduced the specific uptake to virtually zero. At this latter dosage the presence of normal rabbit serum protein at a level of 8 mg per milliliter of irradiated solution reduced the radiation effects to approximately that produced by 10,000 rad.

It is important immediately to minimize radiation damage once the iodination reaction is complete. Such protective measures are the addition of nonspecific inert protein, dilution, then freezing, which are compatible with the intended use of the iodinated protein.

Efficiency of Coupling of Radioactive Iodine to Protein

Data from our own and other laboratories indicate that the macro ICl procedure, when used to couple radioactive iodine to protein at protein levels of a few milligrams and ICl levels of one ICl molecule per 40,000 daltons protein gives labeling levels of 50% or higher in most instances, usually at the 60–70% level. The major exceptions are instances where amounts of ^{131}I in the 100 mCi or higher levels were used, and the total amount of iodine involved was great enough to render the ICl exchange reaction an inefficient one. An extensive use of one adaptation of the macro method for insulin labeling gave preparations with relatively low efficiency of ^{131}I or ^{125}I coupling, but with reproducible production of biologically active insulin metabolized in a normal physiological manner.[13]

The use of the micro method is now standard in our laboratory for 10 mCi or less of radioactive iodine, and amounts of protein ranging from 0.1 mg to 4.0 mg. Labeling efficiencies are similar to the macro method at the

[12] R. W. Helmkamp, R. L. Goodland, W. F. Bale, I. L. Spar, and L. E. Mutschler, "High Specific Activity Iodination of Gamma-Globulin with Iodine-131 Monochloride," Report UR-568, AEC TID-4500 (15th ed.), 1960.
[13] J. L. Izzo, A. M. Roncone, D. L. Helton, and M. J. Izzo, *Arch. Biochem. Biophys.* **198,** 97 (1979).

level of 4 mg of protein, and much better, around 49–60%, when low millicurie levels of ^{125}I are used to label 0.13 mg of protein.

Isotopic Purity of ^{131}I Produced by Fission and from Tellurium

One currently experimental, but potential future routine, use of ^{131}I is as a therapeutic agent attached to tumor-localizing antibodies, either allogeneic or xenogeneic in nature. To diminish the amount of antibody that must be used, and keep low the possibility of allergic reaction, it is desirable to use ^{131}I of as high a specific activity as possible. The most common current method for producing ^{131}I, the irradiation of tellurium with slow neutrons, unavoidably also produces stable ^{127}I and ^{129}I with a 1.6×10^7 year half-life. The highest specific activity ^{131}I currently available contains approximately 100 μg of total iodine per Curie, of which only 8 μg is ^{131}I.

Table I shows the composition of iodine obtained after neutron bombardment of natural tellurium as a function of duration of irradiation and time after activation. Table II shows similar data for iodine produced by slow neutron-induced fission of ^{235}U. Earlier[2] our laboratory was able to obtain from Oak Ridge National Laboratory fission-produced ^{131}I with specific activities as high as 24 μg of total iodine per Curie of ^{131}I.

Inspection of Table I shows that this high specific activity is scarcely possible with ^{131}I obtained from natural tellurium. With sufficient demand,

TABLE I

ISOTOPIC PURITY OF ^{131}I PRODUCED FROM TELLURIUM[a]

Activation time (days)	Time after activation (days)	Percentage of total iodine			Micrograms total iodine per Curie ^{131}I
		^{127}I	^{129}I	^{131}I	
1	2	44	23	33	25
10	2	61	20	19	42
20	2	66	20	14	59
40	2	70	22	8	96
1	8	51	26	23	36
10	8	67	21	12	67
20	8	69	22	9	93
40	8	72	23	5	115

[a] Calculated relative amounts of neutron-induced iodine isotopes present 2 and 8 days after activation of natural tellurium in iodine separated immediately after activation.

TABLE II
ISOTOPIC PURITY OF ^{131}I PRODUCED BY FISSION OF [235]U[a]

Activation time (days)	Time after activation (days)	Percentage of total iodine			Micrograms total iodine per Curie [131]I
		[127]I	[129]I	[131]I	
1	8	0.1	22	78	10
10	8	3.0	34	63	13
20	8	5.0	42	53	15
40	8	8.0	55	37	22
1	32	0.4	68	31	26
10	32	7.0	76	17	47
20	32	10.0	78	12	66
40	32	12.0	81	7	116

[a] Calculated relative amounts of fission product iodine isotopes present 8 and 32 days after bombardment of [235]U in iodine separated immediately after activation.

fission-produced ^{131}I might again become available, or the production of ^{131}I from tellurium highly enriched in [130]Te become economically feasible.

Acknowledgments

This work was supported by USPHS Grant RO1 CA25958 from the National Cancer Institute, NIH, and by the Medical Research Foundation, Inc., Atlanta, Georgia.

[23] Application of High-Performance Liquid Chromatography to Characterize Radiolabeled Peptides for Radioimmunoassay, Biosynthesis, and Microsequence Studies of Polypeptide Hormones

By N. G. SEIDAH and M. CHRÉTIEN

Background

The use of radiolabeled peptides and proteins as tracers in radioimmunoassay (RIA), receptor binding, biosynthesis, and microsequencing studies has gained wide acceptance owing to the ease of preparation, the high specific activities obtainable, and the sensitivity of detection. High-performance liquid chromatography (HPLC) has been used to purify and characterize radiolabeled polypeptides as an efficient and rapid alternative to classical procedures such as polyacrylamide gel electrophoresis, thin-layer chromatography, isoelectric focusing, and gel permeation. This chapter describes HPLC methods used in our laboratory for the past few years for the purification and study of monoiodinated, ^{35}S-, ^{14}C-, and ^{3}H-labeled polypeptides for radioimmunoassay, for biosynthetic characterization of precursor-product relationships and peptide mapping for sequence homologies at lower than picomolar concentrations. Finally, we describe a new method of gel permeation by HPLC that is useful for molecules ranging in molecular weight from 500 to 90,000.

Instrumentation

The basic instrumentation used consists of two pumps (either single-piston or double-piston models), an injector (either manual or automatic), a column, a gradient programmer, a detector, a printer plotter, and a fraction collector. The pumps used in our laboratory are either Waters[1] 6000A or Beckman[2] Model 100A or 110A, all of which have practically very similar performances. A manual injector is preferred for preparative studies, whereas an automatic injector (WISP Model 710)[1] is used for method development and analysis of multiple samples.

The reverse-phase analytical columns used are μ-Bondapak C$_{18}$ (0.39 × 30 cm) (Waters)[1] and 5 μm ultrasphere-ODS (0.46 × 25 cm) (Beckman).[2] For gel permeation, we use a combination of four columns in series obtained from Waters,[1] namely, an I-60, two I-125, and one I-250 (0.78 × 30 cm) column in this order.

Reagents

The basic reagents used consist of reagent grade triethylamine[3] (doubly redistilled), 85% phosphoric[3] or 98% formic[4] acid, and HPLC grade of either acetonitrile or 2-propanol[5] solvents. The method consists of preparing a 1-liter bottle of stock solution of 0.2 M triethylamine phosphate (TEAP) or 0.2 M triethylamine formate (TEAF), pH 3.0, as originally introduced by Rivier[6]; i.e., 25 ml of triethylamine + 950 ml of doubly distilled and deionized H$_2$O + either H$_3$PO$_4$ or HCOOH to bring the pH to

[1] Waters Associates, Inc., Milford, Massachusetts 01757.
[2] Beckman Instruments, Inc., 1117 California Avenue, Palo Alto, California 94304.
[3] Fisher Scientific Co., Fair Lawn, New Jersey 07410.
[4] BDH Chemicals, 6730 Cote de Liesse, Ville St-Laurent, P.Q., Canada H4T 1E3.
[5] Burdick and Jackson Laboratories Inc., Muskegon, Michigan 49442.
[6] J. E. Rivier, *J. Liquid Chromatogr.* **1**, 347 (1978).

3.0, the volume being completed to 1 liter with H_2O. Unless otherwise stated, two bottles are then prepared as follows:

Bottle A: 100 ml of stock solution + 900 ml of H_2O, i.e., 0.02 M TEAP or TEAF, pH 3

Bottle B: 100 ml of stock solution + 900 ml of 2-propanol or CH_3CN

Alternatively, two bottles are prepared as follows:

Bottle A: 1 ml of trifluoroacetic acid[3] (redistilled TFA) + 999 ml of H_2O (0.1% TFA)

Bottle B: 1 ml of TFA + 999 ml of 2-propanol or CH_3CN (0.1% TFA)

These solutions are then filtered under vacuum through Millipore[7] filters of 0.45 μm (type HA) for solution A and 0.5 μm (type FH) for solution B while sonicating for 5 min in a water-bath sonicator. It is advisable to bubble helium continuously in both bottles (e.g., use Biolab[8]-type bottles) to prevent outgassing in the pumps.

Use of HPLC to Purify Monoiodinated Polypeptide Hormones for RIA Studies

Iodination Procedure

All peptides were radioiodinated by the chloramine-T method[9] for a short time to minimize the production of diiodinated derivatives.[10] Iodinations were performed in a final volume of 35 μl of 0.5 M phosphate buffer, pH 7.6, containing 1–5 μg of peptide, 0.4–0.6 mCi of carrier-free $Na^{125}I$, and 2–25 μg of chloramine-T with reaction times of 20–40 sec at room temperature. Iodination was stopped using 40–60 μg of sodium metabisulfite followed immediately either by gel filtration on Sephadex G-25 to remove free iodine or by passage through a Sep-Pak C_{18} cartridge[1,11] as follows: The silanol groups of the Sep-Pak cartridge are first activated by passage of 5 ml, through a syringe, of 100% solution B. This is followed by 30 ml of solution A to equilibrate the Sep-Pak with 100% A. The radioiodination mixture is then deposited on the Sep-Pak cartridge and eluted with 3 times 5-ml washes of 100% A; this should elute all the free ^{125}I from the cartridge. To elute the radiolabeled peptide, usually a 3× wash with 2 ml of B diluted to 30–50% with A should suffice. Counting the eluted fractions in a γ-counter will determine if the iodinated peptide has

---

[7] Millipore Corporation, Bedford, Massachusetts 01730.

[8] Mandel Scientific Co., 143 Dennis Street, Rockwood, Ontario, Canada NOB 2KO.

[9] W. H. Hunter and F. C. Greenwood, *Nature (London)* **194**, 495 (1962).

[10] W. H. Hunter, *Br. Med. Bull.* **30**, 18 (1974).

[11] H. P. J. Bennett, C. A. Browne and S. Solomon, *Proc. Natl. Acad. Sci. U.S.A.* **78**, 4713 (1981).

been eluted. The 6-ml elution volume can then be lyophilized, taking the necessary precautions to avoid contamination of the lyophilizer. Alternatively, it can be dried in a rotary evaporator.

Effect of Chloramine-T Treatment on Elution Position

The native, chloramine-T (CT-) treated and iodinated peptides[12] were analyzed by reverse-phase HPLC on a μ-Bondapak C_{18} column[1] (0.39 × 30 cm) using a single linear gradient elution program of 0 to 50% 2-propanol in TEAP buffer, pH 3.0 (i.e., 0% B/100% A to 55% B/45% A) completed in 75 min and run at a flow rate of 1 ml/min. Protein in the eluent was monitored by optical density at 210 nm, fractions of 0.5 ml were collected, and radioactivity was determined directly in a Beckman Model 300 γ-counter. As illustrated in Fig. 1, CT-treatment of unsubstituted peptides produced a shift to earlier elution volumes on HPLC for all peptides containing methionine,[12] e.g., Met-enkephalin (MET-ENK), α-MSH, β-MSH, γ-MSH, ACTH, and β-endorphin (β-END) (Figs. 1A, E, F, I, J, and K, respectively), while hormones lacking methionine coeluted with CT-treated peptides (denoted by asterisk in Figs. 1B, C, D, G, and H).

Since methionine is readily oxidized by chloramine-T treatment,[13] the change in elution volume for Met-containing peptides most likely reflects the production of peptide hormone bearing the oxidized form of this residue. This is confirmed by coelution of Met-enkephalin, subjected to complete performic acid oxidation, with the CT-treated Met-enkephalin. In contrast, Ci-treated and performic acid-treated lysine-vasopressin (LVP) were clearly separated on HPLC (Fig. 1G), suggesting that the oxidation of cysteine to cysteic acid does not occur under the iodination conditions used. Histidine also appears to be unaffected, as indicated by the coelution of untreated and CT-treated angiotensins I and II (Figs. 1D and C, respectively). Our results do not permit any conclusions regarding the fate of tryptophan, but its susceptibility to oxidation by chloramine-T has been clearly demonstrated.[14] It was observed that increasing the reaction time above 30 sec resulted in marked increase of the relative concentration of oxidized forms[12]; hence a reaction time below 30 sec would be recommended for those peptides containing Met and Trp residues.

[12] N. G. Seidah, M. Dennis, P. Corvol, J. Rochemont, and M. Chrétien, Anal. Biochem. **109**, 185 (1980).

[13] B. H. Stagg, J. M. Temperley, H. Rochman, and S. Morley, Nature (London) **228**, 58 (1970).

[14] C. B. Heward, Y. C. S. Yang, J. F. Ormberg, M. E. Hadley, and V. J. Hruby, Hoppe-Seyler's J. Physiol. Chem. **360**, 1851 (1979).

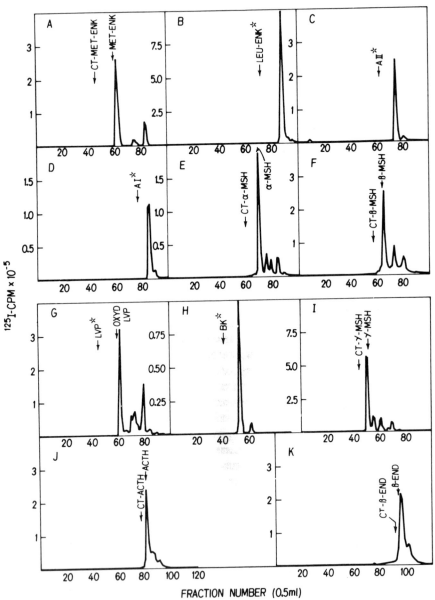

FIG. 1. Elution patterns from high-performance liquid chromatography of radioiodinated peptides. Peptides were radiolabeled by the chloramine-T (CT-) method, and free iodine was removed by Sep-Pak prior to chromatography as described. The asterisk denotes coelution of CT-treated and native peptides. The one-letter notation sequence of the peptides used is as follows: methionine-enkephalin, MET-ENK, YGGFM; leucine-enkephalin, LEU-ENK, YGGFL; angiotensin II, AII, DRVYIHPF; angiotensin I, AI, DRVYIHPFHL; α-melanotropin, α-MSH, acetyl-SYSMEHFRWGKPV-amide; β-melanotropin, β-MSH, DSGPY-

Separation of Iodinated from Unsubstituted Peptides

As illustrated in Fig. 1, iodination resulted in a marked increase in retention time on the C_{18} resin, allowing a clear separation of radiolabeled from CT-oxidized standards for all peptides studied. In addition, several radioactive peaks were resolved in this system. The major radiolabeled species, eluting first, represents monoiodinated peptide, followed by diiodo derivatives and minor contaminants of the peptide. Recoveries of ^{125}I-labeled peptides were greater than 90% in all cases. Careful inspection of Fig. 1, also reveals that monoiodinated and untreated peptides coeluted for those hormones containing residues susceptible to chloramine-T oxidation (Figs. 1A, E, F, I, J, and K). However, since oxidation is near completion in most cases following standard iodination conditions,[12] the monoiodinated derivatives can be recovered with minimal contamination by unlabeled hormone. Indeed, it was shown that under these conditions, HPLC purification allows one to obtain radiolabeled peptide tracers with specific activities approaching closely that of the theoretical maximum for a monoiodinated peptide.[12]

Radioimmunoassay Using HPLC-Purified Peptides

For use in RIA, it is preferred to utilize the TEAF or TFA[11] (see Reagents section) buffer system instead of TEAP, because of the volatility of the former. Under these conditions, the HPLC purified monoiodinated tracer can then be lyophilized to eliminate the TEAF, pH 3.0, or TFA buffer, which could interfere with the RIA procedure. The elution pattern in either TEAF or TEAP was almost identical in most of our studies. To demonstrate the utility of this technique in providing tracers of high specific activity for use in RIA, the antibody titer and sensitivity of a pituitary hormone γ-MSH[12] radioimmunoassay were determined before and after HPLC purification of monoiodinated γ-MSH (Figs. 2A and B). The results indicate that repurification of tracer permitted a threefold increase in antibody titer; i.e., the antibody dilution that produced 40% binding of added ^{125}I-γ-MSH, goes from 1 : 60,000 to 1 : 180,000 (Fig. 2A). Furthermore, as expected this produced an increase in the sensitivity of the γ-MSH RIA (Fig. 2B). In our hands 2-propanol is preferred to CH_3CN as

---

KMEHFRWGSPPKD; lysine-vasopressin, LVP, CYFQNCPKG; bradikinin, BK, RPPGFSPFR; γ-melanotropin, γ-MSH, YVMGHFRWDRFGR; adrenocorticotropin, ACTH, SYSMEHFRWGKPVGKKRRPVKVYPNGAEDESAQAFPLEF; β-endorphin, β-END, YGGFMTSEKSQTPLVTLFKNAIIKNAHKKGQ. In Fig. 1A, CT-MET-ENK and performic acid-oxidized MET-ENK coelute. Reproduced from Seidah *et al.*[12] by courtesy of *Analytical Biochemistry.*

A

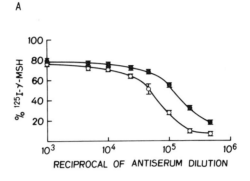

B

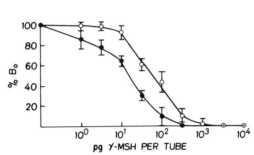

FIG. 2. Radioimmunoassay of γ-melanotropin (γ-MSH) (see legend of Fig. 1 for sequence) using [125I]γ-MSH as tracer purified either by silicic acid only (O——O) or repurified by high-performance liquid chromatography as described in Fig. 1 (●——●). The silicic acid method consists of diluting the iodination reaction mixture with 0.5 ml of heat-denatured human serum, followed by the addition of 50 mg of silicic acid and centrifuging. Free iodine is removed by washing the pellet five times with 1 ml of cold H_2O, and the iodinated peptide was eluted with 0.5 ml of HCl–acetone (60 : 40, v/v). (A) Titer curves: percentage of total added counts per minute of [125I]γ-MSH bound to antibody in the presence of increasing dilutions of anti-γ-MSH antiserum. (B) Standard inhibition curves: percentage B_0 vs amount of cold γ-MSH. Reproduced from Seidah et al.[12] by courtesy of Analytical Biochemistry.

eluting solvent, especially for peptides with a molecular weight greater than 5000.

Conclusion

The HPLC technique described allows routine purification of a wide range of monoiodinated peptide hormones free of unlabeled and diiodo derivatives. The method is fast and simple and provides quantitative recovery of monoiodinated tracers for use in RIA and receptor-binding studies. The oxidation of Met and Trp residues by the iodination procedure might cause problems if the susceptible amino acid residues are necessary for antibody or receptor binding, as has been demonstrated in a number of instances.[13,14] This problem might be circumvented by raising antibodies to the oxidized form of the hormone and pretreating standards and samples with an oxidizing agent prior to RIA, as described for Met-enkephalin[15] or by using biologically active analogs lacking residues susceptible to oxidation for receptor-binding studies.[14]

Coupling HPLC Purification of Peptides with RIA and Microsequencing

The purification of polypeptide hormones and their precursors has been achieved in a number of instances using reverse-phase HPLC.[16,17] In Fig. 3, a typical HPLC fractionation pattern of a human pituitary extract is presented. This method,[17] using a μ-Bondapack C_{18} column and the TEAF–2-propanol eluents described in the Reagents section, allowed the purification of a novel human glycopeptide representing the NH_2-terminal segment of pro-opiomelanocortin, the precursor of the pituitary hormones ACTH and β-endorphin.[18] As shown in this figure, a quantitative radioimmunoassay using an antibody directed against this molecule allowed determination of this peptide in peaks IV, V, and VI. A similar purification scheme was achieved on the reduced and carboxymethylated molecule with [^{14}C]iodoacetamide.[16] Hence the use of the buffer TEAF, pH 3.0, which is lyophilizable, allowed efficient recovery of this peptide from human pituitary extract in a form amenable to sequencing,[17] and that retained biological activity.[17]

In this example, a semipreparative μ-C_{18} column was used (0.78 × 30 cm).[1] This allows processing of up to 100 mg of human pituitary extract per injection, without loss of resolution. In other experiments where a smaller amount of material to be analyzed is available, e.g., less than 5 mg, the use of an analytical (0.39 × 30 cm) column would be recommended. In such a case, use of a fraction collector where fractions of 0.2–0.5 ml can be collected, lyophilized, and then tested by RIA is recommended. The use of 0.02 M TEAF buffer permits the monitoring of the

[15] V. Clement-Jones, P. J. Lowry, L. H. Rees, and G. M. Besser, *Nature* (*London*) **283**, 295 (1980).

[16] N. G. Seidah, R. Routhier, S. Benjannet, N. Larivière, F. Gossard, and M. Chrétien, *J. Chromatogr.* **193**, 291 (1980).

[17] N. G. Seidah, J. Rochemont, J. Hamelin, M. Lis, and M. Chrétien, *J. Biol. Chem.* **256**, 7977 (1981).

[18] M. Chrétien, S. Benjannet, F. Gossard, C. Gianoulakis, P. Crine, M. Lis, and N. G. Seidah, *Can. J. Biochem.* **57**, 1111 (1979).

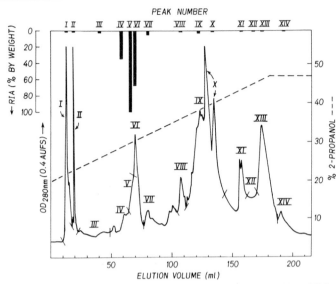

FIG. 3. High-performance liquid chromatography on a 5 μm ultrasphere-ODS column[2] of a human pituitary extract. The 2-propanol linear gradient used is shown as a dashed line (---) on the chromatogram. Material from peaks I–XIV was collected and lyophilized individually. A quantitative radioimmunoassay (RIA) using an antibody directed against the N-terminal portion of pro-opiomelanocortin[17] was performed on a weighed fraction from each peak. The results, expressed as percentage by weight of immunoassayable human material in each fraction, are shown as vertical bars in the upper section of the chromatogram. Reproduced from Seidah *et al.*[17] by courtesy of the *Journal of Biological Chemistry*.

optical density at 235 nm, thereby keeping the detection sensitivity high. If 0.02 M TEAP is used instead, monitoring at 210 nm is possible, but owing to the nonvolatility of this buffer, a great deal of interference with the antigen–antibody reactions could occur,[19] especially if the sample to be analyzed is available in minute quantities and the whole 0.2–0.5 ml fraction must be assayed.

This problem can be resolved[19] by increasing the concentration of buffer used in the RIA (e.g., sodium phosphate) to the highest limit the assay can tolerate (typically 150 mM to 200 mM), and by raising the pH to the highest acceptable value where good antigen–antibody binding is still observed (e.g., use pH 8.0–8.2). Then one can use the TEAP buffer to obtain the desired separation.

Alternatively, use of the volatile TFA system (see Reagents section) allows monitoring of eluents at 210–220 nm. Again a linear gradient from

[19] H. Akil, Y. Ueda, H. L. Lin, and S. J. Watson, *Neuropeptides* **1,** 429 (1981).

0% B to 40–50% B should achieve the separation. However, the resolution in this system could differ from the TEAP or TEAF buffer systems.

Molecular Sieving by HPLC

Methods often used to determine protein molecular weights include centrifugation, gel filtration on cross-linked dextrans or polyacrylamides, and sodium dodecyl sulfate–polyacrylamide gel electrophoresis. Even though reliable, these methods suffer from difficult sample recovery, limited fractionation range and time of separation. Several gel permeation columns suitable for HPLC are now available. These offer alternatives to classical procedures in terms of resolution, recovery, and speed of analysis. Detailed works were published on supports such as TSK-gels PW,[20] Synchropak GP-100,[21] and protein columns.[22,23] Many of these methods rely on the use of nonvolatile buffer eluents of high ionic strength; moreover, the use of denaturing agents, such as guanidine-HCl,[24] proved to be advantageous. Rivier[23] described the use of TEAF and TEAP buffers in gel permeation on protein columns of the type I-125.[1] Although we were able to duplicate Rivier's results, it was clear that the calibration curve was concave downward rather than linear, if the classical K_{av} versus log (M) parameters are used, where M is the molecular weight and

$$K_{av} = (V_e - V_0)/(V_t - V_0) \tag{1}$$

where V_e, V_0, and V_t represent the elution volumes of a retained peptide, the void volume of the column, and the salt position, respectively. This observation was found to be typical of gel permeation by HPLC, and Himmel and Squire[20] suggested the use of $F(v)$ versus $(M)^{1/3}$ defined as

$$F(v) = \frac{(V_e)^{1/3} - (V_0)^{1/3}}{(V_t)^{1/3} - (V_0)^{1/3}} = A_1 M^{1/3} + A_0 \tag{2}$$

where M, V_e, V_0, and V_t retain their meaning, A_1 and A_0 being constants for a given set of columns and elution conditions. Using this approach, it is possible to obtain a straight-line relationship between $F(v)$ and $(M)^{1/3}$ throughout the fractionation range of I-125 columns (500 to 40,000 daltons) using either 0.02 M TEAF or 0.02 M TEAP, pH 3.0. Extension of these results to a series of four protein columns (0.78 × 30 cm each) of

[20] M. E. Himmel and P. G. Squire, *Int. J. Pept. Protein Res.* **17**, 365 (1981).
[21] K. A. Gruber, J. M. Whitaker, and M. Morris, *Anal. Biochem.* **97**, 176 (1979).
[22] R. A. Jenik and J. W. Porter, *Anal. Biochem.* **111**, 184 (1981).
[23] J. A. Rivier, *J. Chromatogr.* **202**, 211 (1980).
[24] N. Ui, *Anal. Biochem.* **97**, 65 (1979).

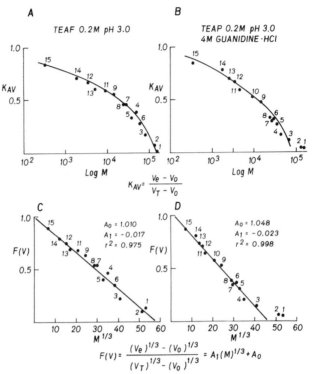

FIG. 4. Relationship between apparent molecular weight (M) and elution volume on protein columns. Data plotted according to K_{av} vs log M [Equation (1); A and B] and according to F(v) vs $M^{1/3}$ [Eq. (2); C and D] using 0.2 M TEAF (A, C) and 0.2 M TEAP–4 M guanidine-HCl (B, D), pH 3, respectively, at a constant flow rate of 1 ml/min. Proteins used were immunoglobulin G (1), bovine serum albumin dimer (2) and monomer (3), ovalbumin (4), myoglobin dimer (5), ovomucoid trypsin inhibitor (6), chymotrypsinogen A (7), cytochrome c dimer (8), α-lactalbumin (9), lima bean trypsin inhibitor (10), ACTH (11), insulin B chain oxidized (12), insulin A chain oxidized (13), γ-MSH (14), and sucrose (15). V_0 and V_t were determined using thyroglobulin and DNP-Gly or sodium azide, respectively. The correlation coefficient (r^2) of the linear regression curves C, D are given.

varying pore sizes is shown in Fig. 4. Here an I-60 column is connected to two I-125 followed by an I-250 column in that order, keeping the length and bore sizes of the connecting tubing to a minimum (e.g., 0.009 inch[1] inner diameter). The elution buffer is 0.2 M TEAF pH 3.0 (stock solution described in the Reagents section). Lower concentrations of TEAF were found to be inadequate, since adsorption of peptides occurred especially on the I-60 column, causing elution of some peptides after the salt volume. The four columns are first equilibrated with 300–500 ml of the 0.2 M

TEAF buffer at 1 ml/min. Once they are equilibrated, they can be used for a long time.

In Fig. 4A, a calibration curve was constructed at a flow rate of 1 ml/min (higher flow rates such as 2 ml/min can be used, with some loss of resolution). Monitoring of eluates was done at 240 nm, due to the absorbance of the 0.2 M TEAF. Alternatively 0.2 M TEAP could be used, and monitoring at 210 nm is possible; however, such a buffer is not volatile, and a desalting step would be necessary for preparative or RIA purposes. As seen in Figs. 4A and 4C, the separation pattern allows a fractionation range of 500–150,000 daltons (i.e., $M^{1/3} = 8$–53) in less than 50 min. The use of Eq. (2) as compared to Eq. (1) allowed a straight-line relationship to be obtained (Fig. 4C), as compared to a concave downward curve (Fig. 4A). However, it was noted that in this buffer system, as well as others,[20] aggregation (mainly dimerization) can occur. To circumvent this problem, the same columns were run using 4 M guanidine-HCl in 0.2 M TEAP, pH 3.0. Using "clean" guanidine-HCl,[25] one could monitor the absorbance at 220 nm and raise the guanidine-HCl concentration to 6 M. With lower-grade batches, absorbance can be read at 280 nm. Again the columns have to be equilibrated with 300–500 ml of this buffer before use at 1 ml/min. The results obtained are shown in Figs. 4B and 4D, where Eqs. (1) and (2) were used to plot the data. It is seen that a better calibration curve is obtained with Eq. (2). The practical range is now from 500 to 90,000 daltons, i.e., $M^{1/3} = 8$–45. It was noted that under these elution conditions the peptide aggregation is drastically minimized. Using this method, molecular weight values were obtained for various polypeptides such as the serine protease tonin[26] ($M_r = 29,839$), human and porcine N-terminal segments of pro-opiomelanocortin[16,27] ($M_r \simeq 15,000$) and bovine *Escherichia coli* enterotoxin ($M_r = 2899$), and all correlated well with values obtained with classical procedures. It should be noted that it is essential to include in the samples an internal marker for V_0 (e.g., thyroglobulin) and V_t (e.g., DNP-Gly) for accurate determination.

This methodology can thus be used to determine molecular weights of unknown peptides in less than 50 min at a 1 ml/min flow rate. Fraction collection (0.2–0.5 ml per fraction) followed by dilution of the collected fractions appropriately with the RIA buffer allows monitoring of the eluates for the presence of a given antigenic determinant. Alternatively, each fraction could be desalted on a Sep-Pak C_{18} (as described above

[25] Schwarz-Mann, Orangeburg, New York 10962.
[26] N. G. Seidah, R. Routhier, M. Caron, M. Chrétien, S. Demassieux, R. Boucher, and J. Genest, *Can. J. Biochem.* **56**, 920 (1978).
[27] N. G. Seidah and M. Chrétien, *Proc. Natl. Acad. Sci. U.S.A.* **78**, 4236 (1981).

under iodenation procedure), lyophilized, and the RIA performed (the guanidine-HCl coming off with the wash step of the Sep-Pak).

As another alternative, the 4 M or 6 M guanidine-HCl could be replaced by a 0.1% sodium dodecyl sulfate buffer, since it was shown that gel permeation columns can be used with that denaturant.[28] After molecular weight separation, protein or peptide can also be identified using rocket immunoelectrophoresis, which permits screening of a large number of fractions at the same time together with microgram quantities of peptides to be detected. Indeed, it was shown that this type of technique can be used in the presence of denaturing agents.[29] This technique even though more difficult than RIA, could prove to be valuable with proteins or peptides exhibiting low solubilities in the absence of denaturing agents.

Radioactively labeled peptides can be eluted from these columns without problems, and hence this system allows the determination of molecular weights of de novo biosynthesized peptides in pulse-chase studies. Chemically labeling peptides, either by radioiodination or other forms of labeling, also would allow the same goal to be achieved. In terms of load, a maximum of about 3–5 mg is possible, and ideally the injection volume should not exceed 500 μl. Therefore, this approach can be used also for preparative purposes where 30–50 mg can easily be processed by repetitive runs in about 1 day.

Peptide Mapping by HPLC of Radioiodinated Polypeptides

Reverse-phase HPLC proved to be very useful when comparative structural studies of polypeptides available in minute quantities were performed. Proteins of interest in biological materials are often present in small amounts, and their detailed analysis is rendered difficult by tedious isolation procedures. This difficulty can be overcome by chemically or metabolically labeling the proteins. These can then be separated, followed by cleavage with specific proteolytic enzymes or chemical agents for further analysis by HPLC.

In the example to be described, we use as a model the matrix or membrane (M) protein of influenza A virus. The M protein, which accounts for 30–40% of total viral proteins,[30] has some characteristics of integrated membrane proteins.[31] To emphasize the potential resolution by

[28] T. Imura, K. Konishi, M. Yokoyama, and K. Konishi, *J. Biochem.* **86**, 639 (1979).

[29] L. D. Lee, H. P. Baden, and C. K. Cheng, *J. Immunol. Methods* **24**, 155 (1978).

[30] P. W. Choppin and R. W. Compans, in "The Influenza Viruses and Influenza" (E. D. Kilbourne, ed.), p. 15. Academic Press, New York, 1975.

[31] S. J. Singer and G. L. Nicholson, *Science* **175**, 720 (1972).

HPLC for comparative studies, membrane proteins from two influenza A virus strains (PR8 and N) were tested, where antigenic variations have been shown to occur on their M proteins.[32] The M proteins from the two strains were isolated[32] and radioiodinated by modification of the iodine monochloride method.[33]

Iodination, initiated by iodine monochloride, was allowed to proceed for 5–10 sec to avoid formation of diiodotyrosine. Labeled proteins were then desalted on Sephadex G-25.[34] The homogeneity of the preparation[33,35] was assessed by electrophoresis on a 15% acrylamide–sodium dodecylsulfate gel, where preparations of the ^{125}I-labeled M protein from the two strains migrated as a single band with apparent molecular weight of 26,000. In preliminary experiments, it was shown that HPLC tryptic peptide mapping of both ^{125}I-labeled M proteins were identical. It was therefore decided to test the digests with *Staphylococcus aureus* V8 protease.[36] Figure 5A depicts the separation of radioiodinated staphylococcal protease digestion products by HPLC using a linear gradient of 2 to 62% CH_3CN on an Altex 5 μm-ODS column[2] with the 0.02 M TEAP, pH 3.0–CH_3CN protocol (see Reagents section). It can be seen that differences in four of the labeled peptides from both M proteins were already apparent (a–d in Fig. 5A). Upon rechromatography of fractions 57–86 (Fig. 5A) using a shallower gradient of 15 to 50% CH_3CN (Fig. 5B), again differences were clearly seen in peak (a) and regions (b) and (c). Recovery was always better than 90% according to this procedure.

Therefore, although this study is by no means exhaustive, it provides evidence that HPLC can be used to advantage for peptide mapping and for comparative studies of proteins available in minute quantities, using a radioiodination labeling procedure and various digestion methods. The use of TEAP–CH_3CN could well be substituted with volatile TEAF–CH_3CN or TEAF–2-propanol if one needs further to characterize the eluted peptides, e.g., by microsequencing or by RIA. It can be envisaged that this powerful method combined with the new hybridoma technology for the production of monoclonal antibodies[37] will add a new dimension to the antigenic topological mapping of membrane antigens present in small amounts in cell extracts, since membrane proteins can be labeled either metabolically or by the lactoperoxidase iodination procedure.

[32] J. Lecomte and J. S. Oxford, *J. Gen. Virol.* **57,** 403 (1981).

[33] R. C. Montelaro and D. P. Bolognesi, *Anal. Biochem.* **99,** 92 (1979).

[34] Pharmacia Fine Chemicals, 2044 St. Regis Blvd, Dorval, P.Q., Canada H9P 1H6.

[35] Most of this work was done in collaboration with Dr. J. Lecompte and A. Darveau of Institut Armand Frappier, Laval, P.Q., Canada.

[36] J. Houmard and G. R. Drapeau, *Proc. Natl. Acad. Sci. U.S.A.* **69,** 3506 (1972).

[37] G. Köhler and C. Milstein, *Nature (London)* **256,** 495 (1975).

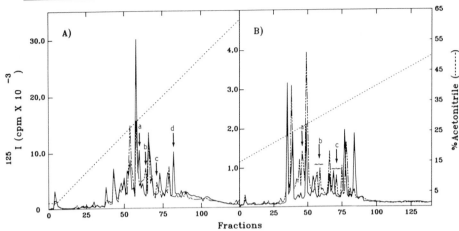

Fig. 5. High-performance liquid chromatography on a 5 μm ultrasphere-ODS column[2] (0.46 × 25 cm), using the TEAP–CH₃CN system, of peptides resulting from digestion with staphylococcal protease V8[36] of the influenza virus protein M obtained from two strains PR8 (——) and N (– – –).[33,35] The digestion was performed at 37° for 2 hr at an enzyme-to-substrate ratio of 1 : 1 (w/w). (A) Elution was done isocratically for 5 min at 2% CH₃CN, followed by a linear gradient of 2 to 62% CH₃CN in 120 min (0.5%/min) at a flow rate of 1 ml/min. Fractions of 1 ml were collected and counted. (B) Fractions 57–86 in Fig. 5A were pooled, lyophilized, and rechromatographed with a shallower gradient of 15 to 50% CH₃CN in 140 min (0.25%/min) at 1 ml/min. Again fractions of 1 ml were collected and counted.

Use of HPLC for Pulse-Chase Studies Characterization of Precursor–Product Relationships

It is now well recognized that many biologically active peptides are initially synthesized on polyribosomes in the form of larger precursor molecules, which subsequently undergo a series of maturation steps leading to the formation of the shorter bioactive entity.[18] In a well studied precursor, the pluripotent pro-opiomelanocortin molecule serves as an initial translation product containing the hormones adrenocorticotropin (ACTH), β-LPH, and β-endorphin and a new N-terminal segment possibly involved in glucocorticoid and mineralocorticoid secretion.[17,27] The usefulness of HPLC is illustrated in the characterization of the maturation products of this precursor molecule obtained following pulse-chase labeling experiments of rat pars intermedia cells with [³H]phenylalanine[38] using sodium dodecyl sulfate–polyacrylamide gel electrophoresis (SDS-PAGE). It was shown[38] that after 30 min of pulse-labeling the cells with [³H]Phe, two major proteins of apparent molecular weights (M_r of 34,000 and 36,000) were synthesized. The analysis of the tryptic fragments from

[38] P. Crine, N. G. Seidah, L. Jeannotte, and M. Chrétien, *Can. J. Biochem.* **58**, 1318 (1980).

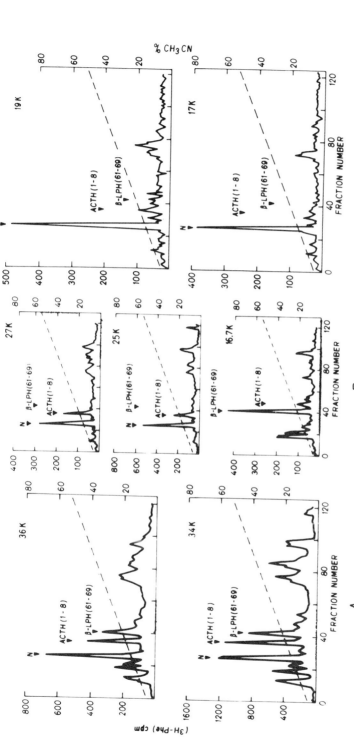

FIG. 6. High-performance liquid chromatography analysis of the tryptic fragments from labeled proteins extracted from rat neurointermediate lobes after a pulse-chase experiment.[38] The [³H]phenylalanine-labeled proteins bands obtained from rat neurointermediate lobes pulsed for 30 min and chased for 0, 30, and 120 min were cut out from the dried sodium dodecyl sulfate–polyacrylamide electrophoresis slab gel,[38] washed extensively with 25% 2-propanol and then 10% methanol, dried by lyophilization, and digested with 5 μg of trypsin for 16 hr at 37°. Aliquots of the digests were oxidized with performic acid and analyzed by HPLC on a μ-Bondapak C₁₈ column (0.39 × 30 cm)¹ using the TEAP–CH₃CN system. A linear gradient of 5 to 70% CH₃CN in 90 min at 1 ml/min was used throughout. Fractions of 0.4 ml were collected and directly assayed for radioactivity. Tryptic maps of (A) the 34,000 and 36,000 daltons (34 K and 36 K) peptides obtained after a 30-min pulse; (B) the 27,000, 25,000, and 16,700 daltons (27 K, 25 K, and 16.7 K) peptides obtained after a 30-min pulse followed by a 30-min chase; and (C) the 19,000 and 17,000 daltons (19 K and 17 K) peptides obtained after a 30-min pulse followed by a 2-hr chase. Reproduced from Crine et al.[38] by courtesy of the Canadian Journal of Biochemistry.

these two proteins by HPLC using the 0.02 M TEAP–CH$_3$CN elution system on a μ-Bondapak C$_{18}$ column (Fig. 6A), showed that they had very similar polypeptide backbones. Both contained the fragments 1–8 of ACTH and 61–69 of β-LPH, in addition to several other fragments, thereby confirming immunoprecipitation studies using antibodies directed against both of these hormones, i.e., ACTH and β-endorphin. After a 30-min chase in nonradioactive medium, it was shown by SDS-PAGE[38] that three new bands appeared with molecular weights of 27,000, 25,000, and 16,700. Tryptic peptide analysis by HPLC (Fig. 6B) showed that the M_r 16,700 band no longer contains the characteristic ACTH 1–8 fragment, but only the β-LPH 61–69 fragment remains. Microsequencing studies of this band showed it to be the intermediate precursor of β-endorphin, i.e., β-lipotropin (β-LPH).[39] The M_r 25,000 and 27,000 bands were shown by microsequencing and peptide mapping to be larger molecular weight inter-mediates containing ACTH at their carboxy terminus, and an N-terminal extension. A longer chase incubation (2 hr) showed that β-LPH is trans-formed in β-endorphin[40] and the M_r 27,000 and 25,000 band forms pep-tides of M_r 19,000 and 17,000 and the ACTH is cleaved to its characteris-tic α-MSH[40] fragment. In Fig. 6C, the tryptic peptide maps of M_r 19,000 and 17,000 forms are shown, and it can be seen that the ACTH 1–8 and β-LPH 61–69 are absent. Microsequencing studies showed that these two peptides have identical N-terminal sequence but could differ in their gly-cosidic linkages.[38] The peptide N in Figs. 6A, B, and C was subsequently shown to be a phenylalanine-containing fragment in the middle portion of these N-terminal peptides.

This brief description of the use of HPLC for mapping newly synthe-sized radiolabeled peptides emphasizes the usefulness of this method for studying the kinetics of maturation and the pathways involved in protein biosynthesis, requiring minute amounts of label (typically 10,000–20,000 cpm). It has the advantage of being nondestructive; i.e., collection of the peptides of interest allows further study by either immunoprecipitation or microsequencing methods, or even degradation with other enzymes for further mapping.

Conclusions

The use of HPLC for characterization of cold and radiolabeled pep-tides and proteins has been demonstrated in a number of examples. The

[39] N. G. Seidah, C. Gianoulakis, P. Crine, M. Lis, S. Benjannet, R. Routhier, and M. Chrétien, *Proc. Natl. Acad. Sci. U.S.A.* **75,** 3153 (1978).

[40] P. Crine, F. Gossard, N. G. Seidah, L. Blanchette, M. Lis, and M. Chrétien, *Proc. Natl. Acad. Sci. U.S.A.* **76,** 5085 (1979).

column most often used is μ-Bondapak C_{18}. The buffer used consists of either TEAP or TEAF, pH 3.0, or 0.1% TFA,[11] and the organic solvent is either CH_3CN or 2-propanol. This simple eluent system can be employed even for gel permeation studies by HPLC. The gradient usually used is the linear one, though concave or convex gradients could be advantageous in some instances. The above-described applications demonstrate the power of this novel technology for rapidly solving difficult biochemical separations. Further advances in column packing materials[41] would further improve on these separations and extend the usefulness of this technique, especially to larger molecular weight proteins that elute better[41] on supports of larger pore size, such as 300–500 Å, as compared to the usually available 100 Å pore size glass beads.

Acknowledgments

This work was supported by a Medical Research Council of Canada Program Grant (PG-2), the National Cancer Institute of Canada, and the National Institutes of Health (NS16315-02). The authors thank Dr. Claude Lazure for his critical reading of this review and for his collaboration on the molecular sieving experiments and Mrs. Diane Marcil for her secretarial assistance.

[41] R. V. Lewis, A. Fallon, S. Stein, K. D. Gibson, and S. Udenfriend, *Anal. Biochem.* **104,** 153 (1980).

[24] Preparation of Stable Radioiodinated Polypeptide Hormones and Proteins Using Polyacrylamide Gel Electrophoresis

By SUSANNE LINDE, BRUNO HANSEN, and ÅKE LERNMARK

Radioactively labeled polypeptide hormones and proteins are widely used as tracers in radioimmunoassays and receptor studies. The peptide or protein is most easily labeled using iodination with ^{125}I or [131]I. The radioactive iodine is substituted in the tyrosine groups of the peptide as monoiodotyrosine or diiodotyrosine, resulting in a heterogeneous mixture of iodine-substituted molecules plus native peptide and unreacted iodide. The iodination mixture has to be fractionated to obtain a well-characterized iodinated product to be used as a tracer. An ideal tracer should retain the full biological activity or immunological reactivity of the native peptide hormone or protein and have a high specific activity and a long shelf life. Studies in our laboratory have shown that it is possible to fractionate

METHODS IN ENZYMOLOGY, VOL. 92

iodinated insulin preparations into homogeneous monoiodoinsulin derivatives[1,2] by polyacrylamide gel electrophoresis using long rods. The same fractionation technique has been useful for the preparation of a number of radioiodinated polypeptide hormones and proteins.[3]

Principle of Iodination

Iodination of a peptide or protein will occur in the tyrosine residues. Iodine substitution in the histidine residues can also take place under certain conditions. In chemical iodination reactions iodide (^{125}I) has to be oxidized to iodine (perhaps I$^+$ is the reactive ion) using, e.g., chloramine-T.[4]

Enzymic iodination using lactoperoxidase is also feasible.[5] Nearly stoichiometric amounts of the oxidizing agent, hydrogen peroxide, are used in this reaction, thus avoiding oxidative destruction of the proteins:

$$I_2 + 2 \text{ Tyr} + H_2O_2 \rightarrow 2 \text{ I–Tyr} + H_2O$$

The substitution with iodine in tyrosine residues may result in monoiodotyrosine or diiodotyrosine, as the reactivity of the individual tyrosine residues depends on availability during the iodination reaction. Tyrosine residues in a protein molecule may attain specific positions allowing better exposure depending on the buffer strength, pH value, or the presence of unfolding agents such as urea. The molar amount of iodine relative to the molar amount of protein and the efficiency of stirring during the reaction will influence the final iodine distribution among different tyrosine residues and the formation of mono- and diiodotyrosines.

Iodination with Lactoperoxidase

The radioactive iodine was obtained from either New England Nuclear (Boston, Massachusetts) as carrier-free Na^{125}I with a specific activity of 629 GBq/mg of I (catalog No. NEZ 033 H) or The Radiochemical Centre (Amersham, U.K.) with a specific activity of 481–629 GBq/mg of I (catalog No. IMS 30). The suppliers provided 37–185 MBq in 1–15 μl of NaOH solution, pH 7–11, in conical glass reaction vials. The ^{125}I$^-$ was routinely transferred from the vial to a polystyrene test tube (NUNC,

[1] J. Gliemann, O. Sonne, S. Linde, and B. Hansen, *Biochem. Biophys. Res. Commun.* **87**, 1183 (1979).
[2] S. Linde, B. Hansen, O. Sonne, J. J. Holst, and J. Gliemann, *Diabetes* **30**, 1 (1981).
[3] S. Linde, B. Hansen, and Å. Lernmark, *Anal. Biochem.* **107**, 165 (1980).
[4] W. M. Hunter and F. C. Greenwood, *Nature (London)* **194**, 495 (1962).
[5] J. I. Thorell and B. G. Johansson, *Biochim. Biophys. Acta* **251**, 363 (1971).

Denmark, catalog No. 341661) permitting adequate stirring during the iodination.

The solution of polypeptide or protein was added and followed immediately by HCl to neutralize the NaOH in the $Na^{125}I$ solution. Alternatively, the $Na^{125}I$ was diluted in buffer before the addition of the polypeptide dissolved in 0.04 M HCl.

The iodination was achieved by adding 0.3–1 mM hydrogen peroxide in distilled water and lactoperoxidase dissolved in buffer. The iodination mixture was magnetically stirred during incubation. Either a commercially available (Bel-Art Products, Pequannock, New Jersey) or a homemade (a piece of a paper clip melted into the thin part of a Pasteur pipette) 2 × 2 mm magnetic stirring bar was used.

Iodination of Insulin (Method A)

$Na^{125}I$, 185 MBq in 15 μl of 0.1 M NaOH from New England Nuclear was transferred to a polystyrene test tube. Thirty microliters of insulin (highly purified porcine insulin from Nordisk Gentofte) suspended at a concentration of 3.3 mg/ml in 0.075 M sodium citrate buffer (pH 5.6) was added, immediately followed by 3 μl of 0.5 M HCl to neutralize the incubation mixture. Two microliters of 1 mM H_2O_2 and 2 μl of lactoperoxidase (0.74 mg/ml) in citrate buffer (pH 5.6) was added twice (at time zero and after 1 min) during continuous stirring. The iodination mixture was incubated for 5 min with stirring. The incubation was stopped by the addition of 56 μl of 0.19 M Tris solution containing 40% sucrose, pH 9.

Iodination of Insulin (Method B)

$Na^{125}I$, 37 MBq in 10 μl of NaOH (pH 7–11) from Amersham was transferred to a polystyrene test tube. Ten microliters of 0.4 M Na_2HPO_4–KH_2PO_4 buffer (pH 7.4) was added followed by 10 μl of insulin (1 mg/ml) in 0.04 M HCl. Five microliters of 0.3 mM H_2O_2 and 5 μl of lactoperoxidase (0.2 mg/ml) in the phosphate buffer (pH 7.4) were then added, and the mixture was incubated for 5 min during continuous stirring. The incubation was stopped by the addition of 40 μl of 40% (w/v) sucrose.

Iodination with Chloramine-T

Radioactive iodine was obtained from the Radiochemical Centre, Amersham, as carrier-free $Na^{125}I$ in dilute NaOH solution, pH 7–11, free from reducing agents. Chloramine-T and sodium disulfite were of analytical grade obtained from Merck.

Iodination of Insulin (Method C)

The iodination reaction was carried out in the reaction vial in which 37 MBq of Na^{125}I in 10 μl of 0.1 N NaOH was supplied. An insulin (highly purified porcine insulin from Nordisk Gentofte) stock solution was prepared by dissolving 1 mg of insulin in 100 μl 0.01 N HCl, which was diluted to 1 mg/ml by the addition of 200 μl of 0.01 N NaOH and 700 μl of 0.5 M sodium phosphate buffer (pH 7.4). Insulin, 10 μl of stock solution corresponding to 10 μg of insulin, was added to the vial followed by 10 μl of chloramine-T (0.88 mg/ml) in 0.5 M sodium phosphate buffer. The iodination mixture was incubated for 45 sec while shaking the vial. The reaction was stopped by the addition of 5 μl of sodium disulfite (4.8 mg/ml) in 0.5 M sodium phosphate buffer.

Iodination of Human Growth Hormone

The iodination reaction was carried out in the reaction vial in which 37 MBq of Na^{125}I in 10 μl of 0.1 N NaOH was supplied. Human growth hormone (highly purified hGH from Nordisk Gentofte) was dissolved in 0.3 M sodium phosphate buffer (pH 7.4) to a concentration of 0.2 mg/ml. Human growth hormone, 50 μl corresponding to 10 μg, was added to the vial followed by 20 μl of chloramine-T (0.03 mg/ml) in 0.3 M sodium phosphate buffer. The iodination mixture was incubated for 3 min while gently shaking the vial. The reaction was stopped by the addition of 5 μl of sodium disulfite (0.2 mg/ml) in 0.3 M sodium phosphate buffer.

Determination of Incorporated Iodine

Incorporated iodine was determined by trichloroacetic acid (TCA) precipitation of the iodination mixture. Immediately after iodination a 0.5-μl aliquot of the iodination mixture was added to 500 μl of 20% trichloroacetic acid. A 500-μl sample of 1% (w/v) human serum albumin in distilled H$_2$O was added, and the test tube was vortexed vigorously before centrifugation (10 min, 1500 g). After centrifugation the supernatant fluid was decanted into another tube, and the precipitate was dissolved in 1 ml of 1 N NaOH. The dissolved precipitate and the supernatant sample, representing unreacted iodide, were counted and the percentage of TCA-precipitable radioactivity was calculated. Separate experiments showed that the presence of cold carrier iodide did not influence the amount of precipitable radioactivity. The efficiency of the iodination procedure expressed as percentage of radioactivity precipitated for porcine insulin varied between 59 and 95% (Table I).

TABLE I
IODINATION YIELD IN DIFFERENT INSULIN IODINATIONS DETERMINED BY
TRICHLOROACETIC ACID PRECIPITATION

Iodination method	Reaction container	Time	Stirring	Percentage of radioactivity precipitated
Method C (chloramine-T)	Vial	45 sec	−	84
Method C (chloramine-T)	Vial	90 sec	−	63
Method B (lactoperoxidase)	Vial	5 min	+	74[a]
Method A and B (lactoperoxidase)	Test tube	5 min	+	95[b]

[a] In four different iodinations the precipitated radioactivity ranged between 66 and 85% (74 ± 8%, mean ± SD).
[b] In nine different iodinations the precipitated radioactivity ranged between 91 and 98% (95 ± 2%, mean ± SD).

The results in Table I demonstrate that effective stirring during the iodination is important. The TCA-precipitable radioactivity was lower and more variable when the iodination reaction was performed in the vial in which the $^{125}I^-$ was delivered. This is exemplified by the highly variable incorporation after chloramine-T iodination without stirring (Table I). The conical shape of the vial makes it difficult to ensure adequate mixing. Iodinations carried out in a round-bottom tube resulted in iodine incorporations exceeding 90%.

The lactoperoxidase iodination was completed within 30 sec. The presence of 6 M urea in the buffer seemed to slow down the reaction time (Table II). The addition of urea to the buffer will unfold the protein, leading to a more equal iodine substitution among the different tyrosine residues.[6–8]

Gel Electrophoresis of Iodination Mixtures

Iodination of the tyrosyl group in a peptide or a protein lowers the pK value of the phenolic hydroxyl in the substituted tyrosine residue. The

[6] L. W. DeZoeten and E. Havinga, *Recl. Trav. Chim. Pays-Bas* **80**, 917 (1961).
[7] A. Massaglia, U. Rosa, G. Rialdi, and A. Rossi, *Biochem. J.* **115**, 11 (1969).
[8] S. Linde, O. Sonne, B. Hansen, and J. Gliemann, *Hoppe-Seyler's Z. Physiol. Chem.* **362**, 573 (1981).

TABLE II
PERCENTAGE OF TRICHLOROACETIC ACID PRECIPITABLE RADIOACTIVITY[a]
AFTER IODINATION OF INSULIN IN DIFFERENT BUFFERS[b]

Iodination time (min)	Citrate, pH 5.6 (method A)	Phosphate, pH 7.4 (method B)	Phosphate–6 M urea, pH 7.8
0.5	91.1	97.0	85.0
1	91.2	97.4	87.6
2	92.5	97.8	—
3	92.6	98.0	93.1
4	92.7	98.0	—
5	92.9	97.3	94.0

[a] A measure of the iodine incorporation.
[b] The iodinations were performed using lactoperoxidase while stirring in a round-bottom tube.

theoretical pK-value of the hydroxyl in noniodinated tyrosine, monoiodo-tyrosine (MIT), and diiodotyrosine (DIT) in proteins is approximately 10.4, 9.1, and 7.9, respectively. These differences prövide the basis for the separation of unlabeled peptide or protein from iodinated molecules using disc electrophoresis in polyacrylamide gels. When the total net charges of insulin and iodinated insulin at the pH values normally used in basic polyacrylamide gel electrophoresis (pH 9)[9] are calculated, it appears that the charge differences are small (Table III). It was necessary, there-fore, to use long (18 cm) polyacrylamide gels.

Electrophoresis Apparatus

An apparatus designed to accommodate 22 cm-long glass tubes with an inner diameter of 5 mm was constructed (Fig. 1). The two buffer containers were made from 2000-ml plastic beakers (diameter 13 cm) cut to a height of 7 cm. In the bottom of the upper buffer container, six holes with a diameter of 15 mm were drilled 15 mm from the edge. The upper buffer container was placed on a stand (23 cm in height) made of plexi-glass. The electrodes, made from a glass tube with a platinum wire in-serted, were mounted centrally in two lids constructed to cover the two buffer containers. The lid for the lower electrode vessel was equipped with six holes (i.d. 8 mm) to ensure a vertical mounting of the six glass tubes.

[9] B. J. Davis, *Ann. N. Y. Acad. Sci.* **121**, 404 (1964).

Reagents and Buffers

Electrode buffer: The buffer used is 0.005 M tris(hydroxymethyl)amino-methane (Tris, Merck p.a) and 0.04 M glycine (Merck p.a), pH 8.3. Prepare fresh from a 10 times concentrated stock solution, stored at 5° for a maximum of 3 months. Each electrode container is filled with 500 ml of electrode buffer.

Tracking dye: A 0.001% bromophenol blue (Merck) solution in distilled water is prepared, and 2 ml are added to 500 ml of electrode buffer in the upper container.

Spacer buffer: A 0.06 M Tris solution is prepared, and the pH is adjusted to 6.7.

Running buffer: A 0.38 M Tris solution is prepared, and the pH is adjusted to 9.15.

Spacer gel (4% w/v) solution: Dissolve 4 g of Cyanogum 41 in 100 ml of spacer buffer and add 100 μl of N,N,N',N'-tetramethylethylene-diamine (TEMED; Koch-Light) and 10 μl of Tween 80 (Fluka AG). Cyanogum 41 is a mixture containing 95% acrylamide and 5% N,N'-methylene bisacrylamide (Acrylogel, BDH; or Gelling Agent, Sigma). Store frozen in 25-ml aliquots for 2–3 months.

Running gel (10% w/v) solution: Dissolve 10 g of Cyanogum 41 in 100 ml of running buffer (before pH adjustment), and add 100 μl of TEMED and 10 μl of Tween 80. Adjust pH with HCl to 9.15 and store frozen in 25-ml aliquots (enough for 6 gels) for 2–3 months.

Ammoniumperoxodisulfate (AP solution): Dissolve 1 g of $(NH_4)_2 S_2O_8$ (Merck p.a) in 4 ml of distilled water. Store at 5° for a maximum of 1 month.

TABLE III
CALCULATED NET CHARGES FOR INSULIN AND IODINATED INSULINS AT pH ABOUT 9

Preparation[a]	Net charge at approx. pH 9
Insulin	−4
Insulin with 1 MIT	−4½
Insulin with 2 MIT	−5
Insulin with 1 DIT	−5
Insulin with 3 MIT	−5½
Insulin with 1 MIT + 1 DIT	−5½

[a] MIT, monoiodotyrosine; DIT, diiodotyrosine.

FIG. 1. (a) Apparatus used for gel electrophoresis of 18–22 cm-long gel rods. (b) View of the electrophoresis apparatus from above to demonstrate fixation of the glass tubes in rubber stoppers.

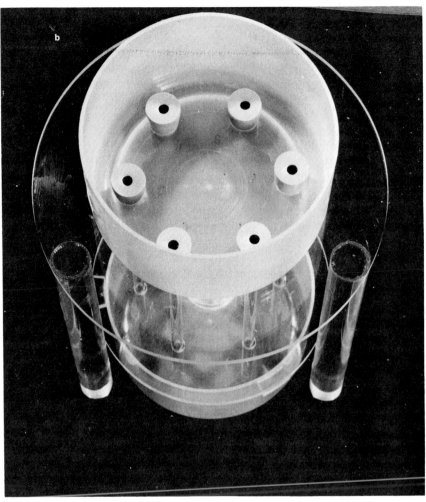

FIG. 1b.

Staining solution: Dissolve 1 g of Coomassie Brilliant Blue G (Serva Blue G, Serva) in 100 ml of distilled water. Store at 5° for a maximum of 1 month.

Staining: Fix the gels in 12.5% TCA for 30–60 min and stain overnight in 1% (w/v) Coomassie Brilliant Blue solution diluted 20 times in 12.5% TCA. Destain the gels in 12.5% TCA. Store in 5% glycerol.

Cleaning solution: "Chromic" acid cleaning solution (Struers A/S, Denmark: 1000 ml of concentrated H_2SO_4 + 40 ml of H_2O + 40 g of

$Na_2Cr_2O_7$) was used to clean the glass tubes before electrophoresis. The glass tubes were placed at least overnight in the cleaning solution followed by washing in distilled water.

Electrophoretic Procedure

Preparation of the Gels. The glass tubes were mounted vertically in a rack and closed at the lower end by inserting them in rubber stoppers. The AP solution (100 μl) was mixed with running gel solution (25 ml), and the tubes were carefully filled to a height of 18 cm avoiding trapping of air bubbles. Immediately after, the gels were overlayered with distilled water (2–3 mm) to ensure a flat and even surface after polymerization, which was completed after 30 min. The water layer was removed and 2–3 cm of spacer gel solution (containing 100 μl of AP solution per 25 ml of spacer gel solution) was layered on top of the running gel and allowed to polymerize for 30 min after water layering. The water was removed, and the rubber stoppers were removed. The lower buffer container was filled with 500 ml of electrode buffer, the lid with the electrode (anode) was mounted, and the glass tubes were mounted vertically in the holes of the upper buffer container inserted into rubber stoppers (size 14/19 mm) with a hole 6 mm in diameter (Fig. 1b). Empty holes in the upper container were closed with rubber stoppers if all six tubes were not used.

Sample Application. As described above, all samples added to the gels contained at least 20% (w/v) sucrose and had a pH above 7 to ensure that the proteins were negatively charged to allow migration toward the anode in the lower buffer container. The sample volume may vary from 10 μl to 400 μl without affecting the final separation. The radioactive samples were added on top of the spacer gels followed by careful layering of electrode buffer containing bromophenol blue filling each tube completely. Finally, the upper buffer container was filled with 500 ml of the same buffer.

Electrophoresis. The upper lid with the cathode was mounted, and the current was switched on (2.5 mA per tube) until the tracking dye, bromophenol blue, was concentrated as a thin band in the spacer gel (10–25 min). The electrophoresis was then continued at 4 mA per tube until the tracking dye was about 0.5 cm from the bottom of the tubes (1.5–2 hr). After electrophoresis the gels were removed by forcing water around the gel with a 22 cm-long hypodermic needle. The water is forced by air pressure using an apparatus schematically shown in Fig. 2.

Slicing the Polyacrylamide Gel

The gels were immediately sliced with a 0.1-mm disposable razor blade into either 1.5- or 2-mm slices using a gel slicer in plexiglass (Fig. 3).

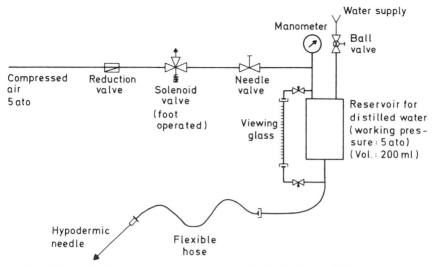

FIG. 2. Pressurized water apparatus constructed (by J. Størling at this institute) to take the gels out of the glass tubes. The hypodermic needle is pushed in between the gel and the inner wall of the glass tube. While the end of the glass tube is closed by the tip of a finger, water is forced out of the needle and the gel is pressed out of the tube.

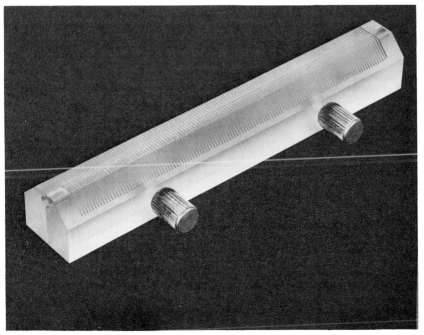

FIG. 3. The gel slicer used for fractionating the gel. Using a razor blade, this slicer allows 18-cm gels to be sliced into 120 1.5-mm slices.

Each slice was placed with a disposable forceps in a test tube and counted for radioactivity in a manual gamma spectrometer (Mølsgaard Medical ApS, Denmark) equipped with a damping device (1000 times damping) in stainless steel. Alternatively, the radioactivity in the gel slices is first eluted in a 0.5–1 ml buffer containing 0.5% albumin, and an aliquot from each eluate is taken for radioactivity determination.

The reproducibility of the gel thickness was 4% as determined by slicing an 18-cm running gel prepared after adding an appropriate amount of radioactivity before polymerization of the gel.

Elution of Tracer from Gel Slices

The iodinated polypeptides or proteins were eluted at 5° in 0.1 M NH_4HCO_3 containing 0.5% human serum albumin (Behringwerke) and adjusted to pH 8.0 with concentrated NH_3-water. The albumin is added to avoid adsorption of labeled material to the walls of the test tubes. The recovery of [125I]monoiodoinsulin eluted from gel slices was determined by measuring the radioactivity in the slice after removing the eluate (Table IV). Elution (with stirring) over 4 hr resulted in more than 80% recovery. However, elution of tracers was most conveniently performed overnight without stirring. If the elution was repeated once more overnight, less than 2% of the radioactivity was left in the gel slices. If stirring was used to increase the elution rate, the eluate turned turbid after about 3 hr of stirring, probably owing to denaturation of albumin. The recovery of different iodinated peptides and proteins was dependent upon their relative molecular mass (Fig. 4). Even the largest protein iodinated (albumin, M_r 68,000) could be eluted overnight at an acceptable yield. Note that the recoveries shown in Fig. 4 are likely to be unique for the running gel used

TABLE IV
ELUTION OF [125I]MONOIODOINSULIN FROM
GEL SLICES

Gel slice size (mm)	Elution volume (μl)	Elution time (hr)	Recovery (%)
1.5	200	2	68
1.5	200	4	82
1.5	125	18	68
1.5	375	18	86
1.5	500	18	89
2	500	18	90

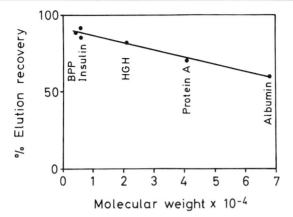

FIG. 4. Recovery of different ^{125}I tracers after elution from gel slices. Gel slices (1.5 or 2 mm) were eluted overnight at 5° in 0.1 M NH₄HCO₃, pH 8.0, containing 0.5% HSA (human serum albumin). In the case of the human [^{125}I]albumin tracer, the elution buffer contained bovine serum albumin. The ^{125}I tracers were bovine pancreatic polypeptide (BPP), insulin (porcine, cod, or lamprey), human growth hormone (hGH), protein A or albumin (human serum albumin).

in the electrophoresis. It has been calculated that the average pore size is about 28 Å at a 10% polyacrylamide concentration (95% acrylamide and 5% bisacrylamide).[10]

Distribution of ^{125}I Radioactivity after Gel Electrophoresis

The distribution of ^{125}I radioactivity after disc electrophoresis of the iodination mixtures of various polypeptide hormones and proteins demonstrates that the tracers are effectively separated from unreacted iodide (Figs. 5–7). The unreacted iodide-125 runs in front of the tracking dye with an R_f value of about 1.2. In all iodinations shown, the position of the unlabeled polypeptide or protein is indicated by an arrow. The positions (R_f value) were determined from a reference gel run in parallel with the sliced gel, separating a sample of the native protein, followed by staining with Coomassie Brilliant Blue. The R_f values were calculated as the migration distance in the running gel toward the anode relative to the migration distance of the tracking dye (bromophenol blue). The radioactivity in the tracking dye region, analyzed by wick chromatography,[11] was found to contain degradation products.

[10] A. H. Gordon, *in* "Laboratory Techniques in Biochemistry and Molecular Biology" (T. S. Work and E. Work, eds.), p. 12. North-Holland Publ., Amsterdam, 1969.
[11] H. Ørskov, *Scand. J. Clin. Lab. Invest.* **20**, 297 (1967).

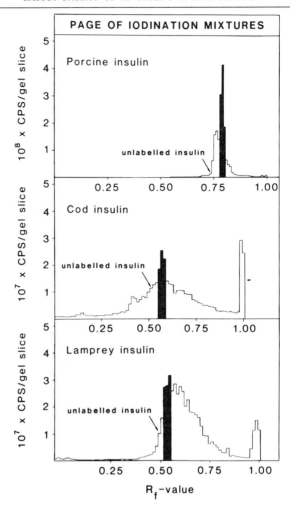

FIG. 5. The distribution of radioactivity in polyacrylamide gels after gel electrophoresis (PAGE) of iodination mixtures. Porcine insulin was iodinated with lactoperoxidase using method B (see text); cod and lamprey insulin, with lactoperoxidase using method A. The gels were fractionated into 1.5-mm (porcine insulin) or 2-mm slices (cod and lamprey insulin). The R_f values of the fractions were calculated relative to the tracking dye (bromophenol blue, R_f 1.00). The filled bars show eluted gel slices.

In iodinations with lactoperoxidase (methods A and B) the enzyme will remain at the top of the running gel with an R_f value lower than 0.06. It will usually remain effectively separated from the iodinated protein. Possible contamination can be avoided by iodination with chloramine-T,

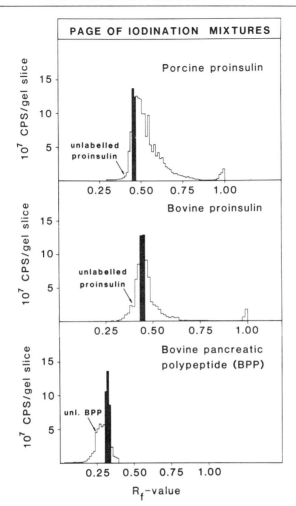

FIG. 6. The distribution of radioactivity in polyacrylamide gel after gel electrophoresis (PAGE) of iodination mixtures. Porcine and bovine proinsulin or bovine pancreatic polypeptide were iodinated with chloramine-T according to method C (see text). The gels were fractionated into 2-mm slices, and the positions of fractions were calculated relative to the tracking dye (bromophenol blue, R_f 1.00). The black bars show eluted gel slices.

as was done with protein A, which had a low R_f value (Fig. 7). The R_f values of various polypeptides and proteins increased by 4–21% after iodination (Table V).

An advantageous feature of the present method is the possibility of fractionating several iodination mixtures from iodinations of different

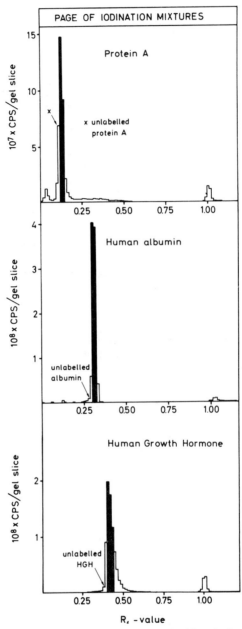

FIG. 7. The distribution of radioactivity in polyacrylamide gel after gel electrophoresis (PAGE) of iodination mixtures. Protein A, human albumin, and human growth hormone were iodinated with chloramine-T according to method C (see text). The gels were fractionated as described in Fig. 6.

TABLE V

Rf Values for Polypeptide and Proteins and the Iodinated Tracers in
10% Polyacrylamide Gel pH 9.1–9.2, Determined Relative to the
Tracking Dye, Bromophenol Blue

Native molecule	R_f	Iodinated molecule	R_f
Porcine insulin	0.73	Tyr A19monoiodoinsulin	0.76
		Tyr A14monoiodoinsulin	0.79
		Diiodoinsulin	0.82
		Triiodoinsulin	0.88
Cod insulin	0.53	Cod iodoinsulin	0.57
Lamprey insulin	0.49	Lamprey iodoinsulin	0.53
Porcine proinsulin	0.38	Porcine iodoproinsulin	0.44
Bovine proinsulin	0.40	Bovine iodoproinsulin	0.46
Pancreatic polypeptide (PP)	0.24	Iodo-PP (peak 1)	0.28
		Iodo-PP (peak 2)	0.32
Protein A	0.12	Iodo-protein A	0.13
Human albumin	0.29	Human iodoalbumin	0.31
Human growth hormone (hGH)	0.39	Iodo-hGH	0.41
Lactoperoxidase	<0.06	Iodide	~1.2

polypeptides or proteins in the same electrophoretic run. Possible cross-contamination (radioactivity) from one gel to another was checked by fractionating an iodination mixture (iodinated insulin) in parallel with an empty gel (no sample applied). The empty gel was sliced, and the radioactivity was measured in the sliced gel. In the gel slices corresponding to the position (R_f value) of the iodinated insulin bands about 0.02% of the total radioactivity in the gel with the iodination mixture was found. In other regions of the empty gel even lower contamination was found, thus allowing the separation of different iodination mixtures at the same time.

Insulin

Porcine Insulin

Iodination of porcine insulin using methods A and B was examined in detail. It was possible to separate different monoiodinated insulins (A14monoiodoinsulin and A19monoiodoinsulin) and diiodoinsulin (Table V). The identification of the monoiodoinsulins and their characterization are given elsewhere.[12,13]

The homogeneity of the tracer with respect to ^{125}I was examined by reelectrophoresis of an aliquot of the eluted tracer (Fig. 8). Visualization

[12] S. Linde and B. Hansen, *Int. J. Pep. Protein Res.* **6,** 157 (1974).
[13] S. Linde and B. Hansen, *Int. J. Pep. Protein Res.* **15,** 495 (1980).

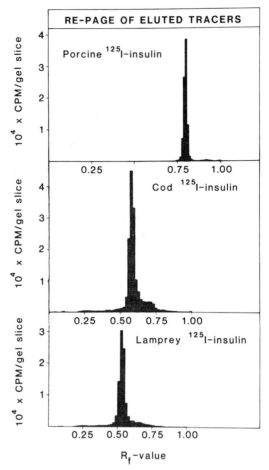

FIG. 8. Homogeneity of porcine, cod, and lamprey ^{125}I-labeled insulins eluted from the gel slices indicated in Fig. 5. The tracers were subjected to a repeated gel electrophoresis (RE-PAGE). The R_f values were calculated relative to the tracking dye (R_f 1.00).

of the reelectrophoresed tracer by staining with Coomassie Brilliant Blue requires at least 0.5 μg of insulin. Stained gels of the reelectrophoresis of gel slices containing A14 and A19monoiodoinsulin (2 gel slices each) from the electrophoretic separation of iodinated porcine insulin (method B) are shown in Fig. 9. Each gel shows only one stained band demonstrating the homogeneity of the tracers.

The insulin tracers, A14 and A19monoiodoinsulin, are routinely used in radioimmunological determinations of insulin.

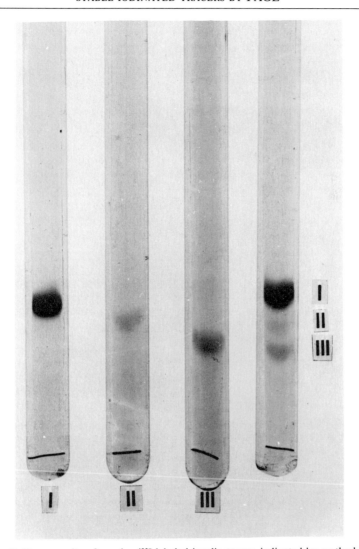

FIG. 9. Homogeneity of porcine ^{125}I-labeled insulin tracer, iodinated by method B (see text) and evaluated after polyacrylamide gel electrophoresis followed by staining with Coomassie Brilliant Blue. The complete iodination mixture analyzed to the right shows three bands corresponding to insulin (I), A19monoiodoinsulin (II), and A14monoiodoinsulin (III) as previously described.[2] Gel slices corresponding to bands I, II, and III from a parallel gel were each reelectrophoresed as shown in the three gels to the left.

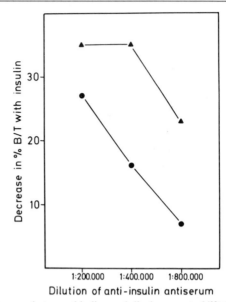

FIG. 10. The difference between binding and displacement of ^{125}I-labeled insulin tracers (method A, see text) with nonlabeled porcine insulin (250 pg/ml) at different anti-insulin antiserum dilutions. The tracers were A14[^{125}I]monoiodoinsulin (●) and A19[^{125}I]monoiodoinsulin (▲).

Figure 10 shows that the displacement of A19monoiodoinsulin with insulin is more effective than the displacement of A14monoiodoinsulin using antiserum D7 (Hagedorn Research Laboratory). The affinity of the two tracers has in fact been found to differ among various insulin antisera.[2] These differences must be considered when optimizing the assay systems.

The *in vitro* receptor binding affinity of A19monoiodoinsulin to adipocytes, hepatocytes, and cultured human lymphocytes was about half that of A14monoiodoinsulin.[1,2] A14monoiodoinsulin of low specific activity (A14[^{125}I,[127]I]monoiodoinsulin) had the same biological activity in isolated rat adipocytes as native insulin, whereas A19[^{125}I,[127]I]monoiodoinsulin was only half as active.[1]

By a combination of polyacrylamide gel electrophoresis and ion-exchange chromatography on QAE-Sephadex, it was possible to isolate all four isomers of [^{125}I]monoiodoinsulin (iodine substituted in Tyr A14, Tyr A19, Tyr B16, or Tyr B26).[8] The apparent binding affinity relative to the A14 isomer to isolated rat adipocytes was 65% for the A19 isomer, 100% for the B16 isomer, and 200% for the B26 isomer.

Storage and Stability

Experiments by gel filtration on Sephadex G-50 have shown that storage of A14monoiodoinsulin tracer at $-20°$ in the elution buffer (0.1 M NH$_4$HCO$_3$ pH 8.0, 0.5% HSA) results in better preservation than storage in a freeze-dried state at $-20°$.[2] Eluted tracers are therefore routinely stored at $-20°$ in 0.1–1 ml aliquots. The stability of stored tracers was also evaluated by paper wick chromatography.[11] In this system the damaged tracer runs with the buffer front, and damage was determined as percentage of the total radioactivity applied to the paper. Tracer damage for A14[^{125}I]monoiodoinsulin was less than 1.3% after 6 months of storage, whereas tracer damage of a [^{125}I]iodoinsulin prepared by gel filtration exhibited 15% tracer damage after 2 months of storage.[2] To test the hypothesis that the stability of monoiodinated insulin was due to the absence of diiodoinsulin, A14 and A19monoiodoinsulins as well as diiodoinsulin were stored for 7 months at $-20°$ after iodination with either method B or C and elution from the polyacrylamide gel slices (Fig. 11). The tracer damage, determined every month, increased from about 0.5% to 1.8% for the monoiodinated tracers, and there was a rapid increase in breakdown products in diiodoinsulin reaching about 15% after 7 months.

It was concluded that diiodoinsulin exhibits a much higher tracer dam-

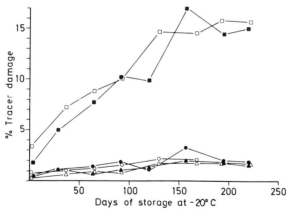

FIG. 11. Tracer damage (percentage of total ^{125}I-radioactivity located at the buffer front after paper wick chromatography) of ^{125}I-labeled insulin tracers stored at $-20°$. Porcine ^{125}I-labeled insulin was prepared by gel electrophoresis after iodination with either lactoperoxidase (method B, open symbols) or chloramine-T (method C, filled symbols). The effect of storage for A14[^{125}I]monoiodoinsulin (○, ●), A19[^{125}I]monoiodoinsulin (△, ▲), and [^{125}I]diiodoinsulin (□, ■) are shown.

age during storage than the monoiodinated tracers. The percentage of radioactivity nonprecipitable in trichloroacetic acid and the percentage of radioactivity nonspecifically bound in radioimmunoassays increased in accordance with the tracer damage (data not shown).

Cod Insulin

Cod insulin (from Dr. Alan Thorpe, Department of Zoology, University of London) was evaluated by gel electrophoresis of a 10-μg sample. Besides the insulin band (R_f 0.53) (Table V) two bands with higher R_f values and 1 band with lower R_f value were demonstrated. The cod insulin was iodinated using method A. The distribution of radioactivity after gel electrophoresis is shown in Fig. 5. The peak of radioactivity was broad, but reelectrophoresis of the three gel slices with maximum radioactivity showed that it was possible to prepare an acceptable tracer despite the presence of impurities in the starting material (Fig. 8). The cod insulin tracer has been suitable for use in radioimmunoassays of fish insulin.[14]

Lamprey Insulin

Lamprey insulin (from Dr. Alan Thorpe, Department of Zoology, University of London) subjected to gel electrophoresis (10 μg) showed more than 30% visually determined impurities with R_f values both higher and lower than the lamprey insulin band (R_f = 0.49) (Table V).

Lamprey insulin was iodinated using method A (Fig. 5). The peak of radioactivity was broad, but the eluted tracer from the slices with highest radioactivity resulted in a nearly homogeneous peak following reelectrophoresis (Fig. 8).

The lamprey tracer was suitable for radioimmunoassays.[14] The iodination of cod and lamprey insulin demonstrates that useful tracers can be prepared by polyacrylamide gel electrophoresis even from crude material.

Proinsulin

Porcine Proinsulin

Gel electrophoresis of 15 μg of porcine proinsulin (from P. Balschmidt, Nordisk Gentofte, Denmark) revealed one major band (R_f value 0.38) (Table V) and a less stained component (R_f value 0.43), the latter

---

[14] A. Thorpe, Department of Zoology, University of London, personal communication, 1979.

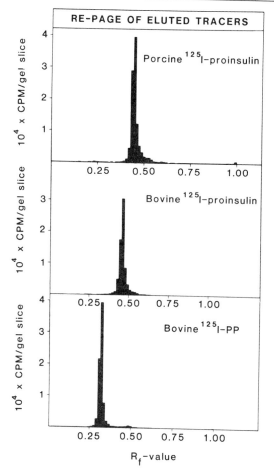

FIG. 12. Homogeneity of porcine and bovine ^{125}I-labeled proinsulin and bovine ^{125}I-labeled pancreatic polypeptide iodinated, purified by gel electrophoresis, and eluted as described in Fig. 6. The eluted tracers were subjected to repeated gel electrophoresis (RE-PAGE), and the distribution of radioactivity was determined after slicing the gel. The position of each fraction was calculated relative to the tracking dye (R_f 1.00).

probably representing deamidated proinsulin. Porcine proinsulin was iodinated according to method C (Fig. 6), and the eluted tracer was subjected to a repeated electrophoresis (Fig. 12).

The R_f value for porcine proinsulin was increased from 0.38 to 0.47 by changing the pH in the running gel to 8.8.[3]

The separation between unlabeled molecules and iodinated molecules is more efficient if the R_f values are high. It is possible, therefore, that iodinated proinsulin could be separated into monoiodinated and diiodinated molecules.

The porcine proinsulin tracer is routinely used in a proinsulin radioimmunoassay separating bound and free hormone with ethanol.[15] The tracer can be used with satisfactory results for a period of 3–4 months.[16]

Bovine Proinsulin

Gel electrophoresis of 15 μg of bovine proinsulin (NOVO Research Institute, Denmark) showed a major band with R_f 0.40 (Table V) together with an additional band on the cathodal side and one or two minor bands on the anodal side.

Bovine proinsulin was iodinated according to method C, and the distribution of radioactivity after gel electrophoresis is shown in Fig. 6 and the electrophoresis of the eluted tracer in Fig. 12.

The bovine proinsulin tracer was used to determine bovine proinsulin by radioimmunoassay separating bound and free hormones with ethanol.[15] The tracer has been used without deterioration over a period of at least 4 months.[16]

Pancreatic Polypeptide

Bovine Pancreatic Polypeptide

Gel electrophoresis of 10 μg of bovine pancreatic polypeptide (BPP) (a gift from Dr. R. Chance, Lilly Research Laboratories, Eli Lilly Co., Indianapolis, Indiana, lot 615-D63-188-7), showed one major band with R_f value 0.24 (Table V). The BPP was iodinated according to methods B and C. There was no difference between the two methods in the distribution of radioactivity after gel electrophoresis. Iodination of BPP resulted in two peaks of radioactivity (Fig. 6). The second peak had the same R_f value after reelectrophoresis and was used as tracer for BPP.[3]

The BPP-tracer was used in the radioimmunoassay of BPP separating free and bound hormone with activated charcoal.[17] The three radioimmunoassay standard curves (Fig. 13) from the bovine pancreatic polypeptide assay using a rabbit anti-BPP serum (lot 615-R110-146-10, kindly provided by Dr. R. Chance) show the effect of storage of ^{125}I-labeled bovine pancre-

[15] L. G. Heding, in "Labelled Proteins in Tracer Studies" (L. Donato, C. Milhaud, and J. Sirchis, eds.), p. 345. Euratom, Brussels, 1966.

[16] P. Nilsson, personal communication, 1981.

[17] T. W. Schwartz, J. F. Rehfeld, F. Stadil, L. I. Larsson, R. E. Chance, and N. Moon, *Lancet* 1, 1102 (1976).

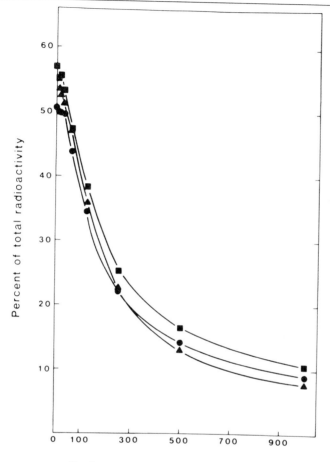

Bovine pancreatic polypeptide (pg/ml)

FIG. 13. Effects of storage of ^{125}I-labeled bovine pancreatic polypeptide (BPP) on the standard curve for radioimmunoassay of BPP. The BPP was labeled as described in Fig. 6. Storage was at −20° for 1 week (▲——▲), 3 months (■——■), and 7 months (●——●). Rabbit anti-BPP was obtained from Dr. R. Chance.

atic polypeptide. It was concluded that the tracer can be used without deterioration for up to 7 months.

Porcine Pancreatic Polypeptide

Tracers of porcine pancreatic polypeptide (PPP) (a gift from J. Ramlau, Nordisk Gentofte, Denmark) were prepared by method C. The same assay system as for BPP was used in radioimmunological determinations

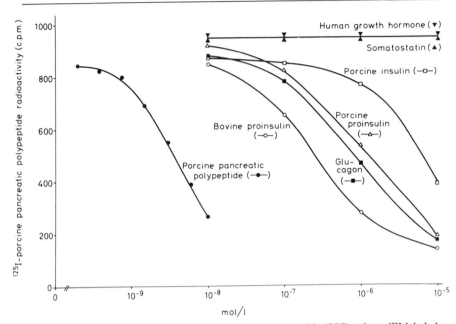

FIG. 14. Radioimmunoassay of porcine pancreatic polypeptide (PPP) using a ^{125}I-labeled PPP tracer prepared by method C (see text). The PP was detected as a contaminant present in bovine proinsulin (○), glucagon (■), porcine proinsulin (△), or porcine insulin (□), all hormones isolated from pancreatic glands, but not in purified human growth hormone (▼) or synthetic somatostatin (▲).

of PPP. Iodinated PPP was specifically displaced by nonradioactive PP (Fig. 14), but not by human growth hormone or synthetic somatostatin. ^{125}I-labeled PPP was suitable for study of the binding of PP to suspensions of living rat and dog hepatocytes.[18]

Human Growth Hormone

Gel electrophoresis of 10 μg of highly purified human growth hormone (hGH) (obtained from Dr. Kim R. Hejnæs, Nordisk Gentofte, Denmark) showed a major band with an R_f value of 0.39 (Table V) plus a minor band, presumably representing deamidated hGH. hGH was iodinated according to methods B and C, and no difference in distribution of radioactivity after gel electrophoresis of the iodination mixture was observed (Fig. 7).

The tracer was used in a radioimmunoassay separating antibody-

[18] V. Bonnevie-Nielsen, K. S. Polonsky, J. J. Jaspan, A. H. Rubenstein, T. W. Schwartz, and H. S. Tager, *Proc. Natl. Acad. Sci. U.S.A.* **79**, 2167 (1982).

bound and free tracer with a *Staphylococcus aureus* immunosorbent.[3] A comparison of the standard curves using the two tracers showed that they were equally good.[3]

Human Albumin

Human albumin [KABI, (R)] was iodinated according to method C.[19] The distribution of radioactivity after gel electrophoresis is shown in Fig. 7. The peak of radioactivity has an R_f value of 0.31 compared to 0.29 for the native albumin (Table V). The human albumin tracer was used in a radioimmunoassay for albumin in human urine.[20] The tracer could be used for at least 4 months.[19]

Protein A

Gel electrophoresis of protein A (Pharmacia, Uppsala, Sweden) showed a major band with R_f value 0.12 and a minor band with a lower R_f value of 0.04 (Table V). Iodination of protein A was performed according to method C (Fig. 7). A peak of radioactivity corresponding to protein A is observed together with a small peak of radioactivity corresponding to the minor component with R_f value 0.04. The low R_f value of protein A in 10% polyacrylamide gels does not allow an effective separation between native and labeled protein A. By changing the polyacrylamide gel concentration to 7.5%, it was possible to increase the R_f value of native protein A to 0.26 and thereby improve the separation of native from labeled protein A.[21] Elution of gel slices with peak activities of iodinated protein A for 72 hr resulted in 90% recovery of protein tracers with a specific radioactivity of nearly 20 μCi/μg. ^{125}I-Labeled protein A is used in radioligand assays of cell-bound antibodies[22] and is useful in screening clones of lymphocyte–myeloma cell hybrids producing monoclonal antibodies.[23] The electrophoretically purified ^{125}I-labeled protein A was stable for up to 60 days when stored at −20°.

[19] J. Sandahl, Steno Memorial Hospital, Denmark, personal communication, 1980.
[20] D. W. Miles, C. E. Mogensen, and H. J. G. Gundersen, *Scand. J. Clin. Lab. Invest.* **26**, 5 (1970).
[21] P. Nilsson and T. Dyrberg, manuscript in preparation.
[22] K. Welsh, G. Dorval, and H. Wigzell, *Nature (London)* **254**, 67 (1975).
[23] G. Köhler and C. Milstein, *Nature (London)* **256**, 495 (1975).

[25] Noncentrifugation Immunoassays: Novel Systems

By Michael Cais

The separation of free and bound fractions is probably the most crucial step in an immunoassay for a given set of reagents. In general, procedures for separating free from bound labeled antigen exploit physicochemical or immunochemical differences between the free and bound fractions, or both. The most widely used procedures[1-3] are based on adsorption (charcoal, silicates, cellulose, ion-exchange resins); fractional precipitation (ammonium sulfate, ethanol, dioxane, polyethylene glycol); immunoprecipitation (double antibody); partition (chromatography, electrophoresis, gel filtration, ultracentrifugation); solid-phase reagents (first or second antibody bound covalently to solid supports, coated tubes or beads); polymerized antisera. The selection of any particular procedure is determined by consideration of many interrelated factors, and a careful study of the separation method is required in order to achieve optimal assay results. However, one feature that is common to many of the above techniques is the need for a centrifugation step to effect aggregation of the suspended solid particles, followed by a decantation (or suction) step to separate physically the solid and liquid phases. Some of the solid support techniques (coated tubes and magnetic particles) do away with centrifugation but still require a decantation (or suction) step. A feasible solvent separation method for immunoassays[4] and a novel system for noncentrifugation, nondecantation solid–liquid separations in immunoassays have been reported.[5]

Solvent Extraction Separation

Materials

Organic solvents, water immiscible, must be of analytical grade or purified by distillation. For assay extraction operations, all solvents must

[1] W. H. Daughaday and L. Jacobs, *in* "Principles of Competitive Binding Assays" (W. D. Odell and W. H. Daughaday, eds.), pp. 303–316, Lippincott, Philadelphia, 1971.

[2] J. G. Ratcliffe, *Br. Med. Bull.* **30,** 32 (1974).

[3] E. H. D. Cameron, S. G. Hillier, and K. Griffith, eds., "Steroid Immunoassays," pp. 207–228. Alpha Omega, Cardiff, Wales, 1975.

[4] M. Cais and M. Shimoni, *Ann. Clin. Biochem.* **18,** 317 (1981).

[5] M. Cais and M. Shimoni, *Ann. Clin. Biochem.* **18,** 324 (1981).

be previously saturated with assay buffer. The radioimmunoassay (RIA) components, antibodies, standards, and radioisotope tracers are the same as those used in regular RIA procedures. Polyethylene, polypropylene, or glass test tubes must be used, since these are not affected by organic solvents.

Selection of Assay Solvent

In order to find a suitable solvent for a particular assay a two-step procedure is necessary[4,6]: (a) determine the extraction efficiency of solvents for the respective antigen in assay buffer; (b) solvents found to extract 90% or more of the antigen from the assay buffer are then tested as separating reagents in the assay protocol in a parallel run of the same assay using one of the regular separation reagents. This will indicate whether the solvent does or does not interfere with the immunological reaction and will help determine the solvent of choice for optimization of the assay.

Procedure for Determination of Extraction Ability

1. For each solvent to be tested, label a group of 4 test tubes alongside 3 tubes for "total counts" determination.
2. Add to each test tube the radioactive tracer to be used in the RIA. For example, add 0.1 ml of tracer solution containing 0.1 μCi.
3. Add to each test tube assay buffer to obtain a total volume as in the RIA test.
4. Add to each test tube (in each group of 4) the solvent to be tested (the recommended solvent volume is twice that of the total aqueous volume in the assay).
5. Cap each test tube, vortex for 20–30 sec, and let stand for 5–10 min to obtain spontaneous separation of the aqueous and organic phases.
6. After phase separation, transfer with a pipette half the volume of the aqueous phase for radioactivity counting: (a) if the tracer is a β-emitting radioisotope transfer the aqueous phase to a counting vial, add scintillation liquid, mix well and count; (b) with γ-emitting tracers, an aliquot of either the solvent or the aqueous phase can be removed for counting.
7. To calculate the extraction efficiency for each solvent, use the counts obtained in the "total counts" tubes and the counts ob-

[6] M. Shimoni, D.Sc. Thesis, Technion-Israel Institute of Technology, Haifa, 1982.

tained after the solvent extraction in the following equation:

$$\% \text{ extraction} = \frac{\text{total counts} - \text{counts in aqueous phase}}{\text{total counts}} \times 100$$

Note: Make sure to take the same volume for "total counts" determination as for the extracted aqueous phase. Thus, if half the aqueous phase is transferred for counting in step 6(a), transfer also half the volume of the "total counts" tubes.

8. Calculate the mean percentage of extraction for each solvent tested (average the results in each group of 4 tubes).
9. If the calculated mean percentage of extraction is 90% or more, test the solvent in the Assay Procedure below.

Assay Procedure with ^{3}H-Labeled Haptens

1. Set up and mark as many tubes, in duplicate, as are required for the RIA system, including tubes for "total counts," blank, and bound zero (B_0).
2. Add 0.1 ml of the ^{3}H-labeled tracer solution to all test tubes.
3. Add 0.1 ml of each standard (calibrator) to the "standards" tubes, 0.1 ml of the analyte to the "unknowns" tubes, and 0.1 ml of assay buffer to "total counts," blank, and B_0 test tubes.
4. Add 0.1 ml of appropriately diluted antiserum to all test tubes except the "total counts" and blank tubes, to which one adds 0.1 ml of assay buffer instead of the antiserum. Mix well all the test tubes.
5. Incubate all test tubes as recommended in the specific RIA protocol.
6. After incubation, add 0.6 ml of solvent (selected in procedure for determination of extraction ability, above) to all test tubes, except the "total counts" tubes.
7. Cap all the test tubes, vortex for 20–30 sec, and let stand for spontaneous phase separation.
8. Transfer from each tube a fixed volume (e.g., 0.2 ml) of the aqueous phase to the counting vials and add scintillation liquid. Mix well and count the radioactivity.
9. Calculate the percentage of B/B_0 from the following equation

$$\% \ B/B_0 = 100 \times \frac{\text{analyte tube (cpm)} - \text{blank tube (cpm)}}{\text{bound zero (cpm)} - \text{blank tube (cpm)}}$$

for $\% \ B/T$ use total counts tube instead of bound zero in above equation.

Note that (a) the lower aqueous phase (0.2 ml), can be removed with a pipette without interference from the upper solvent phase; (b) examples of suitable solvents,[4] generally found not to interfere with the immunological reactions, are methyl isobutyl ketone, *tert*-amyl alcohol, *tert*-butylmethyl ether. These solvents can be used on their own or as paired mixtures between them, in various ratios; (c) all solvents must be saturated with assay buffer prior to use in the assay; (d) the use of a multiple-tube Vortex facilitates the mixing and extraction steps.

Assay Procedure with ^{125}I-Labeled Haptens

With γ-emitting tracers, the assay procedure can be simplified by eliminating the aliquot transfer step for counting the radioactivity at the end of the RIA protocol. This can be achieved in one of the following ways.[4,6]

1. Copper tube adapters can be mounted on the test tubes so that one of the two liquid phases (as desired) can be shielded, and only the unshielded fraction (upper or lower) would be exposed to the detector crystal in the counting well.
2. A Lidex mixer-separator disposable device is used that allows for the immunological reaction, solvent extraction, and physical separation of phases to be performed in the same tube, which is then taken directly to the counter.[4]

Solid–liquid Noncentrifugation Separation

The novel separation system described by Cais and Shimoni[5] comprises, briefly, the following components (Fig. 1): a reaction test tube (A); a separator piston (B), which at its upper end is shaped in the form of a collecting container (E) and at its lower end it has a cavity (F) in which is fitted a suitable membrane (M), the latter being held tightly in place by a ring disk (D). The sealing element (O) on the piston allows the latter to fit snugly into and slide along the inner wall of the test tube. On operation, as the separator piston is pushed down into the tube, the liquid phase is aspirated through the orifice bore of the ring disk (D), then through the axial channel (C), finally to be accumulated in the upper collecting container. The solid particles are separated from the liquid by the membrane barrier. Upon completion of the piston movement, all the solid is retained at the bottom of the test tube. This physical separation between the solid and liquid phases in a single tube has many obvious advantages. One of them is that, with γ-emitting labels, the hermetically closed tube can be placed in the counter either straight up or upside down, enabling the analyst to count either the liquid or the solid phase, or both. With β-emitting labels it is possible to transfer the liquid phase directly to a minivial connected to the collecting container through a special attachment replacing the stopper (S).

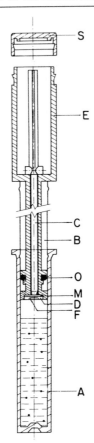

FIG. 1. Lidex unit for noncentrifugation separation of free and bound fractions in immu-noassays. The letters A–F, O, M, S are explained in the text.

The operation of the separator piston described above can be carried out automatically and simultaneously on 40–60 tubes in about 3–4 min with a simple electricity-operated instrument.

Assay Procedure

For the purpose of illustrating the Lidex methodology in a solid–liquid separation system, we shall describe the assay procedure[7] in the specific and quantitive determination in serum or plasma of human chorionic go-nadotropin (hCG), a 38,000-dalton polypeptide hormone normally pro-

---

[7] M. Shimoni, A. Gepstein, and M. Bassat, unpublished results, 1981.

duced by the human placenta. To achieve an assay specific for hCG, the antiserum employed in this example must be raised against the β subunit of hCG.

A comparison of results obtained with Lidex technology and with the classical centrifugation method is illustrated in Fig. 2 for 45 clinical samples assayed in parallel experiments with a commercial kit supplied by Serono Diagnostics (Switzerland). The standard curves of the two methods were fully superimposable.

Reagents and Materials Required

Anti-hCG-β serum
[125]I-hCG tracer
Precalibrated standards with the following hCG concentrations: 0, 3, 6, 12, 25, 50, 10, 200 mIU/ml
Human male serum free of hCG
Polyethylene glycol/double antibody (PEG/DAB) solution containing goat antirabbit γ-globulin in 8% phosphate-buffered polyethylene glycol, pH 7.5
Assay phosphate buffer, pH 7.5
Disposable Lidex plastic test tubes and separators

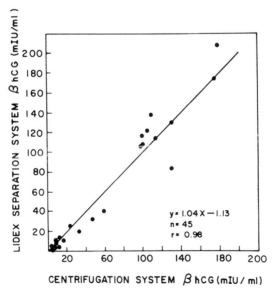

FIG. 2. β-hCG values of 45 clinical samples assayed in parallel experiments with the Lidex separation system and by the centrifugation method.

FIG. 3. The Pressomat (Lidex Corporation) for automatic separation of free and bound fractions with the Lidex methodology. (a) Position at start of operation; (b) position at end of separation process; pressing plate returns automatically to the starting portion.

Lidex test tube racks
Pressomat 301A instrument (Fig. 3)
Automatic micropipettes with disposable tips
Automatic pipettor for addition of PEG/DAB solution.

Specimen Collection and Preparation

Serum: Allow blood to clot at room temperature, centrifuge for 15 min, and collect the serum.

Plasma: Use an anticoagulant such as heparin, centrifuge for 15 min, and collect the plasma.

Dilution for assay: Specimens for pregnant patients should be diluted before assay as follows: weeks 1–2 of pregnancy—undiluted or diluted up to 1 : 4; weeks 3–4—dilute 1 : 5 up to 1 : 100; from week 5 on—dilute 1 : 100 to 1 : 1000. Serum free of hCG is used for dilution up to 1 : 5. Further dilution can be obtained with buffer; when buffer dilutions are assayed, additional hCG-free serum is added to the assay tube to equalize the protein concentrations.

Preparation of Test Tubes. For each assay prepare five groups of tubes: two "total counts" tubes; two nonspecific binding (NSB) tubes (no antiserum added); four B_0 tubes (zero concentration of antigen); two standard tubes for each concentration of standard; two sample tubes for each clinical sample.

Pipetting and Incubation Steps. The following volumes (expressed in milliliters are recommended as optimal for the Lidex System. Reagents have to be adjusted to be used with the recommended volumes. The Lidex test tubes must be used with the Lidex tube racks.

1. Pipette 0.1 ml of zero hCG standard into B_0 tubes.
2. Pipette 0.1 ml of each standard or of each clinical sample into the respective tubes.
3. Pipette 0.1 ml of serum free of hCG into the following tubes: NSB, B_0, standards, and samples diluted more than 1:5.
4. Pipette 0.2 ml of buffer into NSB tubes and 0.1 ml of buffer into samples that are undiluted or diluted up to 1:5.
5. Pipette 0.1 ml of ^{125}I-labeled hCG into all tubes.
6. Pipette 0.1 ml of anti hCG-β serum into all tubes except the NSB and total counts tubes.
7. Mix and incubate at room temperature for 18–24 hr.
8. Mix the PEG/DAB solution by gentle inversion and add 1.5 ml to all tubes except total counts tubes.
9. Gently insert a Lidex piston separator into the top of each tube, except total counts tubes. The caps on the separator should be in place, but not sealed.
10. Keeping the Lidex units in place with any light flat surface, gently mix the reaction mixtures by inverting, by hand, the rack and tubes two or three times (this eliminates the need for vortexing).
11. Place the racks and tubes in the Pressomat and start the instrument. Separation is complete in 3–4 min. The Pressomat has a delay time of several seconds after the caps on the separators have been hermetically sealed, and then the pressing plate automatically moves up to the starting position.
12. Remove the racks from the Pressomat and take out the sealed Lidex unit for counting. The liquid phase will now be trapped in the upper container of the separator, and the precipitate fraction, to be counted, remains at the bottom of the test tube. No decanting is necessary. Place the sealed unit right side up in the γ counter to count the precipitate. If desired, the radioactivity in the upper liquid phase can be counted by placing the Lidex unit upside down in the counter. Count the radioactivity in all tubes for 1 min (or longer depending on the efficiency of the counter).

13. After counting, the sealed Lidex unit is discarded according to required regulatory disposal procedures.
14. Calculation of the results:
    (i) Calculate the mean counts for each group of tubes.
    (ii) Calculate the percentage of binding in absence of antigen, using the formula

$$\frac{\text{standard (or sample) mean counts} - \text{NSB counts}}{B_0 \text{ mean counts} - \text{NSB counts}} \times 100$$

15. To obtain the standard curve, draw a graph of percentage of binding of each standard (y axis) against its hormone concentration (x axis), preferably using semilog paper. The hormone concentration of the samples is obtained by interpolating its percentage of binding on the standard curve.

Concluding Remarks

This chapter describes, inter alia, a new technology for the separation of free from bound fractions in immunoassays using a specially designed test tube and separator device designated LIDEX (acronym for LIquiD EXtraction) separator. The Lidex separator performs the physical separation of the partitioned tracer between the solid and liquid phases, in a closed unit without the need for centrifugation and decantation, and the sealed unit permits the counting of radioactivity in either the free or the bound fraction. The potential and advantages of the method have been described by Cais and Shimoni,[5] and on the basis of extensive studies they have shown that the new Lidex separation technology comprises the following desirable features.

1. It completely, or very nearly so, separates bound and free fractions with a wide margin for error in the conditions used for separation.
2. It does not interfere with the primary antigen–antibody binding reaction.
3. It is simple, easy, and rapid to use.
4. It is inexpensive and uses reagents and equipment that are readily available.
5. It is not affected by plasma or serum.
6. All manipulations are performed in a single tube-separator device.
7. It is highly suitable for automation.
8. It is applicable to a wide range of antigens.
9. The methodology and design of the separator device practically eliminate potential contact with the radioactive reaction mixture, thus ensuring maximum safety from radiation hazards.

[26] Use of Activated Thiol-Sepharose in a Separation Method for Enzyme Immunoassay

By KANEFUSA KATO

In most of the current enzyme immunoassays, except for homogeneous enzyme immunoassay,[1] a batchwise solid-phase method is employed to separate the bound label from the unbound,[2] because it is simple and accurate. However, assays with the solid-phase separation technique have often encountered problems due to nonspecific sample interference,[3] particularly when sensitive assays are applied to the direct estimation of substances in serum.

To overcome these problems, a covalent chromatographic column-separation method based on the thiol–disulfide interchange reaction[4] has been developed.[5–8] This chapter describes the column-separation method applied to the enzyme immunoassay of thyroid hormones, to (anti-insulin) antibodies, and to secretory immunoglobulin A in serum.

Principle of the Separation Method

A mixture of the antibody-bound form of enzyme-labeled antigens and the unbound form is passed through a small column (bed volume = 0.1 ml, Fig. 1) of (anti-IgG) IgG- (or protein A-) coupled Sepharose, in which the (anti-IgG) antibodies (or protein A) have been coupled with disulfide bonds by means of the thiol–disulfide interchange reaction. After washing off unbound label, the antibody-bound label can be eluted with buffer containing excess thiol groups, which split the disulfide bonds between the (anti-IgG) antibodies (or protein A) and the Sepharose derivatives. From the enzyme activity in the eluate, the amounts of antigens or antibodies can be determined.

[1] K. F. Rubenstein, R. S. Schneider, and E. F. Ullman, *Biochem. Biophys. Res. Commun.* **37**, 846 (1972).
[2] A. J. O'Beirne and H. R. Cooper, *J. Histochem. Cytochem.* **27**, 1148 (1979).
[3] K. Kato, Y. Umeda, F. Suzuki, and A. Kosaka, *J. Appl. Biochem.* **1**, 479 (1979).
[4] K. Brocklehurst, J. Carlsson, M. P. J. Kierstan, and E. M. Crook, this series, Vol. 34 [66].
[5] Y. Umeda, F. Suzuki, A. Kosaka, and K. Kato, *Clin. Chim. Acta* **107**, 267 (1980).
[6] R. Yamamoto, Y. Umeda, A. Kosaka, and K. Kato, *J. Biochem. (Tokyo)* **89**, 223 (1981).
[7] R. Yamamoto, S. Hattori, T. Inukai, A. Matsuura, K. Yamashita, A. Kosaka, and K. Kato, *Clin. Chem.* **27**, 1721 (1981).
[8] K. Kato, R. Yamamoto, Y. Umeda, and A. Kosaka, *J. Appl. Biochem.* **3**, 75 (1981).

METHODS IN ENZYMOLOGY, VOL. 92

Preparation of (anti-IgG) Antibody-Coupled Activated Thiol–Sepharose

Immunoglobulin G (IgG) fractions of the antiserum are reduced with 2-mercaptoethylamine, and the reduced IgG, which contains thiol groups in the molecule, can be coupled to the activated thiol–Sepharose.

Reagents

(anti-IgG) Serum, 10 ml
NaCl, 0.15 M
Sodium phosphate buffer, 17.5 mM, pH 6.3
$(NH_4)_2SO_4$
DEAE-Cellulose
Sodium acetate buffer, 0.1 M, pH 5.0, containing 5 mM EDTA
2-Mercaptoethylamine, 1 M
Sephadex G-25 medium
Activated thiol-Sepharose 4B (Pharmacia)
Tris-HCl buffer, 0.1 M, pH 7.5, containing 0.3 M KCl, 1 mM EDTA, and 0.1% NaN$_3$

Procedures. Dilute the (anti-IgG) serum (10 ml) twice with 0.15 M NaCl, and add 6.26 g of solid $(NH_4)_2SO_4$ to the solution at 0° with stirring. Collect the precipitate by centrifugation (6000 g, 10 min), dissolve the precipitate in 10 ml of the sodium phosphate buffer, and dialyze the extract against the same buffer overnight at room temperature. After removing the insoluble proteins by centrifugation, apply the extract to a column (1 × 10 cm) of DEAE-cellulose that has been equilibrated with the phosphate buffer. Pool the IgG fractions passed through the column, which can be monitored with the absorbance of the eluate at 280 nm ($E_{280}^{1\%} = 15$). Concentrate the IgG fractions (to about 6 mg/ml), and dialyze against sodium acetate buffer, pH 5.0.

Mix the dialyzed IgG fraction (5 ml) with 0.4 ml of 1 M 2-mercaptoethylamine (final concentration = 75 mM), and incubate the mixture at 37° for 90 min. Apply the mixture to a column (1.5 × 40 cm) of Sephadex G-25, equilibrated with the acetate buffer, to separate the reduced IgG, which should be eluted in the exclusion volume of the column. Thiol groups in the reduced IgG fractions (about 30 mg in 10 ml) can be determined by the method of Grassetti and Murray,[9] and the IgG contains 6–7 mol equivalents of thiol groups per molecule.

Apply the reduced IgG fractions immediately to a column (0.9 × 15 cm) of activated thiol–Sepharose 4B (prepared with 3 g of the lyophylized Sepahrose), equilibrated with the acetate buffer, at room temperature

[9] D. R. Grassetti and J. F. Murray, Jr., *Arch. Biochem. Biophys.* **119**, 41 (1960).

(15–25°) at a flow rate of 10–15 ml/hr. Wash the column with the acetate buffer and then with the Tris-HCl buffer. Most of the IgG fractions applied should be adsorbed in the column, which can be estimated by measuring the absorbance of the eluate at 280 and 343 nm.[4] Store the antibody-immobilized Sepharose at 4°.

Choice of Antiserum. When the (anti-IgG) antibody-coupled Sepharose is employed as a separation method in the competitive immunoassay, it is highly recommended to use a species-specific (anti-IgG) serum (or at least antiserum with little cross-reactivity with IgG in the serum sample), otherwise the efficiency with which the antibody-bound label is separated will decrease. The (anti-IgG) antibody can be replaced by protein A from *Staphylococcus aureus* for the determination of IgG antibodies. For the assay of antibodies (immunoglobulins) of specified classes, corresponding (anti-immunoglobulin) antibody should be used for preparing the Sepharose column.

Capacity of the (anti-IgG) Antibody–Sepharose Column. The rabbit (anti-guinea pig IgG) antibody-coupled Sepharose was packed in the small column (Fig. 1), and the capacity of the column to bind guinea pig IgG was examined. The guinea pig IgG labeled with β-D-galactosidase from *Escherichia coli* by use of N,N'-o-phenylenedimaleimide (6 milliunits, expressed in enzyme activity, containing less than 0.1 μg of IgG) was mixed with various amounts of nonspecific guinea pig IgG in a volume of 0.5 ml with 0.01 M sodium phosphate buffer, pH 7.0, containing 0.3 M NaCl, 1 mM MgCl$_2$, 0.5% (w/v) gelatin (Difco), 0.1% bovine serum albumin (fraction V), 0.1% NaN$_3$ and 2 mM N-ethylmaleimide (buffer G),

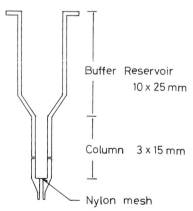

Buffer Reservoir
10 x 25 mm

Column   3 x 15 mm

Nylon mesh

FIG. 1. Plastic minicolumn.

TABLE I

CAPACITY OF THE (ANTI-GUINEA PIG IgG)
ANTIBODY–SEPHAROSE COLUMN AND RECOVERY OF THE LABEL
FROM THE COLUMN[a,b]

Guinea pig IgG added (μg)	β-D-Galactosidase activity (milliunits)		Recovery from the column[c] (%)
	A	B	
0	0.612[d]	3.42[d]	63.5
1	0.762	3.32	63.3
3	0.815	3.35	64.6
10	1.09	2.43	51.5

[a] From Yamamoto et al.,[6] with permission.
[b] Six milliunits of the enzyme-labeled guinea pig IgG and indicated amounts of unlabeled IgG were mixed and applied to the (anti-guinea pig) antibody–Sepharose column. β-D-Galactosidase activities passed through (A) and eluted by dithiothreitol (B) were assayed.
[c] B/(6 − A) × 100.
[d] Means of duplicate assays.

and applied to the (anti-guinea pig IgG) antibody-coupled Sepharose column at room temperature at a flow rate of 3 ml/hr. The column was washed twice with 1 ml of buffer G, and the labeled IgG bound to the column was eluted with 2 ml of 0.01 M sodium phosphate buffer, pH 8.0, containing 0.1 M NaCl, 1 mM MgCl$_2$, 0.1% bovine serum albumin, 0.1% NaN$_3$, and 25 mM dithiothreitol (buffer S). The enzyme activity in the wash and the eluate was assayed with 4-methylumbelliferyl-β-D-galactoside as substrate.[6] As shown in Table I, inclusion of up to 3 μg of guinea pig IgG in the mixture did not result in a marked increase or decrease in the enzyme activity in the wash and in the eluate, respectively, indicating that the (anti-guinea pig IgG)–Sepharose column could trap about 3 μg of guinea pig IgG under the conditions. The recovery of the labeled IgG bound in the column to the eluate was calculated to be about 60%.

The capacity of the column and the recovery of the bound label from the column depended on the lot of antisera used and the preparation of the antibody-coupled Sepharose, and ranged from 3 to 30 μg and 45 to 65%, respectively. The apparently low recovery of the label activity from the column was due in part to the inhibition of β-D-galactosidase activity by dithiothreitol carried with the eluate into the assay medium. The enzyme activity of the labeled IgG might be inhibited sterically when it forms an immune complex with the (anti-IgG) IgG.

Preparation of Protein A-Coupled Activated Thiol–Sepharose

Protein A from *Staphylococcus aureus* is allowed to react with *S*-acetylmercaptosuccinic anhydride,[10] and the mercaptosuccinylated protein A can be coupled to activated thiol–Sepharose 4B.

Reagents

Protein A, 5 mg (Pharmacia)
Crystalline bovine serum albumin
Sodium phosphate buffer, 0.1 *M*, pH 7.0
S-Acetylmercaptosuccinic anhydride
NaOH, 1 *N*
Hydroxylamine, 0.5 *M*, pH 7.0
Sephadex G-25 medium
Sodium acetate buffer, 0.1 *M*, pH 5.0 containing 5 m*M* EDTA
Activated thiol–Sepharose 4B
Tris-HCl buffer, 0.1 *M*, pH 7.5, containing 0.3 *M* KCl, 1 m*M* EDTA, and 0.1% NaN$_3$

Procedures. Dissolve 5 mg each of protein A and bovine serum albumin in 2 ml of the sodium phosphate buffer, pH 7.0, and dialyze against the same buffer. Add 20 mg of *S*-acetylmercaptosuccinic anhydride to the mixture at room temperature with stirring. Maintain the pH of the mixture at 6.7–7.2 by adding 1 *N* NaOH during the succinylation reaction (30 min). Then, add 1 ml of the hydroxylamine solution to the mixture (to remove the acetyl groups of succinylated proteins), and incubate the mixture at 30° for 30 min. Apply the mixture to a column (1.5 × 40 cm) of Sephadex G-25, equilibrated with the acetate buffer (pH 5.0), to isolate the mercaptosuccinylated proteins. Pool the fractions containing thiolated proteins (about 10 ml, detected by the absorbance at 280 nm). Thiol groups in this solution can be estimated as described above (about 5 × 10^{-5} *M*). Apply the thiolated proteins immediately to the column (0.9 × 15 cm) of activated thiol–Sepharose 4B as described in the coupling of (anti-IgG) antibody. Most of the proteins should be adsorbed in the column. Store the protein A-coupled Sepharose at 4°.

Bovine serum albumin is added for two reasons: for diffuse immobilization of protein A on the Sepharose, and for easy detection of the fractions containing protein A after the column chromatographies, since the absorption coefficient ($E_{280}^{1\%} = 1.65$)[11] of protein A is low. The reactivity of protein A with IgG is found not to decrease after succinylation, but to

[10] I. M. Klotz and R. E. Heiney, *Arch. Biochem. Biophys.* **96,** 605 (1962).
[11] J. Sjöquist, B. Meloun, and H. Hjelm, *Eur. J. Biochem.* **29,** 572 (1972).

TABLE II
ASSAY OF PROTEIN A IN FRACTIONS FOR ITS COUPLING TO ACTIVATED
THIOL-SEPHAROSE 4B[a]

Fractions	Volume (ml)	Protein A[b] (mg)	Recovery[c] (%)
Dialyzed	2	4.9	98
Succinylated	2.2	6.2	124
Deacetylated	3.2	5.7	114
Sephadex G-25	10	5.9	118
Passed through the activated thiol–Sepharose column	15	0.25	5

[a] From Kato et al.,[8] with permission.
[b] Five milligrams of protein A were treated.
[c] B/5 × 100.

increase to 120–140% of the untreated protein (Table II), when assessed with a two-site assay system for protein A.[8]

Two-site Assay of Protein A. To estimate the potency of protein A to bind IgG after the chemical modifications, and the efficiency of coupling of protein A to the Sepharose, a sensitive two-site assay for protein A has been developed[8] by using the fact that the protein A has two sites to bind to the Fc portion of IgG.[11] The assay system is composed of silicone rubber solid-phase (3 mm × 4 mm) with noncovalently immobilized human IgG and human IgG labeled with β-D-galactosidase from *Escherichia coli*.

Various amounts of protein A in 0.2 ml of 0.01 M sodium phosphate buffer, pH 7.0, containing 0.1 M NaCl, 1 mM MgCl$_2$, 0.1% bovine serum albumin, and 0.1% NaN$_3$ (buffer A) are incubated with a piece of the human IgG-solid-phase at 30° with shaking. After 4 hr, the reaction mixture is removed by aspiration, and the piece is washed twice with buffer A. The piece is then incubated with the galactosidase-labeled human IgG (3 milliunits/0.15 ml of buffer A) at 4° overnight. The piece is washed with buffer A, and the enzyme activity bound to each piece can be assayed with 4-methylumbelliferyl-β-D-galactoside as described by Kato et al.[12] Figure 2 shows a standard curve for the assay of protein A.

Capacity of the Protein A-Sepharose Column. Capacity of the column (0.1 ml) to trap human IgG was determined, as described in that of the (anti-IgG) antibody-Sepharose column, with human IgG labeled with galactosidase. As shown in Table III, there was little effect of added human IgG up to 30 μg on the enzyme activity of labeled human IgG in the wash

[12] K. Kato, Y. Umeda, F. Suzuki, D. Hayashi, and A. Kosaka, *Clin. Chem.* **25**, 1306 (1979).

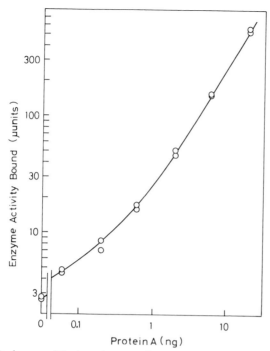

FIG. 2. Standard curve of the two-site assay for protein A from *Staphylococcus aureus*. From Kato *et al.*,[8] with permission.

TABLE III

CAPACITY OF THE PROTEIN A–SEPHAROSE COLUMN AND
RECOVERY OF THE LABEL FROM THE COLUMN[a,b]

Human IgG added (μg)	β-D-Galactosidase activity (milliunits)		Recovery from the column[c] (%)
	A	B	
0	0.12[d]	3.75[d]	63.8
3	0.22	3.66	63.3
10	0.25	3.56	61.9
30	0.37	3.42	60.7
100	0.73	2.96	56.2

[a] From Kato *et al.*,[8] with permission.
[b] Six milliunits of the enzyme-labeled human IgG and indicated amounts of unlabeled IgG were mixed and applied to the protein A-Sepharose column. β-D-Galactosidase activities passed through (A) and eluted by dithiothreitol (B) were assayed.
[c] B/(6 − A) × 100.
[d] Means of duplicate assays.

and in the eluate, indicating that the protein A–Sepharose column could trap about 30 μg of human IgG under the conditions. The recovery of labeled IgG bound in the column to the eluate was about 60% as observed in the case of (anti-guinea pig IgG) antibody–Sepharose column. Similar results were obtained in the experiments with nonspecific guinea pig IgG or rabbit IgG and labeled human IgG.

Application of the Separation Method in Enzyme Immunoassay Assay of Thyroxine and Triiodothyronine[7]

Reagents

Buffer G: Sodium phosphate buffer, 0.01 M, pH 7.0, containing 0.3 M NaCl, 1 mM MgCl$_2$, 0.5% gelatin, 0.1% bovine serum albumin, 0.1% NaN$_3$, and 2 mM N-ethylmaleimide

Buffer S: Sodium phosphate buffer, 0.01 M, pH 8.0, containing 0.1 M NaCl, 1 mM MgCl$_2$, 0.1% bovine serum albumin, 0.1% NaN$_3$, and 25 mM dithiothreitol

Buffer GP: Buffer G with sodium phosphate concentration increased to 0.25 M

Goat (anti-rabbit IgG) antibody-coupled activated thiol–Sepharose 4B

Antisera to thyroxine (T$_4$) and triiodothyronine (T$_3$), antisera for radioimmunoassay (RIA) grade, Calbiochem-Behring (produced in rabbit)

Antigens labeled with β-D-galactosidase, T$_4$, or T$_3$ coupled to the enzyme with 4-(maleimidomethyl) cyclohexane-1-carboxylic acid succinimide ester.[7]

Standard sera, T$_4$, or T$_3$ (Sigma), dissolved in 10 mM NaOH and diluted with human serum freed of T$_4$ and T$_3$ by charcoal treatment

o-Nitrophenyl-β-D-galactoside, 4.5 mg/ml

Na$_2$CO$_3$, 1 M

NaOH, 0.2 M

Procedures. Mix 50 μl of standard or sample serum with 50 μl of 0.2 M NaOH, and incubate at 37° for 20 min. Then, add to the mixture 0.5 ml of the (anti-T$_4$) or (anti-T$_3$) serum diluted with buffer G (containing antibodies corresponding to the amounts for one RIA assay), and 0.2 ml of the labeled antigen (3 milliunits for T$_4$ and 2.5 milliunits for T$_3$, expressed as units of enzyme activity; 1 unit = 1 μmol of o-nitrophenol per minute under the conditions described below). Incubate at 37° for 1 hr, and apply the reaction medium at room temperature at a flow rate of 3 ml/hr to the column of (anti-rabbit IgG) antibody-coupled Sepharose, equilibrated

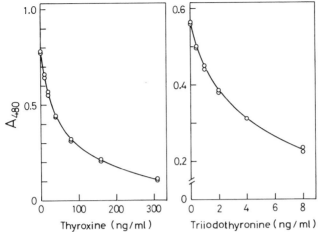

FIG. 3. Standard curve for the assay of thyroid hormones. From Yamamoto et al.,[7] with permission.

with buffer G. Wash the column twice with 1 ml of buffer G, and elute the label with 1 ml of buffer S. (It is recomended to elute the label with 2 ml of the buffer, because most of the label is not always eluted in 1 ml of buffer. There is some variation of elutability among batches of antibody-coupled Sepharose.) Preincubate the eluate (1 ml) at 37°, and start the enzyme reaction by adding 0.25 ml of o-nitrophenyl-β-D-galactoside solution. Incubate at 37° for 1 hr, and stop the reaction by adding 0.25 ml of 1 M Na_2CO_3. Measure the absorbance at 420 nm. The amounts of enzyme activity can be calculated by utilizing the molecular extinction coefficient of o-nitrophenol (21,300) at 420 nm.[13] Figure 3 shows standard curves for the assay of T_4 and T_3.

The separation method has been successfully applied to the competitive enzyme immunoassay of insulin[5] and prolactin[14] in human serum.

Assay of (Anti-insulin) Antibody[8]

Reagents

Buffer G
Buffer S

---

[13] K. M. Jones, in "Data for Biochemical Research" (R. M. C. Dawson, D. C. Eliott, W. H. Eliott, and K. M. Jones, eds.), p. 436. Oxford Univ. Press, London and New York, 1969.

[14] K. Ishikawa, O. Narita, H. Noguchi, and K. Kato, *Clin. Chim. Acta* **121,** 181 (1982).

Insulin labeled with β-D-galactosidase, porcine insulin coupled to the enzyme with N,N'-o-phenylenedimaleimide,[15] diluted with buffer G

Standard (anti-insulin) serum, guinea pig (anti bovine-insulin) serum (containing antibodies capable of binding 0.8 unit of insulin per milliliter; Miles Laboratories), diluted (30- to 10,000-fold) with normal fasted human serum

Protein A-coupled activated thiol–Sepharose 4B

4-Methylumbelliferyl-β-D-galactoside, 0.3 mM

4-Methylumbelliferone, 1×10^{-7} M and 1×10^{-6} M, pH 10.3

Glycine–NaOH buffer, 0.1 M, pH 10.3

Procedures. Dilute standard or sample serum 100-fold with buffer G. Mix each of the diluted samples (0.1 ml) with 0.4 ml of insulin labeled with the enzyme (containing 6 milliunits). Incubate at 30° for 60 min. Apply the mixture to the protein A-coupled Sepharose column. Wash the column with buffer G (2 ml), and elute the label with buffer S (2 ml) as described in the assay for thyroid hormones. Preincubate 0.2-ml aliquots of the eluate for 5 min, and start the enzyme reaction by adding 0.1 ml of 4-methylum-belliferyl-β-D-galactoside solution at 30°. After 20 min, stop the reaction by adding 2.5 ml of the glycine–NaOH buffer. Measure the fluorescence intensity of 4-methylumbelliferone liberated in the reaction mixture with a fluorospectrophotometer at 360 nm, and 450 nm for excitation and emission analysis, respectively, against the standard 4-methylumbelliferone.

Figure 4 shows a standard curve for the assay of (anti-insulin) antibody and detection of the antibody in a serum sample from an autoimmune patient producing (anti-insulin) antibodies.

Comments. Because of the presence of relatively large amounts of the labeled insulin (6 milliunits contained at least 15 microunits of insulin) in the reaction medium, the presence of free insulin in serum sample in amounts up to 100 microunits/ml does not interfere in the assay. But the (anti-insulin) antibodies already bound with endogenous insulin in serum cannot theoretically be determined, unless the exchange reaction between the antibody-bound insulin and the labeled insulin could occur during the incubation.

An advantage of the use of protein A for the assay of antibodies in the column-separation method is that it might be possible for the assay of antibodies in various animal species to be determined quantitatively against a standard antiserum raised in a given animal species. When the (anti-IgG) antibody–Sepharose column is used for the assay of antibodies, the assay has to be carried out with the standard antiserum from the same species.

[15] K. Kato, Y. Haruyama, Y. Hamaguchi, and E. Ishikawa, *J. Biochem.* (*Tokyo*) **84**, 93 (1978).

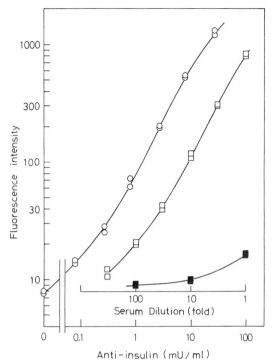

FIG. 4. Standard curve for the assay of (anti-insulin) antibody. Guinea pig (anti-bovine insulin) serum, containing antibodies capable of binding 0.8 unit of insulin per milliliter, was used as a standard serum (–O–) tentatively defined as 0.8 U anti-insulin/ml. A serum sample from an autoimmune patient producing (anti-insulin) antibody was subjected to the assay without (–□–) or with (–■–) 40 milliunits of insulin. Fluorescence intensity was 100 = 1 × 10^{-7} M 4-methylumbelliferone. From Kato et al.,[8] with permission.

Assay of Secretory Immunoglobulin A[16]

Secretory immunoglobulin A (SIgA), which is composed of dimeric immunoglobulin A (IgA) and a secretory component (SC), can be determined specifically with the present separation method by use of the Sepharose column with immobilized antibody to IgA (α-chain) and enzyme-labeled (anti-SC) antibody.

Reagents

Buffer G
Buffer S
Rabbit (anti-SC) antibodies labeled with enzyme. We used rabbit (anti-SC), antibody Fab′ fragments labeled with β-D-galactosi-

---

[16] Y. Ishiguro and K. Kato, unpublished data, 1981.

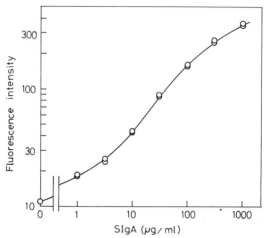

FIG. 5. Standard curve for the assay of secretory immunoglobulin A (SIgA). Fluorescence intensity was $100 = 1 \times 10^{-7}\ M$ 4-methylumbelliferone.

dase,[17] which can be prepared with the IgG fractions from the antiserum by use of N,N'-o-phenylenedimaleimide.[18]

Rabbit (anti-human α-chain) antibody-coupled activated thiol–Sepharose 4B. The IgG fractions of rabbit (anti-human α-chain) serum can be coupled to the Sepharose.

Standard SIgA, purified human SIgA (Boehringwerke) diluted with buffer G

4-Methylumbelliferyl-β-D-galactoside, 0.3 mM

4-Methylumbelliferone, 1×10^{-7} and $1 \times 10^{-6}\ M$, pH 10.3

Glycine–NaOH buffer, 0.1 M, pH 10.3

Procedures. Mix 0.1 ml of standard SIgA or 0.1 ml of serum sample diluted 100-fold with buffer G, with 0.4 ml of the labeled (anti-SC) antibody (6 milliunits in buffer G). Incubate the mixture at 30° for 2 hr. Apply the reaction mixture to the column of (anti-α-chain) antibody-coupled Sepharose. Wash the column with buffer G (2 ml), elute the label with buffer S (2 ml), and measure the enzyme activity in the eluate with 4-methylumbelliferyl-β-D-galactoside as substrate, as described above. Figure 5 shows a standard curve for the assay of SIgA.

Comments. The (anti-α-chain) antibody-Sepharose column prepared with commercially available antisera can trap most of the IgA in 1 μl of human serum.[16] Free SC levels in serum are usually lower than those of SIgA, and 6 milliunits of the labeled (anti-SC) antibodies are sufficient

[17] Y. Ishiguro, K. Kato, and T. Ito, *Clin. Chim. Acta* **116**, 237 (1981).

[18] K. Kato, H. Fukui, Y. Hamaguchi, and E. Ishikawa, *J. Immunol.* **116**, 1554 (1976).

when compared to the amounts of both free and bound SC in 1 μl of serum.

It is highly recommended to use the (anti-SC) serum absorbed with IgA, and to use both antisera (to SC and to α-chain) produced in the same animal species; otherwise SIgA-independent reactions of the enzyme-labeled antibodies with the antibodies coupled on the Sepharose might disturb the assay.

Sensitive and specific enzyme immunoassay systems for SIgA have been developed with the solid-phase separation method using Microtiter plates[19] or silicone rubber pieces[17] as support materials for immobilization of the (anti-SC) or (anti-α chain) antibody. However, when those assay systems are employed for determination of SIgA in serum, high levels of IgA in serum samples interfere with the assay by competitive or nonspecific binding on the solid phase, and precise levels of SIgA in serum cannot be determined with those assay systems.

Labeling of Antigens or Antibodies with β-D-Galactosidase

The enzyme-labeled compounds used in the assays described in this chapter were prepared based on the reaction between thiol groups[20] of β-D-galactosidase from *Escherichia coli* (EC 3.2.1.23) and the maleimide residues introduced to the antigens or antibodies with the bifunctional coupling reagents, N,N'-o-phenylenedimaleimide,[18] 4-maleimidomethylcyclohexyl-1-carboxylic acid succinimide ester,[21] and m-malimidoben-zoyl-N-hydroxysuccinimide ester.[22]

Preparation of Enzyme-Labeled Guinea Pig-IgG with N,N'-o-Phenylenedimaleimide. Guinea pig IgG (8 mg in 2 ml of 0.1 M sodium acetate buffer, pH 5.0) was incubated with 15 mM 2-mercaptoethylamine at 37° for 90 min. The reduced IgG, separated by Sephadex G-25 column chromatography (pH 5.0), was found to contain 1.4 mol equivalent thiol groups per molecule. The IgG (4 ml, A_{280} = 2.5, pH 5.0) was mixed with 2 ml of a saturated solution of N,N'-o-phenylenedimaleimide (Aldrich) in acetate buffer, pH 5.0. The mixture was incubated at 30° for 20 min and applied to a Sephadex G-25 column, equilibrated with 0.1 M sodium phosphate buffer, pH 6.3, to separate the maleimide-IgG (6 ml, A_{280} = 1.4)

---

[19] Å. S. Åkerlund, L. Å. Hanson, S. Aklstedt, and B. Carlsson, *Scand. J. Immunol.* **6**, 1275 (1977).

[20] K. Wallenfels, B. Muller-Hill, D. Dabich, C. Streffer, and R. Weil, *Biochem. Z.* **340**, 41 (1964).

[21] S. Yoshitake, Y. Yamada, E. Ishikawa, and R. Masseyeff, *Eur. J. Biochem.* **101**, 395 (1979).

[22] T. Kitagawa and T. Aikawa, *J. Biochem. (Tokyo)* **79**, 233 (1976).

from the excess dimaleimide. The maleimide IgG (6 ml) was then mixed with 0.75 mg of β-D-galactosidase (Boehringer-Mannheim), and the mixture was left standing at 4° overnight. The mixture was applied to a Sepharose 6B column, which had been equilibrated with 0.01 M sodium phosphate buffer, pH 7.0, containing 0.1 M NaCl, 1 mM MgCl$_2$, 0.1% bovine serum albumin, and 0.1% NaN$_3$ (buffer A). Fractions containing the major peak of enzyme activity were pooled and used as labeled guinea pig IgG.

Preparation of Enzyme-Labeled Insulin with N,N'-o-Phenylenedimaleimide. Five milliliters (about 5 mg) of porcine insulin (Actrapid from Novo Industri A/S, diluted twice with 0.1 M sodium phosphate buffer, pH 6.3.) was allowed to react with 3 mg of S-acetylmercaptosuccinic anhydride with continuous stirring at room temperature for 30 min. Then, 1 ml of 0.5 M hydroxylamine, pH 7.0, was added to the mixture. The mixture was incubated at 30° for 20 min, and applied to a Sephadex G-25 column (pH 6.3) to separate the thiolated insulin (6 ml, $A_{280} = 0.4$). This insulin fraction was added dropwise to 3 ml of the saturated solution (pH 6.3) of N,N'-o-phenylenedimaleimide with stirring. The mixture was incubated at 30° for 20 min and applied to a Sephadex G-25 column (pH 6.3) to separate the maleimide–insulin fraction. Three-milliliter aliquots of the *fraction ($A_{280} = 0.2$)* were mixed with 0.6 mg of β-D-galactosidase. The mixture was left standing at 4° overnight and applied to the Sepharose 6B column as described above.

Preparation of Enzyme-Labeled (anti-SC) Antibody Fab' Fragments with N,N'-o-Phenylenedimaleimide. IgG fractions (20 mg) of rabbit (anti-SC) serum (Behringwerke) were digested with 0.8 mg of pepsin from porcine stomach mucosa (Sigma) to obtain the F(ab')$_2$ fragments (10 mg) of the antibody as described by Kato et al.[23] The F(ab')$_2$ fragments, about 10 mg in 2 ml of sodium acetate buffer (0.1 M, pH 5.0), were mixed with 0.3 ml of freshly prepared 0.1 M 2-mercaptoethylamine. The mixture was incubated at 37° for 90 min and passed through a Sephadex G-25 column (pH 5.0) to separate the Fab' fragments (about 8 mg in 5 ml, estimated from the value $E_{280}^{1\%} = 14.8^{24}$) from excess 2-mercaptoethylamine. The Fab' fragments (5 ml) were immediately added dropwise to 2 ml of the saturated dimaleimide solution in acetate buffer (pH 5.0) with stirring. The mixture was incubated at 30° for 20 min, and applied to a Sephadex G-25 column (pH 6.3) to separate the maleimide–Fab' fragments. This fraction (8 ml, $A_{280} = 1.2$) was mixed with 2 mg of β-D-galactosidase, and the mixture was incubated at 4° overnight. Forty microliters each of 2-mer-

[23] K. Kato, Y. Hamaguchi, H. Fukui, and E. Ishikawa, *FEBS Lett.* **56**, 370 (1975).
[24] W. J. Mandy and A. Nisonoff, *J. Biol. Chem.* **238**, 206 (1963).

captoethylamine (0.1 M) and bovine serum albumin (5%) were added to the mixture to stop the reaction and to stabilize the enzyme activity. The mixture was applied to the Sepharose 6B column to separate the enzyme-labeled antibody Fab' fragments from free Fab' as described above.

Preparation of Enzyme-Labeled T_4 or T_3 with 4-(Maleimidomethyl) Cyclohexyl-1-Carboxylic Acid Succinimide Ester. Fifty microliters of T_4 or T_3 solution (1 mg/ml, in N, N-dimethylformamide) were mixed with $5\mu l$ of the ester solution (2 mg/ml, in N,N-dimethylformamide; Zieben Chemical Co., Ltd., Tokyo) and 0.1 ml of sodium phosphate buffer (0.1 M, pH 7.0), and the mixture was incubated at 30°. After 90 min, the reaction was storped by adding 10 μl of 1 M glycine. Then 0.2 mg of β-D-galactosidase in 0.8 ml of sodium phosphate buffer (0.1 M, pH 7.0) was added to the above solution, and the mixture was incubated at 30°. After 30 min, 10 μl of 0.1 M 2-mercaptoethylamine were added to block remaining maleimide residues. The reaction mixture was applied to a column of Sephacryl S-300 (Pharmacia), equilibrated with sodium phosphate buffer (0.1 M, pH 7.0). The peak fractions of β-D-galactosidase activity were collected, and stored supplemented with bovine serum albumin (0.1%), $MgCl_2$ (1 mM), and NaN_3 (0.1%). The conjugates contained 4–5 mol of T_4 or T_3 per mole of enzyme.[7]

Preparation of Enzyme-Labeled Human IgG with m-Maleimidobenzoyl-N-Hydroxysuccinimide Ester. One milliliter of human IgG solution (1 mg/ml, in 0.1 M sodium phosphate buffer, pH 7.0) was mixed with 50 μl of the ester solution (1 mg/ml, in dioxane; Pierce Chemical Co.), and the mixture was incubated at 30°. After 30 min, the mixture was applied to a Sephadex G-25 column to seperate the maleimide–IgG (2 ml, A_{280} = 0.5). This fraction was mixed with 0.3 mg of β-D-galactosidase and left standing at 4° overnight. The mixture was then applied to the Sepharose 6B column as described above.

[27] Affinity Exclusion: A New Method for the Separation of Free and Bound Fractions in Enzyme Immunoassay

By B. TEROUANNE and J. C. NICOLAS

Assay systems involving the use of antigens, haptens, or antibodies labeled with an enzyme have been used for the measurement of substances in biological fluids.[1,2] Apart from homogeneous enzyme immu-

---

[1] H. Van Hell, J. A. M. Brands, and A. H. W. M. Schuurs, *Clin. Chim. Acta* **91**, 309 (1979).

[2] F. Dray, J. Andrieu, and F. Renaud, *Biochim. Biophys. Acta* **403**, 131 (1975).

noassay,[3] these methods require the separation of free and bound labeled material. This separation can be achieved using insolubilized antibodies or antigens, but whichever absorbent is used, the bound fraction is retained on the matrix and enzyme activity associated with these fractions must be determined in a heterogeneous medium. Efforts have already been made to decrease the time needed for one assay and to determine enzyme activity in solution.[4] For these purposes, we propose affinity chromatography for rapid separation of antigen-enzyme bound to antibodies from free conjugate and its applications in enzyme immunoassay.

We describe enzyme immunoassays for human placental lactogen (hPL) and progesterone, which use Δ^5-3-ketosteroid isomerase as a marker and affinity chromatography on Ultrogel estradiol for the separation of the free and bound fractions.

Principle

In affinity chromatography it has been well established that retention of enzymes is largely dependent on the insolubilized ligand concentration, and the ligand is generally randomly distributed in the gel.

As bound and free fractions are very different in size, reticulated gels are available in which the high molecular weight fraction (bound fraction) is excluded, while the free fraction can enter the gel.

An affinity absorbent using this particular gel as a matrix can specifically retain the free fraction, which can enter the gel and interact with the ligand, while the bound fraction has limited access to the ligand, owing to its higher molecular weight, and is thus not retained.

This technique combines affinity chromatography and penetration of the gel in the same step; consequently we consider "affinity exclusion" to be an apt name for this new method of separation.

Preparation of Glutathione Estradiol Absorbent

17β-Bromoacetoxyestradiol was obtained by esterification of estradiol-17β with bromoacetic acid using the dicyclohexyl carbodiimide method as described by Pons.[6]

[3] K. E. Rubenstein, R. Schneider, and E. F. Ullman, *Biochem. Biophys. Res. Commun.* **47,** 846 (1972).

[4] Y. Umeda, F. Suzuki, A. Kosaka, and K. Kato, *Clin. Chim. Acta* **107,** 267 (1980).

[5] T. T. Ngo and H. M. Lenhoff, *FEBS Lett.* **116,** 285 (1980).

[6] M. Pons, J. Marchand, and A. Crastes de Paulet, *C. R. Hebd. Seances Acad. Sci. Ser. C* **283,** 507 (1976).

One gram of estradiol-17β and 2 g of bromacetic acid were dissolved in 2 ml of dimethylformamide, after cooling to 4°. Then 2 g of N,N'-dicyclo-hexylcarbodiimide were added; after 5 min of reaction, 150 μl of pyridine were added.

After 6 hr at 4°, dicyclohexylurea was precipitated by the addition of 15 ml of ethyl ether; the precipitate was removed by filtration and washed with 20 ml of ethyl ether. The filtrates were pooled and evaporated. The residue was dissolved in 5 ml of benzene and applied to a silica gel column (2 × 20 cm). 3- and 17β-bromoacetoxyestradiol were separated by elution with benzene and benzene-ethyl ether with an increasing concentration of ethyl ether.

Sixty milligrams of 17β-bromoacetoxyestradiol dissolved in 1 ml of dimethyl formamide were incubated with 300 mg of reduced glutathione at pH 8.5 for 6 hr; 20 ml of water were added, and the mixture was extracted twice with 50 ml of ethyl ether. The aqueous phase was applied to a Dowex 1-×2 column (1.5 × 7 cm), previously equilibrated with 0.2 M citrate (pH 6.7). The column was washed with the same buffer, then with 0.2 M citrate (pH 2) until absorbance at 254 nm was almost zero. The glutathione steroid was eluted with 0.2 M citrate (pH 2) containing 50% ethanol. After elution, the pH of the solution was adjusted to pH 9 with 1 M NaOH.

One hundred milliliters of glutathione estradiol solution (3×10^{-3} M), pH 9, were incubated with 100 ml of Ultrogel (AcA-44 or AcA-34) or Sepharose 4B activated by cyanogen bromide. The gel was washed thoroughly with water. To determine the concentration of insolubilized estradiol, 1 g of packed gel was incubated with 1 N hydroxylamine, pH 9, for 16 hr. The estradiol concentration in the supernatant was determined using the enzymic method of Nicolas et al.[7]

Enzyme Preparation

Δ^5-3-Ketosteroid isomerase (EC 5.3.3.1) was prepared by chromatography on DEAE-cellulose[8] and affinity chromatography on 5 ml of Sepharose estradiol-17β. Affinity chromatography was performed in 0.1 M phosphate buffer, pH 7.2. Under these conditions 1 mg of isomerase was retained on 1 ml of Sepharose estradiol. The enzyme was readily eluted with 2 ml of water.

[7] J. C. Nicolas, A. M. Boussioux, B. Descomps, and A. Crastes de Paulet, *Clin. Chim. Acta* **92,** 1 (1979).

[8] A. M. Benson, A. J. Suruda, R. Show, and P. Talalay, *Biochim. Biophys. Acta* **348,** 317 (1974).

Antigen Enzyme Preparation

Conjugates were prepared by the intermolecular disulfide interchange reaction or by alkylation of the thiolated enzyme with bromoacetoxy hapten.

Preparation of the Thiolated Enzyme. Isomerase (1.3 mg) was incubated with 3.5 mg of *S*-acetylmercaptosuccinic anhydride for 2 hr at pH 7.2 and at room temperature. The resulting solution was dialyzed against 200 ml of 0.1 M phosphate buffer, pH 7.2, then against 100 ml of 0.1 M hydroxylamine, pH 8, 10^{-2} M dithiothreitol and finally against 2 liters of 0.1 M Tris-HCl, pH 8.

Seven to nine thiol residues were incorporated per mole of enzyme.

Coupling by Intermolecular Disulfide Interchange. Thiols of the isomerase was activated at pH 8 by 10^{-3} M 5,5'-dithiobisnitrobenzoate. Excess reagent was removed by affinity chromatography on 1 ml of estradiol-17β Sepharose.

The antigen, in this case 2 mg of hPL, was thiolated by reaction with 1.3 mg of methyl 4-mercaptobutyrimidate at pH 9 for 4 hr. After dialysis we determined that 2 thiol residues had been incorporated per mole of hPL. This solution was incubated with the activated enzyme at pH 8. Disulfide formation was shown by the appearance of thionitrobenzoate in the solution, which was determined spectrophotometrically at 412 nm. After completion of the reaction, excess hormone was separated from the enzyme fraction by affinity chromatography.

Coupling by Reaction with Bromoacetoxy Hapten. The thiolated enzyme was also used for the preparation of progesterone conjugate. 11α-Bromoacetoxyprogesterone was prepared by esterification of 11α-hydroxyprogesterone by bromoacetic acid in the presence of dicyclohexyl carbodiimide. The product was purified by chromatography on silica gel. Ten milligrams of 11α-bromoacetoxyprogesterone were allowed to react with 0.5 mg of thiolated enzyme for 16 hr at pH 8.2 and at room temperature, and excess steroid was removed by gel filtration on Sephadex G-25. Two moles of steroid were bound per mole of enzyme.

Affinity Absorption of Enzyme Conjugate

One millimeter of antigen–enzyme solution (isomerase–progesterone or isomerase–hPL) was incubated with an excess of specific antibodies. These solutions were applied to 0.1 ml of estradiol absorbent using AcA-44, AcA-34, and Sepharose as the matrix. The insolubilized estradiol concentration in the absorbent was in the range 0.8×10^{-4} to 2×10^{-4} M. Filtration occurred in less than 1 min, and enzyme activity was determined

TABLE I

PERCENTAGE OF hPL-ISOMERASE ACTIVITY RETAINED ON VARIOUS ABSORBENTS

Absorbents	Estradiol Ultrogel AcA-44 (%)	Estradiol Ultrogel AcA-34 (%)	Estradiol Sepharose 4B (%)
hPL-isomerase	98	99	99
hPL-isomerase + hPL antiserum	5	48	94
hPL-isomerase + hPL antiserum + antirabbit γ-globulin serum	3	6	90

in the effluent. Tables I and II show the retention of enzyme activity on these three absorbents. Retention of isomerase antigen was complete in each case, while the product of the reaction with specific antibodies was completely retained only with estradiol–Sepharose 4B. With estradiol–Ultrogel the retention of the bound fraction was reduced. In Table I we see that in the case of the immune complex formed with hPL-isomerase, only 5% of enzyme activity was retained on estradiol Ultrogel AcA-44 whereas with the complex formed between progesterone isomerase and a specific progesterone antiserum, 58% of enzyme activity was retained on estradiol ACA44. The molecular weight of the complex has to be increased by reaction with second antiserum directed against the first antibody in order to reduce this retention (Tables I and II).

Under these conditions, variations in flow rate between 0.2 ml and 1 ml per minute had no effect on retention of the enzyme conjugates when 0.1 ml of gel was used (Table III).

TABLE II

PERCENTAGE OF PROGESTERONE–ISOMERASE ACTIVITY RETAINED ON VARIOUS ABSORBENTS

Absorbents	Estradiol Ultrogel AcA-44 (%)	Estradiol Ultrogel AcA-34 (%)	Estradiol Sepharose 4B (%)
Progesterone-isomerase	99	99	99
Progesterone-isomerase + progesterone antiserum	58	76	99
Progesterone-isomerase + progesterone antiserum + antirabbit γ-globulin serum	5	8	90

TABLE III
EFFECT OF FLOW RATE ON THE $\Delta^{5\text{-}4}$-3-KETOSTEROID ISOMERASE
CONJUGATE RETENTION ON 17β-ESTRADIOL ULTROGEL AcA-44

	Flow rate (ml/min)			
Conjugates	0.1	0.2	0.5	1
Isomerase	99%	99%	99%	99%
hPL-isomerase	98%	98%	97%	97%
hPL-isomerase + hPL antiserum + antirabbit γ-globulin serum	6%	4%	3%	3%

Enzyme Immunoassay

All dilutions were carried out in buffer A: 0.1 M phosphate, pH 7.2, 0.126 M NaCl, 1 g of BSA per liter. The immunological reaction was performed in polypropylene tubes containing 0.1 ml of sample, 0.1 ml of specific antibodies, 0.1 ml of second antibody anti-rabbit γ-globulin antibody), and 0.1 ml of antigen–isomerase (0.1 IU). After 1 hr of incubation, the free antigen–enzyme was specifically retained by filtration on estradiol Ultrogel. These filtrations were performed in a polypropylene column (diameter 4 mm; Bio-Rad) containing 0.1 ml of estradiol absorbent equilibrated against buffer A. The solution was poured onto the column, then the gel and the tube were washed with 1 ml of buffer A. The enzyme activity of the filtrate was determined spectrophotometrically at 248 nm on 200-μl aliquots. The reaction was carried out in 1 ml of phosphate buffer, 0.03 M, pH 7.2, containing 10^{-4} M Δ^5-androstene-3,17-dione.

Figures 1 and 2 show examples of displacement curves obtained with enzyme–progesterone and enzyme–hPL, respectively. Under the conditions used, the assay detected 30 pg of progesterone and 1 ng of hPL.

The reproducibility of the affinity method for separating bound antibody from free conjugate was tested by repeated measurements of the maximum binding of a progesterone–isomerase conjugate. We found for B_0/T a mean of 42.2% (8 filtrations) with a standard deviation of 1.3% and a coefficient of variation of 3.2%. Affinity exclusion gives results similar to those of radioimmunoassay or enzyme immunoassay using an immunoabsorbent. The new method is quicker, since the separation requires only filtration, and the enzyme is determined directly in 200-μl aliquots. The enzyme activity ranges from 10 to 200 milliunits of absorbance per minute.

In the method described, the binding of antibody induces an increase in size that makes rapid separation of the antigen–isomerase conjugate from the antibody-bound antigen–enzyme complex possible.

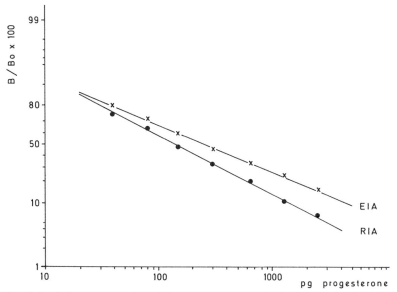

FIG. 1. Logit-log representation of the progesterone standard curve obtained by affinity-exclusion enzyme immunoassay ($\times$——$\times$) and by radioimmunoassay ($\bullet$——$\bullet$). B/B_0 is the percentage of bound enzyme or radioactivity with and without progesterone added.

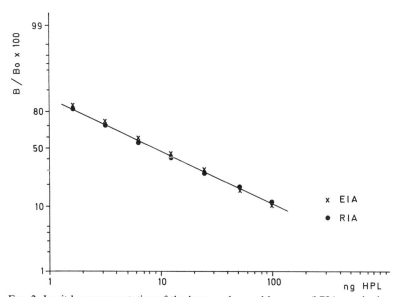

FIG. 2. Logit-log representation of the human placental lactogen (hPL) standard curve obtained by affinity-exclusion enzyme immunoassay ($\times$——$\times$) and by radioimmunoassay ($\bullet$——$\bullet$). B/B_0 is the percentage of bound enzyme or radioactivity with and without human placental lactogen added.

Two conditions that must be satisfied for the system to be operative are that (a) the antigen-isomerase conjugate (free fraction) should enter the gel beads and bind the ligand; (b) the antibody-bound antigen–isomerase conjugate (bound fraction) should be excluded from the gel and readily eluted.

The method is not limited by pore size, since a series of gels of different exclusion properties are commercially available. It is usually possible to select an insoluble matrix to provide an affinity column satisfying both the above conditions. Moreover, it is possible to increase considerably the size of the antibody–antigen–isomerase complex by using a second antiserum directed against the first antibody; large complexes result from the polyvalent binding of several second antibody molecules to the same complex. Complete exclusion of the complexes was obtained by this method in the case of progesterone (Table II).

Finally, analysis of very large antigens is also possible by using the reverse system, in which the antibody labeled by isomerase is retained on an absorbent of smaller pore size, whereas the labeled antibody bound to the large antigen is excluded.

As described above, the separation system used is suitable for the determination of antigen with molecular weights ranging from that of steroids to protein hormones.

The affinity-exclusion method of separating the antigen–enzyme from the fraction bound by antibodies offers several advantages and allows complete automation of enzyme immunoassay. All reactions are performed in solution, and separation is easy and needs only minimum instrumentation. The enzyme activity bound by the antibodies is directly determined in the effluent.

[28] Use of Chromatography Tubes in the Separation of Bound and Free Fractions in Radioimmunoassay

By D. B. WAGNER, J. FEINGERS, D. INBAR,
A. J. PICK, Y. TAMIR, and O. WEISS

An important factor in a good radioimmunoassay (RIA) is the method by which the bound and free fractions are separated so that the radioactivity of either or both can be accurately measured. Regardless of the method of separation chosen, it must be reproducible, simple to perform, and economically feasible.[1] A large variety of specific separation methods

[1] D. S. Skelley, L. P. Brown, and P. K. Besch, Clin. Chem. 19 (2), 146 (1973).

(e.g., solid-supported antibody, second antibody precipitation) as well as nonspecific ones (charcoal, polyethylene glycol, ion-exchange resins) have been described, and some have found use in commercial RIA kits, both manual and automated. Chromatographic, nonspecific separation techniques have not acquired the popularity shared by other methods, most probably because they are rather laborious, time-consuming, and not inexpensive.[2-4] A chromatographic separation method that is fast, very convenient, reproducible, and inexpensive is the subject of this chapter.

In this method, separation between the bound and free fraction is achieved by passive adsorption on small, disposable chromatography columns (termed chromatography tubes) packed with a powered dry adsorbent. This technique is based on a modified form of inverted dry-column chromatography[5]: serum sample (or standard), labeled tracer and antibody are incubated in a regular manner in a test tube. At the end of the incubation, a chromatography tube is placed upright into each test tube, and the incubated mixture is allowed to be absorbed up into the dry resin. The free fraction is trapped by resin at the very bottom of the tube, while the bound fraction (having little or no affinity to the resin) migrates to the top of the tube (Fig. 1). The test tube (containing the chromatography tube) is then placed in a gamma counter. The upper part of the chromatography tube, which contains the bound fraction, remains outside (above) the detector, thus allowing the counting of the free fraction alone. Thus, a separation of the free and bound fraction is achieved without centrifugation, aspiration, or decanting of radioactive solutions. The time needed for complete separation is usually no longer than 5–10 min.

Materials and Methods[6]

Cross-Linked Poly(vinyl Alcohol)

Place distilled water (600 ml) in a 2-liter glass container and warm with stirring to 35–45°. Add poly(vinyl alcohol) (PVA 10 crystalline, water-

[2] H. Orskow, *Scand. J. Clin. Lab. Invest.* **20**, 297 (1967).
[3] N. M. Alexander and J. F. Jennings, *Clin. Chem.* **20** (5), 553 (1974) and references cited therein.
[4] N. M. Alexander and J. F. Jennings, *Clin. Chem.* **20** (10), 1353 (1974) and references cited therein.
[5] V. K. Bhalla, U. R. Nayak, and S. Dev, *J. Chromatogr.* **26**, 54 (1967).
[6] Nonstandard abbreviation and terms used: ANS, 8-anilinonaphthalenesulfonic acid; EDTA, ethylenediaminetetraacetic acid; BSA, bovine serum albumin; hPL, human placental lactogen; PVA, poly(vinyl alcohol); T_3 triiodothyronine; T_4, thyroxine.

Chromatography tube

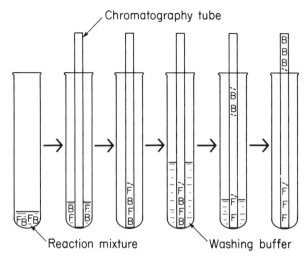

Reaction mixture   Washing buffer

FIG. 1. Schematic representation of the separation of the bound (B) and free (F) fraction using chromatography tubes.

soluble type II from Sigma Chemical Company, St. Louis, Missouri) carefully and slowly to avoid formation of lumps. Stir vigorously until all the PVA is in solution and a heavy foam is formed. Add concentrated hydrochloric acid (5 ml) and glutaraldehyde (10 ml, aqueous, 25%, from Merck, Darmstadt, Germany), and continue the stirring and heating (to 65°) for 20 min. At this point the foam should subside, and a white granular precipitate will appear. Filter the insoluble cross-linked PVA, and wash the white cake with distilled water until the pH of the wash becomes neutral. Wash twice with ethanol, pass through a coarse sieve, and let dry. Sieve the dry powder, and collect the fraction with particle size between 150 and 400 μm.

Preparation of Chromatography Tubes

For Digoxin, Thyroxine, and Triiodothyronine RIA. Cut polystyrene tubing (o.d. 7 mm, i.d. 5 mm) into 90 mm-long pieces. Place filter paper (Whatman No. 41) on a hot plate, and press the end of the polystyrene tube to the surface of the heated filter paper. The filter paper will thus become attached to the tube. The excess paper should be cut and trimmed. Fill the tubes with cross-linked PVA with continuous vibration to ensure thorough packing. The upper end of the tubes may then be sealed with a second filter paper disk, porous foam rubber, sponge or glass wool plug. The amount of cross-linked PVA should be 250 ± 20 mg/

tube. An alternative absorbent for digoxin RIA is dry, fine, acid-washed sea sand. Wash the sand with distilled water until the pH of the washing is approximately 7.0; then dry for 3 hr at 150°, sieve, and collect the fraction between 150 and 250 μm. Use approximately 2.5 g of sand for each chromatography tube.

For Triiodothyronine Uptake Test. Prepare polystyrene tubes sealed with a filter paper disk as for digoxin, thyroxine, triiodothyronine RIA. Place silica gel (silica gel 60, 230–400 mesh from Merck, Darmstadt, Germany) into the tubes to a height of 1.6 cm, followed by a second layer of dry, clean sea sand, and seal the upper end as above.

For hPL. Prepare chromatography tubes as in triiodothyronine uptake test, filled with silica gel only.

Preparation of Buffers

Buffer A: Dissolve disodium EDTA (0.6 g) and disodium phosphate (anhydrous, 1.2 g) in distilled water (to 1000 ml).

Buffer B: Dissolve 2-amino-2-hydroxymethyl-1,3-propanediol (9.8 g), maleic acid (4.1 g), and disodium EDTA (1.5 g) in distilled water (to 1000 ml).

Buffer C: Dissolve disodium phosphate (anhydrous, 12.0 g), disodium EDTA (6.0 g), and sodium azide (0.2 g) in distilled water (to 1000 ml).

Buffer D: Dissolve citric acid (anhydrous, 14.4 g), disodium phosphate pentahydrate (43.7 g) and formaldehyde (370 g/1, 2.7 ml) in distilled water (to 1000 ml).

Buffer E: Dissolve 2-amino-2-hydroxymethyl-1,3-propanediol (9.66 g), maleic acid (8.0 g), disodium EDTA (1.41 g), and Tween 20 (4.0 ml) in distilled water (to 1000 ml).

Buffer F: Dissolve sodium salicylate (7.5 g) in sodium hydroxide, (0.1 N) to 1000 ml.

Radioactive Tracers

Digoxin: ^{125}I-labeled digoxin–histamine conjugate with a specific activity of about 2500 μCi/μg is prepared according to Blazey[7] and diluted with buffer A to approximately 500,000 cpm/ml.

T_4 and T_3: [^{125}I]T_4 and [^{125}I]T_3 obtained from a number of commercial sources (e.g., The Radiochemical Centre, Amersham, U.K.).

hPL: ^{125}I-hPL can be obtained from New England Nuclear, Boston, Massachusetts.

[7] N. D. Blazey, *Clin. Chim. Acta* **80,** 403 (1977).

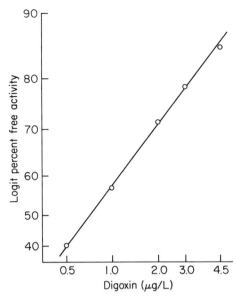

FIG. 2. Standard curve for digoxin radioimmunoassay.

Preparation of Test Tubes Containing Premeasured Lyophilized T_4 or T_3 Tracers

Prepare a solution of [^{125}I]thyroxine (prepared by the chloramine-T method) in water containing PVA (2.5 g/liter) and ANS (4.0 g/liter). Accurately dispense 50-μl aliquots to disposable polypropylene test tubes and lyophilize. For T_3, use the same procedure, but increase the PVA concentration to 5 g/liter and decrease the ANS concentration to 0.5 g/liter. Lyophilize 100-μl aliquots (100,000 cpm for T_3).

Assay Procedures

Digoxin. Incubate for 30 min at room temperature a mixture of antidigoxin antiserum (properly diluted in buffer A), [^{125}I]digoxin (200 μl, approximately 100,000 cpm), and digoxin standard or serum in 75 × 12 mm test tubes. Place a chromatography tube in each test tube and let the reaction mixture be absorbed into the dry absorbent (5–10 min). Without removal of the chromatography tubes, add 1 ml of buffer A (600 μl if sand is used as an absorber) to each test tube, and let it ascend into the chromatography tube. Without removal of the chromatography tubes, place the test tubes in a gamma counter. The radioactivity measured for each test tube represents the "free" fraction and can be used for construction of a standard curve in the usual manner (Fig. 2).

Thyroxine. Add 25 μl of serum or standard to a test tube containing lyophilized [^{125}I]T$_4$, followed by antithroxine antiserum diluted with buffer B (100 μl). Incubate at room temperature for 30 min. Insert a chromatography tube and continue as described for digoxin. A typical standard curve is shown in Fig. 3.

Triiodothyronine. Add 50 μl of serum or standard to a test tube containing lyophilized [^{125}I]T$_3$, followed by 300 μl of antitriiodothyronine antiserum, diluted with buffer C. Incubate for 2 hr at room temperature. Insert a chromatography tube, let the reaction mixture ascend into the dry absorbent; add 0.8 ml of buffer C to the test tube, and allow it, too, to be absorbed. Continue as in procedure for digoxin. The standard curve is shown in Fig. 4.

Triiodothyronine Uptake. Place serum sample (or control, 20 μl) in a test tube (12 × 75 mm). Add [^{125}I]T$_3$ in buffer D (200 μl, about 100,000 cpm). Swirl the test tubes gently, and insert chromatography tubes as above with the silica gel end down. Allow the content of the test tube to ascend into the chromatography tube. Add buffer D (0.5 ml) to each test tube, and continue as in digoxin procedure. Calculate the uptake ratio using the formula

$$\text{Uptake ratio} = \frac{\text{radioactivity of sample}}{\text{radioactivity of euthyroid standard serum}}$$

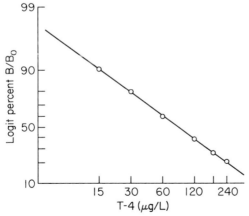

FIG. 3. Standard curve for total thyroxine radioimmunoassay. The same curve was obtained for the neonatal thyroxine RIA.

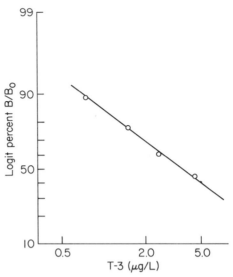

FIG. 4. Standard curve for triiodothyronine (T_3) radioimmunoassay.

Neonatal T_4. Use 1/8 inch paper punch to cut two circular blood-stained disks of either standard T_4 blood or sample, and place the two disks in a test tube. Add [^{125}I]T_4 in buffer F (100 μl), and incubate at room temperature for 1 hr. Add buffer E (100 μl) to each test tube, followed by T_4 antiserum (diluted in buffer, B, 100 μl). Shake gently, and incubate at room temperature for 30 min. Insert a chromatography tube to each test tube; after absorption of the content, add buffer B (500 μl). Continue as in digoxin procedure.

Human Placental Lactogen. Incubate a mixture of anti-hPL antiserum (diluted with buffer A, 50 μl), hPL standard or serum sample (diluted 1:50 with buffer A containing 0.5% BSA, 100 μl) for 40 min at room temperature. Add [^{125}I]hPL (diluted in buffer A containing 0.5% BSA, 50 μl, approximately 40,000 cpm). Incubate for an additional 40 min. Continue as in procedure for triiodothyronine uptake. The standard curve is shown in Fig. 5.

Results

Separation. Very good separation of the antibody-bound from the free fraction was observed in all the RIA tests. A typical example is shown in Fig. 6.

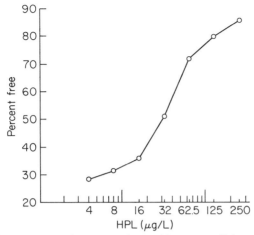

Fig. 5. Standard curve for human placental lactogen radioimmunoassay.

Precision. Three pooled serum samples were assayed in 20 replicates, each on three separate days. An estimate of the interassay precision was calculated for each serum. The results in Table I show that all coefficient of variation (CV) values were within the clinically acceptable range. No

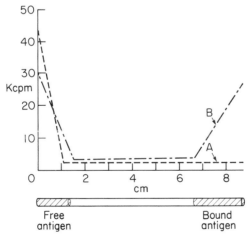

Fig. 6. Position of free and bound tracer in a chromatography tube (digoxin assay). Curve A (– – –), radioactive tracer only, no antibody; curve B (– – – –), 0.5 μg/liter standard and antibody. The chromatographic distribution of the free and bound tracer was obtained by first freezing a chromatography tube, then cutting it, while frozen, into 1.0-cm pieces, and measuring the radioactivity of each piece.

TABLE I
INTERASSAY PRECISION RESULTS

Analyte	Mean (μg/liter)	%CV[a]	N
Digoxin, using PVA[a] chromatography tubes	1.05	6.5	59
	1.94	5.8	60
	3.90	6.0	59
Digoxin, using sand chromatography tubes	1.02	6.8	60
	2.09	5.6	59
	3.75	3.7	59
Thyroxine	27	13.0	60
	72	5.5	60
	163	5.6	60
Neonatal T_4	28.8	23.9	60
	64.7	18.8	58
	140	14.64	59
Triiodothyronine	0.9	12.1	60
	1.5	6.0	60
	3.1	4.7	60
Triiodothyronine uptake (ratio values)	0.60	4.2	60
	0.92	4.4	60
	1.7	3.3	60

[a] CV, coefficient of variance; PVA, poly(vinyl alcohol).

TABLE II
CORRELATION OF THE CHROMATOGRAPHY TUBE TEST WITH COMMERCIAL RADIOIMMUNOASSAY KITS[a]

Analyte/reference kit	Correlation	Intercept	Slope
Digoxin (PVA)/Gammacoat	0.993	−0.089	0.94
Digoxin (PVA)/NEN	0.979	0.186	1.06
Digoxin (sand)/Gammacoat	0.99	−0.036	0.94
Digoxin (sand)/NEN	0.973	0.254	1.06
Thyroxine/Seralute T_4	0.993	5.2	0.92
Thyroxine/Gammacoat T_4	0.972	16.2	0.94
Triiodothyronine/Seralute T_3	0.976	0.046	1.21
Triiodothyronine/Gammacoat T_3	0.991	0.27	1.07
Triiodothyronine uptake/Trilute	0.96	−0.20	1.20
Triiodothyronine uptake/Gammacoat	0.89	−0.10	1.10
Triiodothyronine uptake/NML	0.97	−0.31	1.20

[a] The following commercial RIA kits were used in the accuracy evaluation: Gammacoat from Clinical Assays, Cambridge, Massachusetts; New England Nuclear (NEN), Boston, Massachusetts; Nuclear Medical Lab. (NML), Dallas, Texas; Seralute T3, Seralute T4, and Trilute from Ames Co., Elkhart, Indiana.

TABLE III
MINIMAL DETECTABLE DOSE

Analyte	Sensitivity (μg/liter)
Digoxin (on PVA)	0.4
Digoxin (on sand)	0.3
Thyroxine	3.5
Triiodothyronine	0.3
hPL	4.0

interassay precision studies were performed on the hPL test. Two control sera (0.9 μg/ml and 8.8 μg/ml) were assayed in 20 replicates each, and the intra-assay results were 6.4% CV and 1.9% CV respectively.

Accuracy. Results obtained for 50–60 clinical sera were compared by linear regression with those obtained for the same sera using commercial RIA kits. The correlation values are summarized in Table II. No such study was carried out for the neonatal-T_4 test, as no reliable commercial kit was available at the time this chromatography tube assay was developed. In order to establish a normal range for the assay, samples from 1000 newborns were assayed. Based on this study, the value of 70 μg/liter was suggested as the lower normal limit, the mean value being 135 μg/liter.

Sensitivity. The minimal detectable dose was defined as the lowest concentration for which the percentage of the free activity differs significantly (95% confidence level) from that of antigen-free serum. The results are summarized in Table III. The sensitivity in the hPL assay could be increased, by proper dilutions of the reagents, to 0.05 μg/liter.

Recovery. Recovery was tested by assaying serum samples before and after spiking with known amounts of antigens. The percentage recovery observed was between 94% and 113%.

Discussion

Although complete studies for RIA only of low molecular weight antigens, using chromatography tubes, have been described here, this technique can apparently be applied to the RIA of higher molecular weight antigens as well. Preliminary studies on the RIA for hPL were very promising, demonstrating high sensitivity and precision.

The dry adsorbents that can be used in the chromatography tubes are not limited to those described here. In fact, the only serious limitation is the wet volume of the adsorbent: A high degree of swelling on contact with water causes the "explosion" of the chromatography tube by which most of its content is lost. This was the case with the cross-linked dextran-type resins, such as Sephadex.

It was somewhat surprising to find that plain sand performed, at least in the digoxin RIA, as a very selective adsorbent. In fact, the analytical results obtained by using sand-filled chromatography tubes were indistinguishable from those filled with the more expensive cross-linked PVA. To the best of our knowledge, this is the only (published) report on the use of plain (underivatized) sand for the separation of the "bound" and "free" fraction in immunoassay.

The case of the T_3 uptake test was found to be a special one. Using chromatography tubes filled with one adsorbent (cross-linked PVA, silica gel) only, resulted in poor reproducibility. It was found that the separation oetween the thyroid-binding proteins and other TBG-bound tracer and the free one was dependent on the ascending rate of the reaction mixture up into the chromatography tube. At a low rate a considerable amount of the bound fraction was found in the lower part of the chromatography tube. Because these ascending rates do vary slightly from one chromatography tube to another, the result was poor reproducibility. It appeared that too long a contact of the TBG-bound fraction with the active adsorbent brought about partial "stripping" of the tracer, which, on becoming free, was adsorbed in the lower part, instead of moving upward. By reducing the amount of the silica gel to a minimum, the contact time of the bound fraction with the adsorbent was brought down to a few seconds—too short a time for "stripping." The sand in these chromatography tubes acted as a capillary "pump" only, taking no part in the actual separation.

The preparation of the chromatography tubes as outlined here is the one developed for commercial manufacturing, using automated, rather complex machinery. It should be emphasized that this is not the only method. In fact, chromatography tubes made manually from glass tubes, using glass wool as a support for the adsorbent, gave the same performance as the machine-made tubes.

Also, the T_3 and T_4 RIA can be performed in the more traditional manner without the use of premeasured, lyophilized radioactive tracers. These premeasured tracers were developed in order to reduce the number of accurate pipetting steps to a minimum.

[29] Enzyme-Linked Immunoelectrotransfer Blot Techniques (EITB) for Studying the Specificities of Antigens and Antibodies Separated by Gel Electrophoresis

By VICTOR C. W. TSANG, JOSE M. PERALTA, and A. RAY SIMONS

The enzyme-linked immunoelectrotransfer blot technique (EITB) combines the high resolving power of gradient sodium dodecyl sulfate--polyacrylamide gel electrophoresis (SDS–PAGE) and the high sensitivity of the enzyme-linked immunosorbent assay (ELISA) to produce an extremely powerful qualitative tool for studying antigen–antibody pairs. Using this procedure, antigens electrophoretically resolved on SDS–PAGE are transferred onto nitrocellulose or diazo sheets and identified by ELISA methods.

The general use of transfer blotting electrophoretically separated biological molecules began wth the DNA–RNA hybridization studies of Southern,[1] with a technique now known as the "Southern blot." The usefulness of the blotting technique was further advanced when Towbin et al.[2] demonstrated the feasibility of electrophoretically blotting SDS–PAGE-resolved protein molecules onto nitrocellulose sheets. The possibilities of detecting antigenic activities in SDS–PAGE-resolved molecules by a peroxidase-type ELISA test were suggested by Van Raamsdonk et al.[3] Subsequent marriage of the electrotransfer blot and ELISA techniques produced versions of the present EITB. The blotting media included both nitrocellulose and diazo papers.[4-11] Although the diazo papers couple SDS–protein complexes covalently and probably have higher capacities, we have selected nitrocellulose papers as our

[1] E. M. Southern, *J. Mol. Biol.* **98**, 503 (1975).

[2] H. Towbin, T. Staehelin, and J. Gordon, *Proc. Natl. Acad. Sci. U.S.A.* **76**, 4350 (1979).

[3] W. Van Raamsdonk, C. W. Pool, and C. Heyting, *J. Immunol. Methods.* **17**, 337 (1977).

[4] B. Bowen, J. Steinberg, U. K. Laemmli, and H. Weintraub, *Nucleic Acids Res.* **8**, 1 (1980).

[5] E. J. Stellwag, and A. E. Dahlberg, *Nucleic Acids Res.* **8**, 299 (1980).

[6] J. Renart, J. Reiser, and G. R. Stark, *Proc. Natl. Acad. Sci. U.S.A.* **76**, 3116 (1979).

[7] M. Bittner, P. Kupterer, and C. F. Morris, *Anal. Biochem.* **102**, 459 (1980).

[8] P. T. Desgeorges, P. Ambroise-Thomas, P. Falemga, and J. C. Renversez, *Ann. Biol. Clin.* (*Paris*) **38**, 361 (1980).

[9] H. K. Hochkeppel, U. Menge, and J. Collins, *Nature* (*London*) **291**, 500 (1981).

[10] J. Symington, M. Green, and K. Brackmann, *Proc. Natl. Acad. Sci. U.S.A.* **78**, 177 (1981).

[11] V. C. W. Tsang, Y. Tao, L. Qiu, and H. Xue, *J. Parasitol.* **68** (in press).

principal blotting medium because of their relatively lower background and expense.

Methodology

General Comments about the Technique

The EITB is conducted in three stages: (a) the antigen mixture—which can be extremely complex, such as solubilized whole cells—is first resolved by gel electrophoresis (two-dimensional gels can also be used[10]); (b) the resolved gel is then electrophoretically blotted onto nitrocellulose sheets; and (c) the blotted nitrocellulose is then developed by ELISA.

All SDS–PAGE chemicals (unless specified otherwise), including molecular weight markers, are obtained from Bio-Rad Laboratories (Richmond, California).[12] Gels are cast and run in a $160 \times 200 \times 0.8$-mm vertical slab system from Pharmacia Fine Chemicals (Piscataway, New Jersey). The gel system we use contains a gradient resolving gel (3.3% to 20%) and a 3% stacking gel. Buffer systems are discontinuous. The gels are cast at room temperature. Electrophoresis is carried on at slightly below room temperature (13–18°) in a water-jacketed chamber with running tap water as the coolant.

Electrotransfer blotting is performed at 4° (cold room) in a "Trans-Blot" cell from Bio-Rad Laboratories. The power supply is a regluated dc power source capable of generating 90 Vdc at 1.0 amp.

SDS–PAGE (3% Stacking, 3.3–20% Gradient Resolving)

BUFFERS AND SOLUTION

The following buffers are similar to system J4179 of Jovin et al.[13] as modified by Neville et al.[14] This is a discontinuous buffer system that allows for stacking and unstacking SDS–protein complexes.

Lower gel buffer (4×): 205.6 g of Tris (Schwarz-Mann, Inc. Spring Valley, New York) are titrated with 2 N HCl to pH 9.18; add H_2O to bring the final volume to 1000 ml.

---

[12] The use of trade names and commercial sources is for identification only and does not constitute endorsement by the Public Health Service or the U.S. Department of Health and Human Services.

[13] T. K. Jovin, M. L. Dante, and A. Chrambach, "Multiphasic Buffer Systems Output," Federal Scientific and Technical Information, U.S. Dept. of Commerce, PB 196085-196091, Springfield, Virginia, 1971.

[14] D. M. Neville, Jr., and H. Glossman, this series, Vol. 32, p. 91.

Lower reservoir buffer: Lower gel buffer diluted 1 : 4 with H_2O

Upper gel buffer (4×): 2.62 g of Tris, titrated with $2M$ H_2SO_4 to pH 6.14, H_2O to 100 ml. Tris does not buffer well at pH 6.14; therefore, extreme caution must be exercised when titrating.

Upper reservoir buffer (1×): 2.47 g of boric acid; 4.92 g of Tris; 10 ml of 10% SDS in H_2O (to be added last); H_2O to 1 liter. The pH of this solution usually is 8.64; however, this will vary somewhat with different electrodes. The actual weights of the components are more important.

Acrylamide/bis stock solutions

Resolving gel (40% T : 1% C): 40 g of acrylamide monomers; 1 g of N,N'-methylene-bisacrylamide, H_2O to 100 ml; filter through 0.22-μm filter. This solution must be used within 1 month.

Stacking gel, 4× (3% T : 0.3% C): 3.0 g of acrylamide, 0.3 g of N,N'-methylene-bisacrylamide, H_2O to 25 ml; filter through 0.22-μm filter. Stable for 1 month.

Catalysts

Ammonium persulfate (AP), 6 mg/ml in H_2O, fresh daily for resolving gel. This solution should be diluted 1 : 10 with H_2O for use in the stacking gel.

N,N,N',N'-Tetramethylethylenediamine (TEMED), 75 μl/ml, in H_2O, fresh daily. For the stacking gel, this solution is diluted 1 : 4 with H_2O.

"Pushing" Solution

Glycerol, 50% with a trace of bromophenol blue in H_2O; stable for 6 months

GEL CASTING

Resolving Gel. The gradient gel is formed (usually 2 at a time) by mixing equal parts of "heavy" and "light" gel solutions in a gradient generator. The "light" gel solution is placed in the front chamber. The gels are cast in a Pharmacia GSC-2 slab-gel casting apparatus, by pumping the gradient (light on top) from bottom to top with a 60-ml H_2O overlay. The gel slabs are formed between glass plates measuring 177(width) × 200(length) × 3(height) mm with 10 × 200 × 0.8-mm plastic spacers. The effective running gel measures 155 × 170 × 0.8 mm and has an approximately 10-mm stacking gel on top. The final volume of gel gradient solution is pushed into place with the 50% glycerol pushing solution. The "heavy" and "light" gels are made by mixing the following solutions in the order listed.

Components	"Light" gel (3.3%)	"Heavy" gel (20%)
H_2O	20 ml	7 ml
Lower gel buffer (4×)	8 ml	8 ml
Resolving gel acrylamide/bis	2.5 ml	16 ml
TEMED	0.5 ml	0.5 ml
Ammonium persulfate (AP)	1.7 ml	0.3 ml

Polymerization time for this gradient is usually 30–45 min. Different speeds of polymerization have been encountered with different batches of chemicals. The concentration of the AP can be altered to produce the desirable polymerization time. A larger amount of AP is used for the "light" gel solution in order to ensure a top-to-bottom polymerization sequence. If the "heavy" gel at the bottom is allowed to polymerize first, the heat of reaction would create a thermal inductive turbulence and perturb the gel gradient.

Stacking Gel. Mix 3 ml of each of the following solutions in the order listed: upper gel buffer (4×), stacking gel acrylamide/bis solution, TEMED (1:4 diluted), AP (1:10 diluted).

The H_2O overlay on top of the now polymerized running gel is removed by suction. An 0.8 mm-thick sample-spacer comb is inserted, leaving a 10-mm stacking gel length between the bottom(s) of the sample well(s) and the top of the resolving gel. The stacking gel mixture is injected with a 22-gauge needle with the sample spacer in place. Polymerization is usually completed in 15 min.

Completed gels with sample-spacer combs removed can be wrapped in plastic wraps and stored at 4° for up to 1 month.

SAMPLE TREATMENT

Antigen samples in K^+-free buffers are treated with a 2.5% final concentration of SDS. We generally use a 10% SDS, 9 M urea, in 0.01 M Tris-HCl, pH 8.00, solution. A sufficient amount of the 10% SDS solution is added to protein to give a final 2.5% (SDS) and 0.2 to 1 $\mu g/\mu l$ (protein) in the final treated sample. To achieve these concentrations, 0.05 M Tris-HCl, pH 8.00, may be used to dilute the sample. Samples are then heated at 65° for 15 min in an H_2O bath. Tracking dye is added at 10 μl per 100 μl of treated sample. The composition of the tracking dye is 50 mg of bromophenol blue, 8 ml of glycerol, 1 ml of 0.5 M Tris-HCl, pH 8.00, 1 ml of H_2O.

The amount of sample applied is usually 5–10 μg of total protein in 10–100 μl of total volume for each 8.0-mm width sample lane. Although SDS-treated antigens are generally stable at $-20°$ for periods of up to 1 month, we prefer to use freshly treated samples not over 2 days old.

ELECTROPHORESIS

One or two slabs of gel are placed in the electrophoretic chamber. About 200 ml of top reservoir buffer are added. Sample wells are rinsed by flushing with top reservoir buffer. Samples are applied in measured volumes with a pipetter (e.g., the Excalibur, Ulster Scientific, Inc., Highland, New York), which is fitted with a cutoff pipette tip and a blunt 1.5 inch, 22-gauge disposable syringe needle.

SDS–protein complexes are stacked into the stacking gel at 10 mA per slab with the positive terminal at the bottom. When all samples have entered the stacking gel, the current is increased to 20 mA. Power limits for two gels is set at 20 W. Electrophoresis is continued until the bromophenol blue dye reaches the bottom of the resolving gel. This usually requires 2.5–3 hr.

Staining-Resolved Protein Antigens

In most cases SDS–PAGE gels are stained with the silver stain in our laboratory. This procedure is that of Merril et al.[15] with minor modifications. This method of staining proteins is superior to that of both Coomassie Blue and Amido Black, but is limited to 0.8-mm or thinner gels. The silver stain is capable of detecting 0.1 μg of a single protein band (bovine serum albumin).

Solutions

45% Methanol, 12% acetic acid (MetOH/HAc)
10% Ethanol, 5% acetic acid (EtOH/HAc)
$K_2Cr_2O_7$, 0.3 M (100 × Stock); 10 ml of 0.3 M $K_2Cr_2O_7$ + 0.2 ml of conc. HNO_3, H_2O to 1000 ml = working solution.
$AgNO_3$, 1.2 M (100 × Stock); 10 ml of 1.2 M $AgNO_3$, H_2O up to 1000 ml = working solution.
0.28 M Na_2CO_3, 40% HCHO: 29.68 g of Na_2CO_3 + 0.5 ml 40% HCHO, H_2O to 1000 ml; make fresh daily.
Acetic acid, 1% (1% HAc)

[15] C. R. Merril, D. Goldman, S. A. Sedman, and M. H. Ebert, *Science* **211**, 1437 (1981).

Farmer's reducer (1 : 20 diluted)
  Solution A: 37.5 g of $K_3Fe(CN)_6$, H_2O to 500 ml
  Solution B: 480 g of $Na_2S_2O_3$, H_2O to 2000 ml
  Working solution: 2.5 ml of solution A + 10 ml of solution B; H_2O
    to 250 ml

STAINING PROCEDURE

All the following procedures require the gel to be kept at room temperature in constant gentle motion with a rotary (or side-to-side) shaker.

1. Gel is removed from glass plates with stacking gel attached and dropped into 300 ml of MetOH/HAc in a glass baking dish for 30 min with agitation. MetOH/HAc is removed by suction and may be saved to be used again for up to 3 times.
2. Expose to 200 ml of EtOH/HAc 3 times for 10 min each. EtOH/HAc is discarded.
3. Expose to 300 ml of $K_2Cr_2O_7$ solution for 5 min.
4. Wash 4 times for 30 sec with H_2O (300 ml each time).
5. Expose to $AgNO_3$ solution (300 ml) for 30 min. The first 5 min of this treatment is under the light from a 3200°K photoflood lamp, situated approximately 60 cm from the gel. For the remaining 25 min of treatment, the gel is simply shaken in normal room light.
6. After removing the $AgNO_3$, the gel is quickly rinsed twice with 200 ml of the Na_2CO_3 solution. Another 400 ml of the Na_2CO_3 solution are added, and the development of the protein band is carefully watched, with the gel tray on a light box.
7. When the desired background and degree of staining are reached, the reaction is stopped by removing the Na_2CO_3 developer and adding approximately 400 ml of 1% HAc.
8. Reduction of gel surface stains is achieved by exposing the HAc-treated gel to 250 ml of Farmer's reducer (1 : 20) for 45 sec, followed by 10 washes with H_2O. It is important that the gel be washed thoroughly; otherwise the stained bands will fade.
9. Gels are then photographed and slightly dehydrated with 45% methanol in H_2O and stored in plastic bags.

Alternatively, SDS–PAGE gel may be stained with Coomassie Blue (CB). The procedure we use is the "heavy stain" of Neville et al.[14] However, the CB stain is far less sensitive than the Ag stain, and, therefore, 100 μg of total protein per lane in the SDS-PAGE is required.

Enzyme-Linked Immunoelectrotransfer Blot (EITB)

Solutions

Blotting medium: 0.025 M Tris, 0.193 M glycine, 20% methanol (v/v), pH 8.35. Make up 4 liters at a time; store at 4°.

PBS–Tween: 0.01 M Na_2HPO_4/NaH_2PO_4, 0.15 M NaCl, pH 7.20 (PBS), with 0.3% Tween 20 (polyoxyethylene sorbitan monolaurate from Sigma). The PBS can be prepared in large quantities (20–40 liters), put in liter bottles, sterilized, and stored at room temperature until use. The Tween (3 ml per liter of PBS) is added to the PBS slowly with stirring. The PBS–Tween mixture is prepared fresh daily.

Substrate solution: 50 mg of 3,3′-diaminobenzidine (3,3′,4,4′-tetraaminobiphenyl tetrahydrochloride, 97–99% from Sigma), 100 μl of H_2O_2 (3%), PBS pH 7.2 to 100 ml. Make fresh daily.

EITB Procedure

This procedure requires a Trans-Blot cell from Bio-Rad Laboratories.

1. Soak 16- × 21-cm nitrocellulose sheets (BA85 or BA83, from Schleicher & Schuell, Keene, New Hampshire) in blotting medium about 30 min before the SDS–PAGE ends.
2. Place about 300 ml of blotting medium in a tray with a piece of filter paper (16 × 21 cm)
3. Remove SDS-gel and drop into the tray on top of the filter paper. Position gel squarely on top of the wet paper, and slide both out of the blotting medium while gently holding the two together.
4. Place the paper and gel onto one of the Scotch-Brite pads of the Trans-Blot cell, with the gel facing up.
5. Wet the gel surface with 10 ml of blotting medium. Place the wet nitrocellulose sheet directly onto the gel surface. The nitrocellulose sheet should be eased onto the gel slowly, starting with one edge while lowering toward the opposite edge. All air bubbles must be completely removed between gel and nitrocellulose sheet.
6. Place another wet filter paper on top of the nitrocellulose sheet.
7. The last Scotch-Brite pad is now placed on top. Cut away a small piece from one corner of this last pad to show that it is at the nitrocellulose side of the "sandwich."
8. The "sandwich" is completed by placing everything inside the plastic support frame. The frame and its contents are secured with a rubber band.

9. The "sandwich" is slowly lowered into blotting medium in the Trans-blot cell with the nitrocellulose (pad with cut corner) side toward the positive electrode.

10. Electrophoretic transfer of SDS complexes is accomplished in 3 hr, at 4° with a constant 60 Vdc. In some cases, proteins that are difficult to transfer may require 16 hr. The current required to maintain 60 V is usually 200–400 mA. It is essential to perform blotting at 4° to preserve the solution and extend the life of the medium.

11. After electrophoretic transfer, the nitrocellulose sheet is removed. Some stacking gel may stick to the sheet but can be removed by gentle wiping with dry gauze.

12. The gel–nitrocellulose contact surface is marked with a soft "lead" pencil or ball-point pen. A line is gently drawn across the top or bottom of the nitrocellulose sheet. This line will aid in the alignment of cut strips, if the sheet is to be cut (for exposures to different antibody solutions or sera).

13. The marked sheet is washed 4 times for 15 min each with 300 ml of PBS–Tween at 40° in a shaking water bath. Plastic watertight boxes are used.

14. When a single antigen sample is processed (a single antigen sample resolved in a single chamber comb, as bands across the whole SDS–PAGE, and blotted), the nitrocellulose sheet is placed onto a plastic backing sheet (e.g., "Write on Film" for overhead projections, by 3M Co.) and cut into 0.5- to 0.8-cm strips. When multiple samples are resolved by SDS–PAGE and EITB, the sheet cannot be cut into strips and, therefore, is further processed as an intact sheet.

15. Antibody solutions (either sera or monoclonal products) are diluted in PBS–Tween. The concentration varies with the antibody content of the sera or monoclonal product. For sera of schistosomiasis patients, we usually use a 1:200 to 1:1000 dilution. For monoclonal (hybridoma) products, the dilution is usually 1:500 to 1:2500.

16. Intact (uncut) sheets are processed hereafter in plastic trays. Cut strips are processed in a special slotted tray shown in Fig. 1.

17. Nitrocellulose sheets or strips are exposed to diluted antibody solutions for 1 hr or overnight with agitation on a rotary shaker. The volumes required are 100 ml for the intact sheet and 10 ml for each strip in the slotted tray.

18. The sheet or strips are then washed 3 times for 10 min with PBS–Tween at room temperatures.

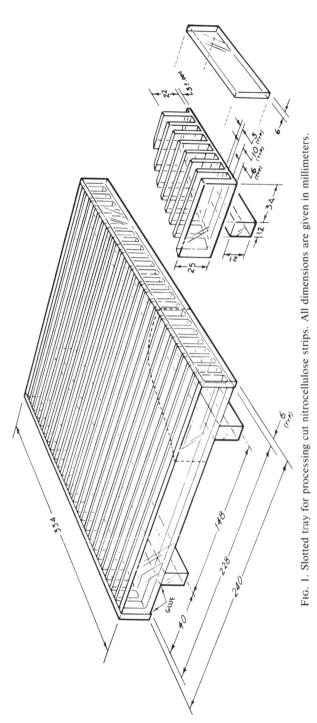

Fig. 1. Slotted tray for processing cut nitrocellulose strips. All dimensions are given in millimeters.

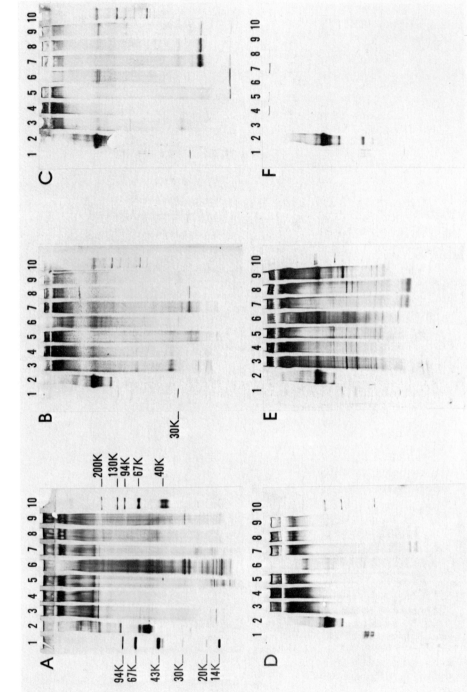

19. The nitrocellulose sheets or strips are exposed to an appropriate solution of horseradish peroxidase (donor:H_2O_2 oxidoreductase, EC 1.11.1.7) labeled anti-antibody (POD-anti-Ig) in PBS–Tween for 1 hr. Commercial preparations of enzyme–antibody conjugates are usually used at a 1 : 1000 dilution. The nitrocellulose are again washed as before.

20. The processed nitrocellulose sheet or strips are then exposed for 5–10 min to the substrate solution. Positive reaction bands usually show up as orange to brown bands within 10 min.

21. The nitrocellulose sheets are then thoroughly rinsed with H_2O.

Photographic Recording of Data. Both Ag-stained gel and EITB nitrocellulose sheets are photographed while wet. The EITB sheets (or strips) may be stored dry, but must be rewet before photography. Photography is performed on Kodak commercial sheet film (4 × 5 inches, No. 6127) under two 3200°K flood lights at an exposure index (E.I.) of 8. The films are developed with Kodak HC-110 dilution B (1 : 31) for 2.25 min at 20°C with continuous agitation. The positive prints are made on Kodabrome "Medium" to "Extra Hard" papers and processed by the Kodak "Ray Print" processor. Since the blot is invariably a mirror image of the SDS–PAGE, it is important to note the polarity of the EITB photograph. For ease of comparison, we usually print the EITB photographs with the negative turned backward, so that the final EITB photos are in the same orientation as that of the SDS–PAGE photos.

Application Notes

Multiple Antigens/Single Antibody Source (Serum). Figure 2 demonstrates the capability of the SDS-PAGE/EITB system in resolving complex antigens. This series of plates also illustrates the power of EITB in distinguishing the antibody specificities and multiplicities of various infection sera. Figure 2A is a silver-stained SDS–PAGE gel of various *Schistosoma mansoni* antigen preparation. Figs. 2B, 2C, 2D, and 2E are the

---

FIG. 2. Sodium dodecyl sulfate–polyacrylamide gel electrophoresis and enzyme-linked immunoelectrotransfer blot (SDS–PAGE/EITB) of multiple antigens from *Schistosoma mansoni*, tested against single sources of antibodies from different patient sera. Lanes: 1, molecular weight markers at 94,000, 67,000, 43,000, 30,000, 20,000, and 14,000; 2, normal human serum; 3–5 microsomal membrane antigens; 6, cytosol antigens; 7, total microsomal antigens; 8, mitochondrial antigens; 9, nuclear antigens; 10, molecular weight markers at 200,000, 130,000, 94,000, 67,000, and 40,000. (A) Ag-stain of the SDS–PAGE gel; (B–D), EITBs with sera from three different North American *S. mansoni* patients; (E) EITB with serum from an Egyptian patient; (F) EITB with a heterologous infection serum (*Trichinella spiralis*).

EITB of sera from patients infected with *S. mansoni*. The sera in Figs. 2B, 2C, and 2D are from North American patients and that of Fig. 2E, from an Egyptian patient. The patients produced antibodies that recognized vastly different antigenic components from the same antigen preparations. Figure 2F is an EITB of a serum from a heterologous infection (*Trichinella spiralis*). Little recognition is demonstrated by this serum to the antigens of *S. mansoni*. The highly reactive band at approximately M_r 160,000 in the normal human serum (NHS) lanes of all the EITB is due to the IgG in the NHS, which is bound by the POD-anti-Ig used in the blotting process.

Single Antigen–Multiple Antibody Source. Figure 3 demonstrates the feasibility of using the cut-strip method to examine the specificities of more than one antibody source against a single antigen SDS–PAGE run. A single cercarial antigen (SCA) is resolved in a single sample well (single comb configuration) on SDS–PAGE and blotted onto a single nitrocellulose sheet. The total antigen protein applied is 100 μg in approximately 200 μl. The nitrocellulose sheet is cut into strips giving approximately 1 μg of total SCA per strip. The strips are then exposed to different antibody sources. These included mouse infection serum (MIS), normal uninfected mouse serum (NMS), hybridoma monoclonal products A, B, C, and D. The different specificities of the antibodies from these sources are evident from their recognition patterns to different components of the SCA antigen preparation. A SDS–PAGE pattern of SCA, as stained by Coomassie Blue with molecular weight markers, is at the left of Fig. 3.

Sensitivity Limits. Figure 4 demonstrates the sensitivity limits of the SDS–PAGE/EITB system. For this purpose, we purified rabbit anti-BSA antibodies by solid-phase immunoabsorption. Pure BSA is resolved by the SDS–PAGE/EITB such that 10, 1.0, 0.1, and 0.01 μg of total BSA is present in each lane, respectively, after blot. The EITB is exposed to pure anti-BSA at a concentration of 0.18 μg/ml. The limit of antigen detection as indicated by Fig. 4 (left column) is at 10 ng of BSA. The antibody sensitivity limit is at 0.01 μg/ml, as demonstrated by the right-hand column of Fig. 4. In this experiment, BSA at 1 μg per lane (strip) is exposed to 1, 0.5, 0.1, 0.05, and 0.01 μg/ml of pure anti-BSA. A concentration of 0.01 μg/ml anti-BSA is sufficient to visualize 1 μg of BSA per strip.

Source of Error

The fact that the antigens in EITB are all treated with SDS, brings forth the theoretical consideration of native configurational integrity of the potential antigenic epitopes. There is the danger that not all the anti-

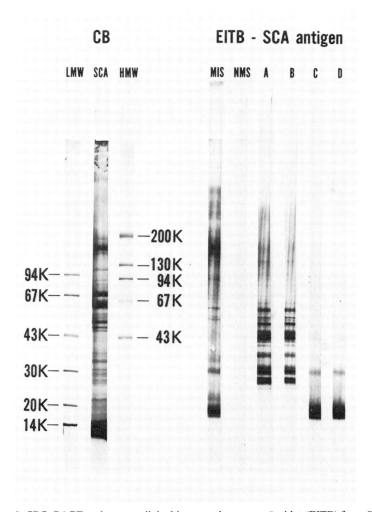

FIG. 3. SDS–PAGE and enzyme-linked immunoelectrotransfer blot (EITB) from *Schistosoma mansoni* soluble cercarial antigen (SCA). The gel is loaded with 100 μg of SCA, and protein bands are visualized with Coomassie Blue (CB) staining. The numbers below the marker represent the molecular weight in thousands of each protein used for calibration. For the EITB, the nitrocellulose strips containing approximately 1 μg of SCA each are incubated with *S. mansoni* mouse infection serum (MIS), normal mouse serum (NMS), and cercarial monoclonal antibodies (A and B are clones derived from the one mother cell line; C and D are clones derived from a second mother cell line).

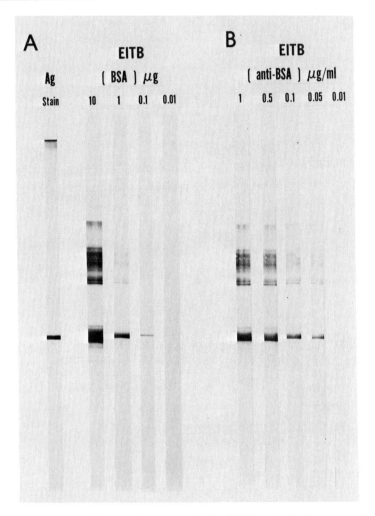

FIG. 4. Enzyme-linked immunoelectrotransfer blot (EITB) using bovine serum albumin (BSA) and rabbit anti-BSA.   (A) Different concentrations of BSA are loaded in the gel and transferred to the nitrocellulose membrane. Rabbit anti-BSA was used at 1 : 1000 dilution. The first lane corresponds to the gel loaded with 1 μg of BSA and strained with silver.   (B) Each strip of the nitrocellulose membrane containing approximately 5 μg of BSA was incubated with different concentrations of rabbit anti-BSA as indicated above the lane.

genic sites can retain the native configuration after SDS treatment to allow recognition by appropriate antibodies. Experiments with IgG, BSA, and other defined antigen, however, would tend to negate this argument. Most of the SDS is presumably removed during washing to allow sufficient refolding of proteins into their native configurations to allow antibody recognition. Only further investigation can fully answer these questions.

Resolved antigen bands on both SDS–PAGE gels and their corresponding EITB sheets or strips must be aligned. Using pencil line markers, and careful attention to the methanol levels in the blotting medium greatly aid in the elimination of the band alignment problems. Many of the adjustments required can be made in the photographic process. The methanol level (20%) and 4° running temperature will prevent intragel distortion of the bands.

Acknowledgments

The work of Dr. J. M. Peralta was supported by a fellowship from Coordenacao de Aperfeicoamento de Pessoal de Nivel Superior (CAPES), Brazil.

[30] Quantitative, Single-Tube, Kinetic-Dependent Enzyme-Linked Immunosorbent Assay (k-ELISA)

By VICTOR C. W. TSANG, BRITT C. WILSON, and JOSE M. PERALTA

The numerous applications and advantages of the enzyme-linked immunosorbent assay (ELISA and related variations) have been reviewed extensively.[1–5] These assays are generally conducted in antibody or antigen-coated Microtiter plates or in a series of polystyrene tubes, where serial dilutions of the biological fluids (usually serum) to be tested are made *in situ*. After a designated incubation period, unbound materials are removed by washing, and the bound molecules (antibody or antigen) are allowed to associate with enzyme-labeled indicator antibodies. After removing unbound, excess labeled antibodies, a chromogenic substrate for the enzyme is introduced. The amount of substrate conversion to product

[1] V. C. W. Tsang, B. C. Wilson, and S. E. Maddison, *Clin. Chem.* **26,** 1255 (1980).
[2] V. C. W. Tsang, Y. Tao, and S. E. Maddison, *J. Parasitol.* **67,** 340 (1981).
[3] E. Engvall, this series, Vol. 70, p. 419.
[4] G. B. Wilson, *Clin. Chem.* **22,** 1243 (1976).
[5] S. L. Scharpe, W. M. Cooreman, W. J. Bloom, and G. M. Laekeman, *Clin. Chem.* **22,** 733 (1976).

at the end of a designated period (usually 15–30 min) is determined by the color generated. Data are reported as the highest dilution (or titer) that yields a predetermined color intensity, after enzymic activity has been stopped. The ELISA as performed by these methods, however, has the following drawbacks.

1. Results depend on an endpoint determination in the presence of an activity-stopping agent, which may or may not "freeze" the color development.[6]

2. Multiple dilutions are required.

3. Endpoint color intensity (usually determined as absorbance) is seldom linear with respect to antibody or serum concentration.[7-9]

4. The ELISA is regarded as quantitative; however, when multitube endpoint ELISA is used, enzyme concentration is linear with respect to activity (velocity) only during the brief initial phase of reaction, and then only when the enzyme concentration is much lower than that of the substrate.[10] The amount of change during a fixed interval is measured, instead of the initial velocities of enzyme activities. Unfortunately, most of the reaction times chosen far exceed the initial linear portion, especially with the horseradish peroxidase (POD) system; data derived from the multitube endpoint ELISA must therefore be considered semiquantitative.

The single-cuvette ELISA based on enzyme rate kinetics, described below, yields quantitative linear data on antibody and antigen concentrations.[1,2] Optimum assay conditions, component concentrations, reaction intervals, and sources of error are described, and assay linearity, reproducibility, and sensitivity are demonstrated. Applications of the k-ELISA in quantitating antigen and antibodies in complex[2] and ideal[1] systems are included.

Theory and Rationale

The ELISA of k-ELISA can be dissected into three discrete components, expressed by the following equations, each of which is governed by the law of mass action.

[6] S. L. Bullock and K. W. Walls, *J. Infect. Dis.* **136**, S279 (1977).

[7] E. Engvall and P. Perlmann, *Protides Biol. Fluids* **19**, 553 (1971).

[8] A. J. Pesce, N. Mendoza, I. Boreisha, M. A. Gazuitis, and V. E. Pollak, *Clin. Chem.* **20**, 35 (1974).

[9] A. Voller, A. Bartlett, and D. W. Bidwell, *J. Gen. Virol.* **33**, 165 (1976).

[10] M. Dixon and E. C. Webb, *in* "Enzymes" (2nd ed.), p. 54. Longmans, Green, New York, 1974.

$$Ag + Ab \rightleftharpoons [Ag\text{-}Ab] \tag{1}$$
$$[Ag\text{-}Ab] + Anti\text{-}Ab\text{-}E \rightleftharpoons [Ag\text{-}Ab\text{-}Anti\text{-}Ab\text{-}E] \tag{2}$$
$$E + S \rightleftharpoons [ES] \rightleftharpoons E + P \tag{3}$$

where Ag represents the antigen(s), Ab the antibody(ies), E the bound enzyme, S the substrate, and P the product.

Reaction (1), which contains the only rate-limiting step, yields the antigen–antibody complex [Ag-Ab]. To promote a right-handed shift of this reaction and to assure a single rate-limiting element, the assay ligand (Ag or Ab) must be present in rate-limiting quantities, while the indicating reagent (Ag or Ab) must be in high excess. Thus, in the Ag quantitation configuration, the Ab must be in high excess.

In Reaction (2), the immune complex [Ag-Ab] is made to react with an excess of enzyme-labeled anti-antibody [Anti-Ab-E], yielding the final complex [Ag-Ab-Anti-Ab-E]. The bound enzyme, E, is then quantitated by adding an excess of substrate, S, which is converted to product P, the measured entity in Reaction (3). Reaction (3) is also the basis for the Michaelis–Menten equation of enzyme kinetics, whereby the final amount of bound enzyme is quantitated.

Use of enzyme rate kinetics to quantitate bound enzyme and assay ligand (either antigen or antibody) differentiates the k-ELISA from conventional ELISA, which measures ligand presence by a single-timepoint colorimetric determination of enzyme action in a multidilution series and yields data in terms of titers that are semiquantitative and nonlinear with respect to ligand concentration.[7-9] The k-ELISA, on the other hand, needs only one tube per ligand assay and yields linear data with respect to ligand concentration. No "block" titration is needed to determine the concentration of each reactant to use, as is required in the conventional ELISA. All reactants except the assay ligand are present in high excess. Thus, only one rate-limiting step exists in the total reaction complex [Eqs. (1–3)], that involving the assay ligand, ensuring a linear relationship between assay ligand concentration and reactivity.

Assay Methods

General Comments

The k-ELISA is a solid-phase immunoassay, performed in four stages.

1. Sensitization. Antigen or antibody is adsorbed onto the ethanol-washed polystyrene cuvette surface.
2. Antibody binding. Antibody or antigen solution is added, and antigen–antibody complex is formed on the cuvette wall.

3. Conjugate. Enzyme-labeled anti-antibody conjugate solution is added. Enzyme conjugate molecules will attach wherever a recognized antibody is present.

4. Enzyme quantitation. An excess of substrate is introduced, the rate of enzymic conversion of substrate to product is measured, and the amount of bound enzyme is quantitated.

All reactions occur on the surface of the reaction vessel; therefore, all conditions that can interfere with the "wall phenomena" must be controlled. We use two types of disposable polystyrene reaction vessels—round cuvettes (4.0-ml capacity, Ultra-VU Disposable Cuvette, LKB S7370-15, obtained from Scientific Products, 1750 Stoneridge Dr., Stone Mountain, Georgia)[11] and square cuvettes (1.5-ml capacity, Kartell, Inc., obtained from Fisher Scientific Corporation, Atlanta, Georgia), both with a 1-cm beam path. Polystyrene cuvettes from other sources have been tested and shown to be usable, although binding capacities for antibodies and antigens may differ slightly among different makes of cuvettes. Cuvettes are routinely washed with 95% ethanol and dried before use. All washing steps, including the ethanol pretreatment, are performed with the vacuum-driven "jet-spray" cuvette washer (Fig. 1). Wash solutions remaining in the cuvettes at the end of each wash session are removed by suction. A polyethylene disposable pipette tip is used as vacuum port to minimize scratching of the cuvette walls.

The system commonly employed in our laboratory uses horseradish peroxidase (donor: H_2O_2 oxidoreductase, EC 1.11.1.7) as the enzyme label. The horseradish peroxidase (POD) is covalently conjugated onto an anti-antibody molecule, usually of the IgG type. Conjugates of various antigenic specificities are obtained commercially or prepared in our laboratory in accordance with published methods.[3,12–14] Since the enzyme-specific activities of conjugates vary greatly from lot to lot, efforts should be made to use a single conjugate lot for each comparative study series. The same lot of cuvettes should also be used whenever possible.

Solutions

Washing buffer (PBS–Tween): 0.01 M Na_2HPO_4/NaH_2PO_4, 0.15 M NaCl, pH 7.20 (PBS) with 0.3% Tween 20 (polyoxyethylene sorbitan monolaurate from Sigma). The PBS can be prepared in large quanti-

---

[11] The use of trade names is for identification only and does not constitute endorsement by the Public Health Service or by the U.S. Department of Health and Human Services.
[12] P. K. Nakane and A. Kawaoi, *J. Histochem. Cytochem.* **22**, 1084 (1974).
[13] S. Avrameas and T. Ternyck, *Immunochemistry* **8**, 1175 (1971).
[14] S. Avrameas, *Immunochemistry* **6**, 43 (1969).

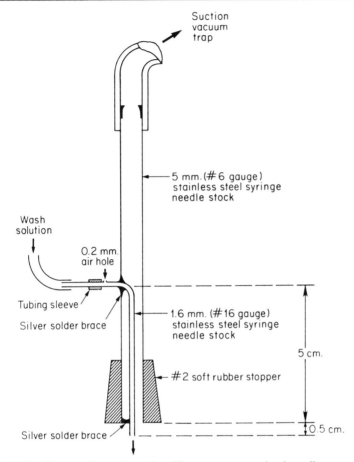

Fig. 1. The "jet-spray" cuvette washer. The vacuum source is a houseline vacuum that generates a negative pressure of 20 inches of Hg. The vacuum pressure alone can maintain a secure seal between the rubber stopper and the lips of the cuvette. The amount of air in the aerated wash solution is regulated by sliding the tubing sleeve over the air hole. Sufficient air is allowed in to create good turbulance in the cuvettes. The washing sequence for each cuvette is 6 sec per round. Three rounds of washing per batch of cuvettes is sufficient to remove all unbound material.

ties (20–40 liters), sterilized in liter bottles, and stored at room temperature until use. The Tween (3 ml per liter of PBS) is added slowly with stirring. The PBS–Tween is prepared fresh daily.

Sensitization buffer (Tris-HCl): 0.05 M Tris, 2.0 mM EDTA, 0.3 M KCl, titrated to pH 8.00 with HCl. This solution can be stored at 4° for up to 2 months. Other buffers can also be used as solvent for the sensitization step. The constraints of ionic strength and pH are usually imposed by the stabilities of the sensitization agent. We have found that this buffer with ionic strength of about 0.05 to 0.5 M NaCl equivalent at pH 4–10 can produce relatively efficient binding.[1]

Conjugate solution. Anti-antibody, covalently conjugated with POD (POD-anti-Ig), is diluted in PBS–Tween. This solution should be prepared fresh, preferably during the antibody–antigen incubation step. The concentration used in our laboratory for commercially obtained conjugate is a 1 : 2000 or 1 : 1000 dilution. This concentration is usually sufficient to maintain an excess of available anti-antibody enzyme conjugate. However, we recommend that the excess level be determined at antibody and antigen saturation levels by a titration series for each batch of conjugate.

Enzyme chromogenic substrate solution and buffer (NaAc/HAc): Dissolve 10 mg of o-dianisidine · 2 HCl (crystalline 3,3′-dimethoxybenzidine · 2 HCl, from Sigma) in 1 ml of absolute methanol and 1 ml of NaAc/HAc pH 5.00 (0.05 M NaCOOH titrated to pH 5.00 with glacial acetic acid; store at −20° until use). Combine 1.66 ml of the o-dianisidine · 2 HCl and 0.1 ml of H_2O_2 (3.0%) and add NaAc/HAc to 25 ml. The substrate should be prepared fresh just prior to use and should be protected from light (place in amber bottle). Glassware used to prepare the substrate solution must be scrupulously clean, free of soap, and preferably rinsed with NaAc/HAc before use.

Instrumentation

Automated. In this mode, in our laboratory, the bound enzyme is quantitated by a LKB 8600 Reaction Rate Analyzer. This instrument is essentially an automatic thermo-controlled spectrophotometer, which is coupled to a data processor and a strip-chart recorder. Up to 100 cuvettes can be input at one time. The cuvettes, containing bound enzyme and 0.75 ml of NaAc/HAc, are automatically fed into the thermal tunnel of the LKB 8600. When warmed (35°), 0.25 ml of chromogenic substrate is machine injected and each cuvette is automatically stirred by vortex at the measuring position. The rate of increase in absorbance at 460 nm (ΔA_{460}) over the first minute of reaction time for each sample is measured and automatically calculated in ΔA_{460}/min units. Long measuring times (up to 10 min) may be set on the LKB 8600 to accommodate slow reactions. For general use we have never encountered the need to extend measuring times to over 3 min. In any event, the data processor is programmed to compute the reaction slopes (rate) from only the straight-line portion of the initial reaction.

Manual or Semiautomatic. In this mode, any good spectrophotometer (either single or dual beam) coupled to a strip-chart recorder and/or a data processor will suffice. In our laboratory, a Series 634 Varian dual-beam spectrophotometer is coupled to a 100-mV chart recorder. The square polystyrene assay cuvette is put in place at the measuring position. A similar but untreated cuvette containing only the substrate solution is in place at the reference position. One milliliter of chromogenic substrate diluted 1 : 4 with NaAc/HAc, pH 5.00, is pumped into the assay cuvette at the measuring position. The reaction rate is recorded over a period of 2 min. The reaction velocity is calculated from the linear slope of the first 1–2 min of reaction. A spectrophotometer equipped to perform enzyme kinetic measurements will greatly simplify the terminal assay step.

Procedure

1. Preparing cuvettes. Prewash all cuvettes in 95% ethanol with the "jet-spray" washer (Fig. 1) and let dry. This can be performed days in advance; washed cuvettes should be covered with plastic wrap. Round cuvettes are used for the automated system and square cuvettes for the manual assay system.

2. Sensitization. Sensitization agent (antigen or antibody) is dissolved or mixed with sensitization buffer, e.g., Tris-HCl. This solution should be very well mixed. One milliliter of the sensitization agent is placed in each of the clean cuvettes. A repeating pipettor should be used. Cuvettes are then placed in a water bath at 37° for 1 hr. The water level in the bath should be just below that of the fluid in the cuvettes. A 2–5 μg/ml protein concentration of most antigens is sufficient for saturating all available binding sites on the cuvettes. We use this concentration of antigen to sensitize cuvettes in all assays for antibody concentrations in patient sera. In this configuration the antigen concentrate at 2–5 μg/ml is in excess, the antibody concentration in the patient sera, used at 5 μl/ml is the rate-limiting step. When used at the antigen rate-limiting configuration, as in comparative antigen assays, the concentration of the sensitizing antigen is generally 0.1–0.5 μg/ml. The corresponding serum concentrations are held at 25–50 μl/ml.

3. Wash. After 1 hr of sensitization cuvettes are washed in PBS/Tween with the "jet-spray" washer (Fig. 1). Each cuvette is washed 3 times for 6 sec per wash. Residual PBS–Tween is removed by suction.

4. Antibody binding. Antibody solutions (e.g., serum) are diluted with PBS–Tween. This is done usually with an automatic diluter, and 1-ml aliquots are dispensed into sensitized cuvette. Incubation again is for 1 hr at 37°. The serum dilution we use for antigen rate-limiting configuration is

usually 25–50 μl/ml. For antibody limiting, antigen excess configurations, as in the routine quantitation of patient sera antibody levels, the serum dilution is usually held at 5 μl/ml in PBS–Tween.

5. Wash as in step 3.

6. Conjugate binding. One milliliter of anti-antibody labeled with POD in PBS/Tween is added to each cuvette. Incubate cuvettes at 37° for 1 hr.

7. Wash as in step 3.

8. Dry cuvette outer surfaces with laboratory wipes.

9. Enzyme quantitation

    a. Automated (also see Instrumentation section): 750 μl of NaAc/ HAc pH 5.00 is added to each cuvette. Cuvettes are placed in the LKB 8600 carrier racks. Enzyme quantitation by rate kinetics (initial velocity of substrate conversion) is automatic, and data are expressed in ΔA_{460}/min.

    b. Manual. Wash solutions are left in the cuvettes after the last wash. Just prior to introduction of substrate solution, with the cuvette already in place at the spectrophotometer measuring position, the residual wash solution is removed by suction. One ml of 1 : 4 diluted chromogenic substrate solution is pumped in *quickly*. The increase of ΔA_{460} is recorded over a period of 1–3 min. The reaction rate is computed for the initial linear portion (usually 1–2 min).

Applications

The k-ELISA is primarily used in our laboratory for two purposes: (*a*) to assay and compare the quantitative reactivities of different antigen preparations; and (*b*) to assay and quantitate the antibody content of serum of patients or experimental animals directed against specific antigens. Examples of these two applications are illustrated below.

Comparative Antigen Assays

In this mode the antigen is the sensitization agent as well as the assay ligand. The antigen concentration, therefore, is the rate-limiting element. We generally sensitize cuvettes with antigens at concentrations of 0.1–0.5 μg/ml. The antibody solution is an infection sera pool diluted at 25 μl/ml in PBS–Tween. Figure 2 illustrates the quantitative nature of the k-ELISA in the antigen assay mode. Three different antigens at various concentrations (from 0.05 μl/ml to 2 μg/ml in Tris-HCl) were used to sensitize cuvettes. They were then exposed to infection serum (25 μl/ml) and POD-anti-Ig conjugate (1 : 1000 dilution) and assayed for bound POD. The data in Fig. 2 show a linear relationship between activity (ΔA_{460}/min) and antigen concentrations over the range of 0–0.5 μg/ml. The plateau

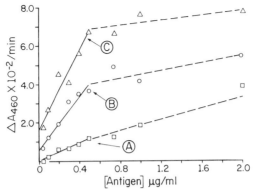

FIG. 2. Quantitative comparison of antigenic activities of different antigens. In this experiment various antigen ligands (curve A, *Schistosoma mansoni* soluble egg antigen; curve B, adult mitochondrial antigen; curve C, adult microsomal antigen) at the concentrations specified on the x axis are used to sensitize curvettes. The rate-limiting element is the antigen concentration. The reacting serum is a pooled infection serum used at 25 μl/ml in PBS–Tween. The linear correlation coefficients (r^2) of all antigen at concentrations between 0 and 0.5 μg/ml are: A, 0.95; B, 0.94; and C, 0.92.

effect at antigen concentrations above 1.0 μg/ml is due to the limitation of available adsorption sites on the cuvette surface. For comparisons of antigen activities, therefore, a concentration in the middle of the linear portions of the curves should be selected. For example, if these three antigens had been assayed once at the 0.3 μg/ml level only, we would have obtained activity units of 0.7, 3.0, and 4.4 (ΔA_{460}/min $\times$ 10^{-2}) for antigens A, B, and C, respectively. These data would then indicate that antigen C is better than both B and A in terms of a specific activity. Quantitatively speaking, in identical 1-mg lots, antigen C would contain a total of 1.47 $\times$ 10^4 units of antigenic activity whereas B and A would contain 1 $\times$ 10^4 and 0.23 $\times$ 10^4 units, respectively.

Another example of antigen quantitation is illustrated by Fig. 3. Here the antigen is normal human IgG. The indicating antibody is POD-anti-human IgG (γ-chain specific), which also serves as the conjugate. There is a linear correlation between concentration and activity from 1.0 ng/ml to 100 ng/ml of IgG (correlation coefficient, r^2 = 0.99). The limit of sensitivity is 10 ng of IgG per milliliter.

Antibody Concentration Assay

In this mode the antibody is the assay ligand; therefore, it should be the only rate-limiting element. To saturate the cuvettes with antigen we generally use a 2–5 μg/ml concentration level for sensitization. Figure 4 illustrates the sensitivity and reproducibility of the k-ELISA in the quanti-

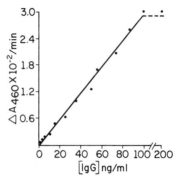

FIG. 3. Assay of IgG. The assay ligand in this experiment is purified normal human IgG. At the concentrations specified on the x axis, IgG is used to sensitize cuvettes. The rate-limiting step is the concentration of IgG. The indicating antibody is horseradish-conjugated anti-IgG used at 1 : 1000 dilution in PBS–Tween. The linear correlation coefficient (r^2) for the 0–100 ng/ml range is 0.99.

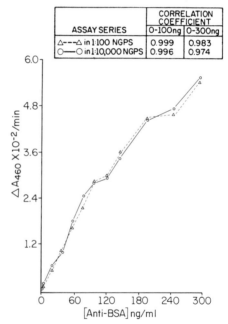

FIG. 4. Reproducibility of the k-ELISA in quantitating specific antibody concentrations. Anti-bovine serum albumin (BSA) antibodies are isolated from immune guinea pig sera by immunoabsorption, and dissolved in 1 : 100 diluted or 1 : 10,000 diluted normal guinea pig serum (NGPS) at concentrations specified on the x axis. These artificial immune sera are assayed 4 days apart. The correlation coefficient (r^2) is computed by linear regression.

tation of antibody. For this experiment two artificial antiserum series were constructed. Specific guinea pig antibodies directed against bovine serum albumin (BSA) were purified by solid-phase immunoabsorption from immune guinea pig sera. Anti-BSA antibodies were dissociated from the immunoabsorption column, and antiserum solutions were made by dissolving specific quantities of the anti-BSA in nonimmune normal guinea pig serum (NGPS) diluted at in PBS–Tween either 1:10,000 or 1:100. The two series of antisera were assayed for their anti-BSA contents on two different days (4 days apart). Figure 4 shows that the k-ELISA is highly reproducible in that the two assay series are superimposable in spite of the fact that they were done 4 days apart and that the two sera series have vastly different concentrations of interfering serum proteins (1:100 vs 1:10,000 NGPS). We also see that the anti-BSA concentrations and k-ELISA activity units maintain good linear correlations ($r^2 \geq 0.9$) and are independent of the concentrations of interfering normal serum proteins.

The linear regression line equations for Fig. 4 can also be used as calibration standard curves for other systems of antigen–antibody assays involving guinea pig serum. As long as the same conjugate is used, the ΔA_{460}/min generated will correspond directly to the actual antibody concentrations present as defined by the line equations of Fig. 4.

An example of an antibody assay with a parasitic antigen (*Schistosoma mansoni* adult microsomal antigen) and the corresponding infection sera is illustrated by Fig. 5. Here, the sensitization agent is a purified microsomal antigen from *S. mansoni*, used at 5 μg/ml in Tris-HCl buffer. Various sera concentrations from two human patients and an uninfected individual were assayed. Figure 5 shows good linear relationship between 1 μl/ml to 15 μl/ml serum concentrations (in PBS–Tween) for the two infection sera (A and B). Although serum from patient A contained approximately twice the amount of antibody serum of patient B, the serum from the uninfected control serum did not show any activity at all. In practice, we use only one dilution per serum, and this is usually held at 5 μl/ml diluted in PBS–Tween.

Sources of Error

The following are some of the most commonly encountered difficulties in this assay. Solutions to the problems are suggested.

1. Erratic results with low activities and reproducibilities. Most likely the cuvettes should have a more thorough prewash with 95% ethanol. Some purely carbohydrate antigens may be poor binders to polystyrene, poly(L-lysine) may be used as a first adsorption layer.

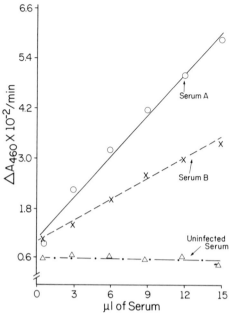

Fig. 5. *Schistosoma mansoni* specific antibody levels in three different sera are quanti-tated by k-ELISA. Cuvettes are sensitized with a microsomal antigen (5 μg/ml) purified from *S. mansoni* adult worms. Activities of both the infection serea A and B are linearly propor-tional to the amount of serum assayed. Thus, in practice only a single concentration (e.g., 5 μl/ml) of each serum need be assayed.

2. Consistent but low activities even with control systems. Check reagent excess levels. Both conjugate and binding agent (antibody or antigen) should be in excess. To determine level of conjugate excess, a series of cuvettes are first fully saturated with both the antigen then the antibody. A serial dilution of the enzyme-conjugate is added and incu-bated as usual. Activity (slope in ΔA_{460}/min) will increase successively with higher concentrations of conjugate. When the saturation level is achieved, the slope curve will plateau. The concentration at which pla-teauing occurs plus an excess margin should be the concentration to use. The same reasoning can be used to determine the excess level for antigen and antibody. This procedure is used for every lot of new conjugate.

3. High background. A series of "blank" controls should be included in every run in addition to the usual normal serum controls. These should include a "no antigen" blank, a "no antibody" blank, a "conjugate only" blank, and a "substrate only" blank. High readings in any of these blanks, especially in the "conjugate only" blank, would indicate insufficient

washing. Increased washing times or vacuum velocity should correct this problem. The Tween 20 used in the PBS–Tween solution should be fresh (less than 3 months old) and maintained at a 0.3% concentration.

4. Decreasing activities in known positive controls. This would indicate a deterioration of the enzyme-conjugate. We normally include two enzymic controls for the conjugate. The diluted working conjugate is further diluted 1 : 2000 in NaAc/HAc pH 5.00; 0.75 ml of this solution is placed in a cuvette and assayed as usual. A drop in enzymic activity should indicate the expiration of the conjugate. Conjugates (after reconstitution, if lyophilized) should be centrifuged at 48,000 g for 20 min and stored in small (100-μl) aliquots with 50% glycerol at $-20°$ (liquid N_2 temperatures are better) until use.

Acknowledgments

The work of Dr. J. M. Peralta was supported by a fellowship from Coordenacao de Aperfeicoamento de Pessoal de Nivel Superior (CAPES), Brazil.

[31] Immunoassay Using Antigen-Coated Plastic Tubes and Radiolabeled or Enzyme-Labeled Protein A

By ADRIAN P. GEE and JOHN J. LANGONE

The procedure used to radiolabel antigens or antibodies for use in radioimmunoassays may result in products that have lost their immunoreactivity or are unstable.[1] To avoid these problems, labeled protein A of *Staphylococcus aureus* (SpA) has been used in immunoassay as a general tracer for IgG antibody. The principles and applications of the basic assay system have been described in Volume 70.[2] Briefly, SpA binds specifically to the Fc region of IgG from rabbits and several other species without inhibiting the antigen–antibody reaction.[3] Since SpA retains its functional activity even when labeled with radionuclides, enzymes, or other tracers, it can be used in immunoassays regardless of antibody specificity. This obviates the need to prepare individual labeled derivatives for each assay. For example, in competition assays with [^{125}I]SpA, ligands supplied in the test sample and immobilized on polyacrylamide beads compete for a lim-

[1] F. C. Greenwood, *in* "Principles of Competitive Protein Binding Assays" (W. E. Odell and W. H. Daughaday, eds.), pp. 288–296. Lippincott, Philadelphia, Pennsylvania, 1971.
[2] J. J. Langone, this series, Vol. 70 [25].
[3] A. Forsgren and J. Sjöquist, *J. Immunol.* **97**, 822 (1966).

ited amount of antibody. Antibody bound to the immobilized antigen can then be determined by incubating the washed beads with excess labeled SpA. Free antigen in the test sample inhibits binding of antibody, and consequently of [^{125}I]SpA, to the solid phase, and its concentration can be determined by reference to a standard inhibition curve.

This method gives reproducible and accurate results, but for each assay system the test ligand must be coupled to the polyacrylamide support, which is repeatedly centrifuged and washed during the assay. Although these procedures are simple, they can become tedious when a large number of samples or different ligands are analyzed.

In 1967 Catt and Tregear[4] found that antibodies to ligands passively adsorbed to plastic surfaces, e.g., polystyrene tubes, could be used in solid-phase radioimmunoassay. The free ligand could easily be separated from the immobilized antibody-bound ligand by simply decanting the fluid phase. Excellent reviews of the principles and applications of the method appear in Volume 73 of this series.[5,6] Until recently, however, the procedure still required the use of labeled antigens specific for each assay. We, therefore, combined the plastic tube technique with the use of [^{125}I]SpA to develop a rapid, simple assay system that can be used to quantify fluid-phase immunoglobulins and a variety of drugs.[7] The antigen or hapten is immobilized onto the surface of plastic tubes either directly by passive adsorption, or indirectly by coupling it to poly-L-lysine which is then adsorbed to the plastic. The antigen on the tube surface then competes with antigen in the fluid-phase test sample to bind a limited quantity of added antibody. Antibody bound to the immobilized ligand is then detected by incubation with excess [^{125}I]SpA. To avoid the problems associated with both the handling and disposal of radioactive material, we have also developed an enzyme immunoassay using alkaline phosphatase-labeled SpA. The principle is the same, although slightly different procedures are used.

Radioimmunoassay of Immunoglobulins

Requirements

Polypropylene tubes, 12 × 75 mm (available from Falcon, Cockeysville, Maryland)

[4] K. J. Catt and G. W. Tregear, *Science* **158,** 1570 (1967).
[5] G. H. Parsons, Jr., this series, Vol. 73 [14].
[6] J. E. Herrmann, this series, Vol. 73 [15].
[7] A. P. Gee and J. J. Langone, *Anal. Biochem.* **116,** 524 (1981).

Purified immunoglobulin (available commercially from many sources, e.g., Cappel Laboratories, Cochranville, Pennsylvania)

Rabbit antiserum to IgA, IgE, or IgM (available commercially from several sources including Cappel Laboratories, Pel-Freeze, Bio-Rad, and Dako Laboratories). The antiserum should contain IgG antibody; whole antiserum is usually satisfactory. Antiserum is not required to assay IgG.

Barbital-buffered saline (BBS-GM), pH 7.4, containing 0.1% gelatin, 0.001 M Mg^{2+} and 0.00015 M Ca^{2+}, prepared as described in Volume 93 [22].

Test sample, e.g., serum, dilutions prepared in buffer

[^{125}I]SpA, prepared as described by Langone,[2] except that 10 μg of protein A (PA) and 0.4 mCi of Bolton-Hunter reagent are used.

Procedure

General. In all the following assays it is important to optimize the test conditions. Each laboratory should determine the kinetics of adsorption of the coating immunoglobulin to the plastic tubes and the amount that should be bound to obtain the best combination of sensitivity and [^{125}I]SpA binding. The conditions described below should, therefore, be used only as guidelines.

Immunoglobulin G. Since IgG binds protein A directly, a one-step procedure can be used. In preliminary experiments with ^{125}I-labeled human IgG and polypropylene tubes, we found the rate of adsorption to be dependent on the concentration of protein offered and the temperature of incubation. In general, the rate of binding was greatest during the first 60–90 min of incubation at 37°, and these conditions were used routinely. Prolonged (up to 32 hr) incubation did not result in saturation of the tube surface with antigen, even at an IgG concentration of 50 μg/ml. We also noted some variability between different batches of tubes and would recommend stockpiling tubes that show good protein adsorption characteristics.

The optimal concentration of IgG for use in the assay is determined as follows.

1. In clean, untreated polypropylene tubes, incubate 1-ml aliquots of BBS-GM containing 11.1, 33.3, or 100 μg of purified human IgG for 1 hr at 37°.

2. Wash the tubes twice with 2 ml of buffer. If not required immediately, they can be stored at 4° for at least 10 days without change in their characteristics in the assay.

3. For each series of coated tubes, a standard curve is established by adding 100 μl of a range of concentrations of purified IgG (0.18–5 μg/ml) or buffer, and 0.9 ml of BBS-GM containing excess [^{125}I]SpA (50,000 cpm for this preparation) to each tube. Duplicate samples should be included. Mix tube contents and incubate for 1 hr at 37°.
4. Decant and discard the contents of the tubes and wash twice with 2 ml of buffer.
5. Drain the tubes well and blot the rims.
6. Determine bound [^{125}I]SpA by counting in a gamma spectrometer for 1 min.
7. Calculate the percentage of inhibition of [^{125}I]SpA binding and plot the standard curves for each series of coated tubes. Typical results in Fig. 1 show that as the amount of IgG used to coat the tubes decreased, the sensitivity of the assay increased, whereas the binding decreased from 11,350 to 3300 cpm. For this IgG preparation and batch of tubes, the optimal combination of sensitivity and binding was achieved using 33.3 μg of IgG per milliliter in the coating step.

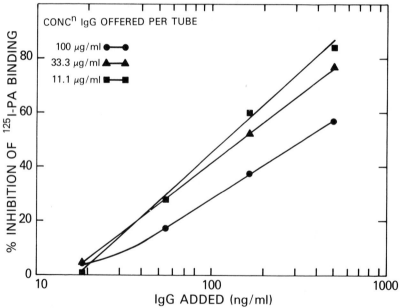

FIG. 1. Effect of amount of IgG used to coat tubes on the sensitivity of IgG radioimmunoassay. In each case excess [^{125}I]SpA (50,000 cpm) was added. Binding was 11,346 cpm (●), 7160 cpm (▲), and 3300 cpm (■). From Gee and Langone.[7]

Once optimal conditions have been established, a large number of tubes can be coated and stored until required. To assay test samples, the same general procedure is followed starting at step 3. In each assay the standard curve is included in addition to duplicates of several dilutions of the test sample. The concentration of IgG in the sample is determined from the standard inhibition curve.

IgM, IgA, and IgE. Since these immunoglobulins react only weakly, if at all, with SpA, a two-step procedure is required. Purified Ig adsorbed to the tube surface competes with Ig in the test sample to bind a limited amount of anti-Ig antiserum. The IgG antibody bound to the Ig ligand on the tube surface is then detected after a second incubation with labeled SpA.

1. Determine the optimal tube coating conditions as described for the IgG assay and prepare a batch of coated tubes. This requires that different concentrations of both Ig for coating and SpA reactive anti-Ig antibody for use in the second step be tested. Control tubes coated with BBS-GM should also be prepared.
2. To each tube add 100 μl of a dilution of a purified immunoglobulin preparation in BBS-GM (see note on IgA assays) for the standard curve; or 100 μl of test sample (inhibitor) diluted in buffer. Add 800 μl of BBS-GM to each tube followed by 100 μl of an optimal dilution (e.g., 1/2000) of anti-immunoglobulin A, M, or E antiserum. The following controls should include: (a) coated tube + buffer + antibody; (b) coated tube + inhibitor + buffer; (c) buffer-coated tube + inhibitor + antibody.
3. Incubate all tubes for 1 hr at 37°. Decant and discard the contents. Wash the tubes twice with buffer and add 1 ml of [^{125}I]SpA in BBS-GM to each (about 50,000 cpm has been found to provide an excess for this dilution of antibody).
4. Incubate the tubes for 1 hr at 37°.
5. Decant and discard the contents. Wash the tubes twice with buffer. Drain well and blot the rims.
6. Determine counts per minute bound by counting for 1 min in a gamma spectrometer. Controls (b) and (c) should bind <600 cpm of [^{125}I]SpA.
7. Using control (a) for the total binding value, calculate the percentage of inhibition of binding for each dilution of the purified immunoglobulin preparation, and plot the standard curve and determine the concentration of immunoglobulin in the test samples.

IgA Assay. Shallow standard curves may be obtained in these assays owing to aggregation of purified IgA. This problem can be overcome by

TABLE I
OPTIMAL CONDITIONS FOR COATED-TUBE RADIOIMMUNOASSAYS[a]

Antigen or hapten	Amount added per tube (μg/ml)	[^{125}I]SpA bound[b] (cpm)	I_{50} (ng/ml)
IgA	30	7200	62
IgE	3	5100	62
IgG	33	7200	150
IgM	33	4400	200
Methotrexate	3.3[c]	6500	175
5-Methyltetrahydrofolate	0.33[c]	5150	1.8

[a] From Gee and Langone.[7]

[b] IgG was analyzed by direct [^{125}I]SpA binding. In the other assays rabbit antisera or purified IgG fractions were used at a final dilution of 1/2000. Binding values have been corrected for background binding of [^{125}I]SpA (60–300 cpm) to antibody-treated tubes.

[c] Given as concentration of poly-L-lysine added.

diluting the standards in BBS-GM containing Tween 80. We found that a final concentration of 0.001% Tween was optimal since at higher concentrations [^{125}I]SpA binding was reduced. Routinely, therefore, all standards and test samples were prepared in BBS-GM containing 0.01% Tween and diluted 10-fold to 1 ml by the addition of BBS-GM and diluted antibody.

IgE Assay. Since the IgE concentration in human serum is low ($\leq$50 ng/ml), undiluted samples often must be analyzed. If the lipid content of the serum is high, the sample may become adsorbed to the tube surface during the incubation with antibody, resulting in high background binding of labeled SpA [control (c)]. This can be reduced either by (*a*) preincubating IgE-coated tubes with BBS-GM supplemented with 0.5% gelatin for 30 min at 30° prior to the addition of the test serum and antibody, or (*b*) absorbing the test serum with SpA–Sepharose, thereby both removing lipid by nonspecific adsorption and IgG by affinity binding. Removal of IgG also eliminates interference by cross-reaction with the anti-IgE antiserum.

The optimal coating conditions, total [^{125}I]SpA cpm bound, and the relative sensitivities of the immunoglobulin radioimmunoassays are summarized in Table I. The sensitivities are given as I_{50} values; i.e., the concentration of inhibitor required to give 50% inhibition of antibody (i.e., [^{125}I]SpA) binding). It should be reemphasized that these data were obtained under specific conditions and are representative. Appropriate conditions must be established in each laboratory depending on the properties of the antibodies and other reagents available. In our hands the intra-assay coefficient of variation was routinely less than 10% and the interassay variation was less than 15%.

Radioimmunoassay of Haptens

Haptens, such as the drugs methotrexate and 5-methyltetrahydrofolate, do not show significant binding to plastic even at high concentrations (50 μg/ml) and after prolonged incubation. Polypropylene and polystyrene can, however, be coated with hapten that has been covalently coupled to poly-L-lysine using carbodiimide. The conjugation procedure is described below.

Hapten Conjugation to Poly-L-Lysine

The ratio of poly-L-lysine to hapten in the conjugate can greatly affect the sensitivity of the immunoassay (Table II). It is therefore important to determine for each hapten the optimal ratio to give the best combination of assay sensitivity and antibody binding.

Requirements

Phosphate buffer, 0.003 M, pH 6.35, throughout
1-Ethyl-3-(3-diethylaminopropyl)carbodiimide (EDAC) (Bio-Rad Laboratories, Richmond, California) at 5 mg/ml in buffer
Poly-L-lysine (average molecular weight 40,000; Sigma Chemical Co., St. Louis, Missouri)
Hapten, dissolved and diluted to the desired concentration in buffer

TABLE II
EFFECT OF METHOTREXATE (MTX)-POLYLYSINE
(PL) CONJUGATE COMPOSITION ON SENSITIVITY
OF THE RADIOIMMUNOASSAY[a]

Conjugate (mole MTX/mole PL)	[125I]SpA bound[b] (cpm)	I_{50} (ng/ml)
7.8	6300	1000
1.98	5890	185
0.16	1790	1

[a] From Gee and Langone.[7]
[b] Corrected for background of 630–730 cpm.

Conjugation Procedure

1. To 1 ml of poly-L-lysine solution add 0.3 ml of hapten and mix
2. Add 1 ml of EDAC dropwise with mixing
3. Allow the mixture to rock for 1 hr at room temperature or overnight at 4°.

The resulting conjugate need not be purified and can be stored for up to 1 month at 4° without loss of antibody-binding activity. One preparation is usually sufficient to coat several thousand tubes. We have not found any advantage to using conjugates freed of unincorporated hapten by chromatography on Sephadex G-25 columns, although this procedure can be used to purify the conjugates to determine the efficiency of the coupling procedure.

Assay Methods

Requirements

Polypropylene tubes, 12 × 75 mm (Falcon, Cockeysville, Maryland)
Purified hapten, for generating the standard curve
Hapten–poly-L-lysine conjugate; see above
Rabbit antiserum to hapten-containing IgG antibody

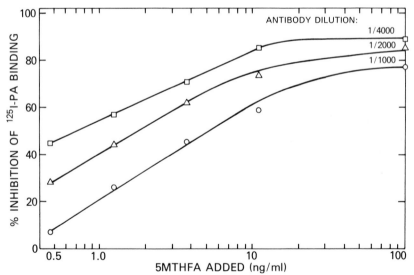

FIG. 2. Effect of antibody concentration on the sensitivity of the radioimmunoassay 5-methyltetrahydrofolate (5MTHFA). All tubes were coated with 5MTHFA conjugate corresponding to 0.33 μg of poly-L-lysine. From Gee and Langone.[7]

Barbital-buffered saline (BBS-GM) prepared as described earlier
Test sample dilutions prepared in buffer
[^{125}I]SpA, prepared as described earlier

Procedure

1. Tubes are coated with the hapten–polylysine conjugate in Hanks' balanced salt solution. The optimal concentration is determined as described for the IgG assay. After the coating step the tubes should be incubated for 30 min at 30° with BBS-GM supplemented with 0.5% gelatin to reduce background binding of [^{125}I]SpA.
2. The remainder of the procedure is as described for the assay of immunoglobulins. The standard curve is generated using dilutions of the unconjugated hapten as the inhibitor. The effect of varying the antibody concentration on sensitivity and total [^{125}I]SpA binding in the assay of 5MTHFA are shown in Fig. 2.

The reproduciibility of the hapten assays was similar to that for immunoglobulins.

Enzyme Immunoassays

Enzyme-labeled SpA can be substituted as the general tracer.

Requirements

Requirements up to the addition of SpA are the same as for the radioimmunoassays.
SpA labeled with alkaline phosphatase (PA-AP) (available from Zymed Laboratories, Burlingame, California), diluted 1 : 500 in BBS-GM
Phosphatase substrate: *p*-nitrophenyl phosphate at 0.67 mg/ml in 0.05 *M* Na$_2$CO$_3$ buffer containing 1 m*M* MgCl$_2$, pH 9.8
"Stopping" reagent: 2 *N* NaOH

Procedure

1. The same procedure as that used for the immunoglobulin and hapten radioimmunoassays is followed up to the addition of labeled SpA.
2. Add 1 ml of 1 : 500 dilution of PA-AP to each tube.
3. Incubate for 1 hr at 37°.
4. Wash the tubes with 4 × 2.5 ml of BBS-GM.
5. Add 1–2 ml of phosphatase substrate to each tube and incubate at 37°.

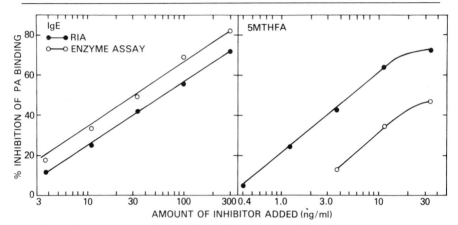

FIG. 3. Comparative sensitivities of radioimmunoassays (●) and enzyme immunoassays (○) for IgE and 5-methyltetrahydrofolate (5MTHFA) under optimal assay conditions. From Gee and Langone.[7]

6. Stop the reaction by adding 0.2 ml of 2 N NaOH to each tube. Mix.
7. Read absorbance at 400 nm.
8. Calculate percentage of inhibition of PA-AP binding, and plot the results as described for the radioimmunoassays

In our hands the enzyme and radioimmunoassays for immunoglobulins were of similar sensitivity, whereas enzyme assays for haptens tended to be less sensitive (Fig. 3). The background binding in tubes coated with BBS-GM and incubated with inhibitor, antibody, and labeled SpA was significantly higher in the enzyme immunoassays ($A_{400 \, nm}$ 0.15–0.2, compared to maximum binding $A_{400 \, nm}$ of 0.6–0.9). This could be reduced slightly by using polystyrene rather than polypropylene tubes.

Conclusions

The plastic tube–SpA tracer immunoassay system has the following advantages over other solid-phase techniques.

1. The disposable plastic tubes can be coated easily and reproducibly and can be stored for long periods without adverse effect.
2. Macromolecules such as immunoglobulins bind directly to the tubes, and haptens can be coupled after conjugation with poly-L-lysine.
3. The hapten conjugates can be prepared rapidly using readily available reagents, and one preparation is sufficient for several thousand assays.

4. Both the enzyme and the radioimmunoassay procedures are simple and fast, since centrifugation of the tubes is unnecessary.
5. Sample volumes of up to 1 ml can be analyzed, e.g., in the IgE assay.
6. Use of [125]I- or alkaline phosphatase-labeled SpA as a general tracer eliminates the need to purify and radiolabel the ligand or antibody.
7. The sensitivity and reproducibility of the method is similar to that reported for other solid-phase techniques.

[32] Colorimetric Immunoassays Using Flavin Adenine Dinucleotide as Label

By DAVID L. MORRIS and ROBERT T. BUCKLER

There has been intense effort to develop immunoassay methods that do not involve the use of radioisotopes. Not only has a wide variety of alternative labels been used successfully, but entirely new approaches to immunoassays have grown out of this work. One of the most important is the homogeneous immunoassay in which the entire assay can be performed without separation of free and labeled antigen from the antigen–antibody complexes formed. On the other hand, when a radiolabel is used, physical separation of the bound and free fractions is obligatory. Materials that have been used successfully as labels in homogeneous immunoassays include, bacteriophages,[1] spin-labeled molecules,[2] enzymes,[3] fluorescent molecules,[4] chemiluminescent molecules,[5] enzyme cofactors,[6] and enzyme substrates.[7] Although the homogeneous assay format does not at present allow detection limits as low as those obtain-

[1] J. Haimovich, E. Hurwitz, N. Novik, and M. Sela, *Biochim. Biophys. Acta* **207**, 115 (1970).
[2] R. K. Leute, E. F. Ullman, A. Goldstein, and L. A. Herzenberg, *Nature (London), New Biol.* **236**, 93 (1972).
[3] K. E. Rubenstein, R. S. Schneider, and E. F. Ullman, *Biochem. Biophys. Res. Commun.* **47**, 846 (1972).
[4] W. B. Dandliker, H. C. Schapiro, J. W. Maduski, R. Alonso, G. A. Feigen, and J. R. Hamrick, *Immunochemistry* **1**, 165 (1964).
[5] H. R. Schroeder, P. O. Vogelhut, R. J. Carrico, R. C. Boguslaski, and R. T. Buckler, *Anal. Chem.* **48**, 1933 (1976).
[6] R. J. Carrico, J. E. Christner, R. C. Boguslaski, and K. K. Yeung, *Anal. Biochem.* **72**, 271 (1976).
[7] J. F. Burd, R. C. Wong, J. E. Feeney, R. J. Carrico, and R. C. Boguslaski, *Clin. Chem.* **23**, 1402 (1977).

METHODS IN ENZYMOLOGY, VOL. 92

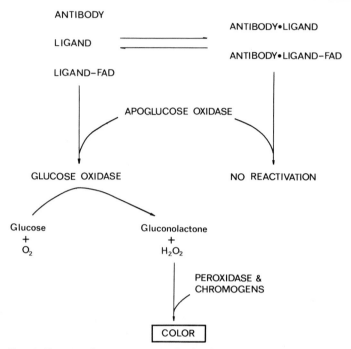

FIG. 1. Concept of apoenzyme reactivation immunoassay system (ARIS).

able with heterogeneous methods and is more susceptible to interferences from the sample matrix, these disadvantages are in many applications offset by the simplicity, convenience, and speed of the former approach.

A colorimetric homogeneous assay has been demonstrated in which a prosthetic group, FAD, is used as a label and detected at very low levels by combination with apoglucose oxidase to generate glucose oxidase activity.[8] This prosthetic group label immunoassay, conveniently known as ARIS (apoenzyme reactivation immunoassay system), is shown diagrammatically in Fig. 1. The ligand–FAD conjugate is able to activate apoglucose oxidase efficiently, but when bound by antibody specific to the ligand this ability is strongly inhibited. Thus, the glucose oxidase activity generated in the assay is related to the concentration of free label in the competitive binding system and hence to the concentration of ligand. The unique substances in this assay system are the FAD-labeled antigens and apoglucose oxidase, and the preparation of these reagents is described in detail.

[8] D. L. Morris, P. B. Ellis, R. J. Carrico, F. M. Yeager, H. R. Schroeder, J. P. Albarella, R. C. Bogusalski, W. E. Hornby, and D. Rawson, *Anal. Chem.* **53,** 658 (1981).

The color-generating reagents described comprise a system modified from that reported by Barham and Trinder.[9,10] However, a wide variety of color-generating systems for detection of hydrogen peroxide is available, and fluorescence, luminescence and electrode methods should be applicable.

Preparation of Apoglucose Oxidase

A method is described for the preparation of apoenzyme on about a 1-g scale. However, the procedure may be readily modified for different batch sizes by proportional changes in reagent volumes and column dimensions.

Aspergillus niger glucose oxidase with a specific activity of at least 150 units/mg should be used (high purity, low catalase grade; Miles Laboratories, Elkhart, Indiana). A 75 mg/ml solution of the enzyme in 30% (w/v) glycerol in 25 mM sodium phosphate, pH 6.0, is prepared. The exact enzyme concentration may be calculated from the absorbance of the solution at 450 nm using the millimolar extinction coefficient 14.1 M^{-1} cm^{-1}.[11]

An acid-dissociation medium is prepared as follows. Glucose oxidase solution (14 ml) containing 1.05 g of enzyme is stirred on a salt–ice bath at 0°, and ice cold 30% (w/v) glycerol in 25 mM phosphate (previously adjusted to pH 1.1 by addition of concentrated H_2SO_4 at room temperature) is added until the pH is 1.7. This should require about an equal volume of reagent. This mixture is then incubated for 30 min at 0°, after which it is loaded onto a column (10 × 15 cm) of BioGel P-10 equilibrated with 30% (w/v) glycerol 25 mM sodium phosphate buffer, pH 1.7 maintained at 4°. The column is eluted with the same medium at a flow rate of 9 ml/min.

The elution of apoenzyme is monitored by its transmittance at 280 nm and collected directly into a stirred suspension of dextran-coated charcoal maintained at 4°. Collection is started when the transmission of the effuent has increased to 70% and is stopped once the transmission has decreased to 70%. The dextran-coated charcoal consists of 8 ml of 0.4 M sodium phosphate, pH 8.0, containing 40 mg of dextran T-70 (Pharmacia Fine Chemicals, Uppsala, Sweden) and 1200 mg of activated charcoal (RIA grade from Schwarz-Mann, Orangeburg, New York). While the protein peak is being collected, the pH is adjusted continually to 7.0 by dropwise addition of 1 N NaOH.

Once the protein has been collected, the charcoal suspension is stirred for 60 min at 4°, after which the charcoal is removed by successive filtra-

[9] D. Barham, and P. Trinder, *Analyst* (*London*) **97,** 142 (1972).
[10] P. Fossati, L. Prencipe, and G. Berti, *Clin. Chem.* **26,** 227 (1980).
[11] B. E. P. Swoboda and V. Massey, *J. Biol. Chem.* **240,** 2209 (1965).

tions. In the first filtration a 47-mm 5.0-μm Millipore filter (GSWP04700) and a 47-mm Millipore prefilter (AP2504700) are used. A second filtration is necessary to remove fines, and a 47-mm 0.45-μm Millipore filter (245-0045) is used. At this stage the volume of apoenzyme is 160–180 ml.

The apoenzyme is most stable when stored in lyophilized form at 4°. The liquid reagent is therefore loaded onto a column (5 × 43 cm) of Sephadex G-25 (coarse) equilibrated at 4° with 0.1 M sodium phosphate, pH 7.0, containing 0.1% (w/v) bovine serum albumin and 5 mg of mannitol per milliliter. The apoprotein is eluted with the same buffer at a flow rate of 8 ml/min and monitored by the change in absorbance at 280 nm using the elution buffer as a blank. Material that is eluted between the 90% transmission limits is pooled and lyophilized. The absorbance of the pooled solution at 280 nm is measured against a blank consisting of the equilibrium medium, and the yield of apoenzyme protein is determined using an absorption at 280 nm of 1.71 for a 1 mg/ml solution. This yield should typically be within the range of 60–70% of the total amount of glucose oxidase starting material.

Apoglucose oxidase in lyophilized form is quite stable and may be maintained at 4° for at least 18 months without loss of activity. When reconstituted into a solution of 0.1 M phosphate buffer, pH 7.0, containing 30% (w/v) glycerol, 0.1% (w/v) bovine serum albumin, and 0.1% (w/v) sodium azide, the apoenzyme (micromolar range) is stable for several weeks. However, gradual loss of activity is observed with storage at 4°, so that typically after 12 months about 40% of the original activity will remain.

The method described for preparation of apoglucose oxidase can be performed with good reproducibility; however, great care must be taken to ensure that contamination with FAD does not occur. Thus, where practical, all materials used in a preparation should be used once and discarded (e.g., BioGel P-10 used for chromatography), and new glassware should be used for manipulation of solutions after elution of the apoenzyme from the column. Where apparatus is reused, it should be thoroughly cleaned with a suitable detergent, such as Isoclean decontaminant (Nuclear Supply & Service Co., Washington, D.C.).

Characterization and Properties of Apoglucose Oxidase

The following properties of the apoenzyme should be determined: the FAD binding site concentration, the residual glucose oxidase activity, and the recoverable glucose oxidase activity.

The FAD binding site concentration is determined by titration of apoglucose oxidase with FAD as originally described by Swoboda.[12]

---

[12] B. E. P. Swoboda, *Biochim. Biophys. Acta* **23**, 365 (1969).

Thus, apoglucose oxidase (50–100 μg/ml) is incubated with a range of FAD concentrations (0–5 μM) in 0.1 M phosphate buffer, pH 7.0, containing 0.1% (w/v) BSA and 0.1% (w/v) sodium azide at room temperature for 16 hr. The activity of the regenerated glucose oxidase can conveniently be measured by use of the colorimetric glucose oxidase assay system described herein. When glucose oxidase activity is plotted vs FAD concentration, a sharply defined equivalence point is obtained. Apoenzyme concentrations are most meaningfully described in terms of the FAD-binding site concentrations, and this is how they are presented throughout this chapter.

The residual glucose oxidase activity is important since this must be sufficiently low to allow sufficient apoenzyme to be incorporated into immunoassays to achieve the desired sensitivity without being hampered by high background color development. The assay procedures described here can be performed satisfactorily when the residual activity is at most 0.010% of the recoverable glucose oxidase activity. This can be calculated from the FAD titration experiment when the residual activity is the activity in the absence of FAD and the recoverable activity is the glucose oxidase activity at the equivalence point.

Organic Synthetic Procedures

N^6-(6-Aminohexyl)FAD

This substance is prepared essentially by the method of Hoard and Ott.[13] A mixture of 473.4 mg (1 mmol) of N^6-(6-trifluoroacetamidohexyl) AMP (prepared by the method of Trayer et al.[14]), 186 mg (1 mmol) of tri-n-butylamine, and 13.5 ml of dry dimethylformamide (DMF) is concentrated at 40° on a rotary evaporator attached to a vacuum pump. The process is repeated twice. The residual gum is then dissolved in 20 ml of dry DMF, and 810.5 mg (5 mmol) of solid 1,1′-carbonyldiimidazole (CDI) is added in one portion. After stirring for 3 hr at room temperature under an inert atmosphere, excess CDI is destroyed by the addition of 405 μl (10 mmol) of methanol. After 12 min the reaction mixture is concentrated on a rotary evaporator attached to a vacuum pump. The residual gum of the imidazolide is three times taken up in 20 ml of DMF and reevaporated to remove the last traces of methanol.

Meanwhile, 684.5 mg (1.5 mmol) of the monotriethylammonium salt of 5′-FMN (prepared by the method of Johnson et al.[15]) is slurried in 13.5 ml

[13] D. E. Hoard, and D. G. Ott, J. Am. Chem. Soc. 87, 1785 (1965).
[14] I. P. Trayer, H. R. Trayer, D. A. P. Small, and R. C. Bottomly, Biochem. J. 139, 609 (1974).
[15] R. D. Johnson, L. LaJohn, and R. J. Carrico, Anal. Biochem. 86, 526 (1978).

of dry dimethyl sulfoxide (DMSO) and combined with 550 mg (1.5 mmol) of tri-*n*-octylamine and 3 ml of dry tetrahydrofuran. After stirring under an inert atmosphere at 45° for 45 min, the mixture is concentrated under reduced pressure. The gummy residue is twice taken up in dry DMSO and reevaporated. It is then taken up in 7 ml of DMSO and added to the imidazolide made in the preceding step. The flask is rinsed with a few milliliters of DMF, and this is added to the reaction. After stirring in the dark for 40 hr, it is concentrated to leave a deep yellow oil that is partitioned between 100 ml of 0.1 *M* triethylammonium bicarbonate and 100 ml of diethyl ether. The aqueous phase is separated, washed with an additional 100 ml of ether, and concentrated to about 10 ml under reduced pressure. This is diluted with 100 ml of ethanol and again concentrated to near dryness. The process is repeated once more. The final solution, about 15 ml, is applied to a column of 150 g of silicic acid equilibrated in 2 liters of 9:1 (v/v) ethanol:1 *M* TEAB. The column is eluted with this solvent at a flow rate of 2 ml/min, and 20-ml fractions are collected. The first yellow fractions contain 4',5'-cyclic FMN, which is closely followed by a mixture of N^6-(6-trifluoroacetamidohexyl)FAD and its cyclic carbonate derivative (cf. Maeda *et al.*[16]). The fractions containing the latter two are pooled and concentrated to about 20 ml. This is diluted to 500 ml with distilled H_2O and concentrated to about 75 ml to remove traces of TEAB. The pH is adjusted to 10.5 with 0.1 *N* NaOH, and the preparation is held at this pH for 8 hr. It is then neutralized to pH 7.0 with 0.1 *N* HCl and concentrated to 20 ml. This solution is applied to a column of BioGel P-2 (90 cm × 5 cm) and eluted with H_2O at a flow rate of 0.5 ml/min. A small amount of N^6-(6-trifluoroacetamidohexyl)FAD elutes first (R_f 0.75, 7:3 ethanol–1 *M* ammonium acetate on silica gel plates) followed by N^6-(6-aminohexyl)FAD (R_f 0.32). The product is salt free and greater than 98% pure (26% yield). It is concentrated to about 10 m*M* and stored at 4° in the presence of 0.02% NaN_3. The concentration of this and other FAD derivatives in aqueous buffer solution at pH 7.0 is calculated using the extinction coefficient at 450 nm of 11.3 mM^{-1} cm^{-1}.[17]

Theophylline-FAD

N^6-(6-Aminohexyl)AMP (2.3 g, 5 mmol) and 1.86 g (7.5 mmol) of the lactam of 8-(3-carboxypropyl)theophylline (prepared by the method of Cook *et al.*[18]) is combined with 10 ml of DMF and 1 ml of triethylamine

[16] M. Maeda, A. D. Patel, and A. Hampton, *Nucleic Acids Res.* **4**, 2843 (1977).
[17] L. G. Whitby, *Biochem. J.* **54**, 437 (1953).
[18] C. E. Cook, M. E. Twine, M. Myers, E. Amerson, J. A. Kapler, and G. F. Taylor, *Res. Commun. Chem. Pathol. Pharmacol.* **13**, 497 (1976).

and stirred for 16 hr at room temperature. Twenty-five grams of silica gel is added to the reaction mixture, and the solvent is removed on a rotary evaporator attached to a vacuum pump. The impregnated adsorbent is placed atop a column of 350 g of silica gel made up in absolute ethanol, and the column is eluted with a linear gradient of 2 liters of ethanol to 2 liters of 7 : 3 (v/v) ethanol–1 M TEAB. When the gradient is exhausted, elution is continued with 7 : 3 ethanol–TEAB. Twenty-milliliter fractions are collected.

Fractions 346–422 are pooled and evaporated to give 1.5 g (37% yield) of the triethylammonium salt of the theophylline–AMP conjugate as a white powder.

The theophylline–AMP conjugate (500 mg, 0.62 mmol) is activated with CDI and allowed to reacted with 5'-FMN as described in the preparation of N^6-(6-aminohexyl)FAD. After 40 hr, the reaction is diluted with 1.5 liters of H_2O and applied at 1 ml/min to a column of DEAE-cellulose (2.5 × 80 cm, acetate form). The column is washed with 500 ml of H_2O, then eluted with a linear gradient of 2 liters of H_2O to 2 liters of 0.6 M ammonium acetate, pH 4.6. The theophylline–FAD conjugate elutes near the end of the gradient and is completely removed by continued elution with 0.6 M ammonium acetate. Ammonium acetate is removed from the conjugate by absorption to and elution from DEAE-cellulose with 0.3 M TEAB. Most of the triethylammonium acetate is then removed by repeated evaporation from H_2O at 30° on a rotary evaporator (30% yield). The structure of this compound is illustrated in Fig. 2.

Stability of FAD Derivatives

Aminohexyl-FAD and theophylline-FAD are stable as solutions in 0.1 M phosphate buffer, pH 7.0, containing 0.1% (w/v) sodium azide. When stored for 6 months in the dark at 37°, room temperature, or 4°, 2 μM solutions are completely stable both with respect to ability to activate apoenzyme and, in the case of theophylline-FAD, in its ability to bind to antibody against theophylline. Lyophilized conjugates prepared from the above solutions are not as stable, and full activity is retained only when they are stored at 4°.

Sodium 3,5-Dichloro-2-Hydroxybenzenesulfonate (DHSA)

In a 5-liter 3-neck flask equipped with a stirrer, thermometer, and dropping funnel is placed 523 g (2.0 mol) of 3,5-dichloro-2-hydroxybenzenesulfonyl chloride (Aldrich Chemical Co., Milwaukee, Wisconsin) and 2.8 liters of distilled H_2O. The slurry is stirred and maintained at 60–70° while a solution of NaOH (160 g, 4 mol) in 400 ml of distilled H_2O is added

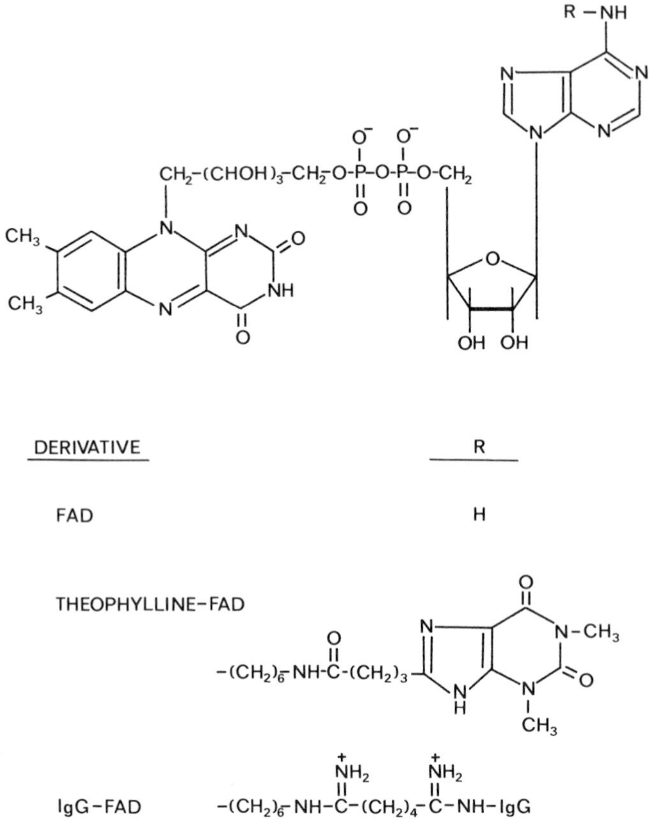

FIG. 2. Structures of flavine adenine dinucleotide (FAD) derivatives.

dropwise over 60 min. After addition is complete, the hot solution is filtered through a pad of diatomaceous earth and the filtrate is cooled in an ice bath at 0–5°. The crystalline product is collected by filtration and washed with 200 ml of cold H_2O. When dry, it amounts to 350 g. Recrystallization from 5 liters of boiling glacial acetic acid gives 330 g (after drying at 100° for 6 hr under vacuum) of DHSA as white crystals, analytically pure.

Preparation of IgG-FAD

Aminohexyl-FAD (4.8 μmol) and 5 μl of triethylamine (30 μmol) are dissolved in 0.9 ml of H_2O. A solution of 24.5 mg of dimethyladipimidate dihydrochloride per milliliter of anhydrous ethanol is prepared immediately before use. Dimethyladipimidate solution (100 μl containing 10 μmol) is then added to the aminohexyl-FAD solution, and the mixture is incubated at room temperature. After 10 min, 40 mg of IgG (0.26 μmol) in 1.0 M potassium carbonate, pH 9.6, is added. The reaction is further incubated for 3 hr at room temperature, after which the solution is chromatographed on a column (2.5 × 50 cm) of Sephadex G-25 (coarse) equilibrated with 0.1 M sodium phosphate buffer, pH 7.0. The first peak with absorbance at 450 nm is collected and dialyzed extensively against 250 volumes of 0.1 M sodium phosphate, pH 7.0; 250 volumes of 0.1 sodium phosphate, pH 7.0, containing 1.0 M NaCl; 25 volumes of 0.1 M sodium phosphate, pH 7.0, for 12 hr each at 4°. Sodium azide [0.1% (w/v)] is included in the buffers. The incorporation of aminohexyl-FAD in the IgG-FAD conjugate may be estimated by determination of the protein content by the Lowry method; and the FAD content from the absorbance at 450 nm using the extinction coefficient 11.3 mM^{-1} cm^{-1}. About 2 mol of FAD are coupled per mole of IgG by this coupling procedure. The structure of the conjugate is illustrated in Fig. 2.

Reagents for Colorimetric Glucose Oxidase Activity Determination

The following reagents are used: sodium 3,5-dichloro-2-hydroxybenzene sulfonate (DHSA) 4-aminoantipyrine (4-AP); glucose, horseradish peroxidase; bovine serum albumin (BSA); sodium phosphate buffer, pH 7.0.

The glucose oxidase activity generated by recombination of FAD derivatives and apoglucose oxidase can be measured *in situ* by a modification of the assay system reported by Fossati et al.[10] Glucose oxidase activity is detected by measurement of the increase in absorbance at 520 nm. The final reagent concentrations are 2.0 mM DHSA, 0.2 mM 4-AP, 0.1 M glucose, 60 μg of peroxidase per milliliter (not less than 60 purpurogallin units per milligram), 1.0% (w/v) BSA, and 0.1 M sodium phosphate buffer. It is recommended that the DHSA and 4-AP should be incorporated into separate liquid reagents in order to prevent slow generation of background color. It has also been found that the commercial sources of purified peroxidase so far investigated (Sigma Co., St. Louis, Missouri; Miles Laboratories, Elkhart, Indiana) contain impurities that destroy both apoenzyme and FAD-label activity. The amount of contaminant activity is also variable from lot-to-lot of the same grade and is present in significant quantity in even the most highly purified preparation of peroxidase. However, the FAD label is completely stable in the presence of peroxidase when 0.1 M phosphate buffer, pH 7.0, is present.

The above assay system may be used also for measurement of the activity of glucose oxidase solutions. Incorporation of 50 μl of a suitably diluted sample of glucose oxidase into 2 ml of the above assay reagents at

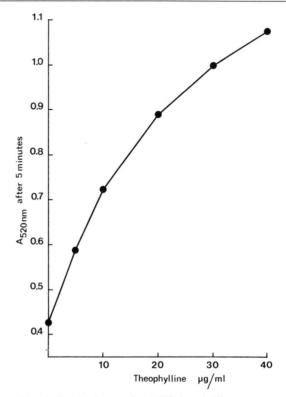

FIG. 3. Standard curve for ARIS theophylline assay.

25° will result in a linear reaction progress curve between 0 and 1.0 absorbance at 520 nm.

Immunoassay for Theophylline

In the case of a low molecular weight antigen, the assay can be performed by allowing the competitive binding reaction, reconstitution of glucose oxidase, and measurement of glucose oxidase activity to proceed simultaneously. In the simplest format the assay reagents are divided between two liquid reagents, which are mixed simultaneously with sample. The following reagent compositions and procedures can be used to perform the assay for theophylline in serum.

Apoenzyme reagent: 4 μM apoglucose oxidase; 4 mM 4-AP; antiserum to theophylline; 30% w/v glycerol; 0.1 M sodium phosphate, pH 7.0

Label reagent: 21 nM theophylline-FAD; 2.1 mM DHSA; 0.105 M glucose; 1.05% (w/v)BSA; 60 μg of peroxidase per milliliter; 0.1 M sodium phosphate, pH 7.0.

Assay Procedure. Dilute serum samples or serum standards 50-fold with 0.1 M sodium phosphate, pH 7.0. Aliquots of diluted sample (50 μl) are placed in disposable plastic cuvettes and mixed with label reagent equilibrated to 25°. Apoenzyme reagent (100 μl) is placed inside cuvette caps, and the reaction is initiated by inserting the cap onto the cuvette and mixing by inversion. The reaction is incubated for 5 min at 25°, after which the absorbance at 520 nm is recorded against label reagent as blank. Figure 3 shows a typical dose-response curve generated using the procedure. The performance characteristics of this assay procedure are shown in the table.

PERFORMANCE OF IMMUNOASSAY FOR
THEOPHYLLINE[a]

A. Precision of Measurement of Theophylline in
   Serum (μg/ml)

Serum concentration	Mean	S	CV (%)
Intra-assay ($n = 20$)			
6.0	6.1	0.4	6.6
12.0	11.8	0.4	3.4
30.0	29.9	0.3	1.0
Inter-assay ($n = 15$)			
2.8	2.8	0.4	13.4
15.4	15.8	1.0	6.4
25.4	32.3	1.7	5.4

B. Correlation of Results of ARIS with Those of
   EMIT. Regression Equation: $Y = A_1X + A_0$

$A_1 = 1.09$
$A_0 = -0.03$
$r = 0.98$
$n = 59$
SE $= 2.0$ μg/ml

[a] CV, coefficient of variation; SE, standard error; ARIS, apoenzyme reactivation immunoassay system.

Immunoassay for Human IgG

The following assay was designed to illustrate the applicability of the ARIS concept to the homogeneous immunoassay of proteins. The reagents and procedures described indicate how the method may be generally used to determine protein concentrations.

Reagents

Samples: Solutions of human IgG were prepared in 0.1 M sodium phosphate buffer, pH 7.0, containing 0.1% (w/v) sodium azide and 0.1% (w/v) bovine serum albumin.

Antiserum solution: Antiserum against human IgG (γ-chain specific), 8 mM DHSA, 0.4 M glucose, and 1.0% bovine serum albumin in 0.1 M sodium phosphate, pH 7.0

Label solution: FAD-IgG conjugate (4.0 nM with respect to FAD content), 0.8 mM 4-AP, 240 μg of peroxidase per milliliter, and 0.1% (w/v) bovine serum albumin in 0.1 M sodium phosphate, pH 7.0

Apoenzyme solution: Apoglucose oxidase (0.4 μM) and 1.0% (w/v) bovine serum albumin in 0.1 M sodium phosphate buffer, pH 7.0

Procedure. It was found that, unlike the procedure used in the ARIS

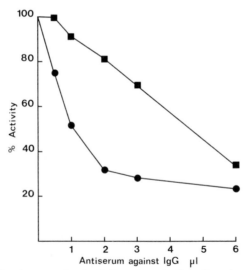

FIG. 4. Effect of antiserum to human IgG on the activation of apoglucose oxidase by IgG-FAD. ●——●, Inhibition in the absence of added IgG; ■——■, inhibition when 50 μl of a solution of 50 μg of IgG per milliliter in 0.1 M phosphate buffer was incorporated into the assay.

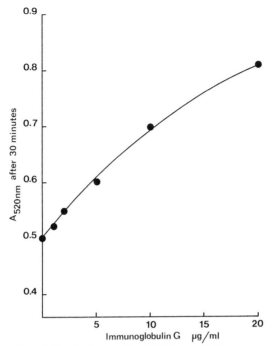

FIG. 5. Standard curve for ARIS human IgG assay.

assay for theophylline, no inhibition could be demonstrated when all the reagents were simultaneously combined. Instead, it was found to be essential to set up the competitive binding incubation with antiserum, FAD-labeled IgG and serum sample prior to addition of apoglucose oxidase. A second incubation is performed after addition of apoenzyme during which simultaneous generation of glucose oxidase and color development occur. Thus, 50 μl of sample, 0.5 ml of antiserum solution, and 0.5 ml of label solution are mixed, and the mixture is incubated for 20 min at room temperature. Apoenzyme solution (1.0 ml) is then added, and the assay is incubated for a further 30 min at 25°. The absorbance at 520 nm is then read against 0.1 M sodium phosphate buffer.

Figure 4 shows the inhibition that may be achieved when different amounts of antiserum against IgG are included in the assay and the reversal of inhibition produced by sample containing 50 μg of human IgG per milliliter. When 15 μl of antiserum was used, 60% inhibition was achieved and about 75% of this inhibition was relieved by the presence of the IgG. The dose response curve shown in Fig. 5 in which 1.5 μl of antiserum was used is representative of what may be achieved by this assay procedure.

[33] Bioluminescent Immunoassays

By Jon Wannlund and Marlene DeLuca

The development of the quantitative radioimmunoassay (RIA) by Yalow and Berson[1] was extremely important for diagnostic methodology in clinical medicine. Since that time RIAs have been reported for many diverse compounds. Many other nonisotopic labels have been used in immunoassays.[2–13] These include fluorescent labels, enzymes, and chemiluminescent labels.

This chapter describes a bioluminescent immunoassay (BIA) for TNT, DNP, and methotrexate. The principles of the assay are the same as for the RIA, differing only in that light rather than radioactivity is the end product. The BIA is comparable in sensitivity to most RIAs.

The procedure involves covalently linking the antigen to firefly luciferase in such a manner that most of the catalytic activity of the enzyme is retained. The luciferase–antigen is then used in a competitive binding assay with increasing amounts of free antigen and a constant amount of antibody. The amount of luciferase–antigen bound is inversely proportional to the amount of free antigen present.

Another procedure is described in which the antigen is linked to glucose-6-phosphate dehydrogenase and the enzyme antigen is used in the binding assay. The advantage of this method is an increased sensitivity due to the large turnover number of the enzyme. The NADH produced is measured with the bacterial luminescent enzymes. The procedure de-

[1] R. S. Yalow and S. A. Berson, *Nature (London)* **184,** 1648 (1959).

[2] J. Haimovich and M. Sela, *Science* **164,** 1279 (1969).

[3] R. K. Leute, E. F. Ullman, A. Goldstein, and L. A. Herzenberg, *Nature (London)* **236,** 93 (1972).

[4] E. Soni and I. Hemmila, *Clin. Chem.* **25,** 353 (1979).

[5] J. S. A. Simpson, A. K. Campbell, M. E. T. Ryall, and J. S. Woodhead, *Nature (London)* **279,** 646 (1979).

[6] S. J. Singer and A. F. Schick, *J. Biophys. Biochem. Cytol.* **9,** 519 (1961).

[7] S. L. Scharpe, W. M. Correman, W. J. Blomme, and G. M. Lackeman, *Clin. Chem.* **22,** (1976).

[8] G. B. Wisdom, *Clin. Chem.* **22,** 1243 (1976).

[9] A. H. A. M. Schuurs and B. K. van Weemen, *Clin. Chim. Acta* **21,** 1 (1977).

[10] G. Felman, P. Druet, J. Bignon, and S. Avrameas, "Immunoenzymatic Techniques." American Elsevier, New York, 1976.

[11] S. B. Pal, "Enzyme Labelled Immunoassay of Hormones and Drugs." de Gruyter, Berlin, 1978.

[12] E. Engvall and A. J. Pesce, *Scand. J. Immunol.* **8,** Suppl. 7 (1978).

[13] B. K. van Weemen and A. H. W. M. Schuurs, *FEBS Lett.* **15,** 232 (1971).

METHODS IN ENZYMOLOGY, VOL. 92

scribed is not a truly competitive binding assay, since the antibody is preincubated with the free antigen followed by the addition of the glucose-6-phosphate dehydrogenase–antigen. The sensitivity of this method for measuring TNT is 10 attomoles.

Procedures

Enzymes

Firefly luciferase is prepared and assayed as described previously.[14] Light intensity can be measured on any one of a number of commercially available instruments.[15] Luciferin can be purchased from Boehringer-Mannheim.

Glucose-6-phosphate dehydrogenase from *Leuconostoc mesenteroides* was purchased from Boehringer-Mannheim. The enzyme is assayed by measuring the amount of NADH produced after a 10-min incubation with NAD and glucose-6-phosphate.[16] The NADH is measured using an immobilized oxidoreductase and bacterial luciferase according to Reactions (1) and (2).

$$NADH + H^+ + FMN \rightleftharpoons NAD^+ + FMNH_2 \tag{1}$$
$$FMNH_2 + RCHO + O_2 \rightleftharpoons FMN + RCOOH + h\nu + H_2O \tag{2}$$

The amount of light obtained is directly proportional to the concentration of NADH when all other substrates are present at saturating concentrations. The NADH : FMN oxidoreductase is purified from *Beneckca harveyi* as described by Jablonski and DeLuca.[17] Luciferase is purified from the same bacteria according to the method of Gunsalus-Miguel *et al.*[18] as modified by Baldwin *et al.*[19]

Immobilization of the Oxidoreductase and Luciferase

One gram of cyanogen bromide-activated Sepharose 4B[20] is added to 3 ml of 0.1 *M* NaHCO₃, pH 8.0, containing per milliliter 5 mg of bacterial

[14] M. DeLuca and W. D. McElroy, this series, Vol. 57, p. 3.
[15] G. L. Picciolo, J. W. Deming, D. A. Nibley, and E. W. Chappelle, this series, Vol. 57, p. 550.
[16] L. A. Decker, "Worthington Enzyme Manual," p. 27. Worthington Biochemical Corporation, New Jersey, 1977.
[17] E. Jablonski and M. DeLuca, *Biochemistry* **16**, 2932 (1977).
[18] A. Gunsalus-Miguel, E. A. Meighen, M. Z. Nicoli, K. H. Nealson, and J. W. Hastings, *J. Biol. Chem.* **247**, 398 (1972).
[19] T. O. Baldwin, M. Z. Nicoli, J. E. Becvar, and J. W. Hastings, *J. Biol. Chem.* **250**, 2763 (1975).
[20] S. C. March, I. Parith, and P. Cuatrecasas, *Anal. Biochem.* **60**, 149 (1974).

luciferase, 25 μg of NADH:FMN oxidoreductase, and 5 mg of bovine serum albumin (BSA). The mixture is stirred in the cold for 16 hr. The Sepharose is then washed with 200 ml of cold 0.1 M sodium phosphate, 0.5 mM dithiothreitol (DTT) pH 7.0, followed by 500 ml of 1 M NaCl, 0.01 M sodium phosphate, 0.5 mM DTT, and finally with 300 ml of 0.1 M sodium phosphate, 0.5 mM DTT pH 7.0. The immobilized enzymes are stored at 4° in 0.1 M sodium phosphate, 0.5 mM DTT, 0.02% sodium azide. One gram of Sepharose–enzymes is suspended in 6 ml final volume of buffer.

Assay of NADH

The reaction mixture contains 500 μl of 0.1 M sodium phosphate, pH 7.0, 10 μl of saturated decanal (prepared freshly every 4 hr), 3×10^{-6} M FMN, and varying amounts of NADH. The reaction is initiated by adding 5 μl of the immobilized enzymes. After mixing, peak light intensity is measured. Light intensity is proportional to NADH concentration from approximately 1 pmol to 1000 pmol. A standard curve with known concentrations of NADH should be run with each preparation of Sepharose–enzymes.

Preparation of Antigen-Enzyme Conjugates

DNP and TNP. Firefly luciferase (1 mg/ml) is incubated with a 5-fold molar excess of 2,4,6-trinitrobenzene sulfonic acid (TNBS) or a 10-fold molar excess of 1-fluoro-2,4-dinitrobenzene (FDNB) in the presence of 2 mM ATP and 10 mM MgCl$_2$ in 0.1 M sodium phosphate, pH 7.8, for 2 hr at room temperature. The sample is then dialyzed against eight 1-liter volumes of 0.1 M sodium phosphate, pH 7.0, 1×10^{-3} M EDTA, and 1×10^{-4} M DTT. The amount of conjugation for DNP–luciferase is determined by the optical density at 340 nm, where DNP has an $A_m = 17,000$. For TNP–luciferase the optical density is measured at 343 nm, where the $A_m = 15,500$. The conjugates prepared in this way have ratios of luciferase to TNP or DNP of approximately 1:1; the enzyme-conjugate retains about 90% of the original catalytic activity.

Methotrexate.[21] Methotrexate, 500 nmol, is incubated with 3.5 μmol of 1-ethyl-3-(3-dimethylaminopropyl)carbodiimide-HCl (EDAC) in 0.3 ml of 0.1 M sodium phosphate, pH 7.0, for 15 min at 25°. Then 2 μmol of ATP and 10 nmol (1 mg) of firefly luciferase are added to give a final volume of 0.5 ml. The mixture is allowed to incubate at 4° for 16 hr. It is dialyzed

[21] J. Wannlund, J. Azari, L. Levine, and M. DeLuca, *Biochem. Biophys. Res. Commun.* **96,** 440 (1980).

against 0.1 M sodium phosphate, pH 7.0, and then passed over a Sephadex G-50 column (1 × 30 cm) to remove any free methotrexate. The amount of bound methotrexate is determined by the absorption at 302 nm, where methotrexate has an A_m of 22,000.[22] The amount of methotrexate linked to luciferase was from 1.3 to 2.0 mol per mole of enzyme.

Glucose-6-phosphate Dehydrogenase. Glucose-6-phosphate dehydrogenase, 100 μg, is incubated with a 10-fold molar excess of FDNB or TNBS in 0.1 M sodium pyrophosphate pH 9.0 for 2 hrs at room temperature. The glucose-6-phosphate dehydrogenase conjugate is then dialyzed against eight 1-liter volumes of 0.1 M sodium phosphate, pH 7.0. Approximately one TNP or DNP is bound per mole of enzyme, with 80–90% of the catalytic activity remaining in the conjugate.

Antibodies

The bioluminescent immunoassay procedure is totally adaptable to all antiserum. The DNP system has been developed using commercially available crude anti-DNP-BSA, obtained from Miles Laboratories. The methotrexate and TNT system use an IgG fraction of antiserum from goat. The goats are immunized with TNP–BSA or methotrexate–hemocyanin according to standard procedures.[23]

The TNP–BSA antigen is synthesized by adding equal volumes of 12 mg of BSA (Calbiochem Behring) per milliliter, 0.02 M sodium borate, pH 9.2, and 12 mg of TNBS per milliliter in H_2O. After incubation at 40° for 8 hr, the mixture is extensively dialyzed against 0.1 M sodium phosphate, pH 8.0. The reaction mixture is then passed through a Sephadex G-25 column to remove any remaining free TNP from the TNP–BSA. The degree of derivatization is determined by OD_{350}, where $A_m = 15,400$. The methotrexate–hemocyanin antigen is synthesized by adding 12.5 mg of methotrexate (obtained from Lederle Laboratories Division, Pearl River, New York), in 0.5 ml of H_2O and 25 mg of hemocyanin dissolved in 2.5 ml of H_2O at pH 8.0, followed by the addition of 25 mg of EDAC. The reaction mixture is incubated at room temperature for 4 hr with continuous stirring, after which the unreacted methotrexate is passed through Sephadex G-50 column.

Immobilization of Antibodies

Sepharose 4B is activated as described by March.[20] Sepharose 4B, 20 g, is suspended in 40 ml of 2 M K_2CO_3, and 2 g of CNBr is dissolved in 1

[22] L. Levine and E. Powers, *Res. Commun. Chem. Pathol. Pharmacol.* **9**, 543.

[23] C. A. Williams and M. W. Chase, eds., "Methods in Immunology and Immunochemistry," Vol. 1. Academic Press, New York, 1967.

ml of acetonitrile. The two are mixed and allowed to react for 5 min with stirring at 4°. The Sepharose is then washed with 1 liter of 0.1 M NaHCO$_3$; 30 mg of antibody are added to 1 g of activated Sepharose in 0.1 M NaHCO$_3$, pH 8.0. The mixture is stirred at 4° for 16 hr. The Sepharose is washed with 100 ml of 0.01 M sodium phosphate, 0.15 M NaCl, pH 7.4; it is then washed with 100 ml of 1 M acetic acid to remove any bound antigen, followed by 100 ml of 0.1 M sodium phosphate pH 7.4, and then with 300 ml of 0.01 M sodium phosphate, 0.15 M NaCl, pH 7.4. The washed Sepharose–antibody is suspended in 0.01 M sodium phosphate, 0.15 M NaCl, pH 7.4 in a ratio of 1 g of Sepharose in 10 ml of buffer. Of the added antibody, 80% is covalently bound to the Sepharose with this procedure.

For the procedure using the glucose-6-phosphate dehydrogenase-antigen, the antibody is immobilized to Sepharose CL-6B, since this material exhibits a much lower nonspecific binding of the enzyme.

Determination of the Binding Capacity of the Immobilized Antibody

Typically, 25 μl of Sepharose-antibody and 100 μl of underivatized Sepharose (1 g/10 ml) are incubated with increasing amounts of antigen–enzyme in 0.5-ml volumes for 3 hr at room temperature. The Sepharose is washed 10 times with 1 ml volumes of 0.1 M sodium phosphate–0.15 M NaCl to remove any unbound enzyme–antigen. The Sepharose is then suspended in buffer and assayed for enzymic activity. Once this stoichiometry has been determined, the amount of Sepharose–antibody to be used is chosen so that the enzyme–antigen bound is readily assayed. For luciferase–antigen this is between 1 and 2 pmol of total binding sites.

Competitive Binding Curves

A constant amount of Sepharose–antibody is incubated with enough luciferase–antigen to saturate approximately 90% of the sites and increasing amounts of free antigen, 0–10 pmol for DNP and TNT. The mixture is incubated for 3 hr at room temperature, followed by washing as described above, and the amount of bound enzyme is assayed. Figure 1 shows a typical competitive binding curve for TNT. The linear range of detection is from 0 to 20 pmol of TNT.

Glucose-6-phosphate Dehydrogenase-TNP

Twenty-five microliters of anti-TNT-Sepharose containing 2 femtomoles of binding sites for TNT, 25 μl of washed Sepharose, and 5–100 μl of TNT (0–100 attomoles) are incubated for 30 min at room temperature.

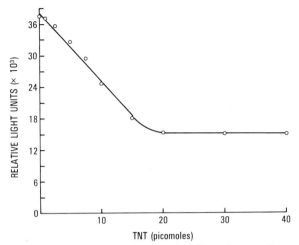

FIG. 1. Competitive binding curve of TNT with TNP–luciferase. Thirty picomoles of TNP–luciferase, 20 μl of Sepharose–antibody, and increasing amounts of free TNT are incubated as described. The Sepharose is washed and assayed for luciferase activity.

After this preincubation, 500 femtomoles of glucose-6-phosphate dehydrogenase–TNP are added. After 2 hr at 23° the Sepharose pellet is washed as described; 100 μl of 0.02 M NAD and 20 μl of 0.2 M glucose 6-phosphate are added to the Sepharose in a final volume of 1 ml of 0.1 M sodium phosphate, pH 7.2. This is allowed to react for 10 min at 23°, and

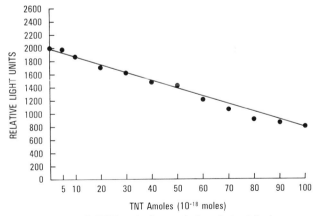

FIG. 2. Binding curve of TNT and glucose-6-phosphate dehydrogenase. Increasing amounts of free TNT were added to 25 μl of Sepharose–antibody. After 30 min 500 femtomoles of glucose-6-phosphate dehydrogenase–TNP was added and allowed to incubate for 2 hr. The amount of enzyme–antigen bound was assayed as described.

the amount of NADH present is determined with the immobilized ox-idoreductase and luciferase. Figure 2 shows typical results obtained with the TNP–glucose-6-phosphate dehydrogenase conjugate and increasing amounts of free TNT[24]; 10 attomoles can be reproducibly detected, and the assay is linear up to 100 attomoles. It is important to note that this is not a truly competitive binding assay, since the free TNT is preincubated with the Sepharose–antibody before the enzyme–TNP is added. This pre-incubation is essential in order to achieve this degree of sensitivity.

Summary

The procedures described here have been developed for only a few antigens at present. It seems very likely that they can be extended to any other antigen of choice. For sensitivities in the picomole range, the luci-ferase–antigen procedure is satisfactory. For increased sensitivity, the glucose-6-phosphate dehydrogenase–antigen is the method of choice. With some minor modifications the sensitivity of this assay can certainly be increased. The lower limits of detection will ultimately be determined by the affinity of the antibody for the antigen, not by the detection system.

Acknowledgments

This work was supported by a contract from the United States Army DAAK70-79-A-0019).

[24] J. Wannlund, H. Egghart, and M. DeLuca, in "Perspectives in Endocrinology and Clini-cal Chemistry" (M. Serio and M. Pazzagli, eds.), pp. 125–128. Raven, New York, 1982.

[34] Immunoassay by Electrochemical Techniques

By Kenneth R. Wehmeyer, Matthew J. Doyle,
H. Brian Halsall, and William R. Heineman

The predominant analytical immunoassay technique is radioimmu-noassay (RIA), in which a radioactive label is attached to an antigen. This method combines the selectivity of an antigen–antibody reaction with the low detection limits of isotopic counting. However, there has been an increasing interest in the development of nonisotopic immunoassay meth-odology. The areas investigated include the use of radicals detected by electron spin resonance,[1] enzymes detected by monitoring substrate turnover,[2] and fluorescent labels detected by fluorometry.[3]

The wide dynamic range and low detection limits of modern electroanalytical techniques, such as differential pulse polarography and stripping voltammetry,[4,5] make labeling the antigen with an electroactive group a potentially useful approach for immunoassay methodology. Two approaches to the development of an electrochemical immunoassay have been examined. The first involves labeling a steroid with a nitro group and detecting it polarographically in a homogeneous assay. The second involves labeling a protein with a metal redox center and detecting it voltammetrically in a heterogeneous assay.

Homogeneous Immunoassay Employing Differential Pulse Polarography

Principle

The procedure involves the determination of haptens employing the detection of an electroactive label by differential pulse polarography (DPP). As in a conventional competitive assay, the unlabeled hapten and the hapten labeled with an electroactive group compete for a limited number of antibody binding sites. The free labeled hapten present at equilibrium is determined by detection of the electrochemical reduction of the electroactive label at a dropping mercury electrode (DME). The cathodic current from the polarographic reduction of the label is proportional to the concentration of unlabeled hapten present. The separation of bound electroactively labeled hapten from free electroactively labeled hapten is not necessary, since the reduction of the antibody-bound labeled hapten is attenuated.

Estriol was chosen as a model compound to demonstrate the assay. Estriol is electroinactive in the potential range -200 mV to -1000 mV vs silver/silver chloride electrode (Ag/AgCl). Estriol labeled with nitro groups in the 2 and 4 positions is electroactive, giving two reduction waves at -422 mV and -481 mV vs Ag/AgCl and was used as the labeled hapten.

[1] R. K. Leute, E. F. Ullman, A. Goldstein, and L. A. Herzenberg, *Nature (London), New Biol.* **236**, 93 (1972).
[2] G. B. Wisdom, *Clin. Chem.* **22**, 1243 (1976).
[3] C. M. O'Donnell and S. C. Suffin, *Anal. Chem.* **51** (1), 33A (1979).
[4] J. B. Flato, *Anal. Chem.* **44** (11), 75A (1972).
[5] T. R. Copeland and R. K. Skogerboe, *Anal. Chem.* **46** (14), 1275A (1974).

Materials

Equipment. Differential pulse polarography was performed using a PARC Model 174A Polarographic Analyzer (Princeton, New Jersey) with a Houston 2000 x-y recorder (Bellaire, Texas), PARC Model 303 static mercury drop electrode, silver–silver chloride (Ag/AgCl) reference electrode and platinum auxiliary electrode. All polarography was performed under the following scan conditions: range of scan, -200 to -1700 mV vs Ag/AgCl; pulse amplitude, 25 mV; scan rate, 10 mV/sec; drop time, 1.0 sec. Solutions were purged for 5–15 min with nitrogen (passed through vanadous chloride deoxygenating towers) and then blanketed with nitrogen while recording polarograms. All solutions containing protein were deoxygenated with a slow N_2 flow rate to avoid denaturation of the protein. Polarograms were repeated three times to obtain average values. The solution volumes used in this study were on the order of 5.0 ml; however, volumes smaller than 1.0 ml can be handled routinely.

Chemicals. Estriol, estradiol, and bovine IgG were of the highest available purity from Sigma Chemical Company. Estrogen specific monoclonal antiserum, amplified in mice as ascites tumors, was a gift from New England Nuclear. The IgG fraction was isolated by use of a Sephacryl 200 (Pharmacia) column using 0.1 *M* potassium phosphate buffer (pH 7.4) as the eluent. The isolated IgG fraction was dialyzed overnight against 0.01 *M* phosphate buffer (pH 7.4) and lyophilized. The lyophilized IgG fraction was reconstituted with deionized water, and this solution was used as the antiserum in this study.

Supporting Electrolyte. All polarograms were recorded in a supporting electrolyte of 0.1 *M* potassium phosphate buffer, pH 7.4. The analysis buffer was purged with N_2 for 1 hr and then pre-electrolyzed over a mercury pool electrode held at -1.3 V vs saturated calomel electrode for at least 48 hr to remove trace metal impurities.

Labeling of Estriol. Estriol was labeled in the 2 and 4 positions by the procedure of Könyves and Olsson.[6] The desired product was purified by silica gel-60 column chromatography, eluting with acetone : chloroform (1 : 1). The 2,4-dinitroestriol (DNE) melted at 225–227°. The DNE was dissolved in 50% ethanol–water (v/v) solution and used as a standard solution.

Procedures and Discussion

Polarography of Phosphate Buffer, Estriol, Antibody, and Dinitroestriol Solutions. Polarograms were recorded of solutions of estrogen-specific antibody, estriol, and phosphate buffer to determine the electro-

---

[6] I. Könyves and A. Olsson, *Acta Chem. Scand.* **18,** 483 (1964).

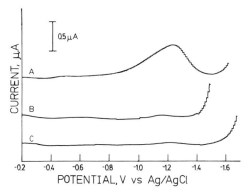

FIG. 1. Differential pulse polarograms of (A) 0.1 M phosphate buffer solution containing 4.6 × 10⁻⁵ M estriol; (B) a 5.0-ml phosphate buffer solution plus 0.10 ml of specific estrogen antibody; and (C) 0.1 M phosphate buffer solution. Taken from K. R. Wehmeyer, H. B. Halsall, and W. R. Heineman, *Clin. Chem.* **28,** 1968 (1982), with permission.

chemical potential window available (Fig. 1). All three solutions were electroinactive in the potential range −200 to −1000 mV vs Ag/AgCl. Thus, a relatively large potential window was available for the observation of the labeled material without interference from these assay components.

A polarogram of DNE was recorded to determine the reduction peak potentials of the electroactive nitro groups. The DNE was electroactive with two distinct reduction waves with peak potentials of −422 mV and −481 mV vs Ag/AgCl (Fig. 2). The peak current was linear as a function of concentration over the range of 60 ng/ml to 3.7 μg/ml.

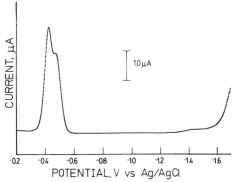

FIG. 2. Differential pulse polarogram of a 0.1 M phosphate buffer solution containing 7.3 × 10⁻⁶ M dinitroestriol, recorded with a 2 mV/sec potential scan rate. Taken from K. R. Wehmeyer, H. R. Halsall, and W. R. Heineman, *Clin. Chem.* **28,** 1968 (1982), with permission.

Interaction of Estrogen Antisera and DNE. The effect of antibody on the reduction of DNE was demonstrated by recording a polarogram of a DNE solution. Successive polarograms were then recorded for the DNE solution following the addition of successive aliquots of the estrogen antibody. The solutions were deoxygenated slowly for 12 min after each aliquot added and before the polarograms were recorded. The resulting plot of peak current vs microliters of antibody added is shown in Fig. 3. The binding of DNE by antibody was demonstrated by the sequential decrease in the reduction peak current with successive aliquots of antisera. The decrease in peak current most likely arises from a sequestering of the electroactive nitro groups from the electrode surface upon binding or from a decrease in the diffusion coefficient of the DNE when bound to the antibody, or both.

Assay of Estriol. In order for an antigen–antibody reaction to serve as a functional analytical method in the competitive mode, it is essential that bound labeled hapten be reversibly displaced from the antibody by unlabeled hapten. A solution containing DNE plus 160 μl of estrogen-specific antibody was slowly deoxygenated for 12 min, and a polarogram was recorded. Successive aliquots of unlabeled estriol were added to the above DNE-Ab solution. After the addition of each aliquot, the solution was deoxygenated slowly for 5 min, and a polarogram was recorded. A plot of peak current vs estriol concentration for successive polarograms is shown in Fig. 4.

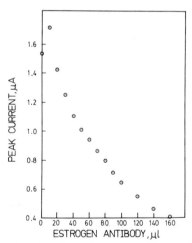

FIG. 3. A plot of the differential pulse polarogram peak currents for a 4.6-ml phosphate buffer solution containing 7.7×10^{-6} *M* dinitroestriol vs microliters of estrogen-specific antibody added to the dinitroestriol solution. Taken from K. R. Wehmeyer, H. B. Halsall, and W. R. Heineman, *Clin. Chem.* **28,** 1968 (1982), with permission.

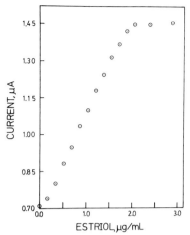

FIG. 4. Plot of differential pulse polarogram peak currents of a 5.5 ml phosphate buffer solution containing dinitroestriol (5.6 μmol/liter) plus 0.15 ml of estrogen-specific antibody vs estriol (μg/ml) added to the solution. Taken from K. R. Wehmeyer, H. B. Halsall, and W. R. Heineman, *Clin. Chem.* **28**, 1968 (1982), with permission.

The reversible displacement of DNE from antibody binding sites by unlabeled estriol is demonstrated by the sequential increase in peak current with successive aliquots of unlabeled estriol. The displacement increases the concentration of free DNE capable of being reduced and correspondingly decreases the concentration of bound DNE. The quantitative detection of this reversible displacement would be the basis of hapten immunoassays by DPP.

Heterogeneous Immunoassay Using Differential Pulse Anodic Stripping
    Voltammetry (DPASV)

Principle

A voltammetric immunoassay (VIA) for human serum albumin (HSA) has also been developed. HSA was chosen as a model antigen for these studies because of its large size (MW 68,460) and biochemical resilience. The larger the macromolecule, the smaller the percentage of antigenic determinants that will be lost upon superficial modification. A bifunctional chelating agent can be covalently coupled to HSA and serve as a site-specific chelon for metal ions. The metal of choice should be normally absent in biological fluids and detectable in an interference-free manner. The metal label can be released from the protein complex and detected electrochemically after a competitive immunoassay.

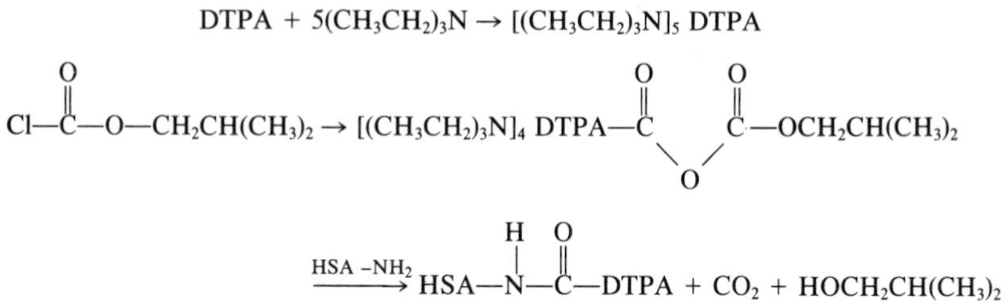

$$DTPA + 5(CH_3CH_2)_3N \rightarrow [(CH_3CH_2)_3N]_5 \; DTPA$$

$$Cl-\overset{\overset{\displaystyle O}{\|}}{C}-O-CH_2CH(CH_3)_2 \rightarrow [(CH_3CH_2)_3N]_4 \; DTPA-\overset{\overset{\displaystyle O}{\|}}{C}\underset{O}{\diagdown \; \diagup}\overset{\overset{\displaystyle O}{\|}}{C}-OCH_2CH(CH_3)_2$$

$$\xrightarrow{HSA-NH_2} HSA-\overset{\overset{\displaystyle H}{|}}{N}-\overset{\overset{\displaystyle O}{\|}}{C}-DTPA + CO_2 + HOCH_2CH(CH_3)_2$$

SCHEME 1. Synthesis of HSA-DTPA.

Labeling Procedure

Diethylenetriamine pentaacetic acid (DTPA) is a bifunctional chelating agent that can be covalently linked to the α and ε amine residues of macromolecules by the formation of an amide bond. A modification of the four-step mixed anhydride procedure of Krejcarek and Tucker[7] was employed. The reaction sequence utilizes reagents common to peptide synthesis and is outlined in Scheme 1.

Buffer and saline solutions were pre-electrolyzed for 24 hr at -1100 mV. Acetonitrile was redistilled and stored over molecular sieves. DTPA (50 mg. Aldrich), H_2O (1.5 ml, deionized), and triethylamine (0.344 ml, Aldrich) were mixed in a lyophilization vessel with gentle hand heating until the DTPA dissolved. The mixture was then lyophilized (4 hr) to a clear glassy residue. The residue was resuspended in 2–4 ml of acetonitrile and cooled in an ice bath for 15 min. The remaining steps were performed at approximately 5°. Isobutylchloroformate (0.056 ml, Sigma) was added, and the solution was stirred for 30 min, during which time a white precipitate of triethylamine hydrochloride formed. This mixture was filtered directly into a cool, stirred HSA solution (0.5 g in 40 ml of 0.1 M NaHCO$_3$, pH 7.57; Sigma). The solution was stirred at 5° for 12 hr and the pH was maintained between 7 and 8. The amount of DTPA coupled to the HSA was kept low so as not to destroy the immunogenicity of the modified protein.

The solution was then dialyzed (12,000 MWCO tubing; Spectrapore) twice against 1 liter of H_2O for 10 hr and once against 1 liter of 0.15 M NaCl (pH 6.70) for 10 hr. Dialysis was followed by ultrafiltration (PM-10 membrane; Diaflo) with 800 ml of 0.15 M NaCl and 500 ml of 0.10 M citrate buffer (pH 2.5) to a final cell volume of 40 ml. This solution may be lyophilized and stored for later use or labeled directly.

---

[7] G. E. Krejcarek and K. L. Tucker, *Biochem. Biophys. Res. Commun.* **77**, 581 (1977).

Indium (In^{3+}) was chosen as the metal ion label because of its high stability constant with DTPA ($K_f = 10^{29}$),[8] its relative absence in biological tissues and fluids, and its very characteristic electrochemical redox behavior. A 17-fold molar excess of In^{3+} was added to the cooled HSA–DTPA (HD) solution, and the mixture was stirred for 35 min in the 0.1 M citrate buffer (pH 2.5). The pH was adjusted to pH 4.5 with 1.0 N NaOH, and the solution was ultrafiltered with 200 ml of 0.15 M NaCl (pH 6.4) and 400 ml of 0.1 M citrate buffer (pH 5.5), then dialyzed once against 1 liter of 0.1 M citrate buffer (pH 5.5) to remove any unbound In^{3+}.

Characterization of the Labeled Complex

The resulting HSA–DTPA–In^{3+} (HDI) complex was examined chromatographically, electrophoretically, immunologically, and spectrophotometrically. Elution of HDI on a Sephacryl-200 column with a PBS–EDTA buffer indicated that the HDI preparation was relatively monodisperse. The electrophoretic mobility of 2-mercaptoethanol-treated HDI was greater than that of native HSA, which is readily accounted for by the greater number of negative charges associated with the DTPA residues of the HDI complex. HDI was shown to retain immunospecificity as a homogeneous precipitin band was obtained by Ouchterlony radial immunodiffusion against anti-HSA. Changes in the average association constant between anti-HSA and HDI remain to be determined.

The molar ratio of In^{3+}: HSA in the HDI complex was determined by plasma emission spectroscopy. A known amount of HDI was nebulized into an argon plasma, and a computer-controlled slew scanning sweep of the 3039.36 Å atomic emmission line of indium was made. The concentration of In^{3+} was determined from calibration curves of standards in similar matrices, and the ratio (In^{3+}: HDI) was calculated. Typically, the In^{3+} content approximated 3–7 mol per mole of HDI. The presence of multiple labels on a single antigen generates an analyte with an inherent amplification factor.

Electrochemical Assessment

All electrochemical studies employed the instrumentation described in the polarography section above. DPP analysis utilized a medium mercury drop (volume = 0.186 μl) with a 1-sec drop time. The potential region between 0.0 V and -1.5 V was scanned negatively at a rate of 10 mV/sec

---

[8] A. E. Martell and R. M. Smith, "Critical Stability Constants," Vol. 1, p. 281. Plenum, New York, 1977.

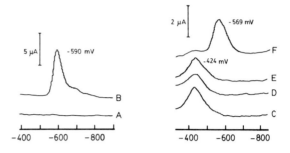

POTENTIAL (mV) vs Ag/AgCl

FIG. 5. Differential pulse polarograms of (A) 0.1 M citrate, pH 5.50, supporting electrolyte; (B) 1.0×10^{-4} M InCl$_3$ in citrate; (C) 1.50×10^{-5} M human serum albumin (HSA); (D) 1.40×10^{-5} M HSA-diethylenetriaminepentaacetic acid (DTPA); (E) 1.08×10^{-5} M HDI (HSA–DTPA–In^{3+} complex); and (F) 1.08×10^{-5} M HDI after acidification with 1.0 M HCl to pH 3.80. Taken from Doyle et al.,[12] with permission.

with a 25 mV (pp) modulation amplitude. Differential pulse anodic stripping voltammetric (DPASV) analysis utilized a large hanging mercury drop (volume = 0.372 μl). A deposition time of 5 min, at a potential of −800 mV, was utilized to preconcentrate the indium. The potential range between −800 mV and 0.0 V was scanned positively at a 2 mV/sec scan rate with a 25 mV (pp) modulation amplitude and a 0.5 sec clock time.

The 0.1 M citrate (pH 5.50) supporting electrolyte was shown to be essentially electroinactive in the potential region scanned (Fig. 5A). Reduction of In^{3+} occurs as a 3-electron process resulting in a pronounced and characteristic cathodic wave at −590 mV (Fig. 5B). HSA, HD, and HDI exhibit reduction waves at approximately −424 mV due to the reduction of surface disulfides of which albumin has seven[9] (Fig. 5C–E). After acidification of the HDI complex, to pH 3.8, a reduction wave appeared at −569 mV characteristic of free In^{3+} (Fig. 5F). No wave at −569 mV appeared when either HSA or HD alone were acidified. The detection limit for the HDI complex by DPP is approximately 1.0 μg/ml (Fig. 6). The response was linear at lower HDI concentrations; however, current attenuation occurred at higher concentrations due to protein adsorption on the electrode surface. Therefore a protocol was developed that involved separating the protein from the analytical solution prior to electrochemical investigation.

[9] M. T. Stankovich and A. J. Bard, J. Electroanal. Chem. Interfacial Electrochem. **86**, 189 (1978).

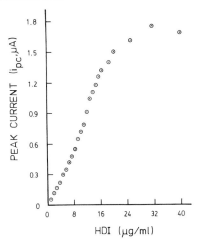

FIG. 6. A plot of the cathodic peak current (μA, for the -490 mV indium wave) vs the total amount of HDI complex (μg/ml) added to 6.0 ml of 0.1 M citrate buffer acidified to pH 1.00. Taken from Doyle et al.,[12] with permission.

Separation and Assay Procedures

Protein A from *Staphylococcus aureus* (SPA, 2 ml, K_b = 1.47 mg/ml, Enzyme Center) reacted with rabbit IgG specific for HSA (2.87 mg; Sigma) has been utilized as an immunoadsorbent for HDI and HSA in a competitive immunoassay. The use of SPA as an immunoadsorbent has been previously discussed in detail[10,11] and will not be addressed further here.

The voltammetric immunoassay protocol is outlined in Fig. 7. Briefly, excess HDI and a limited amount of SPA–Ab were added to a series of HSA (in 0.1 M citrate, pH 7.50) standard solutions, mixed in 12 × 75 mm polystyrene serological tubes, and incubated at 37° for 1 hr. The tubes were centrifuged for 20,000 g/min, and the supernatant was separated and discarded. This supernatant contained the excess protein, and only SPA–Ab-bound antigen remained with the tubes. The pellets were rinsed with 0.1 M citrate (pH 7.50) and resuspended in 0.1 M citrate (pH 1.50), which effected the release of In^{3+} from bound HDI complexes. The tubes were centrifuged a second time for 30,000 g/min, causing the SPA immune complexes to pellet. The supernatant, containing released In^{3+}, was separated, and levels of In^{3+} were determined by DPASV.[12]

[10] S. W. Kessler, *J. Immunol.* **117**, 1482 (1976).
[11] J. M. MacSween and S. L. Eastwood, this series, Vol. 73, p. 245.
[12] M. J. Doyle, H. B. Halsall, and W. R. Heineman, *Anal. Chem.* **54**, 2318 (1982).

VIA PROTOCOL

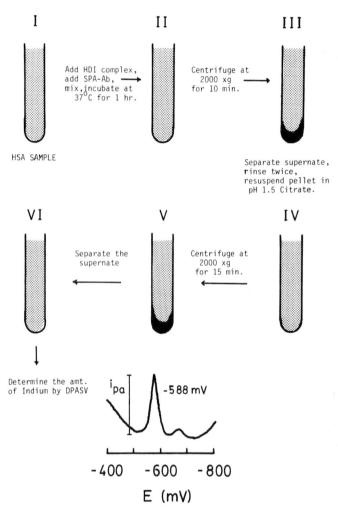

FIG. 7. An outline of the voltammetric immunoassay (VIA) protocol using the SPA–Ab complex as an immunoadsorbent for labeled and unlabeled antigen. DPASV, differential pulse anodic stripping voltammetric analysis. Taken from Doyle et al.,[12] with permission.

Any residual protein remaining in the supernatant would be of such low concentration that attenuation of current would not be a problem. Additionally, the total protein concentration of each tube is identical at this point, only the ratios of bound HSA:HDI differ. Thus, any depression in current response due to adsorption would be equally reflected through-

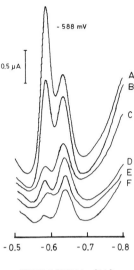

FIG. 8. Differential pulse anodic stripping voltammograms of a typical series of human serum albumin (HSA) standard solutions of increasing concentrations (A–F) following the voltammetric immunoassay procedure. Note that the amount of In^{3+} (−588 mV peak), originating from the HDI complex, decreases as the concentration of unlabeled antigen (HSA) increases. Taken from Doyle et al.,[12] with permission.

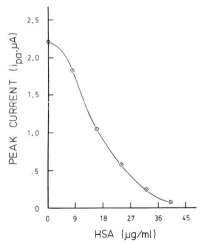

FIG. 9. A plot of the anodic peak current (μA, for the −588 mV indium wave) vs the concentration of human serum albumin (HSA) (μg/ml) in a series of standard solutions. This plot is a typical voltammetric immunoassay standard curve obtained for the model HSA system. Taken from Doyle et al.,[12] with permission.

out the assay. Other approaches designed to remove interfering protein complexes, e.g., reverse-phase chromatography, may also be adapted.

A characteristic anodic wave for released indium appears at -588 mV. The wave at -633 mV also appears in blank sample analysis and is believed to be due to endogenous cadmium (Fig. 8). Since HSA will effectively displace HDI bound to the SPA–Ab adsorbent phase, decreasing amounts of released In^{3+} will be observed as the amount of HSA in the original analytical solution increases. Consequently, a progressive decrease in anodic peak current (i_{p_a}) resulted as HSA standards of increasing concentration were examined (Fig. 8A–F). A plot of peak current (-588 mV wave) vs HSA concentration generated a smooth standard curve, which can be utilized for unknown sample analysis (Fig. 9). A separate standard curve is required for HDI preparations containing different molar ratios of In^{3+} to HSA.

Conclusion

Nonisotopic competitive immunoassay procedures have been investigated employing the electrochemical detection of a labeled hapten by DPP and a metal chelate-labeled antigen by DPASV. In the former case a homogeneous assay could be performed, since the reduction of the bound labeled hapten was attenuated. The detection limit by DPP of the labeled hapten used is limited to the 10^{-6} M to 10^{-7} M range, and the technique's applicability would be restricted to compounds of clinical interest that are present at levels in this range. The metal chelate-labeled antigen approach requires the separation of bound from free labeled antigen but has a much lower limit of detection (10^{-8} M to 10^{-10} M) and would allow the determination of a larger number of compounds of clinical interest.

In summary, model systems employing estriol and HSA have been used to demonstrate the feasibility of using an electrochemically labeled hapten and an electrochemically labeled antigen detected by DPP and DPASV, respectively, as methods of monitoring the competitive and reversible binding of labeled and unlabeled hapten or antigen by a specific antibody.

Acknowledgments

The generous provision of antisera by Dr. Richard Patterson of New England Nuclear is gratefully acknowledged. The authors would also like to acknowledge the Contributions of Drs. C. William Anderson, Julius P. Zodda, and Claude F. Meares. M. J. D. wishes to thank Dr. Michael J. Welch for helpful suggestions concerning the synthesis of the DPTA derivatives. K. R. W. acknowledges support by the University of Cincinnati Research Council summer fellowship. This work was supported by NIH Grant AI-16753 and NSF Grant CHE 79-11872.

[35] Metalloimmunoassay: Principles and Practice

By Michael Cais

The last decade has produced an ever-growing research activity in understanding the role of metal ions in biological systems. Although the human body is only about 3% metals, there is increasing evidence that life processes depend upon these elements far more than this figure would seem to indicate. A particularly strong interest was generated in metallo-biochemical interactions in 1969, when Rosenberg and co-workers reported[1] that platinum coordination complexes exhibited potent antitumor activity. That report engendered a vast amount of research activity aimed at investigating the anticancer properties of platinum complexes. To date, nearly 1000 publications have appeared in the literature on this topic, and the antineoplastic activity of platinum complexes has been demonstrated in many animal tumor-screening systems.[2] In phase II clinical trials, *cis*-dichlorodiammineplatinum(II) in combination with other chemotherapeutic agents has been used with remarkable beneficial results, in particular in the treatment of testicular cancers and of ovarian carcinoma, and this compound has already been released by the Food and Drug Administration for clinical use as an anticancer drug.

A major problem in the application of platinum compounds to antitumor therapy is the apparent lack of target specificity when the platinum compounds are introduced into living systems. It is well known that both biological activity and side effects of drugs can be correlated to their distribution, retention, and excretion in the various organs of animals and humans. The distribution of *cis*-dichlorodiammineplatinum(II) labeled with ^{193m}Pt and ^{195m}Pt has been evaluated in mice, rabbits, and humans.[3-5] Although the rate of clearance in tumor-bearing animals was significantly higher than in control animals, the organ distribution in mice, rats, and rabbits with tumors was similar to that in the control animals. After 40 hr, organs containing most of the radioactivity were the kidney, liver, and the intestine. In all the work carried out so far, no clear distinction has emerged between effects on tumor cells and on normal cells or preferred

[1] B. Rosenberg, L. Van Camp, J. E. Trosko, and V. H. Mansour, *Nature (London)* **222**, 385 (1969).
[2] F. K. V. Leh and W. Wolf, *J. Pharm. Sci.* **65**, 315 (1976).
[3] R. C. Lange, R. P. Spencer, and H. C. Harder, *J. Nucl. Med.* **13**, 328 (1972).
[4] R. C. Lange, R. P. Spencer, and H. C. Harder, *J. Nucl. Med.* **14**, 191 (1973).
[5] P. H. S. Smith and D. M. Taylor, *J. Nucl. Med.* **15**, 349 (1974).

METHODS IN ENZYMOLOGY, VOL. 92

concentration of platinum in tumor tissue versus normal tissue. This lack of selective toxicity of platinum compounds in tumor therapy became in 1973 the subject of a research project in our laboratory, where we had already established some expertise in the field of organometallic and coordination chemistry. In a detailed discussion of the principles of selectivity, Albert[6] emphasized the complexity of the problems involved in the design and/or choice of specific agents. In general, selective agents exert their effects through one or more of the following three principles[6]: (a) selectivity through distribution, which permits use of an agent toxic to both economic and uneconomic cells, provided that it is accumulated exclusively (or very nearly so) by the latter; (b) selectivity through comparative biochemistry, whereby the toxic agents injure a biochemical system that is important only for the uneconomic species; (c) selectivity through comparative cytology, in which the toxic agents interact exclusively with a cell structure that exists only in the uneconomic species. It appeared to us that these principles might be applied to our problem by directing attention to the three major biological systems of high specificity of interaction: antibody–antigen; receptor–hormone; and enzyme–substrate. The concept of targeting a reagent by attaching it to a carrier, which will be recognized selectively by the target site, has formed the basis of much research, in particular with the receptor–hormone system.[7] The studies initiated with the antibody–antigen system resulted in the development of a novel concept for the use of metal atoms in the form of their organometallic and/or coordination complexes as nonradioisotopic labeling agents in biological systems in general and in immunochemical applications in particular. The new method[8] was designated metalloimmunoassay (MIA) by analogy to the well-known procedure of radioimmunoassay (RIA).

General Principle

The general principles of MIA and RIA are the same. In both systems, the basic immunological reaction is that of antibodies (ab) specifically recognizing and binding antigens (ag) to form a strongly bound anti-

[6] A. Albert, "The Selectivity of Drugs." Chapman & Hall, London, 1975.
[7] J. A. Katzenellenbogen, D. F. Heiman, K. E. Carlson, D. W. Payne, and J. E. Lloyd, in "Cytotoxic Estrogens in Hormone Receptive Tumors" (J. Raus, H. Martens, and G. Leclercq, eds.), p. 3. Academic Press, New York, 1980.
[8] M. Cais, S. Dani, Y. Eden, O. Gandolfi, M. Horn, E. E. Isaacs, Y. Josephi, Y. Saar, E. Slovin, and L. Snarsky, *Nature (London)* **270**, 534 (1977).

body–antigen complex (ab · ag). This reaction follows the law of mass action and is described by the equilibrium in Eq. (1)

$$ab + ag \underset{k_d}{\overset{k_a}{\rightleftharpoons}} ab \cdot ag \tag{1}$$

$$ab + ag\text{-M} \underset{k_d'}{\overset{k_a'}{\rightleftharpoons}} ab \cdot ag\text{-M} \tag{2}$$

If the antigen, ag, is replaced by a metal-labeled antigen (metalloantigen), ag-M, it is possible that a similar equilibrium, Eq. (2), will be established provided the metal labeling is carried out in such fashion that there will be as little interference as possible with the antibody–antigen recognition process. In other words, the binding constant (or affinity constant) K for the complex ab · ag should be the same (or very nearly so) as the binding constant K' for the complex ab · ag-M. Under such conditions, if mixtures of variable amounts of antigen [ag] and a constant amount of metalloantigen [ag-M] are allowed to compete for a limited and constant concentration of antibody-binding sites, ab, the reaction mixture, upon equilibrium, will consist of a "free" antigen fraction (unbound ag and ag-M) and a "bound" fraction of antibody–antigen complexes (ab · ag and ab · ag-M). After separation of the two fractions, the amount of metal present in either the "bound" or "free" fraction (or both) can be determined by suitable methods for the detection and quantitative analysis of trace metals. Preparation of a calibration curve plotted for standardized amount of metalloantigen and unlabeled antigen provides the means for determining the concentration of the analyzed substance in unknown samples, by interpolation from the standard curve.

The four major steps in the development of a desired metalloimmunoassay system are (a) production of specific antibodies; (b) synthesis of metalloantigens; (c) selection of techniques for separating the "free" unbound antigens from the antibody–antigen complexes; and (d) choice of analytical method to determine the metal concentration.

A discussion of step a, production of specific antibodies, is beyond the scope of this chapter and the reader is referred to the many literature sources available,[9] including this volume. Step c, a novel system for the separation of "free" and "bound" fractions, which was a spin-off from our metalloimmunoassay studies, is discussed in this volume.[10] The remaining discussion will be limited to steps b and d.

[9] G. E. Abraham, "Handbook of Radioimmunoassay." Decker, New York, 1977; see also references therein.
[10] M. Cais, this volume [25].

Synthesis of Metallohaptens and Metalloantigens

The most important component of MIA, from the novelty aspect of the concept, is the metal-labeling of analyzable haptens and antigens. (The terms metallohapten and metalloantigen will be used interchangeably in the remaining text.)

The considerable research activity in the field of coordination and organometallic chemistry during the past two decades has produced a wealth of information that is pertinent to the subject in hand and readily available in the scientific literature. It is now an established fact that every metal element in the periodic table can be made to react with suitable organic ligands to form either coordination complexes or organometallic derivatives. It is beyond the scope of this report to enter into detailed descriptions of the various types of metallo derivatives that are available, and we shall limit ourselves to presenting some general definitions and concepts pertaining to the topic under discussion. We shall refer to two general classes of compounds that incorporate metal atoms in their molecular structure: (a) compounds in which there is a direct bond between carbon atom(s) of the organic moiety and the metal atom are defined as organometallic compounds, (b) compounds in which the bond between the metal atom and the organic moiety is through a heteroatom present in the ligand are classified as coordination complexes. This is an arbitrary definition, and it should be pointed out that organometallic compounds are a special subgroup of the more general class of coordination complexes. Both types of compounds, organometallic and coordination complexes, are suitable for the synthesis of metallohaptens. The organic moiety of the complex can be any organic compound provided it incorporates a suitable functional group with which one can form a bond to the metal atom. This adds a high degree of versatility to the type of metallohaptens that can be synthesized. The main requirements for an ideal metalloantigen are (a) ready availability in highly purified and stable form at reasonable cost; (b) solubility in aqueous media; (c) that the metal in the label should not occur in significant concentrations in biological fluids; (d) prolonged shelf life, (e) no health hazards; (f) amenability to detection by suitable low-cost, easy-to-operate analytical instruments, preferably of existing technology; (g) high potential sensitivity in assay.

Two general strategies for the synthesis of metallohaptens can be envisaged. One approach would be to introduce a metal atom (or atoms) directly into the hapten, if the structure of the latter is suitable for reaction with metals. In the other approach, one would plan the synthesis of a functionalized metal-containing reagent that could react with the hapten to produce the desired metallohapten. Selected examples will be pre-

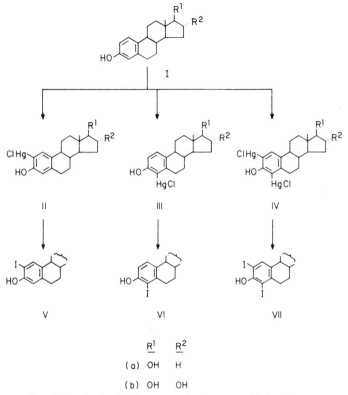

Fig. 1. Synthesis of chloromercuriestrogens and iodoestrogens.

sented to illustrate these two avenues. Additional examples have been published.[11–14]

Synthesis of Mercury-Labeled Estrogens

For an example of the direct introduction of metal atoms into the molecule of the hapten, we investigated the mercuration of the aromatic ring of estradiol and estriol.[12] We were able to obtain the 2-chloromercuri, 4-chloromercuri, and 2,4-(bischloromercuri) derivatives of estradiol and estriol as shown in Fig. 1.

[11] M. Cais, E. Slovin, and L. Snarsky, *J. Organomet. Chem.* **160,** 223 (1978).

[12] M. Cais, *Actual. Chim.* No. 7, 14 (1979).

[13] M. Cais and N. Tirosh, *Bull. Soc. Chim. Belg.* **90,** 27 (1981).

[14] O. Gandolfi, M. Cais, G. Dolcetti, M. Ghedini, and A. Modiano, *Inorg. Chim. Acta* **56,** 127 (1981).

The three chloromercuri derivatives (**II–IV**) could be characterized by their typical nuclear magnetic resonance (NMR) spectrum, in particular that of the aromatic ring protons, which very clearly indicates the position and degree of substitution.[15] A typical reaction procedure is given for the mercuration of estradiol.

Mercuration of Estradiol

A mixture of estradiol (Merck) (1.0 g, 3.67 mmol) and mercuric acetate, $Hg(OAc)_2$ (1.1 g, 3.45 mmol) in methanol (20 ml) is refluxed for 2.5 hr and then stirred at room temperature for 20 hr. The white solid in the reaction mixture is collected by filtration and dissolved in 10 ml of a mixture of methanol–chloroform–dichloromethane (1 : 1 : 1). To this is added a solution of LiCl (400 mg) in methanol (3 ml), and the mixture is stirred at room temperature for 2 hr. The solvent is evaporated under vacuum, and the residue is dissolved in ethyl acetate (25 ml). The solution is extracted with water (3 × 50 ml) to remove inorganic salts, then dried over $MgSO_4$, and the solvent is removed. The resulting white solid product (0.6 g) is a mixture of chloromercuriestradiol derivatives and some unreacted estradiol. This solid product (0.6 g) is refluxed for 2 hr in a mixture of acetic anhydride (3 ml) and dry pyridine (15 ml). After cooling to room temperature, the reaction mixture is added, with stirring, to 200 ml of ice-cold water. The resulting white precipitate is collected by filtration and dried to yield 0.59 of a glassy residue. This is chromatographed on a column of silica (deactivated with 15% water) prepared in benzene. Elution with benzene yields the following pure compounds (in order of elution): 3,17-estradiol diacetate (88 mg); 4-chloromercuri-3,17-estradiol diacetate (190 mg); 2-chloromercuri-3,17-estradiol diacetate (180 mg); 2,4-bischloromercuri-3,17-estradiol diacetate (~20 mg). Hydrolysis of each one of the separated chloromercuriestradiol diacetates produces the respective chloromercuriestradiol derivative, namely 2-chloromercuriestradiol, 4-chloromercuriestradiol, and 2,4-(bischloromercuri)estradiol. A typical hydrolysis experiment is as follows: A mixture of 4-chloromercuriestradiol diacetate (65 mg) in a 5% (by weight) solution of KOH in methanol (20 ml) and LiCl (12 mg) is stirred at room temperature for 1 hr. After evaporation of the methanol under vacuum, the residue is triturated with water; after addition of LiCl (20 mg), the mixture is extracted with chloroform (2 × 25 ml) and then with ethyl acetate (3 × 20 ml). The collected organic solvent phases are dried over $MgSO_4$ and evaporated under vacuum to yield 48 mg of 4-chloromercuriestradiol.

---

[15] M. Cais and Y. Josephi, unpublished results (1977); Y. Josephi, D.Sc. Thesis, Technion-Israel Inst. of Technology, Haifa, Israel, 1979.

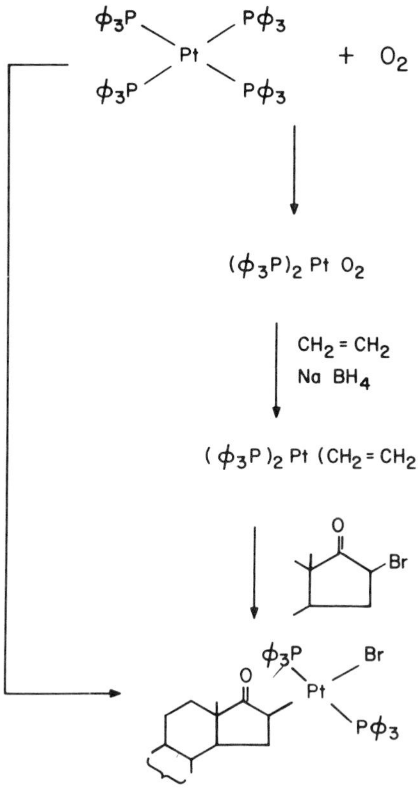

FIG. 2. Synthesis of a platinum-labeled estrone metallohapten.

Analysis. Calculated for $C_{18}H_{23}O_2HgCl$: C, 42.60; H, 4.57%. Found: C, 42.60; H, 4.85%. NMR spectrum (d_8-THF): δ (ppm) [relative to $(CH_3)_4Si$] 6.7–7.3 (2H aromatic, AB system); 0.9 (3H, singlet, angular CH_3) UV spectrum (EtOH) λ_{max} 296 (ε 2400); by comparison: 2-chloromercuriestradiol, λ_{max} 292 (ε 3395); estradiol, λ_{max} 281 (ε 1985). Additional proof of structure is obtained by transformation of the chloromercuriestradiols into their respective iodo derivatives (Fig. 1).

Synthesis of a Platinum-Labeled Estrogen[16]

Another example of a carbon–metal bond synthesis for the direct introduction of a metal atom into a hapten molecule is illustrated for a platinum-labeled estrone derivative (Fig. 2).

---

[16] M. Cais and E. E. Isaacs, unpublished results, 1977; U.S. Patent 4205952 (1980).

A solution of tetrakis(triphenylphosphine)platinum (0.25 g, 0.2 mmol) and 3-acetoxy-16-bromoestrone (0.1 g, 0.26 mmol) in benzene (8 ml) is refluxed under nitrogen for 2 hr. The solvent is removed with a stream of nitrogen, and the residue is washed with several portions (6 × 20 ml) of hot petroleum ether to remove free triphenylphosphine. The remaining solid is dissolved in methanol and filtered; the filtrate is evaporated under vacuum. Recrystallize the residue twice from methylene chloride–hexane to obtain 70 mg (32% yield) of colorless crystals, melting (with decomposition) at 265–267°, and showing a single spot on thin-layer chromatography (TLC) ($R_f \sim 0.57$, silica gel, 2% CH_3OH in $CHCl_3$). The infrared (IR) spectrum (KBr disc) exhibits the following major bands, ν (cm^{-1}): 690(vs), 1000(m), 1090(vs), 1165(m), 1190(m), 1210(s), 1435(vs), 1485(vs), 1745(vs), 1765(s), 2910(s), 3050(s) (absorption intensities, vs = very strong; s = strong; m = medium). The NMR spectrum (CD_2Cl_2) shows δ (ppm): 7.3 (broad, $(C_6H_5)_3P$ and steroid aromatic protons); 2.25 (singlet, CH_3CO); 1.15 (singlet, angular CH_3); 3.0–0.5 (broad resonances due to aliphatic steroid protons.)

The same platinum compound can be prepared by allowing 3-acetoxy-16-bromoestrone to react with bis(triphenylphosphine)ethylene-platinum.[12]

Synthesis of Metalloantigens with Functionalized Metal-Labeling Reagents

In the more general approach to the production of metalloantigens, the scheme requires the synthesis of a metal-containing reagent, which also incorporates a suitable functional group by means of which one can attach the metalloreagent to the antigen or to the modified antigen. For example, if the antigen contains, or can be modified to contain, a carboxylic acid group, one would attempt to synthesize a metalloreagent in which one of the ligands attached to the metal incorporates an amino group available for reaction with the carboxy function of the antigen to form an amide link. In this way the metalloreagent would become the label for that antigen, and the resulting compound would be the metalloantigen used in the metalloimmunoassay for that particular antigen. Obviously, the same metalloreagent could be used also for other antigens containing the required carboxy function. Conversely, if the antigen possesses suitable amino groups (e.g., the ε-amino group of lysine in many proteins) the metalloreagent would be functionalized with a carboxylic acid group to provide a suitable handle for metal-labeling that antigen.

An example of specific metalloantigen synthesis by the above approach is given in the following section and is illustrated schematically in Fig. 3.

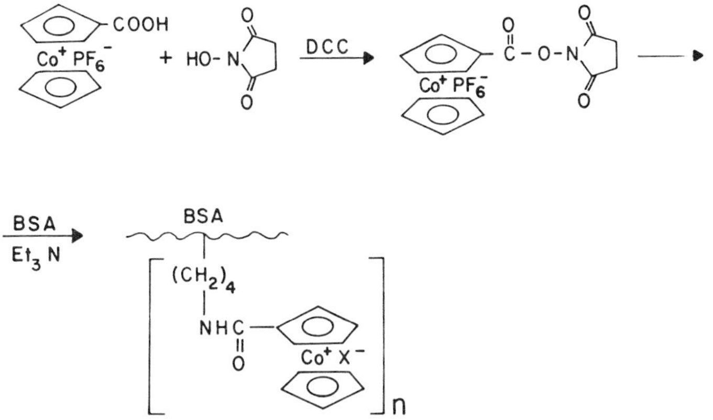

FIG. 3. Synthesis of cobalt-labeled bovine serum albumin.

Synthesis of Cobalt-Labeled Bovine Serum Albumin[17]

Dicyclohexylcarbodiimide (DCC) (130 mg, 0.6 mmol) is added to a stirred solution of 1-carboxycobalticenium hexafluorophosphate[18] (194 mg, 0.5 mmol) and N-hydroxysuccinimide (NHS) (67 mg, 0.6 mmol) in dry acetonitrile (12 ml). After stirring at room temperature for 24 hr, the mixture is filtered (to remove the precipitated dicyclohexylurea), and the filtrate is added in small portions, over a period of 1 hr, to an ice-cooled solution of bovine serum albumin (BSA) (300 mg) in water (35 ml). Simultaneously with the above addition, add triethylamine (225 μl) in small portions, by means of a syringe, to adjust and maintain a pH of 9.5–10 throughout the reaction time. Upon completion of reagents addition, the reaction mixture is stirred at 4° for 24 hr, then dialyzed twice (6 hr) in 2 liters of doubly-distilled water. The dialyzate is lyophilized and dried. Elementary analysis of the lyophilized material (1.93% cobalt) indicates an average of 22 Co gram-atoms per mole of BSA. The Co-labeled BSA prepared as above gave positive ring tests with several rabbit-anti-BSA antisera and moved as a single spot in polyacrylamide gel electrophoresis.

Synthesis of Cobalt-Labeled Protein A[19]

Protein A, which contains 52 lysine residues[20] and has a molecular weight of 42,000, can be labeled with cobalt by the same procedure as that

[17] M. Cais and Y. Altman, unpublished results, 1979.
[18] J. E. Sheats and M. D. Rausch, J. Org. Chem. 35, 3245 (1970).
[19] M. Cais, unpublished results, 1979.
[20] Sjöquist, B. Meloun, and H. Hjelm, Eur. J. Biochem. 29, 572 (1972).

described in the preceding section for BSA. A slightly modified procedure is as follows: A functionalized cobalt complex, 1-carboxy-1'-carbomethoxycobalticenium hexafluorophosphate[17] (46 mg, 0.11 mmol) is allowed to react with NHS (15.5 mg, 0.13 mmol) and DCC (28 mg, 0.13 mmol) in dry acetonitrile (5 ml) with stirring at room temperature for 4 hr and then at 4° for 24 hr. The reaction mixture is filtered, and the filtrate is added portionwise to an ice-cold solution of protein A (Pharmacia) (3.3 mg, 7.85×10^{-5} mmol) in water (3.5 ml). As before, the pH of the reaction mixture is maintained at 9.5–10 by the addition of portions of triethylamine following each addition of the active ester solution (a total of 15 μl of triethylamine is required). The reaction mixture is stirred at 4° for 24 hr, then dialyzed once (4 hr) in 1 liter of saline solution, once (4 hr) in 1 liter of deionized water, and once (18 hr) in 1 liter of deionized water. The dialyzate is centrifuged at 4° for 20 min at 2500 rpm (Damon centrifuge, Model IEC-PR-J), and the supernatant is lyophilized to yield 3.2 mg of cobalt-labeled protein A. Analysis by atomic absorption spectroscopy (see below) showed 2.97% Co content, which corresponds to an average of ~20 Co gram-atoms per mole of protein A.

Synthesis of Cobalt-Labeled Rabbit-Antihuman IgM[19]

Lyophilized rabbit anti-human IgM (heavy chain specific) (Cappel Laboratories, USA) is reconstituted with 7 ml of deionized water (instead of with 2 ml as instructed on manufacturer's label) and allowed to react with the same cobalt complex (21 mg, 4.8×10^{-2} mmol) and under the same reaction conditions as used for the labeling of protein A (see preceding section).

Dialysis (three times in 1 liter of deionized water for 4 hr and twice in 1 liter of deionized water for 18 hr) followed by lyophilization yields 154 mg of cobalt-labeled rabbit anti-human IgM, shown by atomic absorption spectroscopy to contain 0.43% Co.

Analytical Methods in Metalloimmunoassays

The fourth component (see above) in the development of a metalloimmunoassay is the choice of a suitable analytical method to determine the metal concentration in the "bound" and/or "free" phase of the analyte after the separation step. Various methods are now available for metal analysis, such as emission, absorption, and fluorescence spectrometry, microwave excitation emission spectrometry, anodic stripping voltammetry and other electrochemical methods, neutron activation. In our studies so far we have made use only of flameless atomic absorption spectrometry (AAS) employing an electrically heated graphite furnace for the drying, charring, and atomization of the metal-containing sample under analysis. The choice of this method has been governed by several factors, such as relative simplicity of operation, general availability of instruments in clinical chemistry laboratories, the requirement of small aliquots of analyte solution (20–30 μl), and potential high sensitivity of detection (down to picogram levels).

The detection sensitivity of metal atoms in AAS instruments fitted with graphite furnace varies from metal to metal, and it depends on a number of variables connected both with instrumental parameters and the matrix of the assay sample. Therefore it is always necessary to perform a "method development" operation in order to determine the optimal conditions for each assay system. Once this has been done, it is then possible to use the same instrument settings for the required analysis with that particular assay system without having again to go through the procedure of "method development."

The analytical procedure with the graphite furnace AAS involves injection of a small volume of the analytical sample (20–50 μl) into the graphite tube. The solvent is evaporated in the course of the operation, and the metal component of the residue is atomized. The absorption signal recorded is then a measure of the *absolute* quantity of metal element present in the volume of the injected sample. The data presented in the table provide an indication of the type of working range concentrations applicable to AAS. The values in the last column show the approximate metal concentration (ng/ml) required to obtain 0.1 AU (absorption units), keeping in mind that one can read with acceptable precision absorption units (and therefore concentrations) smaller by a factor of 10. Thus, it is clear that one should be able, under optimized conditions, to determine comfortably *metal concentrations* in the range of 3–200 ng/ml. If we now consider that with small hapten molecules, such as steroids, the metal atom constitutes about 10% of the molecular weight of the metallohapten (this value would be higher for heavier elements such as Pt, Au, Hg), the above detection range for the metal components is equivalent to a detection range of 30–2000 ng of metallohapten per milliliter. This range is similar to that reported for other nonradioactive immunoassay methods. Furthermore, the concentration range of tens of nanograms per milliliter is quite adequate for the detection (without prior extraction and concentration operations) of a relatively large number of urinary metabolites, including estriol, morphine, barbiturates, amphetamine, and cocaine.

ATOMIC ABSORPTION SPECTROMETRY-GRAPHITE FURNACE SENSITIVITY FOR
DETECTION OF SOME ELEMENTS[a]

Element	Atomic weight	Sensitivity[b] (ng/0.0044 AU)	Element conc. (ng/20 μl)[c] for 0.1 AU	Element conc. (ng/ml)[c] for 0.1 AU
Au	197	0.030	0.68	34
Co	59	0.080	1.80	91
Cr	52	0.025	0.57	28
Fe	56	0.030	0.68	34
Mn	55	0.008	0.18	9
Mo	96	0.090	2.00	102
Pd	106	0.140	3.20	159
Pt	195	0.090	2.04	102
Rh	103	0.170	3.86	193

[a] Taken from "Analytical Methods Using the HGA Graphite Furnace," Perkin-Elmer Operation Manual, 1974.
[b] By definition, sensitivity expresses the absolute quantity of element required for 0.0044 absorption unit (AU). The values given may vary with the chemical form of the element and instrumental parameters.
[c] Assuming that 20 μl of sample volume were injected into the graphite furnace.

Experimental Considerations

Equipment. An atomic absorption spectrophotometer fitted with electrothermal furnace for drying, charring and atomizing the analyte is recommended. Flame atomic absorption spectrophotometers can be used if the working volumes and metal concentrations in the assay fall within the detection limits of the method. The optimal settings of the instrument are determined in the course of the "method development."

All glassware, disposable plastic tubes and pipette tips have to be thoroughly cleaned by soaking overnight in a 1:1 solution of extra-pure concentrated nitric acid and triply distilled water followed by washing with triply distilled water until the water washings exhibit a neutral pH.

Materials. All reagents used in the preparation of buffers and standards must be checked for absence of labeling-metal as impurity. For phosphate buffers it is preferable to use the potassium salts rather than the sodium salts. The former exhibit less background noise in the AAS signal. Even more advantageous is the use of organic component buffers, such as tris-maleate, pH 7.3 containing 0.025–0.10% polyvinylpyrrolidone (PVP). The latter prevents loss of metal tracer through nonspecific adsorption on the walls of the reaction tubes.

Preparation of Calibration Standards and AAS Calibration Curve.
The procedure will be illustrated with a cobalt-labeled BSA preparation
(Co-BSA) (see above) with a 1.35% cobalt content. Prepare a stock solution of 10.0 mg of Co-BSA in 10 ml of deionized H_2O containing 1% PVP.
The shelf life of this solution, at 4°, is at least 3 months. The calibration
standards are prepared fresh every day by appropriate dilutions from the
stock solution. For example, diluting the above stock solution with 1%
PVP in deionized H_2O 1 : 40; 1 : 80; 1 : 160; 1 : 320; and 1 : 640 gives standards containing cobalt concentrations of 337, 168, 84, 42, and 21 ng/ml,
respectively. For the blank, use the 1% PVP–H_2O solution. The assay
volumes of 20 μl of the blank and the standards are injected in duplicate
(or triplicate) in the graphite chamber of the AAS instrument with suitable
settings, depending on the instrument. In our case we used a Perkin-
Elmer Model instrument with the following settings: drying at 120°, ramp
30 sec, hold 30 sec; charring at 900°, ramp 30 sec, hold 30 sec; atomization
2700°, ramp 3 sec, hold 3 sec. The following readings (means of duplicates) in absorption units (AU) were obtained in a typical experiment:
blank, 0.000; 21 ng/ml, 0.011 ± 0.001; 42 ng/ml, 0.027 ± 0.004; 84 ng/ml,
0.056 ± 0.003; 168 ng/ml, 0.123 ± 0.004; 337 ng/ml, 0.263 ± 0.008. A plot
of AU (y axis) versus the concentrations gave a curve with a linear regression equation $Y = 0.8 \times 10^{-3}x - 8 \times 10^{-3}$ and a correlation factor
$r = 0.996$.

Metalloimmunassay. The protocol for a metalloimmunoassay will follow very closely that for a radioimmunoassay in all the various steps such
as pipetting of reagents, incubation and separation of "free" and
"bound" phases. Just as in RIA, the assay is set up to run the calibration
standards together with the analyte samples using the same instrument
settings throughout. If more than 30 analytical samples are run in the
same assay, it is advisable to check the instrument settings in the course
of the analysis by running at least two of the calibration standards in
between analyte samples.

It is not possible at the present time to give a method of choice for
MIA, just as it is not possible to do so for RIA. We have tried to use many
of the classical separation methods, such as charcoal, double-antibody,
solid-support reagents and columns, with varying degrees of success. The
major difficulty has often been the inadvertent introduction of metal impurities present in the commercial reagents, particularly those used in
the regular techniques for the separation of "free" and "bound"
fractions. This has led us to develop a new separation method for
MIA that subsequently turned out to be of more general applicability.[10]

Concluding Remarks

It should be pointed out that the detection range for metal atoms by AAS, as given above, is that obtainable by current instrument systems. In view of the increasing popularity of AAS, it is reasonable to forecast that in the next few years the AAS instrument companies will produce AA systems with detection capabilities better by at least one order of magnitude. This would mean the possible detection of metallohaptens in the 3–10 ng/ml range, thus bringing MIA very near to the concentration range currently used with many radioimmunoassays. However, in this early stage of development and given the currently commercially available atomic absorption spectrometers, MIA cannot be used for the detection of picogram hapten concentrations, as can be achieved with radioimmunoassays. It is hoped that this is not an unattainable goal, and amplification possibilities are under investigation.

Work with other analytical methods, e.g., polarography, is currently in progress.

The main purpose of this chapter has been to indicate the feasibility of MIA and to hint at some of the many other facets and possibilities that are inherent in it. We hope that this concept, which incorporates an interdisciplinary approach by combining immunochemistry and organic, organometallic, and coordination chemistry, as well as aspects of analytical instrumentation, will stimulate the interest of many other researchers in these fields.

Acknowledgment

My thanks for the work that has been described here are extended to all my collaborators, whose names have been mentioned in the references and footnotes to this chapter.

[36] Fluorescence Fluctuation Immunoassay

By V. B. ELINGS, D. F. NICOLI, and J. BRIGGS

In a homogeneous fluorescent immunoassay[1] one wishes to discriminate between the fluorescence emitted by tagged antigens or antibodies that have reacted with the conjugate species and the fluorescence pro-

---

[1] E. F. Ullman, N. F. Bellet, J. M. Brinkley, and R. F. Zuk, *in* "Immunoassays: Clinical Laboratory Techniques for the 1980s," p. 13. New York, 1980.

duced by those that have not reacted. Previously developed methods that exploit this difference include (*a*) fluorescence quenching, in which the fluorescence intensity is reduced when antigen–antibody binding occurs; and (*b*) fluorescence depolarization,[2] which is a function of the binding. We describe a technique[3,4] in which the reaction occurs on the surfaces of micrometer-sized carrier particles; the relative amount of fluorescence carried by these carrier particles can be determined by measuring the intensity *fluctuations* originating from a small volume δV of the sample that contains a relatively small number of carrier particles. These intensity fluctuations are caused by random variations in the number of carrier particles in the sample volume. If the average number of carrier particles per sample volume equals N, the random root-mean-square (RMS) fluctuation in the number of particles per sample volume will be equal to $\sqrt{N}$.

In order to detect fluorescence intensity fluctuations due to fluctuations in the number of carrier beads per volume δV, one must attempt to make the measurement insensitive to other sources of intensity fluctuations. These include variations in the intensity of the exciting light source, shot noise, fluctuations due to digitization of the analog signal, and fluctuations in the background fluorescence. If one simply scanned through the sample solution, measuring the RMS fluctuation of the intensity, all of the above fluctuations would contribute to the RMS measurement; depending on their magnitude, they might obscure the desired quantity—the magnitude of the number fluctuations due soley to the carrier particles. This would be the result regardless of the number of times one scanned the sample.

As pointed out by Weissman *et al.*,[5] one method of reducing some of these unwanted fluctuations is to employ the technique of *spatial autocorrelation*. One scans through the sample a number of times, thereby obtaining a correlation function that suppresses those fluctuations in fluorescence intensity that are not correlated with the position of the scanning excitation beam. As long as the scanning time is shorter than the characteristic time for diffusion of carrier particles out of a given sample volume δV, number fluctuations of the carrier particles will persist in each sample volume from one scan to the next.

We construct[3] an autocorrelation function $C(n)$ from the measured intensities I_j corresponding to locations j in the solution (using the same

[2] W. B. Dandliker, M. L. Hsu, and W. P. Vanderlaan, *in* "Immunoassays: Clinical Laboratory Techniques for the 1980s," p. 65. Liss, New York, 1980.

[3] D. F. Nicoli, J. Briggs, and V. B. Elings, *Proc. Natl. Acad. Sci. U.S.A.* **77**, 4904 (1980).

[4] J. Briggs, V. B. Elings, and D. F. Nicoli, *Science* **212**, 1266 (1981).

[5] M. Weissman, H. Schindler, and G. Feher, *Proc. Natl. Acad. Sci. U.S.A.* **73**, 2776 (1976).

size volume δV at each location j),

$$C(n) = \sum_j I_j I_{j+n} \tag{1}$$

Here n represents the spacing (i.e., the number of discrete steps in the scan) between the two sample locations, j and $j + n$. In the case of periodic scanning, there exists a value of n, denoted by n_0, for which the sample locations j and $j + n_0$ are identical, for all j. The average in Eq. (1) is taken over many scans of the sample solution. If the fluctuations in the measured intensities are random, i.e., not correlated with sample location, the correlation function will average these fluctuations to zero and simply become equal to the square of the mean intensity for all values of n. If there are intensity fluctuations δI that are correlated with sample position, such as fluctuations in the number of carrier particles in the sample volumes, these fluctuations will not average to zero in $C(n)$ but will instead yield an enhancement equal to $(\delta I)^2$ in $C(n)$ at $n = n_0$, corresponding to one period of the scan. There will also exist spatially correlated fluctuations due to imperfections in the sample cell surfaces and dirt in the solution (which affect the fluorescence excitation efficiency), as well as spatial variations in the level of background fluorescence. To construct the correlation function $C(n)$, one must store the intensity data I so that these values can be retrieved and multiplied by current values of I as each new sample volume is measured. We have accomplished this task using a special-purpose microprocessor-based instrument that is described in a subsequent section.

Theory: Two-Component Immunoassay

Consider an ideal two-component system[3] consisting of fluorescent carrier particles of intensity i_b plus a background of much smaller fluorescent particles of intensity i_f. Subscript b denotes fluorescence *bound* to the carrier particle (owing to antigen–antibody binding), and subscript f denotes fluorescence that is *free* in solution. If in each sample volume δV there are, on average, N_b carrier particles and N_f free fluorescent particles, then the average intensity from each sample volume will be $\bar{I} = N_b i_b + N_f i_f$. The RMS number fluctuations from sample volume to sample volume will be $\sqrt{N_b}$ for the carrier particles and $\sqrt{N_f}$ for the free particles; these fluctuations are statistically independent.

Given these RMS fluctuations in the intensity, one can calculate the value of the correlation function at $n = n_0$,

$$C(n_0) = \overline{((N_b \pm \sqrt{N_b})i_b + (N_f \pm \sqrt{N_f})i_f)^2} \tag{2}$$

where the average is taken over many scans. Terms such as $\pm \overline{N_b \sqrt{N_b}}$, $\pm \overline{N_f \sqrt{N_f}}$ and $\pm \overline{\sqrt{N_b}\sqrt{N_f}}$ will average to zero, resulting in

$$C(n_0) = \bar{I}^2 + N_b i_b^2 + N_f i_f^2 \qquad (3)$$

The correlation function $C(n)$ therefore ideally resembles a flat baseline equal to $\bar{I}^2$ for all n except near $n = n_0$ (one complete scanning period), where there is a peak above the baseline of magnitude $S(n_0)$.

$$S(n_0) = N_b i_b^2 + N_f i_f^2 \qquad (4)$$

The width of the peak in $C(n)$ depends on the density of sampling locations j relative to the dimensions of the sampling volume δV. If there is no spatial overlap of the volumes for locations j and $j + 1$ (all j), then the "peak" will possess no width, consisting of just a single point above the baseline at $n = n_0$. If, on the other hand, there is substantial spatial overlap of the sampling volumes, $C(n)$ will exceed $\bar{I}^2$ for a range of n values centered about n_0.

We have assumed that the lifetimes τ_b and τ_f of the number fluctuations are much longer than the period for one scan, so that the fluctuations persist for at least one scan. If the carrier particles are much brighter than the free particles, with a number density such that $N_b i_b^2 \gg N_f i_f^2$, the peak in the correlation function will be almost entirely due to the carrier particles: i.e., $S(n_0) \approx N_b i_b^2$. Hence, for example, one can determine the bound fluorescent intensity i_b if N_b is known: $i_b \propto \sqrt{S(n_0)}$. Even when the average fluorescent intensity from the free particles ($N_f i_f$) is comparable to that from the carrier particles ($N_b i_b$), the carrier particles can still dominate the correlation peak if they are bright enough. As an example, suppose that on average the carrier particles are each 10 times as bright as each free particle. Also suppose that the contributions from bound and unbound fluorescent sources are equal, i.e., that there are on average 10 times as many free fluorescent particles as carrier particles. Then we have that $N_b i_b^2 / N_f i_f^2 = 10$; thus, the correlation signal is dominated by the bound fluorescence, with only a 10% error in the determination of i_b due to the presence of the free background.

One can suppress the contribution in the correlation peak due to f-type particles if they are small compared to the carrier particles and have a large enough diffusion coefficient. If the mean lifetime of their number fluctuations, τ_f, is substantially less than the time for one complete scan through the sample, then their fluctuations will not persist from one scan to the next and will not give an enhancement in the correlation function. Diffusion causes a number fluctuation in a cubic volume of dimension l to relax exponentially with a lifetime given by

$$\tau = l^2/14D \simeq \frac{1.3\eta l^2}{kT} R_h \tag{5}$$

We have used the Stokes–Einstein equation[6] to relate the diffusion coefficient D to the particle hydrodynamic radius R_h; η is the solvent viscosity, k is Boltzmann's constant, and T is the absolute temperature. This relation provides an estimate of the lifetime of a fluctuation centered in the sample volume. In general there will be a distribution of decay lifetimes due to the spatial distribution of fluctuations; those fluctuations centered near the boundaries of the sample volume will have shorter diffusional lifetimes. Equation (5) therefore provides an upper estimate of the diffusional lifetime τ of a fluctuation. For volume shapes other than a cube, a similar result is obtained with a different numerical factor.

Taking l to be approximately 100 μm (the linear dimension of the sampling volume δV that we employed), we find that the upper limit for the diffusional lifetime of number fluctuations for particles of radius R_h (in micrometers) in water at room temperature is given by $\tau(\sec) \sim 3000\,R_h$. Even for particles as small as a rhodamine 6G dye molecule, which has a molecular weight of 530 and an estimated R_h of 0.7 nm, the upper limit on the fluctuation lifetime is fully 2 sec. Hence, in order to take advantage of diffusion, the scan period must be quite long. Alternatively, the intensity data can be stored over many scans so that one can calculate the correlation function $C(T) = \overline{I(t) \cdot I(t + T)}$, where T is a time delay equal to an integral number of scan periods, with $T > \tau_f$. It should be noted that the fluctuation lifetime of the large carrier particles will be significantly shortened owing to the occurrence of velocity currents in the sample. For instance, although the upper limit for the fluctuation lifetime of a 1 μm carrier particle is 3000 sec, we have not measured fluctuation lifetimes greater than about 10 sec. Very little convection is needed to displace the particles by only 100 μm in 10 sec.

Experimental Apparatus

A schematic diagram of the experimental apparatus[3] is shown in Fig. 1. The major components consist of an argon-ion laser light source, sample cell, and electromechanical translator, illuminating and imaging optics, photomultiplier detector (PMT) and preamplifier, and a microprocessor-based computing autocorrelator and associated oscilloscope display.

[6] B. Chu, "Laser Light Scattering" Academic Press, New York, 1974.

It was convenient to use an argon-ion laser (Spectra Physics, Model 164) to excite the fluorescently tagged molecules in solution, because its strong 488.0 nm (blue) line lies very close to the peak excitation wavelength (485 nm) of fluorescein isothiocyanate (FITC). Furthermore, its highly collimated beam (diameter 1.5 mm) eliminates the need for additional focusing–collimating optics. A laser source was used only because of these convenient characteristics; its coherence property is *not* required for this technique. Any suitably filtered incoherent source, such as an incandescent lamp, would suffice, since only a relatively low beam intensity is required. The laser was typically run at an output power of 80 mW, with the beam attenuated by at least a factor of 10 using neutral density filters to prevent bleaching of the fluorescent labels.

A pinhole P was used to define the diameter of the exciting light beam incident on the sample cell S. The pinhole aperture chosen was 100 μm, conveniently obtained from a 10:1 reduced photographic negative of a 1-mm dot. (Significantly smaller beam sizes could be easily obtained using a focusing lens.) This narrow beam was used to illuminate the sample solu-

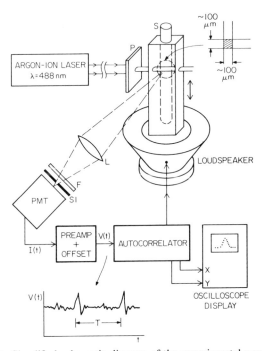

FIG. 1. Simplified schematic diagram of the experimental apparatus.

tion in the cylindrical cell S. The latter was centered vertically in a 1-cm fluorometer cuvette filled with water, which provided index matching to minimize stray scattering due to imperfections in the glass tube surfaces. Two sizes of sample tubes were employed in our experiments: 6-mm diameter disposable culture tubes and 1.35-mm (inside) diameter 100-μl micropipettes. The latter variety were the most effective with respect to reduced convection induced either by thermal gradients or sample motion. A total sample volume of less than 20 μl was easily achieved with the micropipette tubes.

Those FITC-labeled molecules, plus fluorescing "background" impurities, which lie within the cylindrical sample volume defined by P, will emit fluorescent light. Lens L projects this line source of yellow-green light onto slit Sl (photographic negative, width 100 μm) at the face of the PMT detector (EMI type 9865). Hence, the pinhole–slit combination defines an effective detection volume δV of the order of 10^{-6} cm^3. Filter F (Corning No. 3-69) blocks most of the 488-nm light that arrives at lens L owing to scattering from particles in solution or from stray reflections at the glass–water interfaces, leaving only the yellow-green FITC emission plus impurity fluorescence of long wavelengths.

In order to measure the spatial correlation function of fluorescence intensity within the sample solution, the detection volume δV must be scanned within the sample solution. This can be accomplished either by scanning the incident light beam within a stationary sample cell or by moving the sample cell with respect to a stationary light beam. We adopted the latter approach, to avoid scanning the focused image of δV across the PMT photocathode, thereby introducing the risk of periodic signal changes due to spatial nonuniformities in PMT photocathode efficiency. Such spatial variations might contribute to the peak in the correlation function.

Uniform up/down translation of the cylindrical sample cell/cuvette was achieved using a 7-inch loudspeaker as an electromechanical transducer. The latter was driven by a digitally synthesized triangular waveform, consisting of alternating rising and falling staircases of discrete steps [yielding the discrete locations j in Eq. (1)], adjustable in number from 64 to 1024. The scanning period was adjustable from 0.2 to 20 sec, with a maximum peak-to-peak amplitude of translation of 0.8 cm. The time duration of a discrete step in the position of the sample corresponds to the width δt of a correlator "channel," used to form $C(n)$. The values of δt typically employed in these experiments ranged from approximately 2 to 10 msec. The width of the correlation peak is determined by the number of scan steps per translation amplitude. With 64 steps in 8000 μm, each discrete step causes the incident beam to move by 125 μm, resulting in no spatial overlap of successive sample volumes δV and only a single channel, $C(n_0)$, in the correlation peak above the baseline. At the opposite extreme, 1024 steps per 8000 μm, there is substantial overlapping of the volumes and a correlation peak that is approximately 13 channels wide. Signal–noise considerations dictate the optimal degree of overlap (i.e., the number of channels under the peak) for accurate resolution of the peak magnitude $S(n_0)$.

A straightforward operational amplifier circuit was employed to convert the PMT photocurrent to a time-varying voltage. An adjustable gain and variable direct current (dc). Offset were provided to match the fluctuating fluorescence signal optimally to the ± 5 V input range of an 8-bit analog-to-digital (A/D) converter, located in the "front end" of the autocorrelator. An adjustable filter time constant was used to match the signal bandwidth to the step time δt. Only the upper four bits of the A/D converter were used, owing to the relatively large fluctuations in fluorescence intensity encountered in our immunoassay experiments, given the large fluctuations in carrier particle number for the small volume δV utilized. The correlation function $C(n)$ was constructed from 4-bit $\times$ 4-bit multiplications of the real-time intensity with a succession of previous-time values.

In order accurately to determine $C(n)$, the autocorrelator must correctly "match" each sampled fluorescent intensity I_j, corresponding to the jth location in solution, with the appropriate channel number—i.e., with the time base of the correlation function. The instrument performs this synchronization by using the microprocessor both to control the sample cell position and to advance the channel number—i.e., to calculate $C(n)$. Therefore, there can be neither short-term "jitter" nor long-term slippage of the location of δV with respect to the time base, or channel number, of the correlator. This is an important feature in the case of long run times, which is necessary when there are only very weak fluorescence spatial correlations. The linearity of the loudspeaker drive with respect to the driving waveform is irrelevant; only the reproducibility of the sample position for a given programming current is of consequence.

A simplified block diagram of our special-purpose computer-controlled autocorrelator is shown in Fig. 2. It is designed to compute and display $C(n)$ continuously with a built-in programmable long delay, appropriate for the periodic scanning scheme described above. The instrument is normally configured to contain 64 channels, each of which has a capacity of 2^{32} ($\approx 4.3 \times 10^9$) correlated counts. In these respects, it is similar to commercially available 4-bit digital autocorrelators. However, in

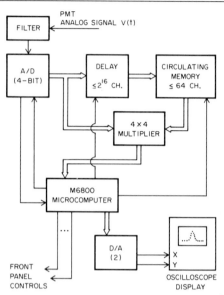

FIG. 2. Simplified block diagram of the microcomputer-controlled autocorrelator. CH., channels.

several significant respects it differs substantially from the latter devices.

Our instrument utilizes a Motorola 6800 microcomputer to obtain 64 channels of $C(n)$ largely with software, thereby eliminating most of the numerous TTL integrated circuits normally required for this task. This simplification is made possible because the shortest channel width δt, 1.75 msec, is very long by normal correlator standards; this interval between successive sampled intensities gives the microcomputer sufficient time to update the contents of 64 channels of $C(n)$ with a set of new products, $I_j I_{n+n}$. By contrast, the measurement of submicron size in solution by photon correlation spectroscopy utilizes much shorter channel widths, in the range 1–100 μsec, which requires fast, parallel digital circuitry. Although there is not sufficient time to construct the autocorrelation function for all n_0 channels, when $n_0 \gg 64$, this is not a limitation since only the value of $C(n)$ near n_0 is of interest in determining the value of the correlation peak, $S(n_0)$.

The instrument incorporates a large block of memory, labeled "Delay" in Fig. 2, in order to center the limited number of computed channels of $C(n)$ about $n = n_0$, thereby covering the range $n = (n_0 - 32)$ to $(n_0 + 31)$. The "4 × 4 Multiplier" multiplies the current intensity I_j by each of the earlier values $I_{j-(n_0+31)}$ through $I_{j-(n_0-32)}$, obtained in sequence from the "Circulating Memory." During the channel interval δt there is sufficient time remaining for the microcomputer to display one channel of $C(n)$ on an $X - Y$ oscilloscope, using a pair of digital-to-analog (D/A) converters. In summary, our design takes advantage of the inherent strengths of the microcomputer—its large memory capacity and ease of data manipulation, as opposed to its limited speed in performing multiplications.

Results of Immunoassay Experiments

The ability of the fluctuation technique to measure the amount of fluorescent label bound to a bead in the presence of high levels of background fluorescence, free in solution, has been demonstrated[3,4] using a variety of idealized systems. Figure 3 plots the relative signal [the square root of the measured correlation peak height $S(n_0)$], proportional to the bound fluorescent intensity I_b, versus the ratio of free (background) to bound fluorescence (I_f/I_b). In these idealized systems, the background resulted from free fluorescent molecules that were added to a suspension of fluorescent beads. The ratio that is plotted along the x axis of Fig. 3 was

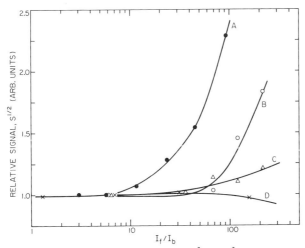

FIG. 3. Relative correlation signal, $S^{1/2}(n_0) \propto (N_b i_b^2 + N_f i_f^2)^{1/2}$, versus the ratio of free fluorescence intensity to bound fluorescence intensity, $I_f/I_b = N_f i_f/N_b i_b$. Curve A, acrylamide beads carrying the complexes Ab_b-Ag-Ab^* plus free Ab^*; curves B and C, 4.3 μm fluorescent polystyrene spheres plus free rhodamine dye molecules, with a scan period of 1.2 sec and 10.2 sec, respectively; curve D, 0.46 μm fluorescent spheres plus free rhodamine with a scan period of 10.2 sec.

varied by controlling the relative concentrations of beads and free fluorophores. The plotted values of this ratio resulted from independent fluorometric measurements of the intensities of the two components at known concentrations.

Let us assume that the x values (I_f/I_b) of the data points in Fig. 3 were achieved by maintaining a fixed concentration of beads of constant fluorescent brightness. Then, if the signal $S^{1/2}(n_0)$ which results from this fluctuation technique is independent of the free fluorescence, the experimental points should form a horizontal line in Fig. 3, independent of I_f/I_b and normalized to 1.0 at small values of I_f/I_b. The observed curves are indeed horizontal at low concentrations of free fluorescence, but generally show an increasing signal at high concentrations of free fluorescence, which results from contributions of the free components to the correlation function. According to our earlier discussion, there are two factors that play the primary role in determining the value along the I_f/I_b axis at which the actual response will deviate from the ideal horizontal response. These factors are the average brightness of the beads relative to that of the free species and the relative lengths of the scan period and the time of randomization of the free species by Brownian motion (or spurious currents) over the sampling volume δV. We now comment on the differences between the four systems represented in Fig. 3 with respect to these two factors.

Curve A of Fig. 3 gives results of measurements[3] using the Immunofluor assay kit for IgG supplied by Bio-Rad Laboratories.[7] These consist of three components: standard concentrations of human IgG (Ag), rabbit antibody to human IgG that has been tagged with FITC (Ab*), and 3- to 5-μm-diameter acrylamide gel beads to which rabbit antibodies have been covalently bound. We shall denote the latter by the symbol Ab_b, the subscript b standing for bound to the bead. These reagents are designed to perform a "sandwich" assay for IgG. That is, the complexes (Ab_b-Ag-Ab*) are formed on the beads in direct proportion to the concentration of Ag. Thus, the bead fluorescence is a direct measure of the analyte concentration. In a homogeneous solution, bead fluorescence may represent a small fraction of the total sample fluorescence, since free Ab* is present in excess. (In a real assay situation, fluorescence from serum components and impurities in the buffer will produce another background component comparable to the signal at low analyte concentrations.)

To produce the data for curve A, after the incubation steps of the standard protocol, the fluorescently tagged beads were isolated from the other components by centrifugation. They were then resuspended, at a

---

[7] Bio-Rad Laboratories, Richmond, California 94804.

fixed concentration, in solutions to which were added serially increasing concentrations of free Ab*. We see from curve A of Fig. 3 that this results in a signal that starts to deviate from the horizontal at a free fluorescent intensity 10 times that of the beads. These data were taken with a scanning period of 1.2 sec, which was presumably short enough to permit the free Ab* to contribute quite efficiently to the correlation peak $S(n_0)$. Therefore, the fact that the signal $S^{1/2}(n_0)$ was flat for $I_f/I_b < 10$ is due nearly entirely to the relative brightness of the beads (i.e., the number of fluorescent complexes per bead). From the magnitude of the increase of the observed signal with increasing I_f/I_b for large values of the latter, we can estimate the average bead brightness. Signal $S^{1/2}(n_0)$ [Eq. (4)] increased by a factor $\sqrt{2}$ when $N_f i_f^2 = N_b i_b^2$. This corresponds to $I_b/I_f = N_f i_f/N_b i_b = (N_f i_f^2/N_b i_b^2)(i_b/i_f) = i_b/i_f$. Hence from Fig. 3 we estimate that the average bead fluorescence is roughly 30 times that of a free Ab* molecule. Since curve A corresponds to an IgG concentration of approximately 33% of the level needed to saturate the signal, we conclude that there were roughly 100 active antibody sites on each carrier particle. There are a wide variety of spurious effects that will influence the signal, including the formation of bead aggregates through the complex Ab_b-Ag-Ab_b, nonspecific binding of Ab* to the beads, and the release of Ag from the beads to form the complexes Ab*-Ag-Ab* in solution. However, curve A shows that the signal is essentially independent of the amount of free Ab* for concentrations less than 10 times the concentration of tagged sites on the beads.

Curves B and C of Fig. 3 summarize the results obtained[3] from idealized systems containing fluorescent polystyrene spheres (Duke Scientific,[8] 4.3-μm diameter) and rhodamine-6G dye molecules. The same 488-nm laser line was used to excite the mixture and the standard fluorescein-matched glass filter used for detection. Although the rhodamine was not being excited near the peak of its absorption spectrum, the level of its fluorescence output was still adequate for detection. In any case, the x values (I_f/I_b) of the points for curves B and C in Fig. 3 resulted from standard fluorometric measurements with these same spectral conditions. For both curves B and C, the bead concentration was fixed and I_f/I_b was adjusted by varying the free rhodamine concentration. The only difference in the two curves was the scanning period: 1.2 sec for curve B and 10.2 sec for curve C. The latter curve is flatter, presumably because the longer correlation time allows more complete randomization of the free rhodamine species between samples. There are two factors that can explain the flatter response of curve B relative to curve A. First, the Duke Scientific beads may be brighter relative to single rhodamine molecules

---

[8] Duke Scientific, Palo Alto, California 94306; Catalog No. 267.

(with excitation at 488 nm) than was the case for the IgG immunobead system of curve A. Second, rhodamine molecules are exceedingly small; their large diffusion coefficient will cause them to randomize more quickly than Ab* molecules.

Smaller fluorescent beads (Dow Chemical,[9] 0.46-μm diameter) were used[4] with free rhodamine to produce curve D. In this case the rhodamine concentration was held fixed and the bead concentration varied to achieve a large range of values of I_f/I_b. The correlation time was again 10.2 sec. This produced a measured signal $S^{1/2}(n_0)$ which decreased with the square-root of the bead concentration [Eq. (4)]. In curve D of Fig. 3 we find that the observed signal is essentially independent of the free background component all the way to $I_f/I_b \simeq 200$. The improvement over curve C is probably due to the fact that the smaller Dow spheres settle less quickly than the Duke particles during the long times employed. Any settling of the beads reduces the bead correlation and thus makes background contributions more significant. These experiments on idealized systems demonstrate the ability of the correlation peak signal accurately to extract the carrier particle fluorescence in a relatively large background of free fluorescence.

For a clinically significant test of a true homogeneous assay, we used the Fluoromatic Gentamicin assay kit produced by Bio-Rad Laboratories.[4] This is a competitive assay with a fixed concentration of FITC-tagged gentamicin [G*] and a variable concentration of untagged gentamicin [G] (i.e., the analyte or standard concentrations to form the standard curve), which compete for a fixed number of antibody sites on the beads. As [G] increases the resulting bead fluorescence will decrease, and vice versa. The correlation peak signal $S^{1/2}(n_0)$ will also decrease as long as it is insensitive to the free fluorescence. Samples to form the standard curve of one measurement were made simply by mixing the G*, G, and bead reagents together and incubating. There was no separation step. The free fluorescence was due to the unbound G* (G + G* must be added in excess of the binding sites) *in addition to* extra dye that Bio-Rad adds to the G* reagent to make it obviously yellow for identification purposes.

Figure 4 shows $S^{1/2}$ plotted[4] as a function of [G] for two ranges of reagent concentrations. The curves are the standard curves for gentamicin determination using our correlation technique on homogeneous samples; they coincide with the standard curve supplied by Bio-Rad, in that a given dilution of the standard results in the same change in the measured signal. Curve B corresponds to solutions that have 10 times the con-

---

[9] Dow Diagnostics, Indianapolis, Indiana 46268.

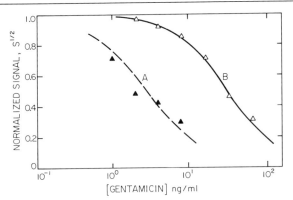

FIG. 4. Normalized correlation signal $S^{1/2}$ versus gentamicin concentration [G] (ng/ml) for a competitive assay, using the "fluoromatic" kit of reagents produced by Bio-Rad Laboratories. Curve B was obtained from reagents that have 10 times the concentrations of G,G* and beads relative to the solutions used to obtain curve A.

centrations of G,G* and beads relative to those of curve A. As expected, the midpoint of curve B is roughly 10 times higher in concentration than the midpoint of curve A. In summary, the assay can resolve gentamicin levels of 1 ng/ml (2.6×10^{-9} M) from a total specimen volume of only 10 μl.

All points in Figs. 3 and 4 represent mean values obtained from several measurements on the same sample. The standard deviation relative to the mean value of S for a given sample decreases with the inverse of the square root of the total sampling time. The scatter of points of curve A (Fig. 3) was greater than that obtained from curve B owing to shorter run times; a typical time for measuring a single mean value on curve B was 20 min. The settling of the beads during long experiments adversely affected the deviation. We used commercially available bead reagents; however, our signal-to-noise ratio would have improved if we had used carrier particles with a greater number of active sites and a density close to that of water to reduce settling.

Summary

The homogeneous fluorescent immunoassay described above allows one to measure the brightness of fluorescently tagged carrier particles that are suspended in a background of free, unbound fluorescent sources. We have demonstrated the feasibility of our technique using a gentamicin competitive assay as well as idealized model systems.

We have seen that the fluctuation–correlation method is able to discriminate against free background sources because each fluorescing particle in solution contributes to the correlation peak [Eq. (4)] with a weighting equal to the *square* of its respective intensity. Hence, a few very bright sources contribute disproportionately to the "signal" relative to many weak ones. To take advantage of this property, one would therefore design an assay that uses relatively larger carrier particles, each of which is capable of binding on the order of 10^3 to 10^4 tagged antibodies or antigens. Unfortunately, the nonlinear dependence of the correlation peak on the brightness of the fluorescing species causes the technique to be perturbed by carrier particle aggregation; the apparent bound fluorescence intensity increases with the extent of aggregation. The latter may be an unavoidable consequence of performing assays using raw blood serum, for example.

The ultimate usefulness of this method will depend on its sensitivity and speed when applied to "real" assays of clinical significance. These characteristics will be influenced by a number of technical details. Given our limited experience with the method thus far, it would appear that its principal drawback is its relatively slow speed. In order to decrease the time needed for a reliable measurement, one must average the *random* fluctuations in the fluorescent intensity to zero more quickly. In principle, this can be accomplished by decreasing the shot noise by collecting a larger fraction of the fluorescent light, and increasing the sampling rate. The method requires rather complicated instrumentation; it is by no means clear that this level of complexity is justified given the realistic level of sensitivity that will be obtained by this technique.

[37] Methods and Applications of Hapten-Sandwich Labeling

By LEON WOFSY

Hapten-sandwich labeling is an indirect procedure: antigen is first localized by treatment with antibody to which hapten groups have been coupled covalently; labeling is achieved with a second-step anti-hapten antibody reagent that bears a desired marker for fluorescence or electron microscopy, autoradiography, or a variety of other purposes (Fig. 1).

Interest in such a labeling methodology was promoted especially by the discovery of differentiation antigens characteristic of subsets of cells that constitute the immune system. Prior to the development of mono-

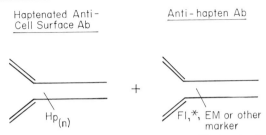

FIG. 1. Hapten-sandwich labeling. A single hapten-modified anti-cell surface antibody preparation may be used in conjunction with its anti-hapten amplifier to permit (a) labeling for fluorescent and electron (transmission and scanning) microscopy; (b) autoradiography; (c) selective killing or supression; (d) isolation of membrane antigens. From Cammisuli and Wofsy.[1]

clonal antibodies, such differentation antigens were defined primarily by low-titer alloantisera. Hapten-sandwich labeling was conceived as a general means of providing amplified visualization of cell surface alloantigens in circumstances where the use of a conventional second-step anti-immunoglobulin is proscribed (e.g., when B lymphocytes are included in the cell population being labeled for the presence of a specific antigen).

A major advantage of the hapten-sandwich procedure is the fact that anti-hapten antibodies and their conjugates with any chosen markers can be purified easily and efficiently by affinity chromatography. The most critical problem, on the other hand, is the requirement that hapten groups be coupled to an immunoglobulin first-layer reagent by means that assure high retention of antibody binding activity.

This chapter describes in detail a procedure by which azophenyl haptens, particularly the azobenzenearsonate (ars) hapten, are coupled to immunoglobulin with a heterobifunctional amidinating reagent.[1] Our laboratory has found that such ars-Ig preparations used in conjuction with markers conjugated to anti-ars antibody provide a simple, effective, and general methodology for specific labeling of cell-surface antigens. We also describe an excellent method that is similar in principle to hapten-sandwich labeling, but uses instead biotin-modified immunoglobulin preparations and markers coupled to avidin.[2,3] Hapten-sandwich and biotin–avidin labeling systems have been reviewed.[4,5] These techniques have been applied usefully with both conventional and monoclonal antibodies.

[1] S. Cammisuli and L. Wofsy, J. Immunol. 117, 1695 (1976).
[2] H. Heitzmann and F. M. Richards, Proc. Natl. Acad. Sci. U.S.A. 71, 3537 (1974).
[3] M. H. Heggeness and J. F. Ash, J. Cell Biol. 73, 783 (1977).
[4] L. Wofsy, C. Henry, and S. Cammisuli, Contemp. Top. Mol. Immunol. 7, 215 (1978).
[5] E. A. Bayer and M. Wilchek, Methods Biochem. Anal. 26, 1 (1980).

Coupling Azophenyl Haptens to Antibodies

Although a number of reagents and procedures have been used to attach a variety of hapten groups covalently to antibodies,[4] the most extensive modification with the least significant loss of antibody activity has been achieved with methyl p-hydroxybenzimidate (HB) as an intermediate to link azophenyl haptens to the amino groups of antibody.[1] As shown in Fig. 2, the aminophenyl derivative of the chosen hapten is first diazotized and azocoupled to HB. After excess diazonium reagent is quenched, the resulting solution is added to antibody so that azophenyl-modified HB can react by way of its imidoester function to amidinate the protein. Amidination reactions are highly specific under mild conditions for amino groups,[6] and even extensive modification of most proteins causes little structural alteration.[7] The procedures described here are applicable with Fab as well as with whole antibody. Starting with diazotized arsanilic acid and HB, at least 15–20 ars-HB groups can be coupled to an antibody with the loss of only about 10% of capacity for binding to an immunoadsorbent antigen column.[1]

Preparation of HB Hydrochloride

HB hydrochloride (Pierce Chemicals, Rockford, Illinois) is easy and inexpensive to prepare by the Pinner synthesis as described by Hunter and Ludwig.[6]

Reagents

p-Hydroxybenzonitrile (Aldrich Chemicals Co., Milwaukee, Wisconsin)
HCl (gas)
Methanol, absolute
Ether, anhydrous

Procedure. Dissolve 50 mmol of p-hydroxybenzonitrile in 75 mmol of absolute methanol. At 0°, under strictly anhydrous conditions, pass HCl through the solution until it is saturated (75 mmol); at saturation, the bubbling of HCl to the surface becomes distinctly more vigorous. An hour after the passage of HCl has been terminated, take up the solid white mass of HB that has crystallized and thoroughly suspend it in cold anhydrous ether. Filter on sintered glass and wash with 100 ml of cold ether. Dry the crystals thoroughly over P_2O_5 under vacuum. HB hydrochloride melts with decomposition at 164°. Store over Drierite at −20°.

[6] M. J. Hunter and M. L. Ludwig, *J. Am. Chem. Soc.* **84,** 3491 (1962).
[7] L. Wofsy and S. J. Singer, *Biochemistry* **2,** 104 (1963).

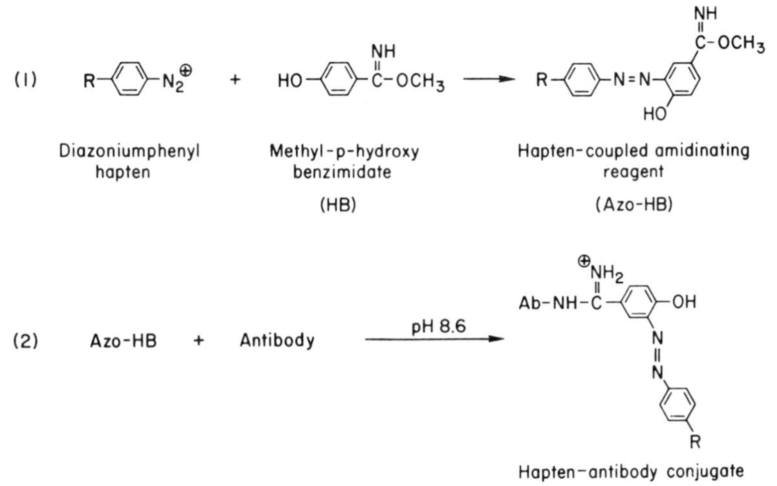

FIG. 2. The preparation of hapten–antibody conjugates with a bifunctional amidinating reagent. From Cammisuli and Wofsy.[1]

Preparation of Ars-HB Modified Antibody

Reagents

Arsanilic acid
HB hydrochloride
Sodium nitrite
Imidazole
Buffers
    Borate, 0.34 M, pH 8.6
    Borate, 0.017 M, pH 8.0
    Phosphate-buffered saline (PBS), 0.01 M phosphate, 0.15 M NaCl, pH 7.2–7.4

Procedure

1. To prepare 12 ml of a 0.25 M diazonium solution, dissolve 3 mmol of arsanilic acid in 8 ml of 1 M HCl on ice. At 0°, with stirring, add 3 mmol of NaNO$_2$ in 4 ml of water. After 15–20 min, the resulting diazonium solution may be reacted with HB or aliquoted, frozen, and stored for several months at −20°.

2. To prepare hapten-azo-HB, the diazonium reagent is allowed to react with HB in a molar excess of 5 : 1 so that no free HB remains to compete with hapten-coupled HB in the subsequent modification of antibody. This is a significant consideration, since the solubility of im-

munoglobulin is sharply diminshed by linkage with even a small number of unmodified HB groups. For a 20-ml stock solution of ars-HB, 0.03 M in initial imidate concentration, add 0.6 mmol of HB to 3 ml of 0.34 M borate buffer adjusted to pH 9.8. (This prevents a drastic drop in pH as diazonium hydrochloride solution is added.) On ice and with stirring, add (in portions) 12 ml of the 0.25 M diazonium solution, maintaining pH 9.2 ± 0.2 with 5 M NaOH. When addition is complete (in about 20 min), allow the reaction to come to room temperature and continue for a total of 2 hr, maintaining pH 9.2 ± 0.2. Add 4 mmol of solid imidazole with stirring and allow mixture to stand for 1 hr to quench residual diazonium. Adjust to pH 8.6 with 2 M HCl and dilute with borate buffer to a final volume of 20 ml. The 0.03 M hapten-azo-HB stock reagent may be aliquoted and stored at −20° for use over several weeks.

3. To modify antibody, allow one volume of a DEAE-purified immunoglobulin preparation, 1.5–7.5 mg/ml in 0.34 M borate, pH 8.6, to react with two volumes of the ars-HB reagent to acheive a solution that is 0.02 M in the initial HB. Allow the reaction to proceed for 15–20 hr at room temperature. Purify the conjugate by passage of the reddish black solution over a BioGel P-60 column equilibrated with 0.017 M borate buffer, pH 8.0. (Passage over Sephadex G-25 or G-50 is not satisfactory for effective separation of the conjugate from azo reaction by-products.) The ars-HB antibody conjugate may be aliquoted and frozen or precipitated in 45% saturated ammonium sulfate and stored at 4°.

4. Before use for labeling cell-surface antigens, adjust the conjugate solution to physiological pH and ionic strength. For samples in 0.017 M borate, add 0.25 volume of 5 × concentrated PBS; samples reconstituted from SAS slurries may be dialyzed directly againt PBS. While ars-HB (and most other charged hapten-azo-HB) antibody conjugates prepared by these procedures remain soluble in PBS, conjugates with some neutral haptens tend to precipitate.[8]

Specifically Purified Anti-hapten Antibody

Procedures for preparation and specific purification of anti-ars and other antibodies against azophenyl haptens have been described in detail[9,10] and closely follow applications of affinity chromatography presented previously in this series, Vol. 34.

[8] E. F. Wallace and L. Wofsy, *J. Immunol. Methods* **25**, 283 (1979).
[9] L. Wofsy and B. Burr, *J. Immunol.* **103**, 380 (1969).
[10] A. H. Good, L. Wofsy, J. Kimura, and C. Henry, *in* "Selected Methods in Cellular Immunology" (B. B. Mishell and S. M. Shiigi, eds.), p. 281. Freeman, San Francisco, 1980.

Biotin–Antibody Conjugates

The association constant of avidin for biotin, 10^{15} M^{-1},[11] exceeds by 7–10 orders of magnitude the range of affinities of most specific antibody–hapten reactions. The fact that both biotin and avidin are commercially available, and that biotin can be linked covalently to proteins or carbohydrates by a variety of relatively convenient methods,[5] has made the biotin-avidin system an attractive and versatile tool in cell biology. Heitzmann and Richards demonstrated its use for labeling biotin-modified membrane sites for electron microscopy with an avidin-ferritin conjugate.[2] Specific labeling of cell surface and intracellular antigens with biotin-modified antibodies and fluorescent avidin is now a fairly common technique.[3,12]

Biotin–antibody conjugates are most effectively prepared with the reagent biotinyl-N-hydroxysuccinimide ester (BOSu),[13] which reacts primarily with amino groups. About seven or eight biotin groups can be coupled per mole of antibody without substantial inactivation.[3] While avidin binds strongly to such biotin–protein conjugates, it should be noted that in comparison to the reaction of avidin with free biotin, the molar valence of four may be effectively reduced and the affinity is lowered dramatically, probably to the range of 10^7–10^{10} M^{-1}.[14] One should therefore not expect all bonds formed in biotin-avidin labeling of cell surface antigens to be irreversible, as might be assumed from the fact that the half-life with free biotin is measured in years.

Preparation of BOSu

BOSu is prepared by coupling biotin and N-hydroxysuccinimide with dicyclohexylcarbodiimide, based on the procedure of Becker *et al.*,[13] as applied by Heitzmann and Richards.[2]

Reagents

Biotin
N-Hydroxysuccinimide
Dicyclohexylcarbodiimide
Procedure. Dissolve the three reagents in dimethylformamide, each at 0.07 M concentration, and allow the reaction to proceed under anhydrous

[11] N. M. Green, *Biochem. J.* **89**, 585 (1963).
[12] L. Y. W. Bourguignon and S. J. Singer, *Proc. Natl. Acad. Sci. U.S.A.* **74**, 5031 (1977).
[13] J. M. Becker, M. Wilchek, and E. Katchalski, *Proc. Natl. Acad. Sci. U.S.A.* **68**, 2604 (1971).
[14] N. M. Green, L. Konieczny, E. J. Toms, and R. C. Valentine, *Biochem. J.* **125**, 781 (1971).

conditions at 50° for 16 hr. Cool the mixture to room temperature and filter off the dicyclohexylurea. Take the filtrate to dryness by rotary evaporation. Crystallize the residue from isopropanol (m.p. 210°).

Coupling Biotin to Antibodies

Biotin-modified antibodies are prepared by reaction with BOSu as described by Heggeness and Ash.[3]

Reagents

BOSu, freshly prepared solution, 1 mg/ml in dimethyl sulfoxide (DMSO)

Buffer: borate, 0.2 M, pH 8.5

Procedure. To 1 ml of a 1 mg/ml solution of DEAE-purified immunoglobulin in borate buffer, add 60–100 μl of fresh BOSu solution. (With monoclonal antibodies, the amount of BOSu that will give optimal modification without significant inactivation may vary within the indicated concentration range.) Allow to react for 4 hr at room temperature. Dialyze or purify by gel filtration.

Coupling Markers to Anti-hapten Antibodies or Avidin

In general, the same procedures are used with either anti-hapten antibody or avidin to couple a wide variety of markers to either ligand. Standard procedures for preparing fluorescent antibody or antibody conjugates with ferritin, hemocyanin, or enzymes apply equally well for avidin conjugates.[5] Here we cite a few procedures that we have found to be most convenient and effective.

Preparation of Fluorescent Avidin

Reagents

Fluorescein isothiocyanate (FITC)

Tetramethylrhodamine isothiocyanate (TRITC)

Buffers

Bicarbonate-buffered saline, 0.5 M, pH 9.2. Mix 10 ml of 0.5 M NaHCO$_3$ in 0.15 M NaCl and 5.8 ml of 0.5 M Na$_2$CO$_3$ in 0.15 M NaCl; adjust to pH 9.2 with 6 M HCl.

Phosphate-buffered saline (PBS): 0.01 M phosphate, 0.15 M NaCl, pH 7.2–7.4

Procedure. To 1 ml of a 10 mg/ml solution of avidin in 0.15 M NaCl, add 0.1 ml of 0.5 M bicarbonate-saline buffer, pH 9.2. For fluorescein modification, add 0.1 ml of FITC, 2 mg/ml in DMSO; for rhodamine, use TRITC in the same manner. Allow to react for 2 hr at room temperature, and purify fluorescent avidin by filtration on Sephadex G-25 in PBS.

Coupling with Glutaraldehyde

Glutaraldehyde is often used in a variety of procedures to couple immunoglobulins or avidin to other protein markers. We have prepared (Fab')$_2$ anti-ars antibody conjugates with keyhole limpet hemocyanin (KLH) for use in several hapten-sandwich applications.

Reagents

Glutaraldehyde, fresh, EM grade
Buffers
    0.1 M Ammonium carbonate, 0.1 M
    Phosphate, 0.1 M, pH 6.8
    Phosphate, 0.05 M, pH 7.5

Procedure

1. To a 1-ml solution containing 5 mg of (Fab')$_2$ anti-ars antibody and 25 mg of KLH, slowly add 0.1 ml of aqueous 0.5% fresh glutaraldehyde with stirring, at room temperature. Allow to react for 1 hr.

2. Terminate the reaction by dialysis against 0.1 M (NH$_4$)$_2$CO$_3$ for 3 hr at 5°. Then dialyze further against 0.05 M phosphate buffer, pH 7.5.

3. Centrifuge at 10,000 rpm for 15 min to remove any precipitate. Chromatograph on a Sepharose 2B column (1 × 40 cm) in 0.05 M phosphate buffer, pH 7.5. Collect the first peak which includes the KLH–antibody conjugates. Samples may be concentrated and stored in 0.1% azide at 4°.

Disulfide-Linked Conjugates

Procedures that use hapten-sandwich techniques have been described for affinity targeting to selected cell surfaces of sealed erythrocyte ghosts, lipsomes, or red cells.[15,16] The method of attachment of antibodies or avidin directly to membranes, or to phosphatidylethanolamine for incorporation into liposomes,[17] employs the bifunctional reagent N-hydroxy-

[15] L. Wofsy, O. Martinez, and W. Godfrey, *Int. Congr. Immunol. 4th*, Abstract 20.85 (1980).
[16] W. Godfrey, B. Doe, E. F. Wallace, B. Bredt, and L. Wofsy, *Exp. Cell Res.* **135**, 137 (1981).
[17] L. D. Leserman, J. Barbet, F. Kourilsky, and J. N. Weinstein, *Nature* (*London*) **288**, 602 (1980).

succinimidyl 3-(2-pyridyldithio)propionate (SPDP). For many purposes, this may represent a distinct improvement over older coupling procedures. SPDP has also been used to link antibodies or avidin to toxins.[18,19]

Coupling with SPDP, developed by Carlsson et al.,[20] adds labile disulfide functions primarily to available amino groups on one or more proteins to be joined. The modified PDP-protein can then react readily with thiol reagents or available sulfhydryls on another protein to form the desired disulfide linkage with the release of pyridine-2-thione.

The following procedures are designed for coupling antibody, Fab', or avidin to red blood cells, but they are readily adapted to preparation of conjugates with other proteins as well as with other types of membranes and surfaces (indicated in our subsequent discussion of hapten-sandwich applications).

Modification of Antibody or Avidin with SPDP

Reagents

SPDP (Pharmacia, Piscataway, New Jersey)
Dithiothreitol (DTT)
Buffer: PBS, pH 7.5
Procedure. Dissolve SPDP in DMSO, 10 mg/ml. To 2–3 mg of protein per milliliter in PBS, add SPDP in a molar ratio of 10 : 1 for antibody or 4 : 1 for avidin. After 20 min at room temperature, pass the reaction solution over Sephadex G-25 to obtain PDP-antibody or PDP-avidin.

The degree of substitution with PDP groups can be determined spectrally, reading the absorbance of the protein solution at 343 nm in the absence and in the presence of 0.02 M DTT; $\varepsilon_M = 8.08 \times 10^3$ M^{-1} cm^{-1} for pyridine-2-thione.[21] Modified avidin can be assayed for retention of biotin-binding activity by methods described previously in this series, Vol. 18A.

Coupling Avidin or Antibody to Red Blood Cells (RBC)

Reagents

SPDP
DTT

---

[18] O. Martinez, E. F. Wallace, and L. Wofsy, *Fed. Proc., Fed. Am. Soc. Exp. Biol.* **39,** 719 (1980).

[19] D. G. Gilliland, J. Steplewski, R. J. Collier, K. F. Mitchell, T. H. Chang, and H. Koprowski, *Proc. Natl. Acad. Sci. U.S.A.* **77,** 4539 (1980).

[20] J. Carlsson, H. Drevin, and R. Axén, *Biochem. J.* **173,** 723 (1978).

[21] T. Stuchbury, M. Shipton, R. Norris, J. P. G. Malthouse, K. Brocklehouse, J. A. Herbert, and H. Suschitzky, *Biochem. J.* **151,** 417 (1975).

Buffers
Sodium carbonate, 0.15 M, pH 9.5
PBS, pH 7.5

Procedure

1. Prepare PDP-RBC by adding 100 μl of SPDP, 50 mg/ml in DMSO, with vortex mixing to a suspension of 1 ml of packed RBC in 9 ml of carbonate buffer. After 20 min at room temperature, wash cells 4 times in PBS.

2. Suspend 1 ml of packed PDP-RBC, freshly prepared, to a volume of 1 ml in PBS. Add 5 ml of 0.1 M DTT, and incubate cells for 20 min at room temperature. After washing 5 times in PBS, allow reduced RBC to react with PDP–protein, 0.2 mg of protein per 0.1 ml of packed RBC-SH in 1–1.5 ml of PBS. Rotate the mixture overnight at room temperature. Wash cells repeatedly in PBS.

Coupling Fab'-anti-ars to RBC

Reduced Fab' (Fab'-SH) may be coupled directly to PDP-RBC. Reduce (Fab')$_2$ anti-ars, 1–3 mg in 0.5 ml of PBS, 2 × 10^{-3} M EDTA, in 2 × 10^{-3} M 2-mercaptoethanol at 37° for 1–2 hr. To remove the reducing agent, pass the solution rapidly over Sephadex G-25 (1 cm × 30 cm), and collect the protein over ice. Mix the reduced protein without delay with PDP-RBC, 0.1 ml of packed cells suspended per volume of PBS containing 0.2 mg of Fab'-SH. Rotate overnight at room temperature, then wash the cells in PBS.

Applications of Hapten-Sandwich Labeling

Hapten-sandwich methods are feasible in virtually all types of procedures to which the more usual indirect or multilayer anti-immunoglobulin techniques[22] have been applied: e.g., labeling antigen with markers for electron microscopy, fluorescence, enzymic and radioisotopic analysis; separating cells on a fluorescence-activated cell sorter (FACS) or by various immunoadsorbent, "panning," and rosetting schemes; isolating specifically labeled membrane components; assaying complement-dependent cytotoxicity; detecting clones of cells producing certain antibodies or factors. In many cases, conventional anti-immunoglobulin procedures and, in recent years, those utilizing staphylococcal protein A,[23] are more convenient and advantageous. Conversely, in cases where recognition by

[22] A. H. Coons, *Int. Rev. Cytol.* **5**, 1 (1956).
[23] J. W. Goding, *J. Immunol. Methods* **20**, 241 (1978).

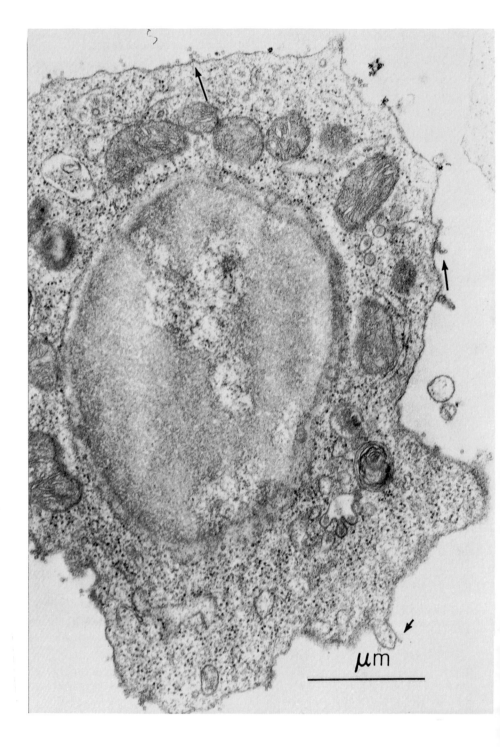

μm

either second-layer anti-Ig or protein A is not narrow enough to meet the needs of a particular experimental design, either anti-hapten antibodies or avidin together with appropriate first-layer conjugates will prove to be most useful. The latter also offer distinct advantages over anti-Ig when it is desirable to have specifically purified second-layer reagents coupled to complex markers (e.g., proteins, cells, liposomes). In singling out particular types of applications below, the focus is on circumstances which may favor choosing the hapten-sandwich technique (generally either the ars-anti-ars or biotin–avidin system).

Labeling Membrane Antigens

Most cell surface antigens that distinguish subsets of leukocytes are not readily visualized directly with fluorescent immunoglobulin. An obvious exception is B cell membrane Ig. Some, but not all, monoclonal antibodies aganist membrane antigens are effective as single-layer reagents coupled to fluorescent or EM markers. In many instances, amplification by indirect labeling is essential or at least preferable. In such cases, first-layer antibodies can be discriminated from B cell Ig by a hapten-sandwich method or by the use of an anti-allotype or (less definitively) an anti-γ specific second-layer reagent. The hapten-sandwich method also permits the use of first-layer ars-Fab and second-layer (Fab')$_2$ anti-ars antibodies in order to avid involvement of macrophage or lymphocyte Fc receptors.[24,25] The electron micrograph in Fig. 3 shows a B lymphocyte labeled with ars-Fab-anti-Ia^k followed by (Fab')$_2$ anti-ars antibody conjugated to KLH. Ars-Fab-anti-Ia^k reagents, in combination with either KLH-conjugated or fluorescent (Fab')$_2$ anti-ars, have been useful in studies detecting Ia antigens on macrophages[24] and T-cell blasts.[26] Anti-ars antibody–KLH conjugates have also been useful in studies of B-cell activation by KLH-primed T cells, since different ars-modified first-layer antibodies can focus the potential antigen target for T cells on any desired B cell membrane receptor.[27,28]

[24] C. Henry, J. R. Goodman, E. Chan, J. Kimura, A. Lucas, and L. Wofsy, *J. Reticuloendothel. Soc.* **26,** 787 (1979).
[25] L. Wofsy, H. O. McDevitt, and C. Henry, *J. Immunol.* **119,** 61 (1977).
[26] C. Henry, B. Doe, J. Kimura, J. North, and L. Wofsy, *Cell. Immunol.* **53,** 125 (1980).
[27] S. Cammisuli, C. Henry, and L. Wofsy, *Eur. J. Immunol.* **8,** 656 (1978).
[28] S. Cammisuli and C. Henry, *Eur. J. Immunol.* **8,** 662 (1978).

FIG. 3. A BALB.K lymphocyte labeled for Ia^k antigens with ars-Fab-anti-Ia^k and (Fab')$_2$ anti-ars–KLH conjugate. Arrows identify the keyhole limpet hemocyanin (KLH) marker. From Henry *et al.*[24]

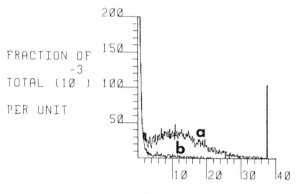

FRACTION OF
TOTAL (10⁻³)
PER UNIT

FLUORESCEIN UNITS

FIG. 4. FACS IV (Becton Dickinson) analysis of CKB mouse spleen cells. Treated with ars-coupled monoclonal anti-Ia^k (specificity 2) antibody and (a) Fl-anti-ars antibodies; (b) anti-ars, biotin-anti-Ig5a, and Fl-avidin. Relative cell number (10,000 total cells) is plotted against fluorescence intensity (linear units).

With monoclonal antibodies that do not fix complement, a highly effective means of achieving cytotoxicity is the use of ars-modified monoclonal with unmodified rabbit anti-ars antibody.[29] FACS analysis shows that monoclonal antibodies of any class or subclass remain effective after ars or biotin modification as first-layer reagents for indirect fluorescence labeling. The choice between the ars and biotin labeling systems in any given case should depend on convenience.

Double Labeling

Hapten-sandwich labeling is particularly valuable when two antigens must be labeled with distinguishable markers. Its use may be essential when detection of one or both of the antigens requires amplification by indirect labeling, since here the use of second-layer anti-Ig reagents presents obvious complications. The requirements for specificity and discrimination for double-labeling are extremely demanding, since none of the ligands (first or second layer) may cross-react with another. This criterion is met successfully when one antigen can be labeled directly and the other is labeled by either ars-anti-ars or biotin-avidin reagents. When double amplification is desirable, both of the latter hapten-sandwich type reagent systems may be used together. Figure 4 shows a FACS analysis of CKB mouse spleen cells labeled with ars-anti-Ia^k(2) and Fl-anti-ars anti-

[29] J. Kimura, C. Metzler, Lee Herzenberg, and L. Wofsy, unpublished observations.

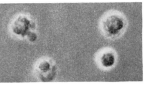

FIG. 5. Double labeling to detect expression of Ia^k antigens on C$_3$H/diSN mouse T-cell blasts. Spleen cells were depleted of B lymphocytes and cultured for 2 days with concanavalin A. Harvested cells were treated with ars-Fab-anti-Ia^k and glut-anti-mouse brain antibody (specific for T lymphocytes), washed, and then labeled with Fl-anti-ars and Rh-anti-glut (glut = glutamate). The photograph shows a field of cells in phase (left), fluorescein (center), and rhodamine (right). All four blasts carry the T-cell label, and two also label as Ia^k positive. From Wallace and Wofsy.[8]

bodies (a); the negative control cells from the same spleen suspension (b) are devoid of fluorescent label even though they have been treated with ars-anti-Ia^k(2) and unmodified anti-ars, followed by an irrelevant (non-binding) biotin–antibody and Fl–avidin. The absence of cross-recognition has also been demonstrated in FACS analysis of the reverse labeling procedure, where an irrelevant ars-antibody with Fl-anti-ars does not stain cells that bind a specific biotin-antibody and unmodified avidin.

Other non-cross-reacting hapten systems have also been used effectively as partners with anti-ars in double labeling.[1,8] For example, an anti-T cell reagent modified with HB-p-azobenzoyl glutamate (glut) has been used with Rh-anti-glut to label T-cell blasts, some of which also are labeled with ars-Fab-anti-Ia^k and Fl-anti-ars (Fig. 5). In general, however, the combination of the ars and biotin reagents is most satisfactory for double labeling.

Specific Delivery of Larger Markers (Vesicles, Cells)

Hapten-sandwich methods are advantageous in several applications aimed at specific attachment of larger markers (e.g., membrane vesicles, selected cell types, beads, virus particles, or microorganisms) to a cell or other surface. Such applications may be involved in studies of membrane fusion, cell–cell interactions, specific delivery of drugs or cytotoxic agents. They may also facilitate procedures for specific cell separation,[30] labeling for scanning electron microscopy,[31] and achieving extra amplification in fluorescence labeling for FACS analysis.

[30] M. Slomich, E. Kwan, L. Wofsy, and C. Henry, in "Selected Methods in Cellular Immunology" (B. B. Mishell and S. M. Shiigi, eds.), p. 212. Freeman, San Francisco, 1980.
[31] M. K. Nemanic, D. P. Carter, D. R. Pitelka, and L. Wofsy, J. Cell Biol. 64, 311 (1975).

Red blood cells that have been conjugated with anti-ars antibody or avidin, prepared as described earlier in the section Disulfide-Linked Conjugates, can readily be converted to sealed erythrocyte ghosts containing fluorescent molecules.[16] Such highly fluorescent vesicles will cluster specifically around subsets of cells labeled with ars- or biotin-modified antibodies (Fig. 6).

Figure 7 shows FACS analyses of C57B1/6 mouse spleen lymphocytes labeled with ars-modified monoclonal anti-Igh-5b followed either by Fl-anti-ars antibody or by anti-ars-coupled RBC vesicles containing fluorescein-modified bovine serum albumin (Fl-BSA). This particular monoclonal antibody is a rather weak reagent for labeling the δ-chain allotype antigens on B cells, and neither Fl-anti-ars antibody nor Fl-avidin is very effective (following appropriately modified first-layer antibody) in resolv-

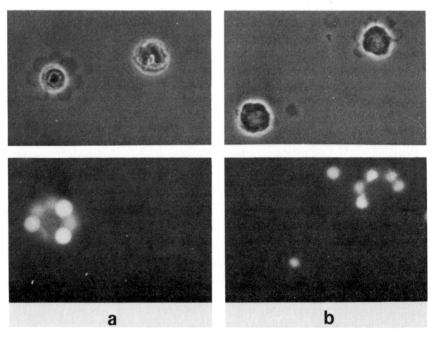

FIG. 6. Targeting of fluorescent sealed red blood cell ghosts to specifically labeled lymphocytes. Each vertically aligned pair of micrographs (a, b) shows the same field in phase (upper) and fluorescence (lower). (a) C57B1/6 cells, when labeled specifically with ars-anti-Igh-5b, formed clusters with Fab'-anti-ars-coupled ghosts. (b) When C3H/Hej mouse spleen cells were treated with biotin-anti-Ia^k (specificity 2), Ia-positive cells formed clusters with avidin-coupled ghosts. From Godfrey et al.[16]

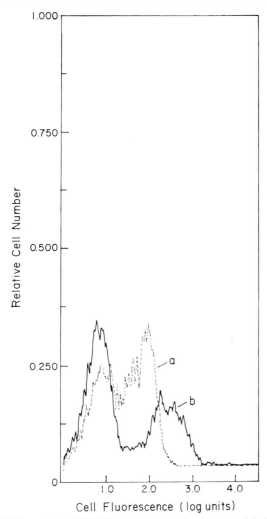

FIG. 7. FACS IV analysis of C57B1/6 mouse spleen lymphocytes labeled with ars-anti-Igh-5b followed by: (1) Fl-anti-ars antibody, or (b) anti-ars-coupled red blood cell ghosts containing Fl-BSA.

ing labeled from unlabeled cells with FACS. Nevertheless, Fig. 7 shows that the specifically targeted fluorescent vesicles permit clear resolution with a greater than threefold average enhancement of fluorescence intensity in the labeled cell population.

Preparation and Targeting of Fluorescent Sealed RBC Ghosts[16]

Protein-conjugated sealed RBC ghosts are prepared by the method of Bodeman and Passow[32] from the corresponding modified RBC previously coupled (see above) with avidin, antibody, or Fab'.

Reagents

Buffers

Sodium barbital, 5.0 mM; MgCl, 0.5 mM; CaCl$_2$, 0.15 mM; pH 7.4

PBS

Minimum Essential Medium (MEM) (Microbiological Associates, Walkersville, Maryland), containing 0.3% BSA, 0.1% sodium azide

Fl-BSA: To prepare Fl-BSA, allow a 1% solution of BSA in 0.15 M sodium bicarbonate, pH 9, to react with a 10 × molar excess of FITC for 2 hr at room temperature; purify by filtration on Sephadex G-25.

NaCl, 1.65 M

Procedure. Suspend 0.2 ml of packed, protein-conjugated RBC with 0.4 ml of Fl-BSA, 3 mg/ml, in PBS. Add 2 ml of barbital buffer, with mixing, at 0°. After 5 min, restore isotonicity by adding 0.2 ml of 1.65 M NaCl. Maintain for another 5 min at 0°, then incubate for 1 hr at 37°. After washing 4 times in PBS, suspend the packed ghosts in an equal volume of MEM.

For targeting to specifically labeled lymphocytes, mix with a pipette 0.1 ml of the fluorescent sealed ghost suspension (1 to 2 × 10^8 ghosts) with 0.1 ml of a suspension of labeled cells containing about 10^7 lymphocytes. Assay the percentage of lymphocytes binding clusters of ghosts (three or more per cell) by phase and fluorescence microscopy. FACS analysis, using a 70-μm nozzle, is readily performed with such cell-ghost suspensions.

[32] H. Bodeman and H. J. Passow, *J. Membr. Biol.* **8,** 1 (1972).

[38] Use of Lectin–Antibody Conjugates for Quantitation and Titration of Antigens and Antibodies

By Jean-Luc Guesdon and Stratis Avrameas

Over the past 10 years, antigen and antibody quantitation by various enzyme immunoassay techniques have been described in many reports,[1–4] and such techniques have been extensively applied in several biochemical fields. Books and reviews on these immunoassay methods have also been published.[5–9] Enzyme immunoassays involve the coupling of antigen or antibody with an enzyme, either by covalent binding or biospecific interaction. In many covalent binding procedures, organic compounds are used to prepare the necessary enzyme antibody or antigen conjugates.[10] Techniques based on biospecific interactions like the antigen–antibody[11–13] or avidin–biotin[14] systems have been successfully applied in various enzyme immunological procedures. We have tested the possibility of using lectin saccharide biospecific interactions for immunoassay.[15] Lectins are proteins found in relatively large amounts in the seeds of various plants,[16] and they are able to interact specifically with carbohydrate moieties. Because these moieties are present on the surface of cells like erythrocytes and in macromolecules like enzymes, lectin–antibody or lectin–antigen conjugates can serve to develop immunoassays using these moieties as markers.

[1] S. Avrameas and B. Guilbert, *C. R. Hebd. Seances Acad. Sci. Ser. D* **273**, 2705 (1971).

[2] E. Engvall and P. Perlmann, *Immunochemistry* **8**, 871 (1971).

[3] B. K. Van Weemen and A. H. W. M. Schuurs, *FEBS Lett.* **15**, 232 (1971).

[4] S. Avrameas and B. Guilbert, *Biochimie* **54**, 837 (1972).

[5] E. Engvall and A. J. Pesce, "Quantitative Enzyme Immunoassay." Blackwell, Oxford, 1978.

[6] A. J. O'Beirne and H. R. Cooper, *J. Histochem. Cytochem.* **27**, 1148 (1979).

[7] M. Oellerich, *J. Clin. Chem. Clin. Biochem.* **18**, 197 (1980).

[8] R. Malvano, "Immunoenzymatic Assay Techniques." Nijhoff, The Hague, 1980.

[9] J. L. Guesdon and S. Avrameas, *Appl. Biochem. Bioeng.* **3**, 27 (1981).

[10] S. Avrameas, T. Ternynck, and J. L. Guesdon, *Scand. J. Immunol.* **8** (Suppl. 7), 7 (1978).

[11] S. Avrameas, *Immunochemistry* **6**, 825 (1969).

[12] T. E. Mason, R. F. Phifer, S. S. Spicer, R. A. Swallow, and R. B. Dreskin, *J. Histochem. Cytochem.* **17**, 563 (1969).

[13] L. A. Sternberger, J. Hardy, J. J. Cuculis, and H. G. Meyer, *J. Histochem. Cytochem.* **18**, 315 (1970).

[14] J. L. Guesdon, T. Ternynck, and S. Avrameas, *J. Histochem. Cytochem.* **27**, 1131 (1979).

[15] J. L. Guesdon and S. Avrameas, *J. Immunol. Methods* **39**, 1 (1980).

[16] H. Lis and N. Sharon, *in* "The Antigens" (M. Sela, ed.), Vol. 4, p. 429. Academic Press, New York, 1977.

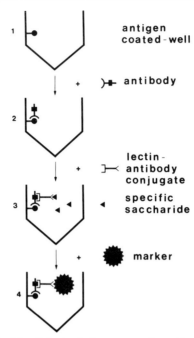

FIG. 1. Principle of the lectin immunotest for antibody titration.

This chapter deals with the preparation of lectin–antibody conjugates and the development of the above type of immunoassay for antigen and antibody titration.

The principle of the use of a lectin–antibody conjugate for antibody titration is schematically illustrated in Fig. 1. An adequate dilution of the serum to be titrated is added to wells coated with the corresponding antigen. After incubation, wells are washed and then filled with lectin-labeled anti-Ig antibody solution, made in saline containing an excess of a saccharide specific to the lectin used. The antigen-combining site of the lectin–antibody conjugate is thus allowed to react with the Ig, but the excess of the saccharide present prevents the lectin-combining sites of the conjugate from reacting with any polysaccharides or glycoproteins present in the reaction medium or on the solid phase. The preparation is washed to eliminate excess saccharide and lectin–antibody conjugate, and a marker substance possessing carbohydrate moieties able to interact specifically with the lectin is added. After additional incubation and washings, the marker is visualized or quantitated by an appropriate procedure. Whatever the marker, the intensity of the response is proportional to the concentration of the antibodies being titrated. Results can be expressed as the serum end point dilution at which antibody activity can no longer be detected.

The antigen is quantitated by a similar procedure. Compared to the above titration technique the differences are that wells are coated with specific antibody directed against the antigen under consideration, and the conjugate is prepared by coupling the same antibody with a lectin. By using a standard solution containing known amounts of antigen, the absolute concentration of the substance under measurement can be obtained. In these lectin immunotests (LIT) the marker can be an enzyme (Enzy LIT), a radioactive glycoconjugate (Radio LIT), or erythrocytes (Erythro LIT).

Materials

Reagents for Lectin Immunotests

Horseradish peroxidase (grade I, RZ = 3; grade II, RZ = 0.6) and *Aspergillus niger* glucose oxidase (grade I) (Boehringer-Mannheim, West Germany)

Escherichia coli K12 β-D-galactosidase (Institute Pasteur, Paris, France)

Agglutinins of *Canavalia ensiformis, Triticum vulgare, Lens culinaris, Ricinus communis, Phaseolus vulgaris*, methyl-α-D-mannoside, *N*-acetylglucosamine, and Ultrogel beads (Pharmindustrie, Villeneuve-La-Garenne, France)

Galactose and *N*-acetylgalactosamine (Sigma, U.S.A.)

Rabbit γ-globulin fraction II (Miles Laboratories, Kankakee, Illinois)

Gelatin (Prolabo, France)

Glutaraldehyde, 25% aqueous solution (Taab Laboratories, Reading, U.K.)

Tween 20 (E. Merck, Darmstadt, West Germany)

Polystyrene plates with U- or V-shaped wells and flat-bottom wells (Greiner, France, and Nunc, Denmark).

Antisera. The rabbit and sheep antisera used in the procedure described here are obtained by hyperimmunizing animals with the corresponding antigen incorporated in complete Freund's adjuvant (Difco Laboratories, Detroit, Michigan) followed by the procedures already described.[17]

---

[17] S. Avrameas, *Immunochemistry* **6**, 43 (1969).

Buffers

PBS: 0.15 M NaCl containing 0.01 M potassium phosphate buffer, pH 7.4

Coating buffer: 0.1 M carbonate–bicarbonate buffer, pH 9.6

Washing buffer: PBS containing 0.1% Tween 20

Incubation buffer: PBS containing 0.1% Tween 20 and 0.3% gelatin

Coupling buffer: 0.1 M potassium phosphate buffer, pH 6.8

Methods

Preparation of Antibodies

The necessary purified antibodies are isolated by immobilizing the corresponding antigens on glutaraldehyde-activated Ultrogel beads, a procedure described in detail elsewhere.[18]

Preparation of Lectin–Antibody Conjugates

Lectins are coupled by the one-step glutaraldehyde procedure.[10,15] For this purpose, lectins and isolated antibodies are dialyzed overnight at 4° against the coupling buffer. Four milligrams of lectin are mixed with 2 mg of antibodies in a final volume of 900 μl. Lectin-specific sugar (100 μl) is added to the mixture to obtain a concentration of 0.1 M; 20 μl of a 1% aqueous glutaraldehyde solution is added while the mixture is stirred. After 3 hr of incubation at room temperature, 50 μl of 2 M glycine solution is added. Two hours later, the mixture is dialyzed against PBS at 4°, left to stand overnight, and centrifuged for 30 min at 3000 g. Finally, a volume of glycerol equal to that of the preparation is added, and the preparation is stored at −20° until use.

Coating of Polystyrene Plates

The necessary immobilized antibody and antigen for the various lectin immunotests are prepared by passive coating of polystyrene plates. This is done by incubating 0.1 ml of antibodies or antigens, diluted in 0.1 M carbonate–bicarbonate buffer, pH 9.6, for 1 hr at 37° and then for 18 hr at 4°. A wide variety of antigens, including DNA, proteins, polysaccharides, and lipopolysaccharides from various sources adhere satisfactorily to the plates by this procedure. Because a prozone phenomenon usually occurs, it is important to determine the optimum concentration of the coating antigen for each new immunological system; this is done by checkerboard

---

[18] J. L. Guesdon and S. Avrameas, *J. Immunol. Methods* **11**, 129 (1976).

titration of the antigen dilution, using one positive and one negative serum. The antigen dilution giving the maximal difference between the positive and the negative serum is used to coat the plates.

Erythrocytes

For the Erythro LIT we usually use sheep red blood cells (SRBC). Before use, they are stored in sterile Alsever's medium and washed about 10 times by centrifugation for 10 min at 600 g with PBS. The SRBC are then suspended in PBS at a concentration such that after complete hemolysis the absorbance measured at 414 nm is between 0.4 and 0.7. Alternatively, glutaraldehyde-treated SRBC can be used instead of normal SRBC. In this case, a 2% suspension of SRBC is treated with 1% glutaraldehyde in cold PBS. The suspension is gently stirred for 1 hr at 4° and then centrifuged at 1500 g for 10 min. The pellet is washed 5 times with PBS and 5 times with distilled water. The suspension can be used for at least 6 months when kept at 4° under sterile conditions. The optimal concentration of SRBC for the assay must be determined for each new batch. Batches are generally used at concentrations ranging from 0.02 to 0.10% (v/v).

Enzyme Assay

Peroxidase is measured by means of a reaction mixture containing *o*-phenylenediamine and hydrogen peroxide.[19] Glucose oxidase activity is measured with a substrate solution containing peroxidase, *o*-phenylenediamine, and D-glucose.[20] The reaction for both enzymes is stopped by adding 3 N HCl, and the optical density is read at 492 nm with a Titertek Multiskan photometer. For β-galactosidase determination, *O*-nitrophenyl-β-D-galactopyranoside is used, and the reaction is stopped by adding 2 M Na$_2$CO$_3$; absorbance is read at 420 nm.[21]

Lectin Immunotest (LIT)

Erythro LIT. For antibody titration by Erythro LIT, 0.1-ml dilutions of the test serum are added to the wells of a plate with U- or V-type wells, coated with the corresponding antigen. After incubation for 2 hr at 37°, the plate is washed by emptying the wells, flooding then with PBS Tween, and emptying then again. This procedure is repeated three times, and the

[19] G. Wolters, L. Kuijpers, J. Kacaki, and A. Schuurs, *J. Clin. Pathol.* **29,** 873 (1976).
[20] H. Labrousse, F. Thron, S. Avrameas, and J. F. Bach, *J. Clin. Lab. Immunol.* **3,** 191 (1980).
[21] J. L. Guesdon, R. Thierry, and S. Avrameas, *J. Allergy Clin. Immunol.* **61,** 23 (1978).

plate is then shaken dry. An 0.1-ml dilution of the lectin anti-Ig antibody conjugate is added. This dilution is made in the incubation buffer containing 0.1 M of one of the following lectin-specific saccharides: methyl-α-D-mannoside for *Canavalia ensiformis* (Con A) and *Lens culinaris* agglutinin (LCA), N-acetylglucosamine for *Triticum vulgare* agglutinin (WGA), D-galactose for *Ricinus communis* agglutinin (RCA), and N-acetylgalactosamine for *Phaseolus vulgaris* agglutinin (PHA). Conjugates are usually used at dilutions of 5–10 μg/ml, except for WGA–antibody conjugate, which gives optimal results at 0.01–0.05 μg/ml. After incubation for 2 hr at 37°, the plate is washed as above and 0.1 ml of SRBC suspension is added. The plate is allowed to incubate at room temperature and is protected from vibrations.

Incubation time usually varies from 4 to 18 hrs, depending on the nature of the substance adsorbed on the plate, the nature and concentration of the erythrocytes, and the shape of the well. At the end of incubation, the erythrocytes in positive wells are uniformly adsorbed on the surface of the well and appear to agglutinate, whereas those in negative wells settle at the bottom of the well (Fig. 2).

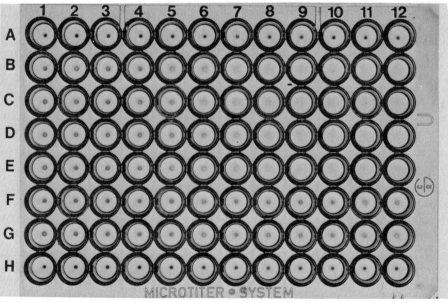

Fig. 2. Photograph of a microplate with U-type wells after incubation with sheep red blood cells. Complete erythrocyte adsorption is visible in wells B_9 to B_{12}, C_9 to C_{12}, D_9 to D_{12}, and E_9 to E_{12}.

False-positive reactions may occur with Erythro LIT and with certain sera tested at weak dilutions (<1/500). In such cases sera should be absorbed twice with a freshly prepared 10% erythrocyte suspension. Note that with some antigens and lectins it is not necessary to add the specific sugar during incubation.[22]

Enzy LIT. Sera are titrated by LIT with enzyme as marker, by a procedure similar to that used for Erythro LIT. The differences are that flat-bottomed micro enzyme-linked immunosorbent assay (ELISA) plates are used instead of plates with U- or V-type wells, and incubation with SRBC is replaced by incubation with the enzyme. The incubation time depends on the affinity of the lectin used for the enzyme marker. With Con A, incubation lasts for 15 min for glucose oxidase and 180 min for peroxidase. After washing, 0.2 ml of substrate solution is added to each well and the enzyme reaction is allowed to proceed at 37°. Note that not all lectins are effective with the enzymes currently used in enzyme immunoassays. Even if an enzyme marker has no carbohydrate moieties, it can nevertheless be used in Enzy LIT after coupling with a polysaccharide or glycoprotein. For example, *Escherichia coli* β-D-galactosidase can be coupled to glucose oxidase for use in Enzy LIT, with glutaraldehyde as coupling agent.[15]

Radio LIT. Antibody quantification by LIT using a radioactive label requires a similar procedure to that described above. Incubation with enzyme is replaced by incubation with a radioactive compound linked to a carbohydrate moiety. The radioactivity associated with the solid phase can be measured in two ways. Either the plate is cut and the radioactivity is measured by placing each well in a radioactivity counter, or the plate is filled with 6 N NaOH to elute the material adsorbed, and the radioactivity in the liquid phase is measured.

Applications

Quantitation of Human α-Fetoprotein (AFP)

Quantitation of AFP by Erythro LIT. Sheep anti-AFP antibody is diluted to 4 μg/ml in coating buffer. Of this solution, 0.1 ml is poured into each well in the U-bottomed polystyrene plate, and the plate is incubated for 1 hr at 37° and then for 18 hr at 4°. The plate is washed three times, wells are filled with 0.1 ml of serum dilutions or standard cord serum, and incubation is allowed to proceed for 90 min at 37°. After further washing,

[22] D. Kovatchev, J. L. Guesdon, D. Ninova, and S. Avrameas, *Ann. Immunol. (Paris)* **132C**, 77 (1981).

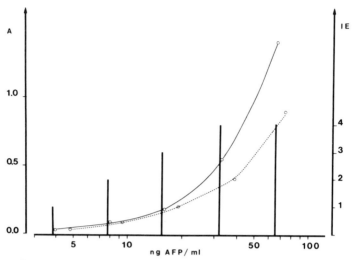

FIG. 3. Quantitation of human α-fetoprotein by enzyme immunoassay ($\cdots$), Enzy Lectin Immunotest (——), and Erythro Lectin Immunotest (vertical lines). Galactosidase antibody conjugate is used for the enzyme immunoassay, and Con A antibody conjugate for the Enzy Lectin Immunotest, with glucose oxidase as marker. In both cases absorbance (A) is a function of the α-fetoprotein concentration. In the present case, the Erythro Lectin Immunotest was carried out with Con A antibody conjugate and SRBC as marker. Results are expressed as the intensity of erythro adsorption (IE).

0.1 ml of Con A-labeled anti-AFP antibody diluted to 5 μg/ml is added to each well in incubation buffer containing 0.1 M methyl α-D-mannoside. The plate is incubated for 90 min at 37°, washed, and then filled with 0.1 ml of SRBC suspension (dilution about 0.08%). Results are scored 3 hr later, and the AFP concentration in each serum sample is determined by comparison with the results obtained with standard cord serum (Fig. 3).

Note: Sera to be tested are absorbed twice with a 10% suspension of SRBC and diluted from 1:2 to 1:256. Cord serum containing a known amount of AFP is diluted to obtain AFP concentrations ranging from 60 μg/ml to 0.5 ng/ml.

Quantitation of α-Fetoprotein by Enzy LIT. α-Fetoprotein is quantified by Enzy LIT by a procedure similar to that described above for Erythro LIT. The difference is that the addition of the SRBC suspension to the wells is replaced by addition of 0.1 ml of glucose oxidase (50 μg/ml) diluted in incubation buffer containing 1 mM Mg Titriplex, CaCl$_2$, and MnCl$_2$. The plate is then incubated for 30 min at 37° and washed three times. The enzyme is measured by adding 0.2 ml of glucose oxidase substrate to each well and allowing the reaction to proceed for 30 min at

37°. At this end point, 50 μl of 3 N HCl are added and the absorbance is measured.

Note: In Enzy LIT it is preferable to use plates with flat-bottomed wells rather than U- or V-shaped wells.

The results obtained by Erythro LIT, Enzy LIT, and conventional sandwich enzyme immunoassay using β-galactosidase-labeled anti-AFP antibody are compared in Fig. 3. The same sensitivity is obtained with all three techniques.

Titration of Antibody to Echinococcus granulosus in Human Sera

Echinococcus granulosus antigen is dissolved in 0.1 M carbonate buffer, pH 9.6, to obtain a final concentration of 10 μg/ml. The antigen is passively adsorbed on a U-bottomed plate by adding 0.1 ml of antigen solution per well and incubating the plate for 1 hr at 37° and then for 18 hr at 4°. The plate is washed to eliminate excess nonbound antigen, and the wells are filled with 0.1 ml of patient sera, at dilutions ranging from 1 : 200 to 1 : 25,600. After incubation for 2 hr at 37°, the plate is washed and each well is filled with 0.1 ml of Con A-labeled anti-human IgG antibody diluted to 5 μg/ml. The plate is further incubated for 2 hr at 37° and then washed and filled with an 0.05% suspension of glutaraldehyde-treated SRBC. The plate is left to stand overnight at room temperature, and the erythro-adsorption pictures obtained are scored as in conventional hemagglutination (Fig. 2). Figure 4 gives an example of the titration of anti-*Echinococ-*

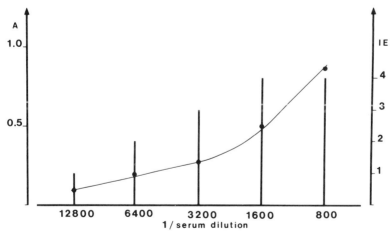

FIG. 4. Titration of anti-*Echinococcus granulosus* antibody in a human serum, either by enzyme immunoassay with phosphatase–antibody conjugate or by the Erythro Lectin Immunotest with Con A antibody conjugate. Absorbance (A) and intensity of erythro adsorption (IE) are a function of serum dilution.

cus granulosus antibody in human sera, using either Con A antibody conjugate followed by addition of SRBC (Erythro LIT) or alkaline phosphatase antibody conjugate followed by addition of substrate.

Concluding Remarks

Lectin immunotests have been successfully used to quantify IgE[15] and α-fetoprotein[22] in human sera and rotavirus antigen in stools,[23] as well as for titration of antibodies against DNA[24] and actin[22] in human sera. Good correlations have been observed between the results obtained with these lectin immunotests and those obtained by other immunological techniques. Furthermore, in these experiments, it was found that the use of lectin-antibody or lectin–antigen conjugates allowed the development of sensitive and reliable immunological techniques. In particular, the use of red blood cells as marker makes the procedure very simple and inexpensive; its detection limit is close to that obtained by conventional enzyme immunoassay. Because of the adaptability of lectin–protein conjugates, they will no doubt be increasingly used for immunological procedures involving labeled antigen or antibody.

[23] J. Prévot and J. L. Guesdon, *Ann. Virol.* (*Paris*) **132E,** 529 (1981).
[24] J. L. Guesdon and S. Avrameas, *Ann. Immunol.* (*Paris*) **131C,** 389 (1980).

[39] Partition Affinity Ligand Assay (PALA) for Quantifying Haptens, Macromolecules, and Whole Cells

By Bo Mattiasson

A crucial point when designing a biochemical binding assay involving two biologically active molecules is the separation of bound from free substance. In a competitive assay a small amount of labeled antigen competes with a small amount of native antigen for a limiting amount of specific antibodies. This situation places high demands on the reproducibility of the binding reaction, on the separation procedure, and on the measuring step. If binding is taking place in a homogeneous solution, the accuracy of the performance of such a step depends on the pipetting accuracy. Today good accuracy can be expected. If solid-phase material takes part in the binding assay, a slightly lower accuracy may be expected. The last step—the measurement—is governed by the accuracy of the equipment used. Here, too, a high degree of accuracy can be ex-

METHODS IN ENZYMOLOGY, VOL. 92

pected. The crucial point is, then, the separation procedure. Various approaches have been tried. In their classical work Yalow and Berson[1] applied electrophoretic separation. This technique proved too tedious and laborious and was later replaced by solid-phase techniques in which one of the reactants is bound to a solid phase.[2,3] After the other reactant in a competitive binding reaction has bound to the solid phase material, the crucial separation process is started. Centrifugation and decantation include several steps that, besides being time- and labor-consuming, also tend to lower the degree of accuracy of the method. Furthermore, if the intervals between the beginning and the end of the separation are long, there is a risk of dissociation of the bound antigen. A fast method must therefore be desirable.

There also have been developed homogeneous enzyme immunoassays for haptens.[4,5] These methods involve no separation steps, since the enzyme activity is read in the incubation solution in which the binding reaction takes place. So far these homogeneous enzyme immunoassays are useful only for haptens.

The term partition affinity ligand assay (PALA) designates binding assays with separation in aqueous two-phase systems.[6] Aqueous two-phase systems have long been applied in the purification of macromolecules, organelles, and cells.[7] The phase systems have a high water content. Each phase contains 80–95% water, and the phases are thus biocompatible. That aqueous polymer solutions are immiscible has long been known,[8] and today a broad spectrum of polymers have been used for preparing aqueous two-phase systems.

According to the general concept of PALA, one of the free molecular entities in the reaction pair studied in the binding reaction passes preferentially to one of the phases, whereas when bound in a complex it can be recovered from the other phase. Binding of the reactant is thus equivalent to a translocation of the molecule across the phase boundary in the subsequent separation step.

The binding reaction can take place in free solution, and the separation is initiated by the addition of a well-mixed phase system followed by

[1] R. S. Yalow and S. A. Berson, *Nature (London)* **184**, 1648 (1959).

[2] L. Wide and J. Porath, *Biochim. Biophys. Acta* **130**, 257 (1966).

[3] K. J. Catt and G. W. Tregear, *Science* **158**, 1570 (1967).

[4] K. Rubinstein, R. Schneider, and E. F. Ullman, *Biochem. Biophys. Res. Commun.* **47**, 846 (1972).

[5] J. F. Burd, R. C. Wong, J. E. Feeney *et al.*, *Clin. Chem.* **23**, 1402 (1977).

[6] B. Mattiasson, *J. Immunol. Methods* **35**, 135 (1980).

[7] P.-Å. Albertsson, "Partition of Cell Particles and Macromolecules." Almqvist & Wiksell, Uppsala, Sweden, 1971.

[8] M. N. Beijerinek, *Zentralbl. Bakteriol.* **2**, 627 (1896).

mixing for a few seconds. Since partition in such a well-mixed system is instantaneous, separation takes place within a few seconds after the addition of the phase system. In a subsequent spontaneous step the phases separate, but this does not influence the partition per se.

The basis of the assay was demonstrated in a rather simple system, digoxin–antidigoxin, where the molecular entities, when free, passed preferentially to different phases; when bound, the antigen was recovered from the antibody-rich phase.[6] However, in many cases both reactants have a predilection for the same phase. In such cases one of the reactants must be modified. This situation was studied for haptens, macromolecules, and cells in competitive assays as well as in direct binding assays.

The use of phase systems also offers good possibilities for direct measurements without further separation. This aspect will be discussed also for radioimmunoassay (RIA) and for enzyme immunoassay (EIA).

This chapter is divided into three sections, for assay of haptens, macromolecules, and cells.

Quantification of Haptens

Haptens are perhaps the group that lends itself most readily to analysis with the PALA technique since there is a marked difference in molecular size between the hapten and the antibody. Provided that there is initially a difference in partition behavior of the free molecular entities, it can be predicted that when bound to the receptor the hapten will move to the phase preferred by the antibody. In some cases, however, when both reactants prefer the same phase, specific precautions are necessary.

Radioimmunoassay of Digoxin

The data largely follow Mattiasson.[6]

Reagents

[125]I-labeled digoxin
Native digoxin in standard sera (0, 0.5, 1, 2, 4, 8, and 16 nmol/liter)
Anti-digoxin antiserum (raised in sheep)
Poly(ethylene glycol)-8000 (PEG-8000)
$MgSO_4 \cdot 7 H_2O$
Tris-HCl buffer, 10 mM, pH 7.5
Procedure. To find out whether the partitioning behavior was favorable, 100 μl of ^{125}I-labeled digoxin were placed in a number of wifuge test tubes. Increasing amounts of antiserum were added before addition of a well-mixed phase system; after phase separation, counts in the top phase and in the bottom phase were counted.

TABLE I[a]

PARTITION OF [^{125}I]DIGOXIN IN A PEG–MAGNESIUM SULFATE
TWO-PHASE SYSTEM[b] AS INFLUENCED BY THE ADDITION OF
ANTI-DIGOXIN ANTISERUM[a]

Antiserum added (dilution 1 : 10) (μl)	Counts per minute found in the top phase (mean value, $n = 2$)	Counts per minute found in the bottom phase (mean value, $n = 2$)
0	10,568	2,614
10	1,263	14,308
20	1,215	14,895
30	1,005	14,702
50	940	15,003

[a] From Mattiasson,[6] with permission.
[b] PEG, 1 ml, 30% (w/w); MgSO$_4$, 1 ml, 30% (w/w).

From such a simple experiment it was clearly seen that binding of digoxin to the antibody changed its partition behavior in the subsequent separation step. Thus, the system would be useful for demonstrating the principle of partition affinity ligand assay (PALA) (Table I).

Fifty microliters of the sample (serum sample or standard serum) were mixed with 100 μl of ^{125}I-labeled digoxin and 100 μl of antiserum (diluted 1 : 25,000 in 10 mM Tris-HCl, pH 7.5) in a test tube. Binding was allowed to take place at room temperature for a predetermined period of time, 15 or 30 min or 24 hrs. After binding had taken place a well mixed phase system consisting of 1 ml of 30% (w/w) PEG-8000 and 3 ml of 30% (w/w) MgSO$_4 \cdot 7$ H$_2$O was added. After mixing on a Vortex mixer for 5 sec, phase separation took place (5–10 min). From each phase an aliquot (50 or 100 μl) was taken out and counted in a gamma counter (LKB Mini-gamma Counter).

When taking an aliquot from the bottom phase, it is important to advance the pipette through the upper phase into the lower phase in order to keep contamination from the upper phase at a minimum. In all these experiments we used Gilson micropipettes with disposable plastic tips. Wifuge tubes were used.

In experiments with 30-min binding, a calibration curve was obtained as shown in Fig. 1; when samples from the local hospital were analyzed with a conventional solid-phase RIA, a correlation (Fig. 2) was found.

The analytical power of such an assay must be related to the partition between the two phases of the participating reactants. This partition behavior is usually expressed as the partition coefficient, $K_{part} \cdot K_{part}$ is defined as the ratio between the concentrations of the substance in the top

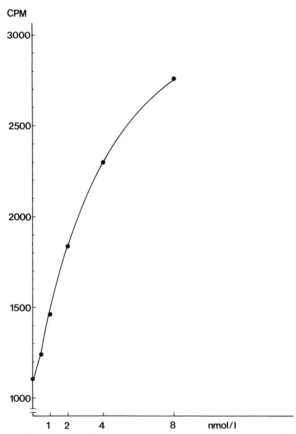

FIG. 1. Calibration curve for digoxin. Counts per minute were measured in the top phase. Reproduced from Mattiasson,[6] with permission.

and bottom phases, respectively. To improve this partition behavior, either of the two reactants may be modified. Since haptens are small molecules, we assumed that modification would probably disturb their ability to bind to the antibodies, especially since large substitutents had to be coupled to them in order to cause the antigen–antibody complexes to partition to the same phase as the modified antigen, not to the same phase as the free antibody.

Instead, the antibody was modified. Much has been written about chemical modification of proteins, especially immobilization of proteins.[9,10] To improve the tendency of the antibody to stay in the bottom

[9] See this series, Vol. 44.

[10] T. M. S. Chang, ed., "Biomedical Applications of Immobilized Enzymes and Proteins," Vols. 1–2. Plenum, New York, 1977.

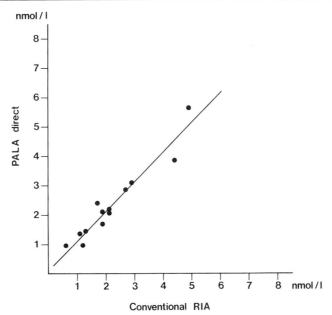

FIG. 2. Correlation between the results obtained with the partition affinity ligand assay (PALA) method and conventional solid-phase radioimmunoassay (RIA).

phase, it was made more hydrophilic by coupling it to small particles of carbohydrate origin, either Sephadex or very small agarose beads. Coupling was performed according to the CNBr method.[11]

Five grams of swollen particles were placed in a beaker with 10 ml of water. The pH was adjusted to 11.0, and a solution of CNBr (25 mg/ml; 5 ml) was added. The pH was kept constant by additions of 2 M NaOH. The titration was maintained for 8 min before the spheres were separated from the activating solution (Sephadex on a glass filter, and the small agarose beads by centrifugation). The beads were washed in cold 0.1 M NaHCO$_3$ and then transferred to a test tube containing 10 ml of NaHCO$_3$ with the antiserum. Coupling proceeded for 16 hr at 4°. The use of this antibody preparation improved the performance of the assay in that it improved the asymmetric behavior of partition.

Binding Assays Using Concanavalin A

These assays have been essentially described by Mattiasson and Ling.[12]

---

[11] R. Axén, J. Porath, and S. Ernback, *Nature* (*London*) **214**, 1302 (1967).
[12] B. Mattiasson and T. G. I. Ling, *J. Immunol. Methods* **38**, 217 (1980).

The lectin–carbohydrate interactions provide a good model for studying binding reactions. Lectins are protein molecules with specific binding properties for carbohydrates. Many lectins with different specificities are commercially available. Lectins are much cheaper than antibodies. A lectin–carbohydrate model system including specific binding is therefore a good model system to study.

The lectin concanavalin A (Con A) is specific for α-D-mannopyranosides and α-D-glucopyranosides. It has been shown to bind several of the glucoenzymes most commonly used in enzyme technology, e.g., glucose oxidase and peroxidase.

The binding assay consisted simply of binding Con A to horseradish peroxidase, thereby using peroxidase as a naturally occurring enzyme-labeled carbohydrate.

The phase system used was the same as in the preceding example. The partition coefficients for Con A and horseradish peroxidase (HRPO) were 0.031 and 0.063, respectively. Obviously such a reactant pair could not be used in a binding assay before modification had been carried out. We initially chose to modify Con A by making it more hydrophobic and thereby translocating it to the top phase. In theory several groups of substances may be used to achieve such effects, e.g., phospholipids, aliphatic chains, and polymers of hydrophobic character. It is known from the literature on affinity partitioning that modification of ligands with poly(ethylene glycol) has been useful for changing a partition.[13] In the cases on record only small molecules have been modified and only moderate changes achieved. If, however, a macromolecule is modified, several PEG chains can be bound to each macromolecule and thus result in a higher factor of change in partition behavior. With the use of monomethoxy-PEG, such modifications can be made without the risk of cross-linking the protein molecules. Monomethoxy-PEG (MPEG) was activated according to the methods of Abuchowski et al.[14] (Fig. 3).

In a typical reaction, 50 mg of activated MPEG (M_r 5000) and 8 mg of Con A were mixed in 1 ml of 0.1 M triethanolamine-HCl buffer, pH 9.4. After coupling for 1 hr at room temperature, the reaction was terminated by addition of 1.0 M glycine–NaOH, pH 9.4. Alternatively, the whole reaction mixture was dialyzed overnight at 4° against the coupling buffer or simply left at 4° overnight. Since no marked differences were observed in the subsequent analyses, the latter alternative was used.

After modification Con A had a K_{part} of about 80, i.e., a change in partition constant of $80/0.031 \approx 2500$ times. Furthermore, when HRPO

[13] V. P. Shanbhag and G. Johansson, *Biochim. Biophys. Acta* **61,** 1141 (1974).
[14] A. Abuchowski, T. van Es, N. C. Palczuk, and F. F. Davis, *J. Biol. Chem.* **252,** 3578 (1977).

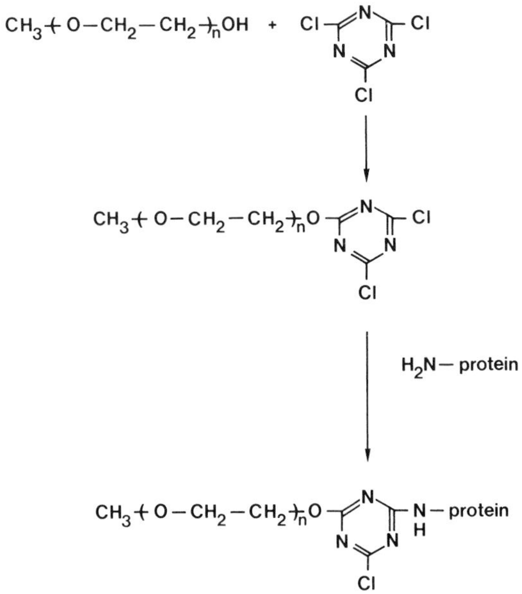

FIG. 3. Reaction scheme for activation of monomethoxy poly(ethylene glycol).

was bound to the modified Con A it was partitioned to the top phase. The degree of binding could be determined, thus, by measuring the distribution of the enzyme between the phases. When a constant amount of enzyme was used analysis of only one of the phases was sufficient. The analyses were performed by taking an aliquot (200 μl; if too much enzyme is present, decrease the amount and add buffer) of the phase (usually the top phase) and adding 650 μl of Tris-HCl buffer, 0.1 M, pH 7.00, and the substrates phenol, 4-aminoantipyrine, and hydrogen peroxide from 20 times stronger stock solutions to final concentrations of 14 mM, 0.8 mM, and 1 mM, respectively. The enzyme activity was read by following the reaction at 510 nm with a spectrophotometer.

As an alternative assay procedure, the substrates may be mixed directly into the phase system, so that when phase partitioning takes place the enzyme reaction is continuously going on. In this manner a kinetic measurement can be made provided the mixture, after addition of phases and the subsequent mixing, is placed in a spectrophotometer cuvette. The spectrophotometer must be used in such a way that the light beam can be made to pass through only one of the phases. Furthermore, in such a direct assay the possibility of an asymmetric partition of substrates in the phase system must be borne in mind.

Binding Assays Involving Con A and HRPO

The enzyme was used as a labeled carbohydrate and could be used as such in competitive binding assays to quantify other carbohydrates.

Assay of Methyl-α-Mannopyranoside. An aqueous two-phase system consisting of 13.5% (w/w) PEG-4000 and 13.5% (w/w) $MgSO_4 \cdot 7 H_2O$ and 10 mM sodium phosphate buffer pH 7.00 was used.

Peroxidase (final concentration 2.5 nM) was mixed with the carbohydrate to be quantified and then MPEG-Con A was added (final concentration 1.8 μM, based on protein content, but with reduced binding capacity due to modification). Incubation in a total volume of 100 μl was allowed to continue for 10 min before addition of 900 μl of well mixed phase system. After mixing on a Vortex mixer, separation was allowed for 10 min before the enzyme content was analyzed according to procedures described in the preceding section. The results of such an assay of methyl-α-D-mannopyranoside when both phases were analyzed separately is shown in Fig. 4. Recordings of kinetic measurements of the same system are given in Fig. 5.

Analysis of Polymeric Carbohydrates. The experimental set-up was the same as that used previously except for change of monomeric sugars

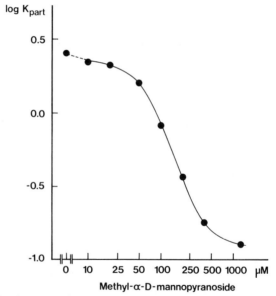

FIG. 4. Calibration curve for methyl-α-D-mannopyranoside. Log K_{part} of the enzyme activity is plotted versus the logarithm of the carbohydrate concentration. Reproduced from Mattiasson and Ling,[12] with permission.

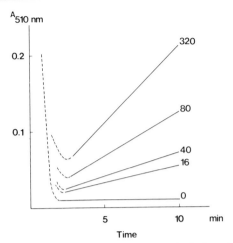

FIG. 5. Time course for a system where all reactants, including phase components and substrates, were mixed. Immediately after mixing, the enzyme activity in the bottom phase was measured. The different micromolar concentrations of methyl-α-D-mannopyranoside are indicated on the graph. The initial period of decreasing absorbance is due to phase separation. Reproduced from Mattiasson and Ling,[12] with permission.

to Dextran T40 (M_r 40,000). This assay proved to be very sensitive, since Dextran T40 could be detected down to 90 nmol/liter.

Estimation of Binding Constants. The experimental conditions used were the same as those described under the analysis of monomeric sugars. By keeping the concentration of monomeric sugar constant (3.3 mM) it was possible to study the influence of various sugars on the partition of HRPO in the phase system after the binding reaction (i.e., the binding of HRPO to Con A had taken place) and thus get an estimate on the capacity of individual sugars to compete with the glucoenzyme for binding positions on Con A. The results, expressed as log K_{part} are plotted against literature data on K_{ass} values.[15] (Fig. 6).

Quantification of Macromolecules

In the application of aqueous two-phase systems in analysis, it seems rather easy to design systems for analyses of small molecules and, as will be discussed later in this chapter, for subcellular particles and cells. The most cumbersome group of substances to analyze is the macromolecules. This may be due to the fact that the macromolecules often have properties similar to those of the binding protein, and that the similarity in size

[15] C. Borrebaeck and B. Mattiasson, *Eur. J. Biochem.* **107,** 67 (1980).

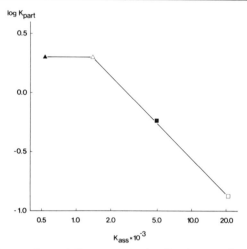

FIG. 6. The influence of association constants, log K_{ass}, between free carbohydrate (3.3 mM) and concanavalin A on the partition constant for horseradish peroxidase in a system as described in the text. Reproduced from Mattiasson and Ling,[12] with permission.

requires that one of the two reactants have a very strong affinity for the other phase if it is to translocate the bound reactant to it.

Concanavalin A

Competitive assays where macromolecules to be assayed compete with a modified form of the molecule for binding to the other reactant is a possibility. This was tried in the Con A case, for example, where, under identical experimental conditions given in the preceding section, MPEG–Con A in a constant concentration was mixed with native Con A before addition of the glucoenzyme horseradish peroxidase (as an analog to enzyme-labeled glucoconjugates). The results of such an assay are given in Fig. 7.

β_2-Microglobulin[16]

β_2-Microglobulin was chosen as a model in the investigation of a competitive binding assay using ^{125}I-labeled antigen. Furthermore, since the antibodies and the β_2-microglobulin moved to the same phase, one of the reactants had to be modified. The structural similarities between antibod-

---

[16] T. G. I. Ling and B. Mattiasson, Poster presented at the 4th Int. Symp. on Affinity Chromatography and Related Techniques, Veldhoven, The Netherlands, June 22–26, (1981) posterabstract B6.

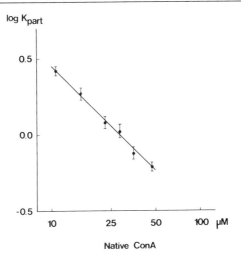

FIG. 7. Standard curve for determination of native concanavalin A. Log K_{part} for peroxidase is plotted versus the log of concentration of native concanavalin A present. Reproduced from Mattiasson and Ling,[12] with permission.

ies and β_2-microglobulin made it reasonable to assume that the two molecules would behave very similarly in the phase system and thus be a good model for testing the value of the system for quantifying proteins. K_{part} for the native molecules was β_2-microglobulin, 0.50, and anti-β_2-microglobulin antibody, 0.31.

Direct MPEG modification of the antibody gave an antibody preparation with poor binding activity; for this reason, rather high concentrations of antisera had to be used for each assay. The system operated as expected, and a calibration curve within the concentration region of clinical interest was obtained. The rather low binding activity of the modified antibodies was probably due to steric hindrance caused by the PEG chains bound to the antibody molecule hampering the binding.

Instead of direct modification of the specific antibody, a modified support with an analog to an antibody against the primary antibody was used. Cells of the bacterium *Staphylococcus aureus* contain on their surfaces a protein, called protein A, with the ability to bind to the Fc part of the human immunoglobulin G (IgG) molecules of subgroups I, II, and IV.[17] By modifying these cells, already killed by heat and formaldehyde, according to the procedure given later in the section on quantification of *Staphylococcus aureus* cells, it was possible to get a preparation separating to the top phase. A preparation of modified cells was treated with

[17] H. Hjelm, *Scand. J. Immunol.* **4**, 633 (1975).

antisera against β_2-microglobulin. Binding was allowed to take place before the cells were washed once. Also IgG-carrying cells partitioned to the top phase.

An assay was performed by mixing ^{125}I-labeled β_2-microglobulin with the sample containing native β_2-microglobulin. Modified cells of *Staphylococcus aureus* carrying specific antibodies against β_2-microglobulin were then added. After incubation for 20 min, a phase system of 13.5% (w/w) PEG-4000 and 13.5% (w/w) $MgSO_4 \cdot 7\ H_2O$ and 10 mM Tris-HCl, pH 7.00, was added. After phase separation, aliquots from both phases were analyzed for radioactivity. The results of such an assay are shown in Fig. 8.

This assay demonstrates the possibility of using an antibody against the antigen-specific antibody when modification of the antigen-specific antibody seriously interferes with the binding reaction. Furthermore, the use of modified *Staphylococcus aureus* cells to carry the specific antibodies eliminates the necessity of modifying antibodies as soon as an assay is to be set up. The modified *S. aureus* cells are, then, a general reagent useful for assays when no spontaneous asymmetric distribution occurs.

Quantification of Cells

Quick diagnostic methods for testing microorganisms are of interest in microbiology, since most established procedures include culturing of

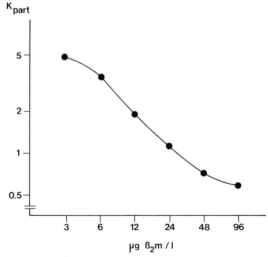

FIG. 8. Calibration curve for β_2-microglobulin. Reproduced from T. G. I. Ling and B. Mattiasson (unpublished).

cells, which is a time- and labor-consuming step.[18,19] The usefulness of aqueous two-phase systems for partition of cells and organelles in preparative biochemical work is well documented.[7] This section discusses model studies on quantification of bacteria and yeasts.

Quantification of Staphylococcus aureus Strain Cowan I[20]

This binding assay, even if it is based on binding of IgG to the bacteria, is not an immunological binding assay. The cells normally carry a layer of protein A, and this molecule has the unqiue ability specifically to bind human IgG subclasses I, II, and IV, via their Fc regions. The assay is thus based upon interactions between protein A and IgG.

The PEG modification is done with activated monomethoxy PEG obtained as described earlier. The assay may be performed in a direct or an indirect way.

Direct Binding Assay of Staphylococcus aureus Cells. In this assay was studied the binding of MPEG-modified ^{125}I-labeled IgG to cells of *Staphylococcus aureus*. The ^{125}I-labeled MPEG-modified IgG partitions spontaneously to the PEG-rich top phase in the phase system used: 400 μl of 10% (w/w) PEG-6000 and 450 μl of 15% (w/w) Dextran T250 diluted with 100 μl of incubation solution or, alternatively, 0.1 M phosphate buffer, pH 7.00. Cells of *S. aureus* pass to the bottom phase. It could thus be concluded that when bound to large bacterial cells the relatively small MPEG-IgG would, on a subsequent partition, move to the same phase as the cells, i.e., the dextran-rich bottom phase. A predetermined amount of MPEG-derivatized ^{125}I-labeled IgG molecules was mixed with the sample containing *S. aureus* cells to be assayed. The mixture was diluted with 0.1 M phosphate buffer pH 7.00, to a final volume of 100 μl. The mixture was carefully mixed and incubated for 30 min at room temperature. Separation was achieved by addition of the well-mixed phase system described above, and, after careful mixing using a Vortex mixer for 10 sec, phase separation was allowed to continue for 2 hr at room temperature. A 200-μl aliquot was taken from the PEG-rich top phase for counting in a LKB-minigamma counter.

The results of this direct binding assay were that, according to the basic principle for PALA, there must be an asymmetric distribution of the reactants between the two phases. This was achieved by modifying the IgG molecules. In their native form the cells as well as the immunoglobu-

[18] M. W. Rytel, *in* "Rapid Diagnosis in Infectious Disease" (M. W. Rytel, ed.), pp. 1–5. CRC Press, Boca Raton, Florida, 1979.

[19] L. A. Chitwood, M. B. Jennings, and H. D. Riley, *Appl. Microbiol.* **18**, 193 (1969).

[20] B. Mattiasson, T. G. I. Ling, and M. Ramstorp, *J. Immunol. Methods* **41**, 105 (1981).

lins are recovered from the dextran-rich bottom phase. The degree of modification of the IgG molecules must be such that it clearly partitions to the top phase. To achieve this, a series of identical ^{125}I-labeled IgG preparations were modified to a varying extent with MPEG. When these modified preparations were tested in the two-phase system, it was found that the partition constant improved with the extent of the modification. On the other hand, the specific interactions utilized in the binding assay must be taken into account. Such an assay was performed, in which a predetermined number of *Staphylococcus aureus* cells were incubated with a predetermined amount of ^{125}I-labeled IgG molecules modified with various amounts of MPEG. After a 30-min incubation, phase separation took place; 400 μl of the top phase were taken out, and the radioactivity was measured. The outcome of such an assay is shown in Fig. 9. It is thus obvious that approximately 1 mg of MPEG per milligram of IgG is optimal if both the binding to protein A on *S. aureus* and partition behavior are to be taken into consideration.

In the binding assay, 1.2×10^{-13} mol of MPEG-derivatized [125]-labeled IgG molecules were incubated with various numbers of native *Staphylo-*

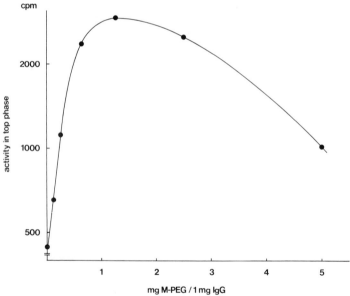

FIG. 9. Determination of the optimal degree of MPEG-modification of IgG in a direct binding assay of *Staphylococcus aureus*. The diagram shows the activity in the top phase as a function of milligrams of MPEG added per milligram of IgG. Reproduced from Mattiasson *et al.,*[20] with permission.

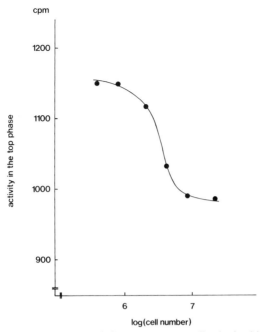

FIG. 10. Calibration curve for *Staphylococcus aureus* cells obtained in a direct-binding assay. The diagram shows the activity in the top phase as a function of the number of cells present in the sample. Reproduced from Mattiasson *et al.*,[20] with permission.

coccus aureus cells for 30 min. The reaction mixture was then partitioned for 30 min in a phase system consisting of 1 ml of 30% (w/w) PEG-4000 and 3 ml of 30% (w/w) $MgSO_4 \cdot 7 H_2O$, 10 mM sodium phosphate, pH 7.00. A 200-μl sample was taken from the top phase for measurement of its activity. The results of such an assay are shown in Fig. 10. It is clear from the figure that a direct binding assay provides a very sensitive system that is useful only within a narrow concentration range—an assay that, from a practical point of view, is of limited value. To improve the sensitivity and the operational concentration range, a competitive binding assay was tried.

Competitive Binding Assay. In this assay competition took place between MPEG-derivatized cells and native bacterial cells for binding ^{125}I-labeled IgG. The idea was that native cells and labeled IgG would be recovered from the bottom phase, whereas modified cells as well as modified cells carrying ^{125}I-labeled MPEG-modified IgG would partition to the top phase. The MPEG-modified cells were prepared according to the following procedure. Two milligrams of activated MPEG were added to

3×10^9 cells in 1 ml of 0.1 M triethanolamine-HCl, 1 M NaCl (pH 9.4). The cell suspension was continuously agitated at room temperature for 1 hr and finally stored at 4°.

A crucial point when setting up such a competitive assay is to ensure that the binding capacity still available on the modified cell preparation is sufficient. Therefore, varying amounts of ^{125}I-labeled IgG were incubated with a constant amount of MPEG-modified *Staphylococcus aureus* cells. After binding and subsequent phase separation, an aliquot was taken from the top phase for analysis in a gamma counter. A titration curve obtained after such an experiment is shown in Fig. 11. Since a linear correlation was found when using a constant number of cells and an increasing amount of the IgG in the incubation, the cells evidently possess enough binding capacity. In the competitive binding assay the concentration of ^{125}I-labeled IgG and the number of MPEG cells were kept constant and the number of native cells added was varied. In Fig. 12 it is shown that the sensitivity of this assay is far better than that of the direct-binding assay. Furthermore, the concentration range over which it is operable is also wider.

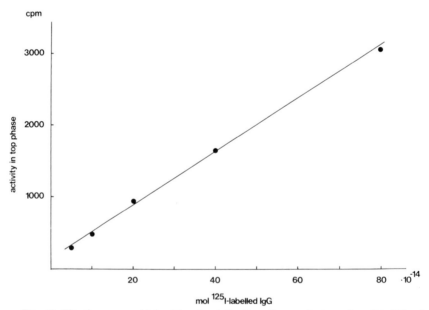

FIG. 11. Titration curve obtained by direct titration of a constant number, 6×10^7, of MPEG-derivatized *Staphylococcus aureus* cells with ^{125}I-labeled IgG. The diagram shows the activity in the top phase when increasing amounts of ^{125}I-labeled IgG were incubated with the cells. Reproduced from Mattiasson *et al.*,[20] with permission.

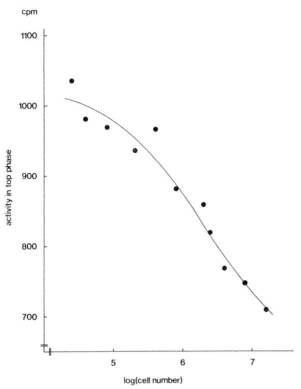

FIG. 12. Calibration curve obtained for *Staphylococcus aureus* cells obtained in a competitive binding assay. Reproduced from Mattiasson *et al.*,[20] with permission.

Assay of Yeast Cells with the Aid of ^{125}I-Labeled Con A[21]

Yeast cells have been shown to partition to the bottom phase in an aqueous two-phase system of PEG-6000 and $MgSO_4$. As discussed earlier, Con A also favors the bottom phase. Since Con A, when modified in such a way as to pass to the top phase loses part of its binding ability, and since there is a marked difference in size between the entities in the binding reaction it was unrealistic to set up an assay based on modified Con A. Instead, a competitive assay was tried. MPEG-5000 (M_r 5000) was used for modification of yeast cells. The MPEG was activated as described earlier in this chapter. Coupling to the cells was performed by incubating 4 ml of a yeast cell suspension (9×10^6 cells/ml) in 0.1 M

[21] B. Mattiasson, T. G. I. Ling, J. Nilsson, and M. Dürholt, *in* "Lectins—Biology, Biochemistry and Clinical Biochemistry" vol. 2. (T. C. Bøg-Hansen, ed.), p. 573. de Gruyter, Berlin, 1982.

TABLE II

PARTITION OF ^{125}I-LABELED CON A IN BINDING ASSAYS USING YEAST CELLS MODIFIED
WITH DIFFERENT AMOUNTS OF MONOMETHOXY-PEG[a]

	Amount of MPEG added	Radioactivity (cpm)		
Number of cells	(mg)	Top phase	Bottom phase	Interface
9×10^6	10	5600	43,400	11,900
9×10^6	100	8900	6,300	45,700
Con A only	—	5100	55,800	0

[a] From Mattiasson et al.,[21] with permission.

triethanolamine buffer, pH 9.4, containing 0.1 M NaCl with 10 or 100 mg of activated MPEG. Coupling was allowed to continue for 1 hr at room temperature before a fivefold molecular excess of glycine was added to stop the reaction.

To evaluate the properties of the two modified cell preparations obtained, cells were incubated with a constant amount of ^{125}I-labeled Con A. It is clear from Table II that the two preparations differed substantially and that the interface plays an important role in this system. It can also be seen that there was a marked difference between counts in the bottom phase and at the interface, whereas only very small effects were observed in the top phase. It can thus be concluded that the assay should be read by measuring the bottom phase and that, even though the interface plays an important role, it may not impair the development of a successful binding assay.

Several competitive binding assays can measure down to 10^5 cells. So could the assay described here. To improve the sensitivity further, a two-step incubation was tried. First, cells and labeled lectin in 0.1 M Tris-HCl, 1 mM MgCl$_2$, CaCl$_2$, and MnCl$_2$, pH 7.2, were incubated in a total volume of 500 μl for 30 min at room temperature before addition of MPEG-modified yeast cells in excess (225,000). After incubation for 3 min, a well-mixed phase system consisting of 1.5 ml of PEG-6000, 30% (w/w), and 1.5 ml of MgSO$_4 \cdot 7$ H$_2$O, 30% (w/w), was added. After phase separation had been allowed to continue for 10 min, aliquots of 500 μl were taken both from the top and the bottom phase, and the radioactivity was measured with a LKB Minigamma counter.

The incubation times in this two-step procedure were chosen so that there should be a good chance of binding of ^{125}I-labeled con A to the native cells in the first step before the addition of the modified cells in

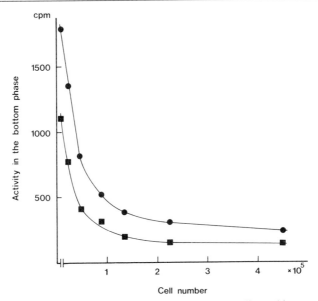

FIG. 13. Influence of the amount of MPEG-modified yeast cells used in an assay and the incubation time in the second step (●, 3 min; ■, 30 min) on the distribution of Con A in the phase system. After separation, the amount of ^{125}I-labeled Con A present in the bottom phase was determined. Reproduced from Mattiasson et al.,[21] with permission.

excess. The duration of this second step should be such that these cells could bind all Con A not bound in the first step, but not be so long that the risk of dissociation of already bound Con A to the native cells would result in transfer of labels from the native to the modified cells. The incubation times used were decided by an initial experiment, where the binding of ^{125}I-labeled Con A to modified yeast cells was studied at different time intervals (Fig. 13).

The result of the two-step incubation assay is shown in Fig. 14. As expected, the sensitivity was substantially improved. Furthermore, it is clearly seen that reading the top phase gave poor results, whereas analysis of aliquots from the bottom phase was more successful.

It should also be mentioned that great care was taken in all the assays where $MgSO_4$ was used as the bottom phase. It can be argued that such concentrated Mg^{2+} solution as that used in the experiments reported here would dissociate the immunochemical reactants. However, owing to the extreme speed at which the separation occurs, no such effects were observed under the experimental conditions used.

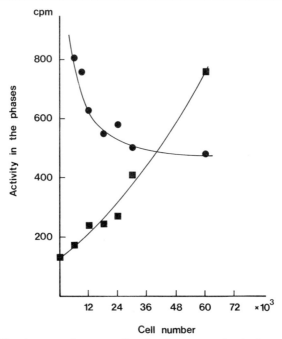

FIG. 14. Calibration curve for yeast cells using the two-step incubation procedure. First, native cells were incubated with ^{125}I-labeled Con A; in the second step, 225,000 MPEG-modified cells were added. After separation, the radioactivity was determined in the top (●) and in bottom (■) phases.

Direct RIA Measurements

As indicated in the Introduction, aqueous two-phase systems can be used for separation and also for permitting the enzymic reaction to take place within the phase system in enzyme immunoassays. This approach was demonstrated in the case of Con A–carbohydrate interactions.

Homogeneous enzyme immunoassays based on modification of the enzyme activity by the immunological reaction have been reported.[4,5] Modulating enzyme activities to perform assays has often been demonstrated, but the development of RIA with no separation has received only little attention in the literature.[22,23] The PALA principle was tested in an effort to develop RIA procedures with no distinct separation step when bound isotope is pipetted into a specific counting vessel. The basic principle behind this direct PALA is demonstrated in Fig. 15. In many gamma

[22] H. Eriksson, B. Mattiasson, and J. I. Thorell, *J. Immunol. Methods* **42**, 105 (1981).
[23] A. Pick and D. Wagner, *J. Immunol. Methods* **32**, 275 (1980).

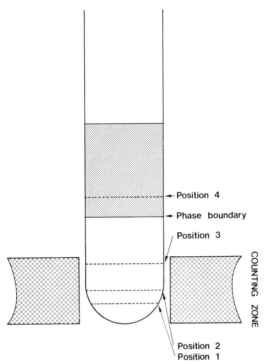

FIG. 15. Schematic presentation of the basis for direct partition affinity ligand assay (PALA) when using γ-emitting isotopes. Positions 1–4 indicate the different measuring positions in the LKB Minigamma counter (position 1 = 0 mm up from the bottom; 2 = 10 mm up, 3 = 20 mm up, and 4 = 30 mm up).

counters the crystals are designed in a fashion similar to that in the LKB gamma counter we used. The basic principle is that on partition the isotope in the top phase is transported out of the counting zone. The assays as such may be performed as described for, e.g., digoxin.[6,24] The only modification is to secure a good physical separation, so that the top phase falls outside the counting zone of the gamma counter.

Assay of Triiodothyronine (T_3) in a Direct PALA Procedure[25]

Materials

Phadebas T_3-RIA kit (standard sera and ^{125}I-labeled T_3 were used)
Antiserum against T_3 raised in sheep

[24] B. Mattiasson, H. Eriksson, and J. Nilsson, *Clin. Chim. Acta,* in press.
[25] H. Eriksson, J. Nilsson, and B. Mattiasson, *Appl. Biochem. Biotechnol.,* in press.

Barbital buffer, 0.075 M, pH 8.6, containing 0.25% (w/v) bovine serum albumin

Test tubes: 12 × 75 mm, purchased from Cerbo, Trollhättan, Sweden

Procedure. Fifty microliters of T_3-containing serum and 100 μl of [^{125}I]T_3 from the commercial kit were mixed with 100 μl of antiserum diluted in 0.075 M barbital buffer, pH 8.6, containing 0.25% (w/v) bovine serum albumin. The sample was incubated for 1 hr before a well mixed phase system [3 ml, 30% (w/w) $MgSO_4 \cdot 7 H_2O$ + 1 ml 30% (w/w) PEG-4000] was added. After separation for 15 min, the test tubes were placed in an LKB Minigamma counter. A calibration curve according to Fig. 16 was obtained, and on analysis of serum samples from the local hospital a coefficient of correlation when comparing the results to those of conventional RIA was 0.96 ($y = 0.77$, $x = 0.25$) (Fig. 17).

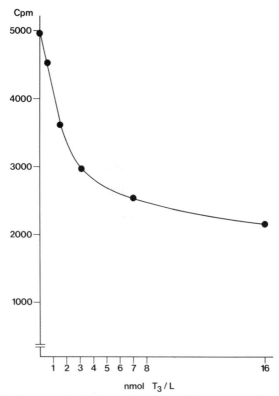

FIG. 16. Calibration curve for T_3. Reproduced from Eriksson *et al.*,[25] with permission.

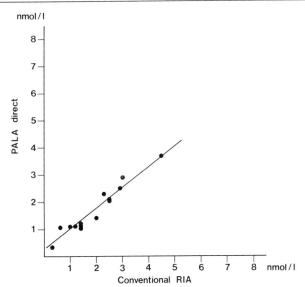

FIG. 17. Correlation of results obtained with a conventional radioimmunoassay (RIA) of 14 human serum samples of T_3 and with the direct partition affinity ligand assay (PALA) technique. Reproduced from Eriksson *et al.*,[25] with permission.

Concluding Remarks

The assays described here cover small molecules, macromolecules, and cell particles. The basis for separation is mainly differences in hydrophobicity, either naturally occurring or chemically induced, but also other possibilities are available.

The separating power of aqueous two-phase systems can be divided into contributions by several factors, e.g., hydrophobicity, hydrophilicity, and charge. A generalized formula has been set up.[7]

$$\ln K_{part} = \ln K_{el} + \ln K_{hydrophob} + \ln K_{hydrophil} + \ln K_{conform}$$

where the subscripts el, hydrophob, hydrophil, and conform denote partition coefficients due to electrical, hydrophobic, hydrophilic, and conformational effects, respectively.[7] In the case of cells the situation is more complicated.

It is, in principle, possible to design analytical procedures based on separation in aqueous two-phase systems owing to any one of the factors in the above equation.

Hydrophobicity has been found to be the best for most of the applications tested. However, in the case of digoxin, the antibody could be made to pass to the bottom phase even better by coupling to hydrophilic carbo-hydrates.

In an attempt to use differences in charges, a phase system of DEAE-dextran-500 and PEG-6000 in Li_2SO_4 was tested.[26] The system studied was bovine serum albumin and FITC-labeled bovine serum albumin as well as antibodies. By changing pH around pI for either of the proteins, it was possible to change the K_{part} by a factor of 2–3. Thus it was shown that assays with partition based on differences in charges were possible and that such assays could work in well standardized conditions. However, when serum samples were introduced, the partition behavior was no longer predictable owing to changes in the buffering capacity. Such as-says were thus of no practical value.

In many studies it was possible to change the partition behavior of proteins by addition of various ions to the phase system. When setting up the conditions, it might be of great importance to choose the right salt medium.

The influence on the interfacial potential, and thus on partition, is different for the halogens

$$F^- < Cl^- < Br^- < I^-$$

and for the alkaline chlorides in the following order:

$$Li^+ < NH_4^+ < Na^+ = Cs^+ < K^+$$

These differences have been used when designing the analytical systems discussed earlier in this chapter.

In conclusion, binding assays involving separation using an aqueous two-phase system seem to be applicable to almost any system. The exper-imental procedure is very simple with few manual steps. These character-stics make PALA a competitive candidate when automation of binding assays is considered.

[26] T. G. I. Ling and B. Mattiasson, unpublished observations.

[40] Preparation and Applications of Multivalent Antibodies with Dual Specificity

By V. GHEȚIE and I. MORARU

General Remarks

Antibody molecules of dual specificity can be obtained by combining univalent Fab'[1] fragments from antibodies of different specificity.[1a,2] When reconstitution takes place in a mixture of Fab', some of the $F(ab)_2'$ molecules so formed are hybrids with double specificity. For instance, the random recombination of two univalent fragments of pepsin-treated antibodies (anti-A and anti-B) yields antibodies anti-A/anti-A, anti-B/anti-B and anti-A/anti-B (hybrid). Since the anti-A/anti-B hybrid is univalent in respect to both antigens, the bivalent anti-A/anti-A and anti-B/anti-B recombinants are able to interfere in the binding of anti-A/anti-B hybrid to the corresponding antigens. The unwanted recombinants can be discarded from the mixture by successive reactions with the corresponding antigens, followed by acid dissociation of the immunoprecipitates.[3]

In order to prepare multivalent hybrid antibody with double specificity, we resorted to the combination of two antibodies with different specificity by means of protein A of *Staphylococcus aureus* (SpA), able to link the IgG molecules through their Fc region.[4] Rabbit IgG treated with SpA forms a soluble complex with a molar ratio of two molecules of IgG per molecule of SpA and with molecular composition $(IgG\text{-}SpA\text{-}IgG)_2$.[5–8] Linkage of IgG antibodies with two different specificities (anti-A and anti-B) by SpA is thought to occur at random, yielding a multivalent hybrid

---

[1] Abbreviations: BRBC, bovine red blood cells; CRBC, chicken red blood cells; Fab, Fab fragment of human IgG; Fer, ferritin; HRP, horseradish peroxidase; HSA, human serum albumin; mIg, mouse immunoglobulin; MSL, mouse spleen lymphocyte; MT, mouse thymocyte; PBS, phosphate-buffered saline; RFC, rosette-forming cells; SpA, protein A of *Staphylococcus aureus;* SRBC, sheep red blood cells.

[1a] A. Nisonoff and M. M. Rivers, *Arch. Biochem. Biophys.* **93,** 460 (1961).

[2] U. Hämmerling, T. Aoki, E. A. De Harven, and L. J. Old, *J. Exp. Med.* **128,** 1461 (1968).

[3] U. Hämmerling, *in* "Methods in Immunology and Immunochemistry" (C. A. Williams and M. W. Chase, eds.), Vol. 5, p. 464. Academic Press, New York, 1976.

[4] J. Sjöquist, A. Forsgren, G. T. Gustafson, and G. Stålenheim, *Cold Spring Harbor Symp. Quant. Biol.* **32,** 341 (1967).

[5] G. Moța, V. Gheție, and J. Sjöquist, *Immunochemistry* **15,** 639 (1978).

[6] J. J. Langone, M. D. P. Boyle, and T. Borsos, *J. Immunol.* **121,** 327 (1978).

[7] J. J. Langone, M. D. P. Boyle, and T. Borsos, *J. Immunol.* **121,** 333 (1978).

[8] S. Mihăescu, A. Sulica, J. Sjöquist, and V. Gheție, *Rev. Roum. Biochim.* **16,** 57 (1979).

antibody complex (anti-A/SpA/anti-B)$_2$ mixed with other antibody complexes consisting of (anti-A/SpA/anti-A)$_2$ and (anti-B/SpA/anti-B)$_2$. The last two unwanted antibody complexes cannot be removed by successive reaction with the corresponding antigens, since acid dissociation of the antigen–antibody complex would also break the linkage between SpA and the two IgG antibody molecules of the hybrid complex.

To avoid this experimental difficulty, we found a different way to prepare hybrid SpA–antibody complexes with double specificity, free of any unwanted monospecific SpA–antibody complexes. By the reaction of rabbit IgG antibody (i.e., anti-A) with an excess of SpA, a soluble complex consisting of one IgG molecule linked to one SpA molecule (IgG anti-A/SpA) was obtained. This complex was able to further attach another molecule of IgG antibody (anti-B), yielding a multivalent hybrid antibody with molecular composition (IgG anti-A/SpA/IgG anti-B)$_2$ (Fig. 1). Using this reaction, several multivalent hybrid antibodies were prepared.

The preparation of hybrid antibody complexes is restricted to rabbit IgG, since the immunoglobulins of other species do not react with SpA (fowl, horse, etc.) or are precipitated by it (mouse, human, etc.).[9] In order to extend the use of hybrid antibodies to antigens reacting only with mouse (i.e., Thy 1.2 antigens) or human (i.e., HLA) alloantibodies, we succeeded by means of SpA to link without precipitation two antibody partners, one deriving from rabbit (nonprecipitating with SpA) and the other originating from mouse or human (precipitating with SpA). Thus we obtained a soluble interspecies hybrid antibody complex with dual specificity.

Procedure for the Preparation of Hybrid Antribody[10]

Rabbit IgG with specificity for antigen A was mixed with SpA at a molar ratio of 1:7. Both reagents were dissolved in 0.05 M NaCl solution. The molecular weight of IgG and SpA were considered to be 150,000 and 42,000.[11] The final concentrations of IgG and SpA were 1 mg/ml and 2 mg/ml, respectively. After incubation for 1 hr at 37°, the mixture was brought to pH 5.5, in 0.02 M acetate buffer, by two ultrafiltrations on Diaflo membrane P-30. The mixture was then chromatographed on CM-Sephadex C-50 equilibrated with 0.02 M acetate buffer, pH 5.5 (Fig. 2).

The unreacted SpA was eluted with this buffer and recovered by affinity chromatography on IgG-Sepharose 4B[12] (and re-used in other experiments).

[9] G. Kronvall, U. S. Seal, J. Finstad, and R. C. Williams, Jr., *J. Immunol.* **104,** 140 (1970).
[10] V. Gheţie and G. Moţa, *Mol. Immunol.* **17,** 395 (1980).
[11] I. Björk, B. A. Peterson, and J. Sjöquist, *Eur. J. Biochem.* **29,** 57 (1972).
[12] H. Hjelm, K. Hjelm, and J. Sjöquist, *FEBS Lett.* **28,** 73 (1972).

1. AN ANTIBODY IS REACTED WITH AN EXCESS OF PROTEIN A (SpA)

   AND THE NON-REACTED SpA IS DISCARDED

2. THE ANTIBODY/SpA COMPLEX IS FURTHER REACTED WITH ANOTHER

   ANTIBODY OF DIFFERENT SPECIFICITY

3. THE HYBRID ANTIBODY COMPLEX IS RECOVERED IN ITS DIMERIC

   FORM (MW 669,000)

FIG. 1. Preparation of multivalent antibodies with dual specificity. Schematic representation of the steps involved in the preparation.

The adsorbed IgG anti-A/SpA complex was eluted with double-concentrated phosphate-buffered saline (PBS), pH 7.4; after adding an equal volume of distilled water, the solution was concentrated to 4 mg/ml.

No free SpA or IgG were detected in this complex by electrophoresis in polyacrylamide gel. The IgG/SpA molar ratio of the complex eluted from CM-Sephadex was 1.06 ± 0.08, and its molecular weight, established by gel filtration on a calibrated Sepharose 6B column (in 0.1 M Tris-HCl buffer, pH 8.0, with 0.2 M NaCl), was 190,000 (Fig. 3a). These results show that the complex IgG-anti-A/SpA has a molecular composition of one IgG molecule to one SpA molecule.

To the IgG anti-A/SpA solution concentrated to 4 mg/ml, an equal volume of a solution containing the same concentration of rabbit IgG with specificity for antigen B was then added, and the mixture was incubated for 1 hr at 37°. The solution was gel filtered on Sepharose 6B. The peak

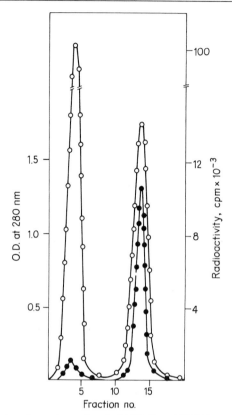

FIG. 2. CM-Sephadex chromatography of rabbit IgG anti-SRBC treated with an excess of radiolabeled SpA. ●——●, Optical density at 280 nm; ○——○ radioactivity. Reproduced from Gheție and Moța,[10] with the permission of *Molecular Immunology*.

containing the hybrid antibody complex (Fig. 3b) was concentrated, dialyzed against PBS, and kept frozen at $-20°$ for months without alteration of antibody activity.

As seen in Fig. 3b the hybrid antibody complex was eluted from Sepharose 6B as a single peak in a volume corresponding to that of a molecular weight of 669,000. The IgG/SpA molar ratio of this complex was 1.92 ± 0.07. This molar ratio along with the high molecular weight shows that the complex is composed of 4 molecules of IgG and 2 molecules of SpA with the molecular formula (IgG anti-A/SpA/IgG anti-B)$_2$. Since a complex with the molecular formula of IgG anti-A/SpA/IgG anti-B (M_r 342,000) was not identified after adding IgG anti-B to IgG anti-A/SpA complex, it is possible that IgG anti-A/SpA/IgG anti-B is able to dimerize instantly, yielding only the (IgG anti-A/SpA/IgG anti-B)$_2$ complex.

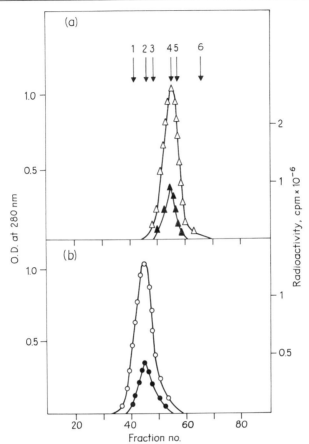

FIG. 3. Gel filtration of rabbit ^{125}I-labeled IgG anti-SRBC/SpA complex and of rabbit (^{125}I-labeled IgG anti-SRBC/SpA/IgG anti-BRBC)$_2$ complex on Sepharose 6B. (a) IgG anti-SRBC/SpA complex. △——△, Optical density (OD) at 280 nm; ▲——▲, radioactivity. (b) IgG anti-SRBC/SpA and IgG anti-BRBC mixture. ○——○, at 280 nm; ●——●, radioactivity. 1, IgM; 2, thyroglobulin; 3, secretory IgA; 4, IgD; 5, rabbit IgG; 6, human serum albumin. Reproduced from Gheție and Moța,[10] with the permission of *Molecular Immunology*.

If instead of rabbit IgG anti-B antibody, mouse or human IgG anti-B was used, a soluble interspecies hybrid antibody complex was prepared.[13] The molecular weight of (rabbit IgG anti-A/SpA/mouse or human IgG anti-B)$_2$ complex (669,000) and its IgG/SpA molar ratio (1.85 ± 0.08) have clearly indicated that mouse or human IgG anti-B was fully integrated into the rabbit IgG anti-A/SpA complex.

[13] G. Moța and V. Gheție, *Mol. Immunol.* **18**, 91 (1981).

Reagents

Protein A of *Staphylococcus aureus* (SpA): SpA (Pharmacia, Uppsala) was labeled with $Na^{125}I$ (The Radiochemical Centre, Amersham, U.K.) using the lactoperoxidase technique.[10] SpA concentration was determined spectrophotometrically at 280 nm using an extinction coefficient of 1.65.

Antibodies. Rabbit IgG was isolated from antisera by chromatography on DEAE-cellulose and gel filtration on Sephadex G-200.[5] Some IgGs were radiolabeled with $Na^{125}I$.[10]

Mouse and human IgG were isolated from antisera by SpA-Sepharose 4B affinity chromatography[12] using 3 M NH_4SCN for elution of bound IgG.

Mouse anti-Thy 1.2 serum, extensively absorbed with mouse red blood cells, hepatocytes, and bone marrow, was used. Human anti-HLA B-12 serum was obtained from National Institutes of Health (Bethesda, Maryland).

Rabbit IgG antibodies specific for red blood cells were purified by immunoadsorption on glutaraldehyde-treated red blood cells and elution with 2.5 M NaI. Rabbit IgG antibodies specific for mouse immunoglobulin (mIg), human IgG or its Fab fragment, horseradish peroxidase (HRP), ferritin (Fer), and human serum albumin (HSA) were isolated by elution with 0.1 M glycine-HCl buffer pH 1.8 from the appropriate immunoadsorbent prepared by polymerization of proteins with glutaraldehyde.[10] The antibody content of IgG preparations varied between 66 and 85%.

Note: The hybrid antibody complexes can be prepared starting from purified antibody preparations (e.g., anti-SRBC) or even from the IgG fraction of the antisera (e.g., anti-Thy 1.2).

Random hybrid antibody complexes prepared by mixing two different IgG antibodies with SpA (in molar ratio 2:1) can also be used in experiments in which the unwanted monospecific complex does not hinder the reaction of the hybrid complex with the corresponding antigens.

Properties of Hybrid Antibody

By electrophoresis in 6% polyacrylamide gel the hybrid antibody complexes prepared with rabbit, human, or mouse IgG showed a narrow banding (Fig. 4), thus indicating that the complexes are homogeneous and free of nonreacted SpA or IgG.

The presence in the same molecule of IgG antibodies originating from two different species or having two different specificities was demonstrated by immunodiffusion. Thus double diffusion with rabbit (IgG anti-HSA/SpA/IgG anti-human Fab)$_2$ complex shows a reaction of complete

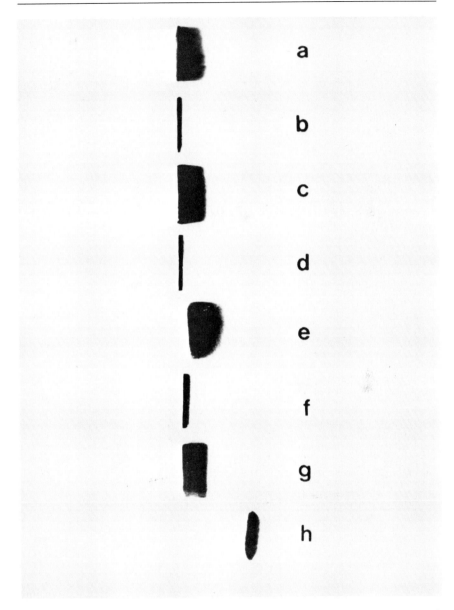

FIG. 4. Electrophoresis in polyacrylamide gel of multivalent antibodies with dual specificity. a, Human IgG; b, (rabbit IgG/SpA/human IgG)$_2$; c, rabbit IgG; d, (rabbit IgG/SpA/mouse IgG)$_2$; e, mouse IgG; f, (rabbit IgG/SpA/rabbit IgG)$_2$; g, rabbit IgG/SpA; h, SpA. Reproduced from Moţa and Gheţie,[13] with the permission of *Molecular Immunology*.

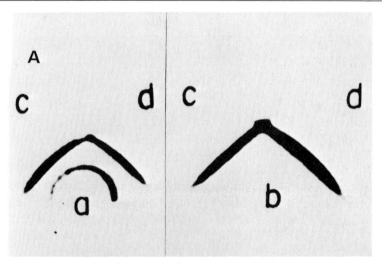

FIG. 5. Immunodiffusion analysis of the hybrid antibody complex. (A) Rabbit hybrid antibody: a, (IgG anti-HSA/SpA/IgG anti-human Fab)₂; b, mixture of IgG anti-HSA and IgG anti-human Fab fragment; c, human Fab fragment; d, HSA. (B) Interspecies hybrid anti-body: a, goat anti-rabbit IgG; b, goat anti-human IgG; c, (rabbit IgG/SpA/human IgG)₂; d, (rabbit IgG/SpA/mouse IgG)₂; e, goat anti-mouse IgG. Reproduced from Gheție and Moța,[10] and Moța and Gheție,[13] with the permission of *Molecular Immunology.*

identity between the precipitin lines of HSA and human Fab fragment, two unrelated antigens, whereas with a mixture of anti-HSA and anti-Fab, a reaction of nonidentity was observed (Fig. 5A). Double diffusion with (rabbit IgG anti-HRP/SpA/mouse IgG anti-Thy 1.2)₂ and (rabbit IgG anti-BRBC/SpA/human IgG anti-HLA B-12)₂ shows a reaction of complete identity between the precipitin lines given by anti-rabbit IgG and anti-mouse IgG or anti-human IgG, respectively (Fig. 5B).

The ability of some hybrid antibody complexes with dual specificity to react with both sheep red blood cells (SRBC) and bovine red blood cells (BRBC) and with SRBC/chicken red blood cells (CRBC) antigens was demonstrated by hemagglutination and hemolysis (Table I). As seen in Table I, the hybrid antibody complex (anti-SRBC/SpA/anti-BRBC)₂ was able to react with both SRBC and BRBC, but not with CRBC, whereas the (anti-SRBC/SpA/anti-CRBC)₂ complex reacted with SRBC and CRBC, but not with BRBC.

Rosette Formation with Hybrid Antibody

Hybrid antibody complexes are able to draw together two distinct cell types, thus enabling rosette formation. BRBC, CRBC, and mouse spleen

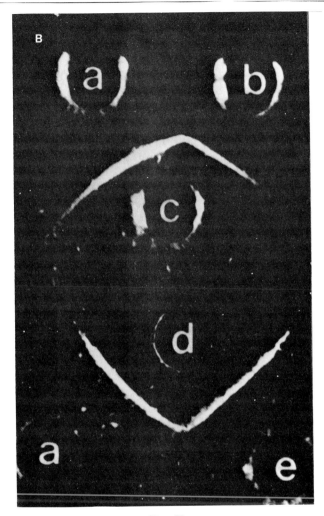

FIG. 5B.

lymphocytes (MSL) treated with hybrid antibody complexes against some
of their cell membrane antigens and SRBC formed rosettes with SRBC
(Fig. 6 and Table II). No rosettes were observed with two cell types if the
hybrid antibody complex did not contain the specific antibodies for both
cells.

The ability of (rabbit IgG anti-HRP/SpA/mouse IgG anti-Thy 1.2)$_2$ to
react with both HRP-coated SRBC and Thy 1.2-bearing lymphocytes was
demonstrated by rosette formation. Mouse spleen lymphocytes or mouse

TABLE I
Hemagglutinating and Hemolytic Activity of Some Hybrid Antibody Complexes with Double Specificity[a]

Ligands	Titer[b] (μg ligand/ml) with		
	SRBC	BRBC[c]	CRBC
Anti-SRBC/SpA/anti-BRBC	0.55	1.46	NA
Anti-SRBC/SpA/anti-CRBC	0.49	NH	0.39
Anti-SRBC/SpA/anti-mIg	0.29	NH	NA
Anti-SRBC	6.64	NH	NA
Anti-BRBC	NA	1.46	NA
Anti-CRBC	NA	NH	3.12

[a] Reproduced from Gheţie and Moţa,[10] with the permission of *Molecular Immunology*.

[b] NH, no hemolysis; NA, no agglutination.

[c] A hemolytic assay was used, since BRBC are not agglutinated by anti-BRBC antibodies.

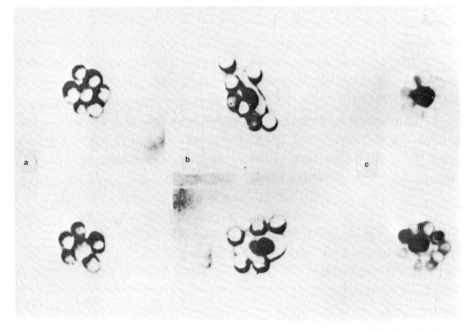

FIG. 6. Rosettes with various cell types formed by cross-linkage with hybrid antibody complexes. (a) Bovine RBC rosetted with sheep RBC after previous treatment of BRBC with (IgG anti-SRBC/SpA/IgG anti-BRBC)$_2$ complex; (b) chicken RBC rosetted with SRBC after previous treatment of CRBC with (IgG anti-SRBC/SpA/IgG anti-CRBC)$_2$; (c) Mouse spleen lymphocytes (MSL) rosetted with SRBC after previous treatment of MSL with (IgG anti-SRBC/SpA/IgG anti-mIg)$_2$. Reproduced from Gheţie and Moţa,[10] with the permission of *Molecular Immunology*.

TABLE II

ABILITY OF HYBRID ANTIBODY COMPLEXES TO ESTABLISH CELL-TO-CELL CONTACTS[a]

Ligands[b]	Protein concentration[c] (μg/ml/10^7 cells)	Percentage of SRBC rosettes formed		
		BRBC	CRBC	MSL
Anti-SRBC/SpA/anti-BRBC	12	85[d]	<1	ND
Anti-SRBC/SpA/anti-CRBC	1.2[e]	<1	48[e]	5
Anti-SRBC/SpA/anti-mIg	6	<1	ND[f]	46[g]
Anti-SRBC + anti-BRBC (without SpA)	1000	0	0	1
None	0	0	0	<1

[a] Reproduced from Gheție and Moța,[10] with the permission of *Molecular Immunology*.

[b] Target cells (BRBC, CRBC, MSL) were treated with the appropriate concentrations of hybrid antibody complexes. After washing, the cell-bound hybrid antibody was able to attach SRBC to target cells.

[c] Lowest hybrid concentration still able to give a maximum percentage of rosettes.

[d] Rosette-forming BRBC determined by UV microscopy using fluorescent BRBC.

[e] Maximum percentage of rosettes not determined due to the agglutination of CRBC by increase of hybrid antibody concentration.

[f] ND, not done.

[g] By staining with fluorescent anti-mIg antibody, the percentage of Ig-bearing MSL was 45. When mouse thymocytes were treated with hybrid antibody complex, no rosettes were formed with SRBC.

thymocytes (MT) treated with hybrid complexes containing antibodies against both Thy 1.2 antigen and HRP or CRBC formed rosettes with the indicator cells (Table III).

The specificity of the reaction was demonstrated by showing that rosette formation between MT treated with (rabbit IgG anti-HRP/SpA/ mouse IgG anti-Thy 1.2)$_2$ complex and HRP-coated SRBC was inhibited if the cells were previously treated with HRP (Fig. 7).

Assay.[10] Cells (5×10^6) were suspended in 0.45 ml of TC-199 medium, and 0.05 ml of hybrid antibody solution (10–25 μg) was added. After 30 min of incubation at room temperature the cells were washed three times with TC-199 medium by centrifugation (10 min at 250 g) and resuspended to 10^6 cells/0.45 ml. To this suspension 0.05 ml of indicator cells (e.g., SRBC) was added (5×10^7 cells). The mixture was centrifuged for 10 min at 250 g and further incubated at 4° for 1 hr. The pellet was then resuspended and about 300 cells were counted. Cells binding more than three indicator cells were considered to be positive rosettes. The percentage of rosettes was calculated according to the following equation: 100 R/R + NR, where R = number of rosetted cells and NR = number of nonrosetted cells.

TABLE III
ABILITY OF INTERSPECIES HYBRID ANTIBODY COMPLEXES WITH DUAL SPECIFICITY TO
DRAW TOGETHER TWO DISTINCT CELL TYPES[a]

Ligands[b]	Percentage of HRP-SRBC rosettes formed		Percentage of CRBC rosettes formed	
	MSL	MT	MSL	MT
Rabbit anti-HRP/SpA/mouse anti-Thy 1.2	28	85	<1	<1
Rabbit anti-CRBC/SpA/mouse anti-Thy 1.2	<1	<1	22	78
Rabbit IgG anti-HRP + mouse IgG anti-Thy 1.2 (without SpA)	<1	<1	ND[c]	ND

[a] Reproduced from Moța and Gheție,[13] with the permission of *Molecular Immunology*.

[b] Target cells (MSL, MT) were treated with hybrid antibody complexes (100/μg/ml per 10^7 cells). After washing, the cell-bound hybrid antibody was able to attach indicator cells (HRP-SRBC, CRBC) to target cells.

[c] ND not done.

Detection of Cell-Surface Antigens by Electron Microscopy Using Hybrid Antibody

Electron microscopy of MSL treated with either (IgG anti-HRP/SpA/ IgG anti-mIg)$_2$ and HRP or (IgG anti-Fer/SpA/IgG anti-mIg)$_2$ and Fer showed a strong and specific staining of lymphocyte membranes.[14,15]

The labeling pattern with HRP on ultrathin sections consisted of electron-dense lines of different lengths on the outer aspect of the plasma membrane suggesting a patchlike distribution. By counting spleen lymphocytes we noted that 53% were labeled. Morphometry of MSL showed that the average labeled fraction of the cell surface was 49% (Fig. 8a).

The labeling pattern with Fer consisted of electron-dense grains situated in a monolayer on the outer aspect of the plasma membrane and separated from the latter by a narrow space approximately three times the thickness of a cell membrane (Fig. 8b). The Fer molecules were found in clusters, and never singly, and appeared on thin sections as segments of labeled plasma membrane separated by marker-free zones (Fig. 8b). These labeled segments have a distribution very similar to that obtained

[14] E. Mandache, E. Moldoveanu, G. Moța, I. Moraru, and V. Gheție, *J. Immunol. Methods* **35,** 33 (1980).

[15] E. Mandache, G. Moța, E. Moldoveanu, I. Moraru, and V. Gheție, *J. Immunol. Methods* **42,** 355 (1981).

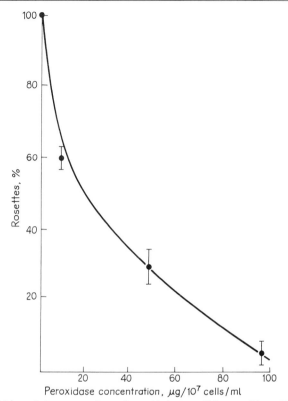

FIG. 7. Inhibition of the rosette formation with horseradish peroxidase (HRP)-SRBC of hybrid-treated mouse thymocytes, (MT) with various concentrations of HRP. 10^7 MT were treated with 75 μg of (rabbit IgG anti-HRP/SpA/mouse IgG anti-Thy 1.2)$_2$ complex. The percentage of rosettes formed without prior incubation with HRP was 80%. Reproduced from Moţa and Gheţie,[13] with the permission of *Molecular Immunology*.

using hybrid antibody with anti-HRP specificity. The only difference was that the plasma membrane segments labeled with Fer were shorter and more numerous than the segments labeled with HRP. The Fer staining gave better resolution and higher accuracy than HRP staining.

By using a mixture of two hybrid antibodies consisting of anti-HRP and anti-Fer IgG linked by SpA to different antibodies directed against two membrane antigens, their simultaneous detection on the cell surface by electron microscopy was easily done (Fig. 9). Thus treatment of human IgG-charged MT with a mixture of hybrid antibody containing (anti-Fer/SpA/anti-Thy 1.2)$_2$ and (anti-HRP/SpA/anti-Fab)$_2$ enabled us to visualize simultaneously on the same cell both Fc receptor (FcR) and Thy 1.2

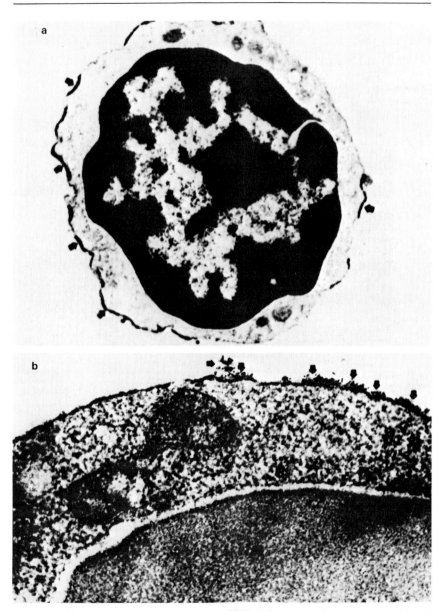

FIG. 8. Electron microscopy of cells treated with multivalent antibodies with dual specificity. (a) Mouse spleen lynphocytes treated with (IgG anti-HRP/SpA/IgG anti-mouse Ig)$_2$ showing peroxidase-positive labeled areas (arrows). ×25,000, (b) MSL treated with (IgG anti-Fer/SpA/IgG anti-mouse Ig)$_2$ showing labeled areas with ferritin in monolayer (vertical arrows). ×75,000. Reproduced from Mandache et al.,[14,15] with the permission of Journal of Immunological Methods.

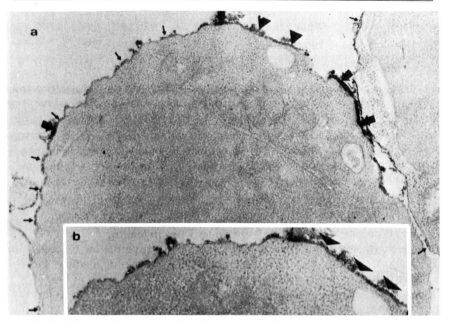

FIG. 9. Electron microscopy of IgG-charged mouse thymocytes treated with (anti-Fer/SpA/anti-Thy 1)₂ and (anti-HRP/SpA/anti-Fab)₂ showing ferritin-labeled segments (thin arrows), peroxidase-labeled segments (thick arrows), and mixed (ferritin and peroxidase)-labeled zone (arrowheads). ×35,000. (b) Detail of mixed labeled zone showing the apposition of ferritin molecules and peroxidase (arrowheads). ×50,000. Reproduced from Mandache et al.,[15] with the permission of *Journal of Immunological Methods*.

alloantigen. These two membrane antigens were identified by two distinct electron microscopic markers, namely, Fer for Thy 1.2 and HRP for FcR. All the investigated cells showed rounded shapes with more or less rough surfaces. Besides the two distinct kinds of labeled areas (rows of ferritin grains and an electron-dense black layer with peroxidase), some thymocytes showed both fine grains and the electron-dense black layer in apposition (Fig. 9). This third kind of labeling occurred with a lower frequency, usually showing small clusters of ferritin spread through a larger peroxidase-labeled zone. Concomitant visualization of FcR and Thy[1.2] antigen on MT clearly shows that their distribution is largely independent and that there is a greater amount of Thy 1.2 antigen present.

The occurrence of mixed labeled areas may be explained by assuming that, after interaction with the corresponding ligands, both FcR and Thy 1.2 are incompletely segregated in the redistribution process. This explanation implies that, on the one hand, the antigens are initially mixed together on the cell surface, at least one of them (Thy 1.2) having a uniform distribution; and, on the other hand, patch formation of both membrane antigens is not hindered by the simultaneous binding of two multivanlent ligands. Thus the mixed Fer- and HRP-labeled areas may result from the countercurrent flow of Thy 1.2 antigen molecules through clusters of FcR molecules. However, the disparity between the size of the hybrid complex and the size of antigenic sites on cell membranes intro- duces some uncertainity as to the actual position and the redistribution pattern of such antigenic markers.

Assay.[14,15] Mouse spleen lymphocytes (10^7) were suspended in 0.15 ml of TC-199 medium containing 0.015 M sodium azide, and 0.05 ml of hy- brid antibody solution (5–25 μg) was added. After 30 min of incubation at 4°, the cells were washed thrice with cold TC-199 medium containing sodium azide and resuspended to 10^7 cells per 0.1 ml. To this suspension, 0.025 ml of HRP (10–20 μg) and/or Fer (25 μg) in phosphate-buffered saline was added, and the mixture was incubated for 30 min at 4°. After repeated washings the cells were pelleted by centrifugation.

Mouse thymocytes (10^7 cells/ml) were treated with 10 μl of human IgG (1 mg). After 30 min of incubation at 4°, the cells were repeatedly washed with cold TC-199 medium containing sodium azide, resuspended at 10^7 cells/0.15 ml, and then processed as described above.

The cell pellet was fixed for 5 min with 1.5% glutaraldehyde in 0.067 M sodium cacodylate buffer, pH 7.4. For uniform fixation the cells were resuspended in the fixing medium and centrifuged. The cell samples treated with hybrid antibody preparation containing anti-Fer antibody were washed three times in sodium cacodylate buffer and postfixed with 1% OsO_4 in the same buffer, pH 7.4, then washed, pelleted again, incor- porated in 1% Difco agar, sampled, and processed for embedding in Epon 812.

All the cell samples treated with hybrid mixtures containing anti-Fer and anti-HRP or only anti-HRP were washed in 0.05 M Tris-HCl buffer, pH 7.4, and incubated for 20 min at room temperature in a medium con- taining 0.05% 3,3'-diaminobenzidine and 0.02% H_2O_2 in 0.05 M Tris-HCl buffer, pH 7.4. The cells were washed, postfixed in 1% OsO_4 in sodium cacodylate buffer, pH 7.4, and again washed and pelleted. The pellet was incorporated in 1% Difco agar at 65° and divided into 1-mm^3 fragments. These were dehydrated an embedded for transmission electron micros- copy.

Cytotoxic Lymphocytes Coated with Hybrid Antibody

Because of its multivalency the hybrid antibody has an increased functional affinity for the cognate membrane antigen and thus allows cell-

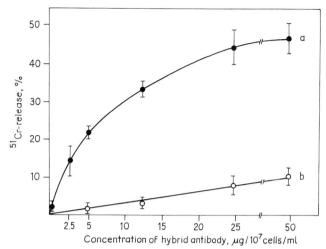

FIG. 10. Killing of CRBC targets by mouse spleen cells treated with multivalent antibodies with dual specificity. Curve a: (IgG anti-Thy 1/SpA/IgG anti-CRBC)₂; Curve b: (IgG/SpA/IgG anti-CRBC)₂, a complex without anti-Thy 1 activity (control). Reproduced Laky *et al.*,[16] with permission of *Molecular Immunology*.

to-cell contact between two unrelated cells (Fig. 6). Since hybrid antibody can be directed both against cells with cytotoxic potential and cells susceptible to be lysed we demonstrated that hybrid antibody can activate normal T lymphocytes to become killers. Thus by coating mouse lymphocytes with hybrid antibody containing rabbit anti-Thy 1 antibody linked to anti-CRBC antibody, we were able to demonstrate that a subpopulation of Thy 1 bearer cells was able to kill specifically CRBC.[16]

Spleen cells from adult AKR mice were separately treated with (IgG anti-Thy 1/SpA/IgG anti-CRBC)₂ and (IgG/SpA/IgG anti-CRBC)₂ complexes (the latter lacking anti-Thy 1 activity); after incubation and washing ^{51}Cr-labeled CRBC were added. Results of several experiments show that spleen cells treated with (IgG anti-Thy 1/SpA/IgG anti-CRBC)₂ complex interacted with and killed the target cells (Fig. 10). The ability to kill CRBC depended on the concentration of (IgG anti-Thy 1/SpA/IgG anti-CRBC)₂ complex. Cells treated with the (IgG/SpA/IgG anti-CRBC)₂ complex (taken as control) had no significant effect in killing cells even at a concentration of 25 μg/10^7 cells. Very similar results were obtained when, instead of spleen cells, purified lymphocytes from adult mice (containing less than 5% monocytes) were used.

[16] M. Laky, G. Moţa, and V. Gheţie, *Mol. Immunol.* **18,** 1029 (1982).

Spleen cells isolated from young (14 days) mice and homozygote nude mice were also treated with IgG antibody/SpA complexes, and their ability to attach and kill CRBC was further assayed (Table IV). As compared to the cytotoxic potential of lymphocytes from adult mice, it can be seen that lymphocytes from nude mice had a higher killing efficiency and the cytolytic effect of lymphocytes from young mice was consistently lower.

The specificity of the CRBC binding and killing reactions by lymphoid cells treated with hybrid antibody having dual, anti-Thy 1 and anti-CRBC, specificity was clearly demonstrated by (a) the ability of Fab anti-Thy 1 antibody to inhibit killing of CRBC targets by spleen cells charged with the Fab antibody prior to hybrid antibody treatment (Fig. 11); (b) the ability of Fab anti-CRBC antibody to inhibit killing of CRBC targets by spleen cells charged with MHA (when CRBC were coated with Fab anti-CRBC antibody) (Fig. 11); (c) the inability of spleen lymphocytes coated with the hybrid antibody complex (IgG anti-Thy 1/Spa/IgG anti-Fer)$_2$ to kill CRBC.

Previous studies have shown that mouse spleen lymphocytes or peritoneal macrophages treated with the (IgG anti-CRBC/SpA/IgG anti-CRBC)$_2$ complex displayed an enhanced ability in rosette formation or in

TABLE IV

BINDING AND KILLING OF CRBC TARGETS BY SPLEEN CELLS ISOLATED FROM YOUNG MICE AND NUDE MICE AFTER TREATMENT WITH IgG ANTIBODY/SpA COMPLEXES[a]

Source of spleen cells	Treatment with IgG/SpA complexes[b]	Rosetting and killing ability of treated cells	
		RFC[c] (%)	CRBC-lysis[d] (%)
Young mice	A	8.1	8.4
	B	1.4	4.2
	C	1.0	2.0
Nude mice	A	25.8	66.2
	B	7.6	20.3
	C	1.0	3.1
Adult mice (control)	A	18.8	44.0
	B	3.5	6.5
	C	1.0	2.1

[a] Reproduced from Laky et al.,[16] with permission of Molecular Immunology.

[b] Treatment performed at a concentration of 25 μg/10^7 cells for (IgG anti-Thy 1/SpA/IgG anti-CRBC)$_2$ (A) and (IgG/SpA/IgG anti-CRBC)$_2$ (B). Without treatment (control)(C).

[c] Percentage of CRBC rosette forming cells (RFC).

[d] Percentage of ^{51}Cr-CRBC release after 18 hr of incubation at 37°.

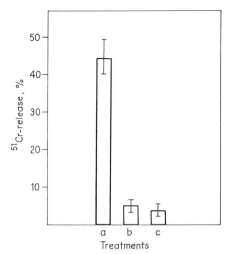

Fig. 11. Inhibition of killing abilities of mouse spleen cells with Fab antibody against Thy 1 and CRBC antigens. Bars: a, cells charged with hybrid antibody (IgG anti-Thy 1/SpA/IgG anti-CRBC)$_2$; b, cells treated with Fab anti-Thy 1 prior to hybrid antibody; c, cells charged with hybrid antibody and CRBC coated with Fab anti-CRBC. Reproduced from Laky *et al.*,[16] with permission from *Molecular Immunology*.

lysis of CRBC as compared to the ability acquired after treatment with free IgG anti-CRBC molecules.[17,18] The enhancement of rosetting and CRBC killing for both cell types was a result of increased activity of the octavalent (IgG anti-CRBC/SpA/IgG anti-CRBC)$_2$ antibody complex.[8] These results suggest that (IgG anti-Thy 1/SpA/IgG anti-CRBC)$_2$ may bind to Thy 1- and FcR-Positive cells either by anti-Thy 1 antibody or by the Fc region of IgG molecule or even by both sites of the complex.

In order to estimate separately the rosetting and cytotoxic efficiency of cells binding IgG antibody/SpA complexes via either the antibody combining sites or the Fc region, we used in the present work a control complex containing one of the IgG partners without antibody activity (IgG/SpA/IgG anti-CRBC)$_2$. Since the results showed that the cells did not gain any significant killing ability after treatment with this control complex (Fig. 10 and Table IV), we considered that the effect of (IgG anti-Thy 1/SpA/IgG anti-CRBC)$_2$ complex was entirely due to its binding with one antibody specificity to the lymphocyte Thy 1 antigen and with the other to CRBC surface antigen(s).[16]

[17] A. Sulica, M. Laky, M. Gherman, V. Gheţie, and J. Sjöquist, *Scand. J. Immunol.* **5**, 1191 (1976).
[18] M. Laky, M. Gherman, V. Gheţie, J. Sjöquist, and A. Sulica, *Scand. J. Immunol.* **7**, 345 (1978).

TABLE V
COMPARISON BETWEEN MONOVALENT AND MULTIVALENT HYBRID ANTIBODY

Preparation, properties, and use	Monovalent hybrid	Multivalent hybrid
Molecular composition	Fab' anti-A/Fab' anti-B	(IgG anti-A/SpA/IgG anti-B)$_2$
Molecular weight	100,000	669,000[a]
Valence	1	4
Affinity	Low	High
Stability	Reducing agents destroy the activity	Low pH destroys the activity
Preparation	Tedious	Easy
Purity	Impurified with anti-A or anti-B and with inactive IgG	Impurified with inactive IgG
Use	Location of surface antigens by electron microscopy	The same use as for monovalent hybrids plus investigation of cell-to-cell interactions

[a] Calculated: 684,000 M_r (taking as molecular weights 150,000 for IgG and 42,000 for SpA).

Normal T lymphocytes can be "instructed" to kill target cells only when cross-linking between CRBC and lymphocytes is achieved by the multivalent hybrid antibody. The cytotoxic cells detected by treatment with hybrid antibody are a preexisting subpopulation of T cells requiring only a link to the target for the expression of their cytotoxic potential. One possible candidate for the killer activity might be NK cells which, similar to the hybrid-treated MSL, have a high cytotoxic activity in nude mice and a low one in young mice as compared to the activity of spleen cells from normal adult mice.[19]

Assay.[16] Mouse spleen cell suspension in TC-199 medium containing 0.015 M sodium azide was treated with different amounts of hybrid antibody preparation (1–25 μg/10^7 cells/ml). Cells were incubated for 1 hr on ice and washed three times, the last washings containing TC-199 medium with 10% fetal calf serum minus sodium azide. The hybrid antibody-treated mouse splenocytes (5 × 10^6 cells/ml) were used as effector cells, and sodium (^{51}Cr) chromate-labeled CRBC (10^5 cells/ml) were used as targets. The cells were maintained in culture for 18 hr at 37°. The ^{51}Cr release was measured in a well-type gamma counter. The release was

[19] R. B. Herberman, J. Y. Djeu, H. D. Kay, J. R. Ortaldo, C. Riccardi, G. D. Bonnard, H. T. Holden, R. Fagnani, A. Santoni, and P. Puccetti, *Immunol. Rev.* **44,** 43 (1979).

expressed as percentage of the total radioactivity contained in 10^5 radiolabeled CRBC. As a control for spontaneous ^{51}Cr release, CRBC were incubated with untreated effector cells or treated with (IgG/SpA/IgG anti-CRBC)$_2$ complex without anti-Thy 1 activity.

Concluding Remarks

As compared with the hybrid antibody prepared by recombination of Fab' fragments,[2,3] the main advantage of the hybrid SpA–antibody complex[10,13] is its simplicity of preparation, purity, and multivalency (Table V). By its multivalency (four combining sites per antibody partner of the complex), the hybrid antibody complexes have an increased functional affinity for the corresponding antigens, as reported for SpA–antibody complexes with a single specificity.[6,8] Hence the gain in avidity allows the hybrid SpA–antibody complexes to draw together not only soluble antigens, but also particulate ones (e.g., cells, bacteria, viruses). Such multivalent hybrid SpA–antibody complexes are useful tools for the study of cell-to-cell interaction and location of cell-surface antigens.

Acknowledgments

Our investigation on the preparation of multivalent hybrid antibody and its exploitation in immunology was supported by grants from the Romanian Academy of Medical Sciences and was done in cooperation with Drs. Gabriela Moța, E. Mandache, Mariana Laky, and Elena Moldoveanu and with the technical assistance of Mrs. Mariana Caralicea, Miss Mariana Nicolae, Mrs. Cornelia Popescu, and Mrs. Petra Stanciu, all from Babeş Institute, Bucharest. We thank Mrs. Mariana Pavlovski for skillful secretarial assistance and for typing the manuscript.

[41] LIGAND: A Computerized Analysis of Ligand Binding Data

By PETER J. MUNSON

In this chapter, a computerized approach for the analysis of data from ligand binding experiments is described.[1,2] The computer program facilitates a systematic, objective data analysis by providing efficient estimates of the binding parameters, evaluating the adequacy of fit of the data, and

[1] P. J. Munson and D. Rodbard, *Anal. Biochem.* **107**, 220 (1980).
[2] P. J. Munson and D. Rodbard, *Endocrinology* **105**, 1377 (1979).

allowing convenient formulation and comparison of a variety of models, so that the best model among these may be chosen on a sound, statistically meaningful basis.

For many decades, progress in the field of enzymology and, more recently, in hormone-receptor studies has relied almost exclusively on graphical methods of analysis of the binding isotherm, without the aid of computerization. With the widening availability of mini-, micro-, and even "personal" computers, many experimentalists are turning to these devices for help in the analysis of their data. Yet some are reluctant to use this tool, arguing that if the experimental result cannot be demonstrated on the appropriate plot then it is probably not credible, or if the data are so "noisy" as to require computerized curve fitting then the experiment is of basically poor quality and the results should not be accepted. However, a seemingly poor experiment, with careful statistical analysis may still yield important information about how to redesign future experiments.

Graphical approaches, though simple, powerful, and well tested, do have some shortcomings. There are many different graphical approaches (Scatchard plot,[3] Lineweaver–Burk plot,[4] Woolf plot,[5] Hill plot[6]), each providing a slightly different view of the same data. Therefore conclusions drawn from an experiment may be influenced by which plot is used for analysis. Many graphical methods rely on a transformation of the data into a straight-line representation. However, a simple straight line will result only under the simplest binding reaction mechanisms. Graphical methods do not generalize easily to more complicated binding models; e.g., consider the difficulty of analyzing the Scatchard plot for two independent classes of binding sites. Furthermore, manually fitting a line to data may allow the investigator unconsciously to disregard (or overemphasize) certain data points in the graph in order to obtain a desired result. Thus, graphical results contain an element of subjectivity.

An appropriate, intelligently used, computerized analysis can overcome many of the difficulties presented by graphical methods. First of all, computers can be heartlessly objective and will always arrive at the same result, given the same data. Second, it is usually possible to generalize a computer program to test more complex binding mechanisms. For example, if a one-site (straight-line Scatchard) model is inadequate, one may attempt to fit a two-site model, using the same computer program. Third, whereas there are three or four graphical coordinate systems in which to view the binding isotherm, there is a unique, "best" coordinate system

[3] G. Scatchard, *Ann. N.Y. Acad. Sci.* **51**, 660 (1949).
[4] H. Lineweaver and D. Burk, *J. Chem. Soc.* **56**, 658 (1934).
[5] J. B. Haldane, *Nature (London)* **179**, 832 (1957).
[6] A. V. Hill, *J. Physiol. (London)* **40**, IV (1910).

that may serve as the basis for the computerized analysis, namely, the system in which the data were originally collected. Graphical transformations that linearize the binding curve also tend to distort and obscure the statistical properties of the data. In the untransformed data (viz. Bound vs Total ligand concentration), virtually all of the measurement error is confined to a single variable, Bound concentration, which is thus suitable as the independent variable in the regression. In the Scatchard plot, for instance, measurement errors appear in both Bound and B/F axes, thus invalidating ordinary linear regression.

Automated computerized analysis should never completely replace graphical methods, however. Graphical presentations enhance and facilitate the investigators intuition in a way that pure numerical computation cannot. Graphs sometimes suggest possible enhancements in the theoretical model describing the binding reaction. For instance, upward curvature in the Scatchard plot suggests heterogeneity or negative cooperativity in binding sites; downward curvature suggests positive cooperativity. The numerical output of the computer is generally not as helpful in this regard. Furthermore, not every set of data requires a careful computer analysis. If there is little or no "noise" in the data and the experiment is well designed, then all valid analysis techniques should give the same results. However, every investigator ought to be familiar with computerized nonlinear least squares, because for certain difficult or critical experiments, a sophisticated, computerized analysis may be essential. Thus, in routine situations where measurement errors have been carefully controlled and the biochemical model is fully validated, graphical techniques will continue to yield adequate and useful results. But on the frontiers of a new discipline, where data are obtained under difficult, noisy situations and the model itself is in question and under development, a computerized analysis approach will often become the method of choice.

The ideal analysis technique is a hybrid of pure statistical calculation and graphical approaches. In fact, the computer can also routinely provide high-resolution, convenient graphical output. Automatic graphical representation of the data with the best fitting curve provides a check on the validity of the statistical data analysis and helps one ensure that the results "make sense." In addition to the objectivity, reliability, and efficiency, there is an important additional advantage of using a computerized analysis method. Many graphical methods place unnecessary constraints on the design of the binding experiment. For instance, a Scatchard plot will be linear only if one plots the "specific" binding of a single class of receptor sites, after measuring and subtracting the "nonspecific" binding in a separate experimental step. On the other hand, using an appropriate computer program, one may fit both specific and nonspecific binding simultaneously, largely avoiding the need for additional "nonspecific" tubes. Computerized analysis thus allows for greater flexibility in the experimental design.

In this chapter we describe unified strategy and computer program for the analysis of a wide variety of experiments involving one or more ligands binding to possibly several classes of binding sites. The general "$n \times m$" model for n ligands binding to m classes of receptor sites is used as a basis for calculation.[7,8] Many experimental designs can then be handled as special cases of this model.

The computer program described here[1] provides a means of obtaining maximally efficient estimates of the parameters of the specified model with their standard errors, and the correlation matrix for the various parameters. From this output one may construct the confidence limits or the joint confidence regions for any pair of parameters. Typically, the high correlation of estimates of K and R make inspection of the bivariate elliptical confidence regions useful and, in some cases, necessary. The parameters are estimated by means of a weighted nonlinear least squares algorithm and thus enjoy a number of important statistical properties.[9]

The use of appropriately weighted nonlinear least squares is an established methodology for estimating the parameters in many situations besides binding reactions. Although this methodology is not the only one available, its wide acceptance in the scientific community, in addition to its statistical properties, argues for its use as a standard method in ligand-binding studies as well.

Finally, we caution users of computerized analysis methods not to attach a greater significance and importance to the output than it is due. Clearly, the output of a computerized analysis can be no better than the input of data and experimental technique. The true significance of affinity values obtained by nonlinear least squares curve fitting still depends heavily on the assumption of thermodynamic ideality of the experimental conditions (true equilibrium, no errors in separating bound and free, bimolecular, reversible binding, etc.). The receptor molecule may not be in true solution; unstirred layers or membrane charges may modify the true binding energy. All these caveats are as true for the computerized analysis as they are for any form of graphical analysis. Nevertheless, even in the absence of one or more of these preconditions, fitting a reasonable biochemical model to the data is generally to be preferred over simply drawing a smooth curve through the data points.

[7] H. A. Feldman, *Anal. Biochem.* **48**, 317 (1972).

[8] H. A. Feldman, D. Rodbard, and D. Levine, *Anal. Biochem.* **45**, 530 (1972).

[9] Y. Bard, "Nonlinear Parameter Estimation." Academic Press, New York, 1974.

We illustrate the use of a computerized analysis strategy with a series of examples of increasing complexity. The analysis makes use of a BASIC language computer program (LIGAND, available from the author upon request) which is adaptable to various mini- and microcomputers now available in many laboratories.

One Labeled Ligand, One Class of Sites—Linear Scatchard Plot

The simplest model for ligand binding is the case of a single ligand reacting with a single homogeneous class of binding sites, frequently represented as a linear Scatchard plot. Here, the investigator wishes to evaluate the affinity, K, and binding capacity R. The apparent simplicity of this case would at first seem to make the payoff of a complex, computerized analysis very small. However, the actual experimental technique frequently contains a subtle complexity making the use of computerized nonlinear least squares worthwhile.

First, we describe the analysis of a typical saturation binding experiment. Increasing doses of labeled ligand are added to a series of tubes, each containing a constant, small amount of tissue or cell preparation containing the receptor of interest. After allowing for the binding to reach equilibrium, the bound and free fractions of labeled ligand are separated, usually via centrifugation or filtration, and the bound concentration is determined, often by counting radioactive disintegrations. Optionally, free ligand concentration may also be determined. In either case, since total ligand concentration is known for each tube, both bound and free concentrations may be calculated for each tube. One typically constructs a plot of a transformation of these data, for example, B/F ratio vs Bound concentration. If the data points fall along a straight line, its slope and intercept may be used as estimates of binding affinity K and capacity R. The analysis is actually more complex in most biological systems, since receptor protein is frequently contaminated by many other proteins, membranes, cells, or organelles that may also bind the ligand with low affinity. In order to obtain a straight-line graph, one must first subtract this nonspecific, nonsaturable binding, which is measured by adding a large (100-fold excess) concentration of unlabeled ligand, thus displacing virtually all the specific bound labeled ligand from the receptor.

In many situations, this simple approach is entirely adequate. For instance, in the analysis of estradiol binding to mammary tumor cells, where a single class of receptors is presumed to exist, the graphical approach appears to yield adequate results.[10] However, there are situations

[10] D. Rodbard, P. J. Munson, and A. K. Thakur, *Cancer* **46**, 2907 (1980).

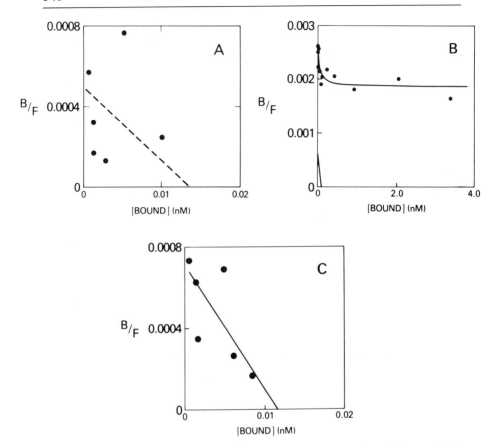

FIG. 1. [³H]Dexamethasone binding to squirrel monkey mononuclear leukocytes.[11] (A) Scatchard plot of original data (six points) after subtraction of nonspecific binding estimated by addition of 100-fold excess unlabeled dexamethasone. Scatter of data is extremely high; existence of specific binding (dashed line) is questionable. (B) Crude binding (six points) and nonspecific binding (six points) are fit simultaneously with mathematical model for one specific site and one nonspecific site. Scatchard plot shows the presence of an extremely large horizontal asymptote at $B/F = 0.002$ (nonspecific binding), but the specific binding component is now clearly recognizable, although it represents less than one-third of the total initial binding. (C) Specific binding, using computer estimate of nonspecific binding, represented in Scatchard plot. Scatter of original six data points is now reduced with the use of a smoothed, consistent estimate of nonspecific binding.

where the process of subtracting nonspecific binding itself introduces substantial amounts of statistical error, resulting in unnecessary scatter of points in the Scatchard plot (Fig. 1A). We illustrate this effect with data from a study of dexamethasone binding to cortisol receptors on mononuclear leukocytes. The supply of white blood cells, taken from the squirrel monkey,[11] was severely limited. Thus, additional replication could not be used to reduce the scatter of the data. The squirrel monkey is considered to be a model for human cortisol resistance because its high circulating cortisol levels suggest that the hormone is somehow ineffective in carrying its message to the target tissue. In this experiment, the investigators wished to determine the binding affinity and capacity for squirrel monkey cortisol receptors, and thus to characterize a possible receptor defect. Owing to the limited number of data points and the large degree of scatter in both the specific and nonspecific binding measurements, the Scatchard plot obtained by presubtracting nonspecific binding first appeared to be useless (Fig. 1A). However, by first reexpressing the data as Bound ligand (labeled or unlabeled) vs Total (either hot or cold) ligand added, and then fitting the parameter for nonspecific binding, N, simultaneously with the affinity, K, and capacity, R, to the experimental data, the method finds reasonably good estimates for K ($6 \times 10^7 M^{-1}$), R ($1 \times 10^{-11} M$) and N (0.002). [See Eq. (8) for a mathematical description of the model.]

Although the estimates of K and R were determined only to within a factor of two, there was no question about the existence of specific binding of dexamethasone. The model including parameters for both specific binding and nonspecific binding (K, R, and N) fit significantly better than the model that allowed only nonspecific (N only) binding. The "extra sum of squares" comparison yielded an F value of 12.17 with 2 and 9 degrees of freedom, which is significantly at the $p = 0.01$ level. The data were next replotted in the Scatchard coordinates together with the computer-fitted curve, this time without substraction of nonspecific (Fig. 1B). The high-affinity binding component (lower left corner) is almost completely obscured by the nonspecific binding. Nevertheless, the overall binding model fits the data reasonably well with a root mean square (RMS) residual error reflecting an average scatter of 8% around the predicted curve for bound ligand concentration, which is quite acceptable. The unusually high background or nonspecific binding resulted from a biological contaminant that could not be removed experimentally. Finally, the data were replotted (Fig. 1C) in the Scatchard plot after subtracting the computer-estimated nonspecific binding component. The reduction in the scatter

[11] G. P. Chrousos et al., Proc. Natl. Acad. Sci. U.S.A. **79**, 2036 (1982).

around the straight line (Fig. 1A vs Fig. 1C) resulted from the use of a single, stable, computer-generated estimate of the nonspecific binding.

The general strategy here was to fit the "raw" data, Bound ligand concentration, with the appropriate mathematical model for a single ligand binding to one class of specific sites and to an additional class of nonsaturable, nonspecific sites (see the last section, the mathematical model). We did not use the Scatchard plot as the basis for computerized analysis, but only for presentation of the data. One of the most common mistakes of early attempts to computerize the analysis of ligand binding was to use simple linear regression in a transformed plot, such as the Scatchard. Some of the pitfalls of this approach are apparent here and have been detailed elsewhere.[1,12–16] Basically, owing to the unusual or bizarre nature of the error structure in the Scatchard plot induced by the data transformations, this simplistic approach can give rise to substantial, 2- to 10-fold estimation errors. Several workers have attempted to compensate for this unusual error structure with the use of a robust or weighted robust regression in the Scatchard plot.[12] While this idea has some merit, it is generally simpler, and certainly more direct, to use weighted nonlinear least squares with the exact mathematical model of the binding reaction, thus taking full advantage of the well-behaved error structure of bound ligand concentration.

Curved Scatchard Plots—One Ligand, Two Classes of Binding Sites

Frequently, some curvature may become evident in the Scatchard plot. Curvature complicates the analysis sufficiently that most investigators turn to the computer for help at this point. Although curvature may in fact be due to a variety of causes (positive or negative cooperativity of receptor sites, experimental artifacts), one frequently assumes, with some biological justification, that curvature is due to the presence of two or more classes of receptors or binding sites. The mathematics involved in analyzing and modeling this situation rapidly becomes quite complex. Unfortunately, because of this complexity, several incorrect, simplified analysis methods are quite widespread in the literature.

First, many investigators may be tempted to ignore the curvature altogether, and fit a straight line to the plot. The resulting affinity estimate will have little true thermodynamic significance, although some will con-

[12] N. A. C. Cressie and D. D. Keightley, *Biometrics* **37**, 235 (1981).
[13] E. E. Baulieu and J.-P. Raynaud, *Eur. J. Biochem.* **13**, 293 (1970).
[14] H. R. Kalbitzer and D. Stehlik, *Z. Naturforsch. C. Biosci.* **34C**, 757 (1979).
[15] J. T. Woosley and T. G. Muldoon, *J. Steroid Biochem.* **8**, 625 (1977).
[16] J. Wahrendorf, *Int. J. Bio-Med. Comput.* **10**, 75 (1979).

sider this to be an "average" affinity of the system. Failure to detect curvature may have serious consequences in that the receptor binding affinity can be significantly over- or underestimated.

Second, the investigator may recognize the curvature but then attempt to fit only the upper "high affinity" segment of the curve with a straight line, and simply ignore the data points in the "low affinity" region. Although in certain restricted circumstances this will yield a satisfactory estimate of the high-affinity component, it is definitely nonoptimal and generally yields misleading results. Similarly, the tendency toward curvature in the Scatchard plot may go undetected simply because data were not collected over a wide enough concentration range. This may explain why many studies of the estrogen receptor in rat uterus report a straight-line Scatchard plot, whereas there is some more recent evidence of curvature at the very low affinity, high concentration region of the curve.[17]

Third, one is tempted to conclude that since the slope of a linear Scatchard plot is an estimate of the binding affinity, the slope of the two "linear" portions of a curved Scatchard plot can be taken as estimates for the high and low affinities, K_1 and K_2 for two classes of sites. In general, this is false; such estimates may contain a two- to fivefold bias. However, if we can estimate the slopes and intercepts of the extreme ends of the curve, it is possible to compensate for this bias using a technique described by Thakur *et al.*[18] The correct estimates of the binding parameters are represented graphically as the asymptotes to the hyperbolic curve. A manual, graphical approach known as curve peeling has been developed[19] for constructing the asymptotes, but in practice it is quite tedious.

The optimal approach to estimation of the binding parameters K_1, R_1, K_2, R_2, for two classes of sites requires use of a computer program. However, not all programs are equal; the investigator must take care that the computer program is designed to give optimal results. It is important not to fit the hyperbola directly to the Scatchard plot, owing to the distorted error structure discussed previously. Rather, the "raw," untransformed data should be fit with the correct biochemical model of two independent classes of sites. Although the algebra involved for this model may at first appear formidable, once the equations have been programmed, the computer can solve them numerically, and thus the weighted least-squares estimates of K_1, R_1, K_2, R_2, and N are easily obtained (see the section The Mathematical Model for details).

[17] K. Kelner and E. Peck, *J. Receptor Res.* **2**, 47 (1981).
[18] A. K. Thakur, M. L. Jaffe, and D. Rodbard, *Anal. Biochem.* **107**, 279 (1980).
[19] H. E. Rosenthal, *Anal. Biochem.* **20**, 525 (1967).

A number of additional difficulties become apparent when programming the computer to perform this analysis. The first is the question of how to detect curvature automatically. If a single class of sites has already been fit to the data (straight Scatchard plot), then one may instruct the computer to look for a pattern in data points around the fitted curve. In Fig. 2, we illustrate this process using artificial data, first fitting it with an (incorrect) one-site model. Here, in the 14 points plotted, there are three runs in the residuals, i.e., 3 groups of points that lie to the left (below) or to the right (above) of the dashed line. If these 14 points were randomly scattered, we would expect roughly 7 such groups or runs above and below this line. Since 3 runs is surprisingly few (see the section Statistical Considerations), we have evidence that a one-site model, even including nonspecific binding, is not a good explanation of the data.

The one-site model (Table I, line 1) requires 3 parameters, K_1, R_1, and N. When a two-site model, requiring 5 parameters is fit to the data, a substantial improvement in the quality of fit results (Fig. 2), reflected in the reduction in RMS error from 8.5% to 6.3% (Table I, lines 2 and 3). In fact, a mathematical consequence of adding parameters to any model is additional flexibility, which results in better fit to the data. But does this degree of improvement in fit reflect true superiority of the two-site model?

A formal statistical test, utilizing the extra-sum-of-squares principle (see under Statistical Considerations) can give the answer. Intuitively, this test provides a penalty according to the number of new parameters added to the model and computes an F value (in this case 6.96), which can be compared to the "critical" value (in this case 6.7 for $F_{2,13}$ at the $p = 0.01$ level). Since the F value is greater than the critical value, we reject the one-site model in favor of the two-site model. The "p value" refers to the probability of drawing this conclusion if in fact there were only a single class of sites.

The general strategy is as follows: If there is a suggestion of lack of fit using one model, we generalize the model (add another class of sites) to attempt a correction. Then, after fitting the expanded model, we determine whether the improvement in fit was substantial enough to warrant the additional complexity. We now consider some of the statistical properties of our parameter estimates. Because we cannot determine the "true" values of parameters exactly in any real biological system, our example is based on a computer-generated experiment, where we have predetermined the values for all the binding parameters. Table I compares the results of fitting each model with the true values. We see that the estimate for the affinities of the correct (two-site) model are well within two standard errors of the true values, whereas, using a one-site model, the affinity of K_1 is underestimated by a factor of two. Looking at the

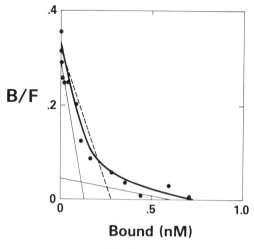

Fig. 2.

parameter estimates for the two-site model, some investigators might argue that the greater than 100% error in the estimates of R_2 and K_2 indicates the suitability of a two-site model.

However, these large standard errors are frequently a consequence of the experimental design, not a measure of the suitability of the model. Large uncertainties in the parameters can be expected if the range of ligand concentration does not span the affinity values we are attempting to measure. Furthermore, even though K_2 and R_2 are poorly defined, their product $K_2 \times R_2$ remains well defined, owing to the large negative correlation in these estimates. Our best overall criterion for choosing a two-site over a one-site model remains the F test described above.

There are some practical considerations involved in computer analysis of a two-site model. Nonlinear regression programs generally require ini-

TABLE I
NONLINEAR REGRESSION PARAMETER ESTIMATES

	K_1 (nM^{-1})	R_1 (nM)	K_2 (nM^{-1})	R_2 (nM)	N	RMS[a] $(\%)$	df[a]	SS
One-site model	1.1 ±0.2	0.28 ±0.04	—	—	0.055 ±0.002	8.5	15	1084
Two-site model	2.1 ±0.7	0.14 ±0.05	0.074 ±0.077	0.62 ±0.34	0.051 ±0.003	6.3	13	516
True values	2.3	0.11	0.11	0.55	0.051	7.0	17	833

[a] RMS, root mean square; df, degrees of freedom; SS, sum of squares.

tial estimates for model parameters, which are then iteratively refined by the program. Appropriate initial values are required for successful completion of the program, and values may be taken from previously determined values or from the Scatchard plot directly. Use of the "limiting slopes" technique[18] to obtain initial values for two-site models has proved to be quite successful. Interestingly, there is a high degree of correlation between the parameters for the low-affinity site and for nonspecific binding. In fact, a curved Scatchard plot can sometimes be adequately straightened by the simple expedient of optimizing the estimate of nonspecific binding. This is an additional virtue of fitting the nonspecific parameter simultaneously with affinity and capacity.

There are several other explanations of curvature in the Scatchard plot to consider. Curvature may be due to a complex ligand–receptor interaction, such as negative cooperativity between neighboring receptor sites. A variety of negative cooperative models have been proposed[20] that may generate curves similar to the two classes of sites model. It is practically impossible to distinguish between the negative cooperativity and heterogeneity of receptors on the basis of equilibrium binding alone. Indeed, under certain circumstances the mathematical formalism for these models becomes identical.[20a] Additionally, there are a number of experimental artifacts that may cause curvature in the Scatchard plot. Underestimation of nonspecific binding may cause a suggestion of a low-affinity class of sites in the Scatchard plot. The presence of endogenous ligand in the tissue preparation, degradation or metabolism of the free ligand, inadequate correction for counter background, impure labeled ligand, misestimated specific activity, contamination of the bound fraction by the free fraction all may influence the curvature and shape of the Scatchard plot.[21,22]

Another cause of curvature may be assay drift; i.e., a continuous slow change in the assay conditions with time may result in systematic deviations from the model. Drift is frequently observed in the practice of routine radioimmunoassays (RIAs), where measurement errors of neighboring samples in an assay will display a substantial degree of serial correlation. This in turn, may be due to minor fluctuations of temperature, pH, etc., with resulting change in binding affinity. One may largely remove the effects of "drift" by randomly ordering the various concentra-

[20] F. W. Dahlquist, this series, Vol. 48 [13].

[20a] J. E. Fletcher, A. A. Spector, and J. D. Ashbrook, *Biochemistry* **9**, 4580 (1970).

[21] P. J. Munson, *J. Receptor Res*. **3** (1983) (in press).

[22] J. M. Boeynaems and J. E. Dumont, "Outlines of Receptor Theory." Elsevier/North-Holland, Amsterdam, 1980.

tions of ligand within the assay. Although this is generally impractical, a quasi-random approach may be effective, i.e., first performing an increasing series, then a decreasing series, of concentrations.

Displacement Curves—One Labeled Ligand, One Class of Sites, One, Two, or More Unlabeled Ligands

Displacement of competition curve analysis is a common methodological tool in pharmacological studies of drug-receptor binding. When only a single radioactive labeled tracer ligand is available, one may measure the degree of inhibition of binding of a labeled (hot) drug or hormone by a possibly large number of unlabeled (cold) ligands. The analysis usually proceeds with a plot of inhibition (or binding) of the label vs dose of unlabeled ligand.

From the estimated dose yielding 50% effect (ED_{50}), one may estimate the relative potency of these unlabeled drugs or hormones. Assuming that all the ligands are competing for a single, homogeneous class of receptor sites, we can also estimate the affinity, K_1, of the labeled ligand or the relative affinity K_2/K_1 of the unlabeled ligand. A simplistic approach for calculating the relative affinities uses the relation:

$$K_2/K_1 \cong ED_{50_1}/ED_{50_2} \tag{1}$$

where ED_{50_1} refers to the ED_{50} for the first ligand, and ED_{50_2} refers to the second. Several workers have recognized that this approach is valid only in certain limiting situations. Therefore, several rules for relating ED_{50} values to binding affinities have been developed. The Cheng–Prusoff–Chou correction[23,24] was developed in the context of enzyme inhibition studies but is also frequently applied in ligand competition studies. This rule relates the ED_{50} of a competitive inhibitor as a function of equilibrium constants and substrate concentration. This equation is applicable only when free ligand is approximately equal to total concentration, i.e., when the B/T ratio is small. This assumption is usually satisfied in enzyme studies, but is frequently violated in ligand–receptor studies. When B/T is small, we have

$$ED_{50_2} \cong (1 + p^*K_1)(1/K_2) \tag{2}$$

Where p^* is the concentration of the first, labeled ligand, ED_{50_2} is the dose of the second ligand inhibiting 50% of the initial binding, K_1 and K_2 are affinity constants of the labeled and unlabeled ligands, respectively.

[23] Y. Cheng and W. H. Prusoff, *Biochem. Pharmocol.* **22**, 3099 (1973).
[24] T.-C. Chou, *Mol. Parmacol.* **10**, 235 (1974).

Analogous to Eq. (2), when the displacer (unlabeled) ligand is identical to the labeled ligand, and assuming $K_1 = K_2$, we have

$$ED_{50_1} \cong 1/K_1 + p^* \tag{3}$$

Combining these two relationships, we may derive values for K_1, K_2, and K_2/K_1 from a pair of displacement curves. Another calculation rule[25] that does not assume that B/T is small, but provides an exact calculation for the relative affinity, is

$$K_2/K_1 = ED_{50_1}/[ED_{50_2} + ED_{50_2}(B/F^*) + ED_{50_1}(B/F^*)] \tag{4}$$

where B/F^* is the bound to free ratio of the labeled ligand at the 50% displacement point, and the ED_{50}s are designated for ligands 1 and 2.

Where the same ligand is used as tracer and competitor (self-displacement), a number of rules are available for converting ED_{50}s to affinity values. The data may also be reexpressed in a Scatchard or Lineweaver–Burk plot, and both the affinity and the capacity, R, can be estimated. If p^*, the labeled ligand concentration, is very large so that all the receptor sites are initially saturated, we see from Eq. (3) that the ED_{50} is solely determined by p^* and is no longer an indication of binding affinity. When p^* is small but receptor concentration, R, is not,

$$ED_{50} \cong 1/K + R/2 \tag{5}$$

where ED_{50} now refers to the dose yielding 50% reduction of the initial B/F ratio.

One can avoid this multiplicity of rules and approximations by fitting the exact biochemical model for two ligands competing for a single class of receptors, a special case of the general n by m model. Although the mathematical model effectively requires solving a set of quadratic equations, the time required for computation is negligible, and the results are valid regardless of the value for tracer concentration p^*, initial binding B/T, receptor occupancy, etc. We illustrate this computerized approach with synthetic, computer-generated data shown in Fig. 3. Here we have a single labeled ligand, L_1, displaced from its receptor by three unlabeled ligands, L_1, L_2, and L_3. L_2 has apparently a lower affinity and L_3 a higher affinity than L_1, the labeled ligand. For purposes of computation, the data shown in Fig. 3 are reexpressed as Bound ligand concentration for the labeled ligand vs Total added. From these data the computer estimates five parameters K_1, K_2, K_3, R, N_1. The last parameter describes the nonspecific binding for the labeled ligand, which would ordinarily be sub-

[25] D. Rodbard and J. E. Lewald, *Acta Endocrinol.* **64** (Suppl. 147), 79 (1970).

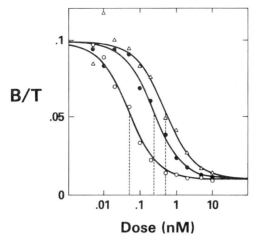

B/T

FIG. 3. Typical displacement curves from experiment involving one labeled ligand (L^*_1) displaced by unlabeled version of same ligand $(L_1, \bullet, \bullet)$, ligand L_2 $(\triangle, \triangle)$ or ligand L_3 $(\bigcirc, \bigcirc)$. Data are artificially generated by adding random "noise" to exact values predicted from mathematical model. ED_{50} values (- - -) for the three curves are conventionally used to calculate affinities and relative potencies for the three ligands. Approximations may be avoided and more accurate results be obtained by fitting the correct mathematical model for three ligands binding to one class of sites with nonspecific binding using nonlinear least squares. From Munson and Rodbard.[1]

tracted from the data before graphical analysis is begun. When using the computer, it is advantageous to fit this parameter to the uncorrected data.

For purposes of comparison, Table II gives the results of estimating the affinities of each of the ligands by three graphical approaches and then

TABLE II
RELATIVE BINDING AFFINITY ESTIMATES USING VARIOUS TECHNIQUES[a]

	K_1 (nM^{-1})	K_2 (nM^{-1})	K_3 (nM^{-1})	R (nM)	N_1	RMS (%)	K_2/K_1	K_3/K_1
Relative ED$_{50}$ method	4.17	1.96	21	—	—	—	0.474	5.05
Cheng–Prusoff–Chou	4.54	2.12	23.7	—	—	—	0.471	5.21
Rodbard–Lewald	—	—	—	—	—	—	0.448	6.33
LIGAND curve fitting	5.7 ±0.6	2.5 ±0.2	37.2 ±3.8	0.020 ±0.002	0.01 ±0.001	7.42	0.431	6.53
True values	5.5	2.3	40	0.02	0.01	7	0.418	7.27

[a] ED_{50} for the labeled ligand, L_1, is 0.24 nM; for L_2, 0.51 nM, and L_3, 0.048 nM. The B/F ratio at the ED_{50} is 0.05. The concentration of labeled ligand, p^*, is 0.02 nM. RMS, root mean square.

by computerized modeling. Compared to the true values, curve-fitting (using LIGAND) provides better estimates for all the parameters than do any of the various rules based on the ED_{50} (Table II, rows 1–3).

The worst error (roughly a factor of 2) is made when estimating the affinity for the most potent ligand, L_3, using either the ED_{50} (Eq. 1) method or with the addition of the Cheng–Prussoff–Chou [Eqs. (2) and (3)] correction. By contrast, the computer estimate of K_3 lies within 10% of the true value. The computer estimates are also accompanied by standard errors, providing the user with an idea of experimental variability. The computer program may also provide measures of the goodness of fit of the model, thus allowing one to detect when the model is inadequate, e.g., when there are actually multiple classes of receptors.

Finally, because the exact model was used as a basis for fitting, the computerized approach is valid for all concentrations, p^*, of labeled ligand, all receptor occupancy levels, all initial binding values, regardless of the relative affinities of labeled and cold ligand. The computerized method described here does not rely on intermediate graphical steps, but estimates the parameters directly, thus making optimum use of the data.

"Bumpy" Displacement Curves—One Labeled Ligand, One Unlabeled, Two or More Classes of Sites

Occasionally, the curve resulting from a competition or displacement experiment will have stepped or terraced appearance suggesting competition at more than one class of receptor sites. These steps in the curve may also overlap, giving rise to a gradual displacement, with a logit-log slope less than 1. In this situation, use of the ED_{50} as a means of characterizing the curves becomes questionable. There follows an example of such a case (Fig. 4).

We illustrate and develop this analysis with a study of the binding of pituitary gonadotropin-releasing hormone (GnRH) to its receptor.[26] Native GnRH, a decapeptide, shows significant low-affinity as well as high-affinity binding to the putative physiological GnRH receptor. Owing to this low-affinity binding as well as substantial degradation of GnRH *in vitro,* valid estimates of the number of high-affinity receptors were difficult to obtain. To remedy these problems, analogs of GnRH substituted in position 6 and 10 of the native decapeptide ([D-Ser(TBu)^6]-des-Gly^{10}-GnRH ethylamide, (D-Ser^6), and [D-Ala^6]-des-Gly^{10}-GnRH ethylamide, (D-Ala^6) were developed. These analogs did not show such rapid degradation *in vitro,* yet apparently had many times the natural potency or biolog-

---

[26] R. N. Clayton, R. A. Shakespeare, J. A. Duncan, and J. C. Marshal, *Endocrinology* **105**, 1369 (1979).

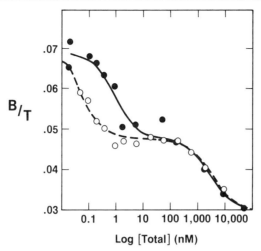

FIG. 4. Displacement of labeled gonadotropin-releasing hormone (GnRH) from pituitary membranes by unlabeled GnRH (●,●) or D-Ser[6] analog (○,○). The curve shows a stepped appearance; thus characterization in terms of the midpoint (ED$_{50}$) is inadequate. Computer fitting of a two-ligand, two-binding site model gives adequate representation of data (——, ---) and gives reliable estimates of affinities of both ligands to the two binding sites.

ical activity of the native peptide. With the development of the analogs of the GnRH came the problem of accurately determining whether or not these analogs bind to the same receptors as do the native hormone and whether these analogs could be used to assess the receptor concentration of the GnRH receptor. In the first series of experiments, GnRH labeled with ^{125}I was displaced from its receptors in bovine pituitary plasma membranes by unlabeled GnRH. The resulting curve (Fig. 4, ●, ●) showed a definite plateau at the midpoint, suggestive of two classes of receptors. [^{125}I] GnRH was also displaced with the new D-Ser[6] GnRH analog (Fig. 4, ○, ○). The graph suggested that two ED$_{50}$s would be required to describe each curve, one for the upper "step," and one for the lower one.

Intuitively, one would like to say that these ED$_{50}$s are indicative of the binding affinities of each ligand for each class of sites. However, it is virtually impossible to extract this information from the graph, owing to the substantial degree of experimental scatter. With the aid of the computer program for modeling two ligands binding to two classes of sites, however, we can make considerable headway in estimating the binding parameters. The mathematical model is most conveniently described with the aid of a diagram (Fig. 5; see The Mathematical Model section for the equations). Here we see the possibility for four binding affinities, K_{11}, K_{12}, K_{21}, and K_{22}; two capacities, R_1 and R_2; and two parameters for nonspe-

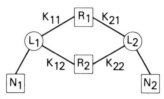

FIG. 5. Diagram of two-ligand (L_1, L_2), two-binding site (R_1, R_2) model with nonspecific binding (N_1, N_2). Four affinities (K_{11}, K_{12}, K_{21}, K_{22}), two binding capacities (R_1, R_2), and two nonspecific parameters (N_1, N_2) must be estimated from the data.

cific binding, N_1 and N_2. As in the case of the simple displacement curves, the data are first reexpressed as Bound GnRH vs Total GnRH added, disregarding the presence or the absence of ^{125}I in the molecule (assuming that iodination has not altered the binding affinity substantially). The parameter values are then adjusted by the computer program to give the best fit to these data.

In Fig. 4 we show the fitted curves superimposed on the data. A reasonably good fit is obtained for this experiment with an RMS error of 4.1%. The parameters for both binding sites are reasonably well determined (Table III). However, there is virtually no information in this experiment about the nonspecific binding of D-Ser[6]; any value of N_2 will given an equivalent fit. Thus, it is probably reasonable to assume that N_2 in roughly equal to N_1.

Thus, even though the traditional displacement curve analysis is unable to adequately handle "bumpy" displacement curves, mathematical modeling with the aid of the computer is quite effective in modeling the data and determining the binding parameters. This same analysis strategy has been successfully applied to the analysis of subtypes of α- and β-adrenergic receptors using a single labeled ligand ([^{3}H]dihydroergocryp-

TABLE III
BINDING PARAMETERS FOR LABELED GONADOTROPIN-RELEASING HORMONE (GnRH)
DISPLACED BY UNLABELED GnRH AND D-Ser[6] ANALOG

	K_1 (nM^{-1})	K_2 (mM^{-1})	R_1 (pM)	R_2 (nM)	N	RMS[a]	df[a]
GnRH	1.26 (35%)	0.43 (41%)	19.4 (31%)	44.5 (45%)	0.031 (5%)	4.1%	19
D-Ser[6]	22.7 (34%)	0.34 (37%)					

[a] RMS, root mean square; df, degrees of freedom.

tine for α receptors and [^{3}H]dihydroalprenolol for β receptors) together with several, partially selective unlabeled ligands.

With the use of a computer modeling it is possible to resolve a non-smooth, nonsteep displacement curve into α_1 and α_2 subtypes of the α-adrenergic receptor; similarly, for the β-adrenergic system. Under certain circumstances, it is possible to resolve and quantify these receptor subtypes even though only partially selective ligands are available.[27–29]

Opiate receptors have been subclassified into two or possibly three classes, μ, δ, and κ on the basis of biological activity. However, many opiate ligands show a moderate to substantial degree of cross-reactivity in the binding to the various receptor classes. Thus, progress in resolving and identifying these classes has been difficult. In the presence of significant cross-reactivity, almost all the simple graphical approaches fail to satisfy the usual assumptions. However, through the use of computerized modeling, Pfeiffer has been able to characterize multiple opiate receptor classes μ, δ, κ in the absence of a truly κ selective ligand.[30] Ultimately, the existence and characterization of subclasses of receptors is facilitated if a pure non-cross-reacting ligand can be found, or if the classes themselves can be physically separated. In the absence of such selective ligands, the best available tool is the computerized analysis of appropriately designed cross-competition studies.

The CORRECTION FACTOR—A New Parameter for the Combination of Results from Several Experiments

Scatchard plots from different replications of the same experiment will often tend to show a similar, but not identical, shape or curvature. Owing to the experimental variability, or possibly the lack of sufficient data of any single experiment, it may be impossible to estimate the binding parameters for two classes of sites, K_1, R_1, K_2, R_2, and N. Nevertheless, over the course of several experiments the distinct impression is given that the curves reflect a consistent biological phenomenon. We may be unable to average the estimated parameter values because a two-site model may fit only some of the experiments, or the parameter values may be so unstable that pooling and averaging the parameters values seems inappropriate. Further, minor changes in experimental design may make it impossible to simply average the raw data. For instance, on the second

[27] A. A. Hancock, A. L. DeLean, and R. J. Lefkowitz, *Mol. Pharmacol.* **16**, 1 (1979).
[28] B. B. Hoffman, A. DeLean, C. L. Wood, D. D. Schocken, and R. J. Lefkowitz, *Life Sci.* **24**, 1739 (1979).
[29] A. DeLean, A. A. Hancock, and R. J. Lefkowitz, *Mol. Pharmacol.* **21**, 5 (1981).
[30] A. Pfeiffer and A. Herz, *Mol. Pharmacol.* **21**, 266 (1982).

or third trial, one may have used more receptor bearing tissue or included more of the low dose points or used different replicates at different dose levels. Further, since the apparent receptor concentration, R, is frequently observed to change with each experiment even when the amount of tissue or protein is controlled, fitting a single "mass action law" model to all the experiments is also inappropriate.

An improved approach utilizing a general nonlinear least squares method is to provide additional parameters, the correction factors, to account for the changing apparent receptor concentrations between experiments. We illustrate their use with data gathered from a study of [^{3}H]aldosterone binding to A-6 epithelial cells derived from toad kidney in continuous culture (data kindly provided by Drs. Watlington and Handler[31]). Data were collected in three separate experiments and plotted in a Scatchard plot, after correction for DNA content of each assay dish.

Each plot appeared nonlinear with some variation in the shape and in the Bound axis intercept, but each plot was suggestive of two classes of binding sites (Fig. 6A). However, there was insufficient information in each single experiment (10 points) to find stable estimates of the five parameters required in the model. In fact, the parameter estimates were essentially indeterminate (see Table IV, rows 1–3). If we are to apply a single binding model to the reaction in all three experiments simultaneously, we must introduce a proportionality factor to scale the data from the second two experiments to the first one. Rather than using a single value such as a B_{max} determination or DNA or protein assay for each experiment, it is better to estimate an optimal value for the correction factor, based on all the data at hand. The scale factor, C, allows for varying receptor concentrations between experiments while requiring the same values for K_1 and K_2 and the same ratio ($R_1 : R_2$) of high-affinity sites to low-affinity sites for every experiment.

The results of this approach are shown in Fig. 6B. Requiring the same "shape" for all three curves still results in an adequate fit with no systematic departures from the model. The average scatter of the points around the fitted curve (RMS error) corresponds to a 19% error in the bound concentration (Table IV, row 4). The values of K_1 and R_1 have approximately 40 and 25% errors, respectively, compared to the nearly 100% errors found when each curve is fitted individually (Table IV, rows 1–3). The error in the estimates of K_2 and R_2 are reduced to roughly 100% (Table IV, row 4). This value may appear unacceptably large at first. But recognizing that the estimates have more nearly a log-normal than a normal distribution, we see that K_2 and R_2 have standard errors less than a factor of 2.

[31] C. O. Watlington, F. M. Perkins, P. J. Munson, and J. S. Handler, *Am. J. Physiol.* **242,** F610 (1982).

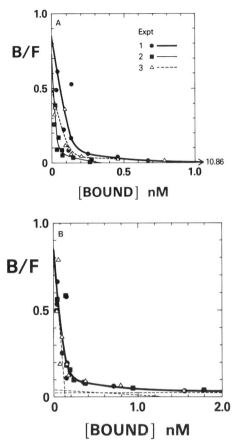

FIG. 6. Scatchard plots for binding of [³H]aldosterone to A-6 toad kidney epithelium in culture from three separate experiments. (A) Values corrected for nonspecific binding. Note systematic change in overall magnitude of binding, but similarity of shape of the curves. (B) After applying correction factors, C_2 and C_3, to account for the changing receptor concentration between experiments, the data points for all three experiments are randomly scattered about a single fitted curve. Dashed lines show components of high affinity, low affinity, and nonspecific binding. From Munson and Rodbard.[1]

Assuming no second class of binding sites (Table IV, row 5) increases the RMS error from 18.9% to 24.5%, with a corresponding "extra sum of squares" F test value of 9.39, which is highly significant. Hence, even though the parameters for the second class of binding sites are poorly determined, we conclude that it is necessary to include them in the model.

Are we justified in combining the data from all three experiments, forcing the curves to have exactly the same shape? This question is an-

TABLE IV

One Ligand, Two Receptors, Three Experiments, Six Fits

Row	Expt. No.	K_1 (nM⁻¹)	R_1 (nmol/ mg DNA)	K_2 (nM⁻¹)	R_2 (nmol/ mg DNA)	N	C_2	C_3	Points	Parameters	df[a]	SS[a]	RMS (%)
(1)	1	6.3 ±5.0	0.70 ±0.30	0.027 ±0.066	6.4 ±11.3	0.020 ±0.006	—	—	10	5	5	2,698	23.2
(2)	2	18.6 ±21.5	0.18 ±0.09	0.01 ±0.11	2.0 ±1.2	0.013 ±0.002	—	—	10	5	5	1,334	16.3
(3)	3	3.9 ±1.8	0.62 ±0.18	0.002 ±0.020	81.0 ±1150	0.001 ±0.100	—	—	10	5	5	1,514	17.4
Total									30	15	15	5,546	19.2
(4)	1–3	6.1 ±2.4	0.64 ±0.16	0.033 ±0.033	6.4 ±4.3	0.020 ±0.003	0.54 ±0.05	0.69 ±0.06	30	7	23	8,300	18.9
(5)	1–3	3.0 ±0.8	1.06 ±0.18	—	—	0.030 ±0.004	0.55 ±0.07	0.70 ±0.09	30	5	25	15,000	24.5
(6)	1–3	5.0 ±3.0	0.53 ±0.20	0.024 ±0.045	5.8 ±5.8	0.014 ±0.003	1.0	1.0	30	5	25	24,580	31.3

[a] df, degrees of freedom; SS, sum of squares; RMS, root mean square.

swered by comparing the fit of three experiments fit individually, estimating 15 parameters (see Table IV, rows 1–3), to the fit of the same three experiments while sharing the K and R values but introducing correction factors C_2 and C_3 for the second and third curves, thus estimating 7 parameters. This comparison (Table IV, rows 1–3, total vs row 4) shows that allowing each curve to have its own set of binding parameters does not give a significantly better fit than pooling the affinities and $R_1 : R_2$ ratio for all three curves. In fact, the average scatter of the data is slightly larger for the individual experiments (RMS = 19.2%) than for the combined fit (RMS = 18.9%). Thus, it is entirely reasonable to attribute the differences in the original three curves to random noise.

What do we gain by adjusting experiments 2 and 3 relative to experiment 1 over and above the original correction for DNA content per cell? That is, is it really necessary to correct these data? Or, would it have been possible simply to pool the data without correction factors? We answer this by refitting the data, this time forcing C_2 and C_3 equal to 1 (Table IV, row 6). Here the RMS error increased significantly from 18.9% to 31.3%. We conclude that in these experiments there was a significant degree of day-to-day variation in receptor content of the cells beyond that measured by DNA content, possibly due to varying receptor number per cell, varying cell size, cell stage, or passage number.

It might have been preferable to perform a single experiment with 28 data points, so that no correction factors would be necessary. Unfortunately, at the time of this experiment it was not yet possible to grow such a large batch of cells. Even this approach would not be optimal, since it sacrifices information about the reproducibility of the result in independent experiments. Through the expedient of pooling results from several experiments and using correction factors, it is possible to make headway in the face of these technical difficulties. Experiments frequently are done in small series, and it is generally useful to analyze the series in its entirity. This is facilitated by fitting these correction factors in a general nonlinear least squares context. Use of the model-fitting strategy employed here provides an efficient way to combine data from multiple experiments.

Modeling of Many Ligands to Numerous Classes of Receptors

In this section, we continue the analysis of the GnRH receptors, begun in the preceding section: Bumpy Displacement Curves. When the analogs of GnRH (D-Ser[6], D-Ala[6]) became available in ^{125}I-iodinated form, it was possible to do a complete cross-displacement study: labeled GnRH displaced by D-Ser[6] analog and labeled D-Ser[6] versus GnRH. Thus, with native GnRH and two analogs, both iodinated and unlabeled, there are nine possible displacement curves. Six of these nine experiments were attempted. Because of the importance of correctly determining the parameter values for the new analogs, the experiments were repeated numerous times in two different laboratories, producing total of 40 displacement curves. The complete set of raw data is displayed in Fig. 7. Scatchard plots were constructed for each individual curve for the homologous (same labeled and unlabeled ligand) experiments (Fig. 8). The data for GnRH were modeled with two classes of binding sites. Therefore, we obtained estimates of binding affinities of GnRH, K_{11}, K_{12}, binding capacities, R_1, R_2, and nonspecific binding, N. Some homologous curves for the analogs were modeled adequately with a single class of sites, but for other curves, two classes were required. Thus, for each curve, estimates of K_{21}, R_1, N_2, and possibly K_{22}, and R_2, were obtained for D-Ser[6], and K_{31}, R_1, K_{32}, R_2, and N_3 for D-Ala[6].

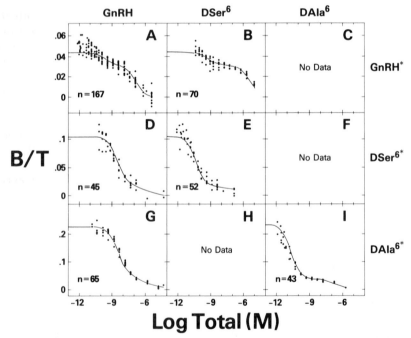

FIG. 7. Data for cross-displacement study of gonadotropin-releasing hormone (GnRH) and two analogs. Labeled ligand is indicated in right margin; unlabeled ligand in the heading. Curves are computer-generated simultaneous fit of the three-ligand, two-binding site model to all of the data. The B/T values have been corrected for nonspecific binding and for variable receptor concentrations between experiments. From Munson and Rodbard.[1]

Next, we wished to combine these estimates from the many replicate experiments. Rather than take the arithmetic averages of the parameter values, it is more appropriate to use the geometric mean, since the distribution of parameter values is generally skewed. However, since a two-site model did not fit some of the individual curves, averaging the parameter values was inappropriate. A better approach is to fit all the data for each ligand simultaneously, with correction factors (Fig. 8). Although this requires a greater computational effort on the part of the machine, it is only marginally more difficult for the user. A correction based on measured protein content was also attempted, but did not completely account for the variability of receptor content between experiments. If experiments using different ligands made use of the same receptor preparation, then the correction factor for these two experiments was "shared," so that only one correction factor was used for each preparation.

The results of the combined analysis of the homologous curves (same labeled and unlabeled ligand) is presented for each ligand (Table V, rows 1–3). Here it appears that the D-Ala^6 and D-Ser^6 analogs were "seeing" significantly more high-affinity receptors than the native GnRH (26 or 37 pM vs 14 pM). Perhaps the synthetic peptides could in fact bind to an additional class of binding sites. At this point in the analysis we could not rule out the hypothesis that GnRH might bind to a completely different protein than any of the modified analogs. These hypotheses are represented diagrammatically in Fig. 9, showing only one analog, D-Ser^6 (L$_2$) versus native GnRH (L$_1$). In model A, we propose that GnRH bind to two classes of receptors, R_1 and R_2, whereas, the analog binds only to the first class. Compare this to model B, where the analog is allowed to bind to a second class of receptor sites. There is already some tentative evidence from the homologous curves favoring the latter model. The larger estimate of receptor number using the subsituted analogs suggested that we consider model D, which allows an additional high-affinity class binding to L$_2$. Finally, we need to rule out model C, where the analog (D-Ser^6) binds to completely different proteins from the native hormone.

Having formalized the competing models, we now proceed to select among them on an objective, statistical basis. We do this by fitting each model to all the data gathered so far, homologous and heterologous curves, comprising some 443 data points, using weighted, nonlinear least squares. Compared to model A, model B explained a significantly greater degree of variability (RMS for A is 7.1%, for B, 6.1%, $F = 86$ with 2 and 246 df; $p < 0.01$). Thus, we rule out model A and conclude that the analogs bind to a second class of sites. Although model C could be entertained on the basis of the homologous curve analysis alone, inclusion of the heterologous curves rules it out on a qualitative basis. That is, since

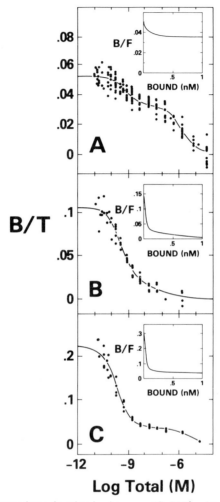

FIG. 8. Binding of gonadotropin-releasing hormone (GnRH) and two analogs to pituitary membranes. Computer-fit displacement curves to homologous (same labeled and unlabeled ligands) data were corrected for nonspecific binding and for optimal correction factor for between-experiment receptor variability. Insets show same curve transformed to Scatchard plot. (A) GnRH; (B) D-Ser[6]; and (C) D-Ala[6]. The curves correspond to parameters given Table V, rows 1–3. From Munson and Rodbard.[1]

each ligand effectively cross-displaces the others, they must compete at the same binding sites. A more subtle issue is the comparison of model B with model D, which has a third class of sites, available only to the substituted analogs. Interestingly, when model D was fit to the complete

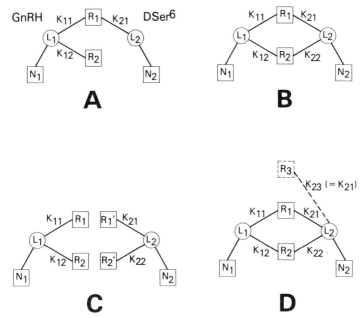

FIG. 9. Alternative binding models for gonadotropin-releasing factor (GnRH)–pituitary receptor study. GnRH (L_1) and only one analog (L_2) are shown for simplicity. (A) Allows only one binding site for the analog (D-Ser[6]). (B) Allows two binding sites for both native GnRH (L_1) and the analog (L_2). (C) Binding sites for GnRH (L_1) are completely distinct from binding sites for D-Ser[6] (L_2). (D) Allows higher binding capacity for D-Ser[6] (L_2) by postulating additional high-affinity receptors (R_3) with same affinity for D-Ser[6] ($K_{23} = K_{21}$), as has the other high-affinity class. From Munson and Rodbard.[1]

data set, the computer optimized value for R_3 was 0. Thus, model D fit no better than B. The increased value of R_1 originally observed from the partial data was judged to be a fortuitous occurrence.

Having established the most suitable model by systematically rejecting alternatives, we can now estimate optimal values for all the parameters shown in lines 4–6 of Table V. Comparison to the estimates made from the individual Scatchard plots shows a marked change in the value of K_1 for GnRH (1.4 to 0.31), and R_1 for GnRH (14 to 42 pM). However, since the latter values are based on the entire set of data, we assume that they are more reliable. Part of the reason for the poor estimation of R_1 for GnRH is revealed in the biphasic nature of the Scatchard plot (Fig. 7A, inset), where the low-affinity component almost completely obliterate the high-affinity component.

Further, extensive "noise" in the low-dose, high-affinity range of the displacement curve for GnRH makes estimation of R_1 based on GnRH

TABLE V
GnRH, D-Ser⁶ Analog, and D-Ala⁶ Analog Binding Parameters[a]

Rows	K_1 (nM^{-1})	K_2 (mM^{-1})	R_1 (pM)	R_2 (nM)	N	RMS (%)	df	Number of experiments
1. GnRH (L_1)	1.4 ±0.4	0.68 ±0.11	14 ±4	44 ±7	0.036 0.001	7.0	150	14
2. D-Ser⁶(L_2)	3.1 ±0.7	20 ±19	26 ±6	0.75 ±0.66	0.070 ±0.003	5.6	43	5
3. D-Ala⁶(L_3)	6.4 ±0.9	0.20 ±0.12	37 ±4	190 ±150	0.023 ±0.007	8.3	35	4
4. GnRH	0.31 ±0.02	0.66 ±0.09						
5. D-Ser⁶	2.0 ±0.2	0.29 ±0.06	42.1 ±3.2	47 ±6.5	0.066 ±0.003	8.1	409	42
6. D-Ala⁶	6.0 ±0.6	0.70 ±0.10						

[a] Rows 1–3: Data from homologous displacement curves. Rows 4–6: Simultaneous analysis of all data (homologous and heterologous displacement curves) from the three ligands. GnRH, gonadotropin-releasing hormone; RMS, root mean square; df, degrees of freedom.

alone unreliable. The uncertainties for the parameters K_2 and R_2 for the D-Ser⁶ and D-Ala⁶ have been reduced from nearly 100% when fit separately to approximately 20% in the simultaneous fit. Thus, by including data involving ligands having different affinity ratios or selectivity for two classes of receptor (R_1 and R_2), a dramatic reduction in the uncertainty of the binding parameters was obtained, compared to experiments involving only a single ligand.

Finally, the overall goodness of fit of the model to the data may be visually evaluated in Fig. 7. The average RMS error for these data points is 8.1%, a value compatible with the RMS for individual curves. The model appears to be an adequate description of the data with the possible exception of the panel A for the homologous GnRH curves. This apparent systematic lack of fit may be related to the known high rate of GnRH degradation *in vitro,* but it is small in overall magnitude, relative to the entire data set.

Statistical Considerations

Goodness-of-Fit Criterion

Throughout our analysis of ligand binding data we have based our choice of model, choice of parameter estimate, determination of curvature, etc., upon a goodness-of-fit criterion of the model to the data. That is, we choose the model that fits the data the best. What does "best" mean? Mathematically speaking, there are a great number of possibilities for this criterion. We may require for instance, that the points be, on the average, close to the line in the Scatchard plot. The parameter values that give us this result will in general be different from the values that cause the line to be close to the points in the double-reciprocal or Lineweaver–Burk plot. Additionally, we may require that the points be not only "close," but scattered randomly about the line in any of these coordinate systems. Which is the appropriate criterion to use? A widely used criterion for "goodness of fit" is least squares. That is, we choose the model such that the sum of the squared distances from each data point to the line is a minimum. Distance in a Scatchard plot might be very different from distance in the Lineweaver–Burk plot, which, in turn, is different from the distance in the Hill plot. Thus, simply specifying that we are using a least-squares criterion is not sufficient. The choice of coordinate system also makes a difference. Is there a "best" coordinate system? The answer, statistically speaking, is yes. We should analyze, as nearly as possible, the original raw data. In most ligand binding studies, this coordinate system is Bound ligand concentration vs Total concentration. In this system, the statistical or measurement errors are predominantly confined to the Bound variable. Total concentration, on the other hand, is generally known with virtually no error.

After some consideration, we see that least-squared errors of bound concentration is not a completely satisfactory criterion for goodness of fit. Since the Bound concentration frequently ranges over several orders of magnitude, an error in measurement of 0.2 nM may have much less significance when measuring in the micromolar range than when trying to measure less than 1 nmol. We may account for the changing significance of measurement errors through the use of statistical weights. Weights should be assigned inversely proportionally to the expected variance of an observation. This variance is not constant across the measurement range, but tends to increase with increasing values of Bound ligand concentration. We may produce an empirical model of the variance of bound (B) thus:

$$\text{Var } (B) = a_0 + a_1B + a_2B^2 \tag{5a}$$

The values for a_0, a_1, and a_2 may be determined empirically by a variety of methods.[32] For instance, the assumption of a constant relative precision of measurement (constant percentage of coefficient of variations) corresponds to setting $a_0 = a_1 = 0$, and $a_2 = 1$. After first fitting a model using unweighted least squares, we may determine predicted values for the

[32] D. Rodbard, R. H. Lenox, H. L. Wray, and D. Ramseth, *Clin. Chem.* **22**, 350 (1976).

Bound concentration at each dose level. These estimates are used in turn to calculate estimates of the variance of the measurement, Var(B), which are then used for weighting in a refined weighted least squares. This process is termed iteratively reweighted least squares. Because K and R do not appear as linear parameters in the binding equations, we make use of the Marquardt–Levenberg algorithm for the solution of the nonlinear least-squares problem.[9] This procedure is somewhat more complicated than linear least squares in that iterative approximations to the final result are required. Unfortunately, in certain circumstances parameter estimates are not obtainable because the iterations do not converge. Often, the problem of lack of convergence may be resolved by trying new starting estimates or by slightly modifying the model.

Alternatives to the least-squares criterion of goodness of fit have been proposed. Some authors argue that robust statistical techniques be used to guard against the undue influence of outlying or erroneous observations.[16] Our experience has been that outlying points can usually be recognized in the appropriate computer-generated graphical display of the data, where they will frequently lie outside the 95% confidence band around the fitted curve. Also, analysis of replicate measurement at a single ligand concentration may be used to reveal outliers when we compare the observed variance of the replicates with the expected variance at this concentration [see Eq. (5a)]. Weighted least squares has good statistical justification, is relatively easy to understand, and hence is widely accepted. Coupled with automated graphical display, it remains an excellent choice for the criteria of goodness of fit.

Precision of Parameter Estimates

If we treat the computer program as a "black box," then the input is the data points and the model specifications (one binding site, two sites, etc.). The output of the "black box" is parameter estimates. Thus, our parameter estimates can be no better than the data we have fed into the computer, a simple fact sometimes overlooked by computer users. However, the computer can give an optimistic estimate of the precision of all of the parameters estimated from the data. This is usually expressed as a standard error of the parameter or as a percentage of coefficient of variation (% CV). These standard errors are only approximate and represent a lower bound on the uncertainty of the results. Standard errors are based on the relative agreement of the data points to the fitted curve; thus noisy data will result in large uncertainties of the parameters. Parameter uncertainty will also result from insufficient data or from essential indeterminacy of the model. Furthermore, various parameter estimates may be highly correlated. For example, even in well designed binding studies an estimate of the affinity, K, and capacity, R, are highly negatively correlated. Thus, a slight overestimation of K is frequently accompanied by a comparable degree of underestimation of R. In some situations there may be too many parameters in the model. Clearly we must have at least one, and preferably three or four, data points per parameter. Lack of data, or too many parameters, or a poor experimental design may result in an "ill-conditioned" model and lack of convergence of the nonlinear least-squares algorithm. Also, if an experiment does not have a wide enough concentration range, especially in the high dose region, then the parameters K_2 and R_2 will likely be ill determined. Careful consideration of the design of the experiment therefore becomes important.

Many investigators are disappointed to learn that their experiments yield apparently poorly determined parameter estimates with a CV of 40–50%. In fact, experience with two-site ligand binding models suggests that this value is not atypical. The error is due not primarily to the scatter of the data points, which is frequently small, but rather to the high correlation of several parameters, which is determined by the experimental design. Even with 100% error, parameter estimates convey significant information, as they are determined to within a factor of 2. Published values for affinity measurements for the same ligand–receptor systems frequently differ by a factor of 5 or more.

Combining or Pooling Experiments

There are at least two sources of errors in experiments, namely, within-experiment error and between-experiment error. Experience in routine radioimmunoassays (RIA) has shown that the latter source is often up to three times larger than the former. Thus, minor changes in buffer systems, temperature, pH, timing of assay procedures between assays contribute significantly to the variability of results. Similarly, in ligand-binding assays we would expect the uncertainty of parameters evaluated on different days to be greater than the expected uncertainty of the same parameter as determined within a given assay.

There are two methods for summarizing data from these series of assays. The simplest approach is to compute the average and standard error of the replicate parameter values, determined on, say, five different replications of the assay. A more sophisticated approach is to fit the same model to all the data simultaneously, allowing for varying receptor concentration with correction factors, if necessary. The standard errors provided by the program LIGAND in the second approach assume homogeneity of variance and independence of each measurement. If the experimental system is stable between experiments, that is, the component of between-experiment variance is relatively small, these standard errors are appropriate. However, when between-experiment error is substantial, and thus the assumption of independence is violated a larger and more realistic estimate of the error is obtained by simply computing the standard deviation and standard error of the replicate determinations of parameter values. Thus, the approach of pooling data, while giving the best possible simultaneous parameter estimates, may give biased underestimates of the standard errors of these same parameters when the series of assays being analyzed has a large component of between-assay error.

Choice of Binding Model

How certain are we of the choice of models? Several criteria may be utilized in choosing the appropriate model. Most important is the weighted RMS error defined as

$$\text{RMS} = [\Sigma w_i(y_i - \hat{y}_i)^2]/[n - p] \tag{6}$$

where y_i is the observed value of bound ligand concentration, $\hat{y}_i$ is the fitted value, and w_i is the statistical weight for this data point, n is the number of data points, and p is the number of estimated parameters.

When two models are being compared, then one may use "extra sum of squares" principle[33] to decide objectively whether the more complex model fits the data significantly better. This test is exact under the assumption of normality and when using linear models. Although we are dealing with *non*linear models, we assume that when standard errors for the parameters are reasonably small, the linear least-squares theory is approximately correct. Thus, with reasonably well-behaved data, the F test may be used to determine whether adding parameters to a model results in a significant improvement in the fit. The F statistic is defined as:

$$F = [(\text{SS}_1 - \text{SS}_2)/(\text{df}_1 - \text{df}_2)]/(\text{SS}_2/\text{df}_2) \tag{7}$$

Here SS_1 and SS_2 are the weighted residual sum of squares for the original and extended model, respectively, df_1 and df_2 are the associated degrees of freedom, i.e., number of data points minus the number of estimated parameters. This calculated F ratio is then compared to a tabulated or critical F value with $(\text{df}_1 - \text{df}_2)$ and df_2 degrees of freedom available in most statistical textbooks.

In addition to this parametric statistical test for evaluating the relative goodness of fit, it is sometimes useful to utilize a nonparametric approach. By looking at the residuals, i.e., differences from data points to

[33] N. R. Draper and H. Smith, "Applied Regression Analysis." Wiley, New York, 1966.

the fitted line, one may determine whether the data points are randomly scattered about the line. One sort of nonrandomness is revealed by the runs test. This test counts the number of runs of the signs of the residuals above or below the fitted line. In a random situation with n data points we expect roughly $n/2$ runs. Thus, the sequence $+++-----++++$ is a sequence of signs that seems nonrandom because it has too few, in this case three, runs. The number of observed runs may be compared with tabulated values to determine whether we have significantly too many or too few runs.[34] Other tests for nonrandomness, such as the MSSD test or the run length test or simple graphs of standardized residuals, may also be useful in this situation.[33]

The Mathematical Model

The mathematical model of n ligands binding to m classes of sites has been developed by Feldman.[7] We have added new parameters (N, C) for nonspecific binding and correction factors. For a single ligand and a single class of binding sites and nonspecific, the implicit set of equations or model defining Bound concentration (B) as a function of Total ligand (T) are

$$B = [KR/(1 + KF) + N]F \qquad \text{(8a)}$$
$$T = B + F \qquad \text{(8b)}$$

where F is free ligand. Although this set of equations may be solved explicitly, it is of no advantage to do so. The numerical solution actually requires less computing time than computing the explicit solution, which involves a square root function. Further, in this form it is easy to see how these equations are generalized. For one ligand binding to a two-receptor class (R_1, R_2) plus nonspecific, the equations become

$$B = [K_1R_1/(1 + K_1F) + K_2R_2/(1 + K_2F) + N]F \qquad \text{(9a)}$$
$$T = B + F \qquad \text{(9b)}$$

The extension of the model to several classes of sites is obvious. For two ligands binding a single class of receptors, the model becomes

$$B_1 = [K_1RF_1/(1 + K_1F_1 + K_2F_2) + N_1]F_1 \qquad \text{(10a)}$$

$$B_2 = [K_2RF_2/(1 + K_1F_1 + K_2F_2) + N_2]F_2 \qquad \text{(10b)}$$
$$T_1 = B_1 + F_1 \qquad \text{(10c)}$$
$$T_2 = B_2 + F_2 \qquad \text{(10d)}$$

---

[34] C. A. Bennett and N. L. Franklin, "Statistical Analysis in Chemistry and the Chemical Industry." Wiley, New York, 1954.

where the subscripts on T, B, F, K refer to ligand L_1 or L_2. With two ligands and two classes of receptors, we require a double subscript on the affinity constant, K. The model is then defined by

$$B_1 = [K_{11}R_1/(1 + K_{11}F_1 + K_{21}F_2)$$
$$+ K_{12}R_2/(1 + K_{12}F_1 + K_{22}F_2) + N_1]F_1 \quad \text{(11a)}$$
$$B_2 = [K_{21}R_1/(1 + K_{11}F_1 + K_{21}F_2)$$
$$+ K_{22}R_2/(1 + K_{12}F_1 + K_{22}F_2) + N_1]F_1 \quad \text{(11b)}$$
$$T_1 = B_1 + F_1 \quad \text{(11c)}$$
$$T_2 = B_2 + F_2 \quad \text{(11d)}$$

Again, the extension to any number of ligands and classes of sites should be clear. The correction factor that adjusts for varying receptor concentration between experiments is introduced as follows. For a one-ligand, one-site model fit to two separate experiments we need one new correction factor (C_2) to adjust the binding of the second experiment relative to the first. The model becomes

$$B_1 = [KR/(1 + KF_1) + N]F_1 \quad \text{(12a)}$$

$$B_2 = [KR/(1 + KF_2) + N]F_2C_2 \quad \text{(12b)}$$
$$T = B_1 + F_1 \quad \text{(12c)}$$
$$T = B_2 + F_2 \quad \text{(12d)}$$

where B_1, F_1 refer to bound and free concentrations in the first experiment, and B_2, F_2 refer to the second. Alternatively if specific and nonspecific binding do not vary proportionately between experiments, we may include separate correction factors, C_2 and D_2 as follows:

$$B_1 = [KR/(1 + KF_1) + N]F_1 \quad \text{(13a)}$$

$$B_2 = \{[KR/(1 + KF_2)]C_2 + ND_2\}F_2 \quad \text{(13b)}$$
$$T = B_1 + F_1 \quad \text{(13c)}$$
$$T = B_2 + F_2 \quad \text{(13d)}$$

Again the extension to multiple ligands and receptor should be obvious, provided one takes care in interpreting the subscripts.

Acknowledgments

The author wishes to acknowledge the contributions of Dr. David Rodbard. Many of the modeling techniques described here were developed with his collaboration. Dr. Ajit Thakur provided a critical review of the manuscript. I also acknowledge Drs. R. N. Clayton, C. Watlington, and G. Chrousos whose data illustrate the use of these techniques. Prototype versions of program LIGAND (SCAFIT) were developed in collaboration with Dr. Andre DeLean. Andrea Kinard provided excellent assistance in preparation of the manuscript.

[42] Two Simple Programs for the Analysis of Data from Enzyme-Linked Immunosorbent Assays (ELISA) on a Programmable Desk-Top Calculator

By D. G. RITCHIE, J. M. NICKERSON, and G. M. FULLER

We have designed two simple programs for use with an inexpensive programmable calculator that rapidly and accurately convert raw data generated from enzyme-linked immunosorbent assays (ELISA) directly into antigen concentration. The first program computes and compares effective doses (ED_{50}) between a standard and each unknown sample assayed. The ED_{50} from the unknown sample is then multiplied by a concentration factor that yields the unknown concentration. The second program linearizes the sigmoidal enzyme-linked immunosorbent assay titration curve using a logit-log transformation of the data in order to compute unknown concentration values. Both programs employ stringent limit conditions to decrease "nonsense" calculations. Data are then processed by a least-squares best-fit linear regression analysis.

While these programs have been designed for use with a competitive fibrinogen ELISA (i.e., absorbance is inversely proportional to the amount of antigen in a well), both may be used with direct (i.e., absorbance α [antigen]) ELISA test systems.

Assay Method

Principle. Enzyme-linked immunoassays are performed in 96-well round-bottom Microtiter plates (Costar, Data Packaging Corp.) coated with monospecific goat anti-rat fibrinogen.[1-4] Serial twofold dilutions of a fibrinogen standard as well as samples containing an unknown amount of fibrinogen are pipetted into the appropriate wells and shaken for 2.5 hr at 25°. Unbound antigen is removed and antigen–enzyme (alkaline phosphatase) conjugate is added. After shaking for another 2.5 hr, the unbound antigen-conjugate is removed and substrate is added to each well. The absorbance of the yellow product, *p*-nitrophenol, is determined on a colorimeter, and the absorbance reading is then converted directly into anti-

[1] E. Engvall and P. Perlman, *J. Immunol.* **109**, 129 (1972).
[2] E. Engvall and A. J. Pesce, *Scand. J. Immunol* **8**, Suppl. 7 (1978).
[3] E. Ishikawa and K. Kato, *Scand. J. Immunol.* **8**, Suppl. 7, 43 (1978).
[4] S.-W. Kwan, G. M. Fuller, M. A. Krautter, J. H. van Bavel, and R. M. Goldblum, *Anal. Biochem.* **83**, 589 (1977).

METHODS IN ENZYMOLOGY, VOL. 92

gen concentration using a TI-59 programmable calculator interfaced to the colorimeter.

Immunosorbent Assay

Reagents

Monospecific goat anti-rat fibrinogen
Antibody-coated Microtiter plates
Fibrinogen–alkaline phosphatase conjugate
PBS–Tween: 0.85% NaCl, 0.05 M sodium phosphate, 0.05% Tween 20, pH 7.1
Tween–saline: 0.85% NaCl, 0.05% Tween 20
p-Nitrophenyl phosphate: 0.1 mg/ml diethanolamine buffer, pH 9.8

Procedure. The immunoassays were performed in 96-well round-bottom Microtiter plates coated with monospecific goat anti-rat fibrinogen. Briefly, 0.2 ml (5 μg/ml) of rat fibrinogen ($E_{1\,cm}^{1\%}$ = 15.9; γ_{max} 280 nm[5]) in PBS–Tween were placed in row A columns 1 and 2. Samples containing fibrinogen (2–14 μg/ml) were then added (0.2 ml/well) in duplicate or triplicate to row A columns 3–12. Two plates were required for 10 samples plus two standards. Serial twofold dilutions in Tween–saline were made to rows B through G. Row H contained only PBS–Tween throughout the course of the assay. The plates were tightly sealed with cellophane tape and shaken for 2.5 hr at 25°. Unbound antigen was then removed by washing with Tween–saline, and antigen–enzyme conjugate (fibrinogen–alkaline phosphatase) was added. The plates were again incubated for 2.5 hr, washed with Tween–saline, then incubated with substrate (p-nitrophenyl phosphate). The enzymic reaction was terminated after 15 min by the addition of 0.025 ml of 2 N NaOH. The absorbance of the yellow product, p-nitrophenol, was measured at 405 nm with an ELISA plate colorimeter (Fisher).

Data Analysis

Hardware. CompuPrint 700 (Artek); TI-59 (Texas Instruments); PC-100C printer.

Program I Operation

When serial twofold dilutions of a 5 μg/ml fibrinogen solution were assayed by the enzyme-linked immunoassay method, a sigmoidal curve

---

[5] R. F. Doolittle, in "The Plasma Proteins" (F. W. Putnam, ed.), 2nd ed., Vol. 2, p. 109. Academic Press, New York, 1975.

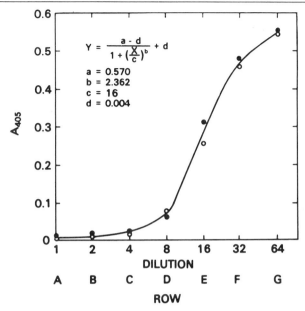

Fig. 1. Standard curve for fibrinogen determination. Plates were coated with a 6 μg of monospecific goat anti-rat fibrinogen per milliliter. Serial twofold dilutions were made (using PBS–Tween) with a rat fibrinogen standard (5.0 μg/ml) from row A through row G. ●, experimental data; ○, theoretical data obtained from Eq. (1); b corresponds to the slope of the logit-log plot (Fig. 2).

relating absorbance to log dilution (i.e., actual concentration) could be drawn through the data (Fig. 1). The general form of the logistic equation that can be used as a model for this relationship may be expressed as

$$Y = (a - d)/[1 + (X/c)^b] + d \tag{1}$$

where Y is the response; X, the arithmetic concentration; a, the response when $X = 0$; d, the response for "infinite" concentration; c, the ED_{50}, i.e., the concentration resulting from a response halfway between a and d; and b, a "slope factor" that determines the steepness of the curve.[6,7] This "slope factor" corresponds to the slope of a logit-log plot (see Fig. 2). It should be emphasized that the absolute concentration of a sample changes the placement of the curve (i.e., to the right or left), but not its shape. Thus, two or more curves may be characterized separately when compared in terms of slopes and ED_{50}s. Program I takes advantage of this fact in the following ways: after reading the absorbances from a known

[6] A. DeLean, P. J. Munson, and D. Rodbard, *Am. J. Physiol.* **235,** E97 (1978).
[7] S. E. Davis, M. L. Jaffe, P. J. Munson, and D. Rodbard, *J. Immunoassay* **1,** 15 (1981).

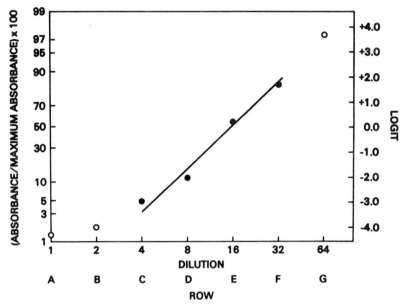

FIG. 2. Logit transformation of absorbances from Fig. 1. Data points from rows A, B, and G (○) were omitted from the linear-regression analysis since they were either less than 3% or greater than 97% of the maximum absorbance (see Discussion). The line describing the binding of enzyme–antigen conjugate was obtained by linear-regression analysis of the four remaining points (●). The correlation coefficient for this line is 0.9906.

fibrinogen standard, the user determines from the program print-out the median absorbance, minimum acceptable slope (MAS), and concentration factor (see Appendix I, step 4.1), then enters these values into the appropriate program storage registers. When reading an unknown, slopes between successive dilutions are calculated automatically, then compared with the slope (MAS) obtained for the standard. Once these two slopes match one another, the ED_{50} of the unknown curve is calculated, then multiplied by a concentration factor to give the concentration of the unknown sample.

Since the user can prescan each column using the ELISA plate reader, rows having absorbances immediately above and below the median absorbance can be readily identified. Included in the program are direct access labels for rows B through F. By using the appropriate address label, the user can move the program pointer directly to that row having an absorbance immediately below the median absorbance and begin data analysis. Used in this way, the program can convert absorbances immediately above and below the median absorbance into concentration, thus avoiding altogether the need to read every well.

When known fibrinogen standards were assayed and then analyzed with this program, a linear relationship ($r = 0.9957$) between actual versus assayed values was obtained (Fig. 3A). The slope of the linear regression curve was 1.17. Within this range of fibrinogen concentrations the maximum slopes obtained from all samples differed from that of the known calibration standard by 3.4–32.5% (15.5 ± 9.3%, mean ± standard deviation; $N = 10$).

The accuracy of this method is not diminished when the maximum slope obtained from an unknown sample differs by large values from that obtained from a calibration standard. This is attested to by the absolute errors derived from duplicate samples having slopes that are different from a known calibration standard by 3.4 vs 32.5%. These duplicate samples had absolute errors of 1.03 and 0.48 μg/ml, respectively. However, since the majority of unknown samples have maximum slopes that are within ± 20% of the calibration standards, we routinely eliminate from analysis all unknown slopes that are less than 80% of that obtained from the standard. This is accomplished automatically with Program I by selecting the appropriate value for a minimum acceptable slope (MAS: see Appendix I, step 4.4). The average absolute error obtained from the fibrinogen values (2 to 14 μg/ml) depicted in Fig. 3A was 0.85 μg/ml. The average absolute error is defined here as the sum of the deviations of the experimental values from the actual values divided by the number of determinations.

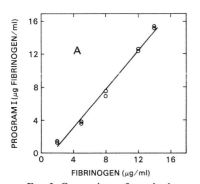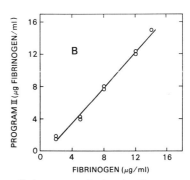

FIG. 3. Comparison of standard curves with known fibrinogen standards using Programs I and II. Using $E_{1\,cm}^{1\%} = 15.9$ for rat fibrinogen, standards containing between 2 and 14 μg/ml were assayed in duplicate. The resultant absorbances were then converted to fibrinogen concentration (relative to a 5 μg/ml standard) using either Program I (panel A) or Program II (panel B). A best-fit linear-regression line has been drawn through each set of data.

Program II Operation

Program II was written so that the logistic model could be utilized in estimating the concentrations of unknowns. Absorbance measurements were transformed into logit units and then plotted as logit versus log dilution. In this plot a straight line can be drawn through data that the logistic model fits. The logit transformation is given in Eq. (2).

$$\text{Logit } (Y) = \ln [Y/(100 - Y)] \tag{2}$$

where Y is the percentage response. In this equation Y is defined as

$$Y = (100) [OD_i/OD_{max}] \tag{3}$$

where OD_i is the sample absorbance at dilution i and OD_{max} is the absorbance at infinite antigen dilution. Equation (2) may be rewritten as a function of OD_i as follows:

$$f(OD_i) = \ln \left[\frac{OD_i}{OD_{max}OD_i} \right]$$

ELISA Program II was designed to obtain a linear regression best-fit analysis from the relationship

$$f(OD_i) = (b) \ln (dilution_i) + a \tag{4}$$

where a is the ordinate intercept and b is the slope. These constants are determined for each regression analysis. A typical logit versus log plot of the data from Fig. 1 is shown in Fig. 2. When the slope, b, from Fig. 2 (obtained from the program print-out for each sample) was entered into Eq. (1), a theoretical sigmoidal curve was obtained that closely approximated the curve obtained from the assay (see Fig. 1). To obtain the slope, the program accepts and averages from one to three absorbance measurements from each row, eliminates values outside of the 3–97% OD_{max} range, then converts the remaining values into logit units. The dilutions of known and unknown samples at which the logit is equal to zero (as calculated by a least-square linear regression subprogram) are then compared to determine the concentration of the unknown.

When the same set of known fibrinogen standards were analyzed by this program, a linear relationship ($r = 0.9977$; slope $= 1.11$) between actual versus experimentally determined values was obtained (Fig. 3B). The average absolute error obtained from Fig. 3B by this method of analysis for rat fibrinogen was 0.46 μg of fibrinogen per milliliter.

Appendix I. Detailed Description of Program I Operation

In Table I, the individual steps for this program are listed. The 494 steps are stored on three sides of two magnetic cards. Programming and storage of the program of the magnetic cards is performed as described by the manufacturer.

Notes

1.1. Partition calculator to 719.29 by pressing 3; 2nd; Op; 17.
2.1. Insert program card side 1 after pressing INV; 2nd; FIX; CLR. Again press CLR and insert card side 2.

TABLE I
PROGRAM I: STEPS 000–494

Step	Code	Step	Code	Step	Code	Step	Code	Step	Code	Step	Code	Step	Code
000	25 CLR	073	06 6	148	85 +	223	69 OP	298	02 2	373	95 =	448	24 24
001	69 OP	074	28 LOG	149	25 CLR	224	05 05	299	69 OP	374	42 STO	449	95 =
002	00 00	075	42 STO	150	36 PGM	225	71 SBR	300	01 C1	375	10 10	450	77 GE
003	03 3	076	18 18	151	01 C1	226	85 +	301	69 OP	376	69 OP	451	04 C4
004	06 6	077	03 3	152	71 SBR	227	25 CLR	302	05 C5	377	06 C6	452	54 54
005	01 1	078	02 2	153	25 CLR	228	36 PGM	303	71 SBR	378	98 ADV	453	92 RTN
006	03 3	079	28 LOG	154	43 RCL	229	01 C1	304	85 +	379	92 RTN	454	95 =
007	03 3	080	42 STO	155	15 15	230	71 SBR	305	25 CLR	380	76 LBL	455	99 PRT
008	00 0	081	19 19	156	32 X:T	231	25 CLR	306	36 PGM	381	33 X2	456	43 RCL
009	03 3	082	06 6	157	43 RCL	232	43 RCL	307	01 C1	382	25 CLR	457	22 22
010	03 3	083	04 4	158	13 13	233	17 17	308	71 SBR	383	91 R/S	458	69 OP
011	02 2	084	28 LOG	159	78 Σ+	234	32 X:T	309	25 CLR	384	55 ÷	459	15 15
012	07 7	085	42 STO	160	43 RCL	235	43 RCL	310	43 RCL	385	01 1	460	95 =
013	69 OP	086	20 20	161	16 16	236	13 13	311	19 19	386	00 C	461	22 INV
014	01 01	087	58 FIX	162	32 X:T	237	78 Σ+	312	32 X:T	387	00 C	462	28 LOG
015	01 1	088	03 C3	163	43 RCL	238	43 RCL	313	43 RCL	388	00 C	463	65 ×
016	07 7	089	71 SBR	164	10 10	239	18 18	314	13 13	389	95 =	464	43 RCL
017	00 0	090	85 +	165	78 Σ+	240	32 X:T	315	78 Σ+	390	42 STO	465	21 21
018	00 0	091	25 CLR	166	43 RCL	241	43 RCL	316	43 RCL	391	11 11	466	95 =
019	00 0	092	69 OP	167	22 22	242	10 10	317	20 20	392	99 PRT	467	99 PRT
020	00 0	093	00 C0	168	11 F	243	78 Σ+	318	32 X:T	393	91 R/S	468	98 ADV
021	00 0	094	03 3	169	25 CLR	244	43 RCL	319	43 RCL	394	55 ÷	469	92 RTN
022	00 0	095	05 5	170	69 OP	245	22 22	320	10 10	395	01 1	470	76 LBL
023	00 0	096	03 3	171	00 C0	246	11 F	321	78 Σ+	396	00 C	471	14 I
024	00 0	097	02 2	172	03 3	247	25 CLR	322	43 RCL	397	00 C	472	61 GTO
025	69 OP	098	04 4	173	05 5	248	69 OP	323	22 22	398	00 C	473	01 C1
026	02 02	099	03 3	174	03 3	249	00 0	324	11 F	399	95 =	474	69 OP
027	69 OP	100	00 C	175	02 2	250	03 3	325	25 CLR	400	42 STO	475	76 LBL
028	05 05	101	00 C	176	04 4	251	05 5	326	81 RST	401	12 12	476	15 E
029	25 CLR	102	01 1	177	03 3	252	03 3	327	76 LBL	402	99 PRT	477	61 GTO
030	58 FIX	103	04 4	178	00 0	253	02 2	328	85 +	403	25 CLR	478	02 C2
031	00 00	104	69 OP	179	00 0	254	04 4	329	25 CLR	404	69 OP	479	08 C8
032	91 R/S	105	01 C1	180	01 1	255	03 3	330	91 R/S	405	00 C0	480	76 LBL
033	99 PRT	106	69 OP	181	06 6	256	00 C	331	55 ÷	406	03 3	481	10 E'
034	98 ADV	107	05 C5	182	69 OP	257	00 C	332	00 1	407	00 C	482	61 GTO
035	22 INV	108	71 SBR	183	01 C1	258	02 2	333	00 C	408	03 3	483	02 C2
036	58 FIX	109	33 X2	184	69 OP	259	01 1	334	00 C	409	00 C	484	47 47
037	25 CLR	110	25 CLR	185	05 05	260	69 OP	335	00 C	410	01 1	485	76 LBL
038	91 R/S	111	36 PGM	186	71 SBR	261	01 C1	336	95 =	411	07 7	486	13 C
039	25 CLR	112	01 C1	187	33 X2	262	69 OP	337	42 STO	412	01 1	487	61 GTO
040	69 OP	113	71 SBR	188	25 CLR	263	05 C5	338	08 C8	413	03 3	488	01 C1
041	00 00	114	25 CLR	189	36 PGM	264	71 SBR	339	99 PRT	414	03 3	489	30 30
042	03 3	115	43 RCL	190	01 C1	265	33 X2	340	91 R/S	415	01 1	490	76 LBL
043	05 5	116	14 14	191	71 SBR	266	25 CLR	341	55 ÷	416	69 OP	491	12 B
044	03 3	117	32 X:T	192	25 CLR	267	36 PGM	342	00 1	417	04 C4	492	61 GTO
045	02 2	118	43 RCL	193	43 RCL	268	01 C1	343	00 C	418	43 RCL	493	00 C0
046	04 4	119	10 10	194	16 16	269	71 SBR	344	00 C	419	11 11	494	92 RTN
047	03 3	120	78 Σ+	195	32 X:T	270	25 CLR	345	00 C	420	85 +		
048	00 0	121	43 RCL	196	43 RCL	271	43 RCL	346	95 =	421	43 RCL		
049	00 0	122	15 15	197	10 10	272	18 18	347	42 STO	422	12 12		
050	01 1	123	32 X:T	198	78 Σ+	273	32 X:T	348	09 C9	423	95 =		
051	03 3	124	43 RCL	199	43 RCL	274	43 RCL	349	99 PRT	424	55 ÷		
052	69 OP	125	13 13	200	17 17	275	10 10	350	25 CLR	425	02 2		
053	01 01	126	78 Σ+	201	32 X:T	276	78 Σ+	351	69 OP	426	95 =		
054	69 OP	127	43 RCL	202	43 RCL	277	43 RCL	352	00 C0	427	42 STO		
055	05 05	128	22 22	203	13 13	278	19 19	353	03 3	428	13 13		
056	01 1	129	11 F	204	78 Σ+	279	32 X:T	354	00 C	429	69 OP		
057	28 LOG	130	25 CLR	205	43 RCL	280	43 RCL	355	03 3	430	06 C6		
058	42 STO	131	69 OP	206	22 22	281	13 13	356	00 C	431	98 ADV		
059	14 14	132	00 00	207	11 F	282	78 Σ+	357	01 1	432	92 RTN		
060	02 2	133	03 3	208	25 CLR	283	43 RCL	358	07 7	433	76 LBL		
061	28 LOG	134	05 5	209	69 OP	284	22 22	359	01 1	434	11 F		
062	42 STO	135	03 3	210	00 C0	285	11 F	360	03 3	435	69 OP		
063	15 15	136	02 2	211	03 3	286	25 CLR	361	03 3	436	12 12		
064	04 4	137	04 4	212	05 5	287	69 OP	362	01 1	437	95 =		
065	28 LOG	138	03 3	213	03 3	288	00 C0	363	69 OP	438	22 INV		
066	42 STO	139	00 C	214	02 2	289	03 3	364	04 C4	439	95 =		
067	16 16	140	00 C	215	04 4	290	05 5	365	43 RCL	440	42 STO		
068	08 8	141	01 1	216	03 3	291	03 3	366	08 C8	441	24 24		
069	28 LOG	142	05 5	217	00 C	292	02 2	367	85 +	442	29 CP		
070	42 STO	143	69 OP	218	00 C	293	04 4	368	43 RCL	443	43 RCL		
071	17 17	144	01 C1	219	01 1	294	03 3	369	09 C9	444	23 23		
072	01 1	145	69 OP	220	07 7	295	00 C	370	95 =	445	32 X:T		
		146	05 05	221	69 OP	296	00 C	371	55 ÷	446	25 CLR		
		147	71 SBR	222	01 C1	297	02 2	372	02 2	447	43 RCL		

2.2. Press CLR insert second card side 1.

2.3. Press RST; R/S to start program.

3.1. Key in sample number, then press R/S.

4.1. Enter concentration factor, median absorbance, and minimum acceptable slope as follows:

4.2. Median absorbance: This is equal to one-half the maximum absorbance obtained from the standard curve. Enter this value and press STO 22.

4.3. Concentration factor (C.F.); for standards enter 1 then press STO 21. For unknown C.F. = [standard]/dilution; where the dilution is obtained from the standard curve. For example, from Fig. 1, the program calculated a dilution of 16.0. Therefore, C.F. $= 5 \div 16 = 0.313$.

4.4. Minimum acceptable slope (MAS): The slope will vary from antigen to antigen. For fibrinogen the slope is 0.4. For most indirect immunoassays, 0.1 can be used initially. Press 0.1; STO 23. Slopes greater than 0.1 will then be printed for each dilution of the standard. From these values the maximum slope is obtained. The MAS = 0.8 × max slope. Enter this value for unknown samples and press STO 23.

4.5. These numbers may all be recorded on tape by pressing TRACE key on printer prior to keying in each of the numbers. Release TRACE before proceeding.

5.1. Before reading absorbances press R/S.

6.1. Scan column to determine where median OD is located. In Fig. 1, rows D and E are closest to the median OD.

6.2. Press appropriate user defined key (i.e., B through E to access rows B through E and 2nd; E, to access row F). For example, for Fig. 1 the user would press D.

6.3. Read adsorbance. When using a MicroELISA Reader press PRINT button on Reader. Absorbances will then be automatically entered into the program. When using a spectrophotometer that is not interfaced to a calculator, absorbance must be manually keyed in as a whole number followed R/S. Be careful not to enter new data unless a 0 appears in calculator display. If new data are accidentally entered while calculations are in progress, the program overloads and must be rerun from the beginning by pressing RST; R/S; R/S and beginning again at row A.

6.4. Continue reading next row. Print-out will include mean absorbance for each row followed by two numbers if the slope is greater than MAS in register 23. The first number is the slope, and the second is either the dilution (when reading the standard) or the final antigen concentration (when reading an unknown).

7.1. To begin reading another sample, press RST; R/S. Be sure storage register 21, 22, and 23 contain the appropriate numbers for either standards or unknowns.

Appendix II. Detailed Description of Program II Operation

In Table II, the individual steps for Program II are listed. The 612 program steps are stored on three sides of two magnetic cards. Programming and storage of the program are performed as described by the manufacturer.

Notes

1.1. Partition calculator to 719.29. Press 3; 2nd; Op; 17.
1.2. Enter the program stored on magnetic cards. Press RST; CLR. Insert card side 1. Press CLR, insert card side 2. Press CLR, insert second card side 3.
2.1. Scan the ELISA plate for the highest single absorbance measurement. Multiply that reading by 1000, and store it in register 21. Example: If the highest absorbance measurement is 0.632, enter 632 and press STO 21. Press RST, CLR.
3.1. To start the program, press R/S, R/S.
3.2. The calculator will then ask "CONC OF STD IS?" Enter from the keyboard the undiluted concentration of the standard solution of antigen. Press R/S.
3.3. The calculator will then print "1 = STD 0 = UNK." Press 1, R/S, if the set of data about to be entered is a standard. Press 0, R/S, if the set of data to be entered is a solution of unknown antigen concentration.
3.4. The calculator will then print "X̄ of 1, 2, or 3 ODS?" Press 1, R/S, if the data to be entered are singlets. Press 2, R/S, if the data are in duplicate. Press, 3 R/S, if the data are in triplicate.
3.5. The calculator will then print "OD IS." Make sure that the Microtiter plate is positioned at the first well to be measured (position A-1), then press the PRINT button on the ELISA colorimeter. The absorbance is automatically entered into the calculator. If the calculator and spectrophotometer are not interfaced, multiply the absorbance by 1000, enter that number, and press R/S. The calculator will then print "OD IS." Enter the next measurement. If samples are assayed as duplicates, read position A-2. The calculator will print the mean of the two absorbance measurements. If data are in triplicates, read well A-3 and the calculator will print the mean of the three absorbances.

TABLE II
Program II: Steps 000–611

Step	Code	Key	Step	Code	Key	Step	Code	Key	Step	Code	Key	Step	Code	Key	Step	Code	Key
000	91	R/S	080	22	22	161	01	1	242	42	STO	323	76	LBL	404	05	05
001	25	CLR	081	42	STO	162	69	OP	243	20	20	324	19	D'	405	69	OP
002	69	OP	082	24	24	163	04	04	244	55	÷	325	07	7	406	12	12
003	00	00	083	25	CLR	164	69	OP	245	53	(	326	32	X:T	407	32	X:T
004	01	1	084	69	OP	165	05	05	246	43	RCL	327	43	RCL	408	99	PRT
005	05	5	085	00	00	166	91	R/S	247	21	21	328	07	07	409	25	CLR
006	00	0	086	00	0	167	99	PRT	248	75	-	329	67	EQ	410	69	OP
007	01	1	087	02	2	168	42	STO	249	43	RCL	330	03	03	411	00	00
008	03	3	088	06	6	169	17	17	250	20	20	331	35	35	412	01	1
009	01	1	089	04	4	170	42	STO	251	54	)	332	61	GTO	413	05	5
010	01	1	090	03	3	171	08	08	252	95	=	333	01	01	414	00	0
011	05	5	091	06	6	172	08	8	253	23	LNX	334	81	81	415	01	1
012	00	0	092	03	3	173	42	STO	254	24	STO	335	76	LBL	416	03	3
013	00	0	093	07	7	174	09	09	255	24	24	336	13	C	417	05	5
014	~69	OP	094	01	1	175	00	0	256	43	RCL	337	00	0	418	03	3
015	01	01	095	06	6	176	42	STO	257	20	20	338	69	OP	419	05	5
016	00	0	096	69	OP	177	07	07	258	55	÷	339	15	15	420	00	0
017	01	1	097	01	01	178	61	GTO	259	43	RCL	340	22	INV	421	00	0
018	02	2	098	00	0	179	01	01	260	21	21	341	23	LNX	422	69	OP
019	01	1	099	01	1	180	92	92	261	95	=	342	42	STO	423	01	01
020	00	0	100	06	6	181	76	LBL	262	42	STO	343	13	13	424	01	1
021	00	0	101	04	4	182	12	B	263	19	19	344	25	CLR	425	05	5
022	03	3	102	04	4	183	00	0	264	02	2	345	69	OP	426	00	0
023	06	6	103	01	1	184	42	STO	265	45	Y^X	346	00	00	427	01	1
024	03	3	104	03	3	185	22	22	266	43	RCL	347	01	1	428	01	1
025	07	7	105	01	1	186	43	RCL	267	07	07	348	06	6	429	07	7
026	69	OP	106	02	2	187	17	17	268	95	=	349	02	2	430	02	2
027	02	02	107	06	6	188	42	STO	269	42	STO	350	04	4	431	01	1
028	01	1	108	69	OP	189	08	08	270	18	18	351	02	2	432	02	2
029	06	6	109	02	02	190	76	LBL	271	23	LNX	352	07	7	433	01	1
030	00	0	110	69	OP	191	11	A	272	42	STO	353	03	3	434	69	OP
031	00	0	111	05	05	192	76	LBL	273	11	11	354	01	1	435	02	02
032	02	2	112	91	R/S	193	14	D	274	69	OP	355	00	0	436	69	OP
033	04	4	113	42	STO	194	68	NOP	275	27	27	356	00	0	437	05	05
034	03	3	114	12	12	195	25	CLR	276	93	.	357	69	OP	438	69	OP
035	06	6	115	99	PRT	196	69	OP	277	00	0	358	01	01	439	13	13
036	07	7	116	25	CLR	197	00	00	278	03	3	359	01	1	440	99	PRT
037	01	1	117	69	OP	198	00	0	279	32	X:T	360	03	3	441	01	1
038	69	OP	118	00	00	199	01	1	280	43	RCL	361	03	3	442	32	X:T
039	03	03	119	06	6	200	01	1	281	19	19	362	07	7	443	43	RCL
040	69	OP	120	07	7	201	06	6	282	22	INV	363	00	0	444	12	12
041	05	05	121	00	0	202	00	0	283	77	GE	364	00	0	445	67	EQ
042	91	R/S	122	00	0	203	00	0	284	03	03	365	02	2	446	05	05
043	42	STO	123	00	0	204	02	2	285	04	04	366	07	7	447	96	96
044	10	10	124	01	1	205	04	4	286	93	.	367	00	0	448	25	CLR
045	99	PRT	125	02	2	206	03	3	287	09	9	368	01	1	449	69	OP
046	76	LBL	126	01	1	207	06	6	288	07	7	369	69	OP	450	00	00
047	15	E	127	00	0	208	69	OP	289	32	X:T	370	02	02	451	04	4
048	00	0	128	00	0	209	01	01	290	43	RCL	371	02	2	452	01	1
049	42	STO	129	69	OP	210	69	OP	291	19	19	372	02	2	453	03	3
050	01	01	130	01	01	211	05	05	292	77	GE	373	02	2	454	01	1
051	42	STO	131	00	0	212	91	R/S	293	03	03	374	04	4	455	02	2
052	02	02	132	02	2	213	99	PRT	294	04	04	375	03	3	456	06	6
053	42	STO	133	05	5	214	44	SUM	295	43	RCL	376	07	7	457	00	0
054	03	03	134	07	7	215	22	22	296	11	11	377	06	6	458	00	0
055	42	STO	135	00	0	216	68	NOP	297	32	X:T	378	04	4	459	01	1
056	04	04	136	03	3	217	97	DSZ	298	43	RCL	379	00	0	460	05	5
057	42	STO	137	05	5	218	08	08	299	24	24	380	01	1	461	69	OP
058	05	05	138	07	7	219	01	01	300	78	Σ+	381	69	OP	462	01	01
059	42	STO	139	00	0	220	90	90	301	61	GTO	382	03	03	463	00	0
060	06	06	140	00	0	221	25	CLR	302	03	03	383	69	OP	464	01	1
061	42	STO	141	69	OP	222	69	OP	303	23	23	384	05	05	465	03	3
062	07	07	142	02	02	223	00	00	304	76	LBL	385	43	RCL	466	01	1
063	42	STO	143	01	1	224	03	3	305	18	C'	386	13	13	467	01	1
064	08	08	144	01	1	225	00	0	306	25	CLR	387	99	PRT	468	05	5
065	42	STO	145	03	3	226	01	1	307	69	OP	388	25	CLR	469	06	6
066	09	09	146	05	5	227	07	7	308	00	00	389	69	OP	470	04	4
067	42	STO	147	00	0	228	01	1	309	00	0	390	00	00	471	00	0
068	11	11	148	00	0	229	03	3	310	01	1	391	03	3	472	00	0
069	42	STO	149	00	0	230	03	3	311	03	3	392	06	6	473	69	OP
070	15	15	150	04	4	231	01	1	312	00	0	393	02	2	474	02	02
071	42	STO	151	00	0	232	69	OP	313	02	2	394	07	7	475	69	OP
072	17	17	152	00	0	233	04	04	314	04	4	395	00	0	476	05	05
073	42	STO	153	69	OP	234	43	RCL	315	03	3	396	01	1	477	43	RCL
074	18	18	154	03	03	235	22	22	316	07	7	397	03	3	478	10	10
075	42	STO	155	01	1	236	55	÷	317	69	OP	398	03	3	479	65	×
076	20	20	156	01	1	237	43	RCL	318	04	04	399	01	1	480	43	RCL
077	42	STO	157	06	6	238	17	17	319	43	RCL	400	07	7	481	13	13
078	19	19	158	03	3	239	95	=	320	19	19	401	69	OP	482	55	÷
079	42	STO	159	06	6	240	69	OP	321	69	OP	402	01	01	483	43	RCL
			160	07	7	241	06	06	322	06	06	403	69	OP	484	14	14

TABLE II (continued)

PROGRAM II (con't.)

485	95	=	566	00	0
486	99	PRT	567	00	0
487	25	CLR	568	03	3
488	69	OP	569	07	7
489	00	00	570	02	2
490	01	1	571	03	3
491	06	6	572	01	1
492	00	0	573	03	3
493	01	1	574	03	3
494	03	3	575	01	1
495	01	1	576	69	OP
496	01	1	577	03	03
497	07	7	578	02	2
498	07	7	579	06	6
499	01	1	580	03	3
500	69	OP	581	06	6
501	01	01	582	07	7
502	00	0	583	03	3
503	02	2	584	07	7
504	06	6	585	03	3
505	04	4	586	00	0
506	04	4	587	00	0
507	05	5	588	69	OP
508	01	1	589	04	04
509	07	7	590	69	OP
510	03	3	591	05	05
511	06	6	592	61	GTO
512	69	OP	593	06	06
513	02	02	594	08	08
514	00	0	595	76	LBL
515	00	0	596	16	A'
516	00	0	597	43	RCL
517	01	1	598	13	13
518	06	6	599	42	STO
519	04	4	600	14	14
520	03	3	601	00	0
521	01	1	602	42	STO
522	00	0	603	13	13
523	01	1	604	61	GTO
524	69	OP	605	00	00
525	03	03	606	47	47
526	69	OP	607	76	LBL
527	05	05	608	17	B'
528	91	R/S	609	68	NOP
529	99	PRT	610	91	R/S
530	42	STO	611	81	RST
531	15	15			
532	00	0			
533	32	X:T			
534	43	RCL			
535	15	15			
536	67	EQ			
537	00	00			
538	47	47			
539	25	CLR			
540	69	OP			
541	00	00			
542	01	1			
543	05	5			
544	02	2			
545	07	7			
546	01	1			
547	07	7			
548	01	1			
549	03	3			
550	03	3			
551	01	1			
552	69	OP			
553	01	01			
554	00	0			
555	00	0			
556	01	1			
557	03	3			
558	03	3			
559	05	5			
560	01	1			
561	07	7			
562	01	1			
563	03	3			
564	69	OP			
565	02	02			

3.6. Be sure to wait until the calculator has finished all calculations before entering data. When "C" appears in the calculator display left-hand corner, the calculator is still processing data. Wait until the "C" is no longer displayed before entering more data.

3.7. Proceed to the next dilutions, in order, rows B through G, and enter those data pairs, singlets, or triplets to be averaged.

3.8. At the end of each set of dilutions the calculator will perform a least-squares analyses of the data. It will print "DILN AT LOGIT = 0." This phrase means the dilution of the sample, calculated by least-squares fit of Eq. (4), where $f(OD_i) = 0$. The dilution at logit = 0 is equivalent to the ED_{50}. This dilution is used in further calculations to determine the sample concentration. The slope of the least-squares fit and the correlation coefficient are also printed. If the sample just entered was an unknown, the calculator will print "UNK CON =" followed by the unknown concentration. The dimensions of the unknown concentration are the same as those of the standard concentration.

3.9. The calculator will print "DONE? 1 = YES 0 = NO." If finished with the program, push 1, R/S. If not finished, press 0, R/S.

4.1. In certain cases, absorbance measurements are omitted from analysis. Data points outside the range of 3–97% of the maximum absorbance are omitted. These data are printed and identified by the calculator. After the calculator prints the mean absorbance, it will print the ratio of the mean absorbance to the maximum absorbance followed by "OMIT."

4.2. Be sure to read all wells in a row. If the ELISA assay makes use of more or less than rows A through G, an adjustment in the program must be made. Change step 325 from 7 to the number of wells used in a single row of dilutions.

4.3. If other than serial twofold dilutions are used in an ELISA assay, Program II may be modified as follows: Change step 264 from 2 to any single integer (3–9 inclusive) that describes the fold dilution.

[43] Determination of Affinity and Specificity of Anti-Hapten Antibodies by Competitive Radioimmunoassay

By ROLF MÜLLER

The emphasis of this chapter is on the use of competitive radioimmunoassay (RIA) to determine specificity and affinity of anti-hapten antibodies,[1] both of which considerably affect the sensitivity of immunological assays. No additional experiments are required for the determination of antibody characteristics once a competitive RIA has been established. The described technique is particularly worthwhile for screening analyses of large numbers of antisera or monoclonal antibodies. In these instances, several characteristics can be obtained by analyzing unpurified antibodies in a single experiment: the average antibody affinity constant (K), the antibody specificity (i.e., the cross-reactivity of the antibodies with other structurally related compounds), the concentration of specific antibodies in, e.g., serum, ascites fluid, or cell culture medium, and the theoretical detection limit of the competitive RIA for a given antibody and a tracer of maximum specific radioactivity.

Several other methods have been described for determination of the affinity of anti-hapten antibodies, including equilibrium dialysis,[2] fluorescence quenching,[3] and fluorescence polarization.[4] These techniques are applicable only to purified or partially purified antibodies. Steward and Petty[5,6] established an ammonium sulfate precipitation method that allows the determination of antibody affinity in whole serum. This method requires extrapolation of a Scatchard plot that, in some instances, deviates from linearity, so that precise determinations of K values may be difficult. These shortcomings can be circumvented by calculating K from the data obtained by competitive RIA as described in this chapter. K is calculated from the molar inhibitor concentration giving 50% inhibition of tracer–antibody binding in a competitive RIA, the molar tracer concentration in the assay, and the amount of tracer bound in the absence of inhibitor. These

[1] R. Müller, *J. Immunol. Methods* **34**, 345 (1980).

[2] H. N. Eisen and F. Karush, *J. Am. Chem. Soc.* **71**, 363 (1965).

[3] S. F. Velick, C. W. Parker, and H. N. Eisen, *Proc. Natl. Acad. Sci. U.S.A.* **46**, 1470 (1960).

[4] W. B. Dandlicker, H. C. Schapiro, J. W. Meduski, R. Alonso, G. A. Feigen, and J. R. Hamricks, *Immunochemistry* **1**, 165 (1965).

[5] M. W. Steward and R. E. Petty, *Immunology* **22**, 747 (1972).

[6] M. W. Steward and R. E. Petty, *Immunology* **23**, 881 (1972).

METHODS IN ENZYMOLOGY, VOL. 92

values can be precisely determined, and so a reliable value for K should be obtained.

Representative examples for the application of this method to the determination of antibody specificity and affinity will be presented, and some parameters by which these antibody characteristics could be influenced will be discussed.

The Competitive RIA

The calculation of K from the data obtained by a competitive RIA requires performance of the assay under the following conditions: (a) the competitive RIA must be carried out under equilibrium conditions; (b) the antibody concentration should be adjusted to ~40–70% tracer-binding in the absence of inhibitor; and (c) a separation technique has to be applied that permits almost complete separation of bound from free tracer (e.g., ammonium sulfate precipitation of the antibodies, double-antibody precipitation or filtration of the antigen–antibody complex). The following standard procedure was used for all our studies concerning the characterization of anti-nucleoside antibodies,[1,7–9] but any other competitive RIA meeting the conditions described above would be equally well suited for the determination of specificity and affinity of a given anti-hapten antibody or antiserum.

Materials

Tris-buffered saline (TBS): 140 mM NaCl; 20 mM Tris-HCl, pH 7.5; 3 mM NaN$_3$

RIA buffer: TBS containing 2% (w/v) bovine serum albumin (BSA; e.g., fraction V, Sigma) and 0.2% (w/v) normal rabbit or bovine IgG (e.g., Calbiochem)

[8-^{3}H]deoxyguanosine ([8-^{3}H]dGuo): 2.3 Ci/mmol (Amersham); diluted in RIA buffer

O^6-Ethyl-[8-^{3}H]guanosine (O^6-Et[8-^{3}H]Guo): 1.4 Ci/mmol (for synthesis see Müller and Rajewsky[7]); diluted in RIA buffer

O^6-Ethyl-[8,5′-^{3}H]deoxyguanosine (O^6-Et[8,5′-^{3}H]dGuo): 14 Ci/mmol (for synthesis see Müller and Rajewsky[7,8]); diluted in RIA buffer

Inhibitors: normal deoxynucleosides (Sigma) and various alkylated nucleic acid components (synthesized as described elsewhere[7–9]); dissolved and diluted in TBS

---

[7] R. Müller and M. F. Rajewsky, *Z. Naturforsch. C: Biosci.* **33C**, 897 (1978).

[8] R. Müller and M. F. Rajewsky, *Cancer Res.* **40**, 887 (1980).

[9] M. F. Rajewsky, R. Müller, J. Adamkiewicz, and W. Drosdziok, *in* "Carcinogenesis: Fundamental Mechanisms and Environmental Effects" (B. Pullman *et al.*, eds.), p. 207. Reidel, Dordrecht, 1980.

Anti-nucleoside sera: produced in rabbits after immunization with the respective ribonucleoside covalently bound to a carrier protein (detailed elsewhere[1,10,11]); diluted in RIA buffer

Ammonium sulfate: aqueous solution saturated at room temperature; pH adjusted to 7.0 with diluted ammonium hydroxide

Scintillation fluid: Rotiszint 22 (Roth, Karlsruhe, Germany)

Method. The RIA was carried out in 1.5-ml Eppendorf reaction tubes. Each sample contained 50 μl of tracer (3000–10^4 dpm), 100 μl of inhibitor solution (natural or alkylated nucleic acid components at various concentrations), and 50 μl of anti-nucleoside serum at a dilution giving 50–60% of tracer–antibody binding in the absence of inhibitor. The appropriate serum dilution has been determined prior to performance of the competitive RIA using the same conditions described above except that 100 μl of TBS had been added to all samples (various antiserum dilutions) instead of inhibitor solution. After incubation for 2 hr at room temperature (equilibrium; i.e., no change of inhibition values observed upon further incubation), 200 μl of saturated ammonium sulfate solution were added. Ten minutes later the samples were centrifuged for 3 min at 10,000 g. The ^{3}H-radioactivity in 200 μl of the supernatant was determined in 10 ml of scintillation fluid in a liquid scintillation spectrometer. The inhibition of tracer-antibody binding was calculated by the following formula:

$$\text{Inhibition (\%)} = (\text{dpm}_1 - \text{dpm}_2)/(\text{dpm}_3 - \text{dpm}_2) \times 100$$

where dpm_1 = disintegrations per minute in the supernatant of inhibitor-containing sample (test sample); dpm_2 = dpm in the supernatant of sample without inhibitor (0% inhibition); dpm_3 = dpm in the supernatant of sample without antibodies (taken as the value for 100% inhibition).

Determination of Antibody Specificity

The competitive RIA can be used for the determination of antibody specificity, i.e., the cross-reactivity of antibodies with compounds structurally related to the hapten used for immunization. In this assay, the competition of the tracer (i.e., the radioactively labeled hapten against which the antibodies are directed) and various inhibitors [i.e., unlabeled haptens; either the same hapten used as the tracer (I_{Tr}) or any other hapten whose cross-reactivity with the antibodies is to be determined (I_x)] for the antibody binding sites is measured. The radioactivity (tracer) bound by the antibodies is determined for the respective inhibitor concentration, and the calculated inhibition of tracer–antibody binding (see the

[10] R. Müller and M. F. Rajewsky, *J. Cancer Res. Clin. Oncol.* **102**, 99 (1981).
[11] B. F. Erlanger and S. M. Beiser, *Proc. Natl. Acad. Sci. U.S.A.* **52**, 68 (1964).

preceding section on the Competitive RIA) is plotted as a function of the inhibitor concentration. As an example, Fig. 1 demonstrates the inhibition curves for various normal and alkylated nucleosides in a competitive RIA developed for the quantification of the carcinogen-modified DNA component O^6-ethyl-2'-deoxyguanosine (O^6-EtdGuo).[7–9] The cross-reactivity of the antibodies can then be determined by comparing the concentrations of I_{Tr} and I_x required for 50% inhibition of tracer–antibody binding (Table I). In those instances where inhibition values of 50% cannot be obtained (e.g., in the case of very low cross-reactivity; see Fig. 1 and Table I, normal deoxynucleosides), an estimation of the order of cross-reactivity is possible by comparing lower inhibition values, e.g., 10–20%, which, however, usually gives less precise results than the comparison of 50% inhibition values owing to the unlinearity of the inhibition curves in the range of low inhibition values. Table I shows that (a) the antibodies exhibit the highest affinity for O^6-EtdGuo (since all other haptens require higher concentrations for 50% inhibition of tracer–antibody binding); (b) the antibodies recognize the size of the alkyl group [affinity for O^6-ethylguanosine (O^6-EtGuo) > O^6-n-butylguanosine (O^6-BuGuo) > O^6-

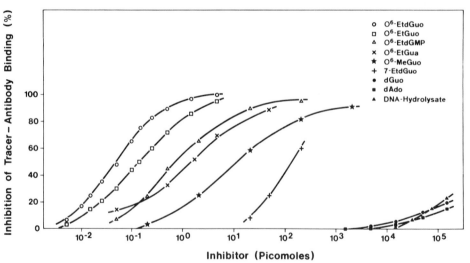

FIG. 1. Inhibition of tracer–antibody binding by various alkylated and natural nucleic acid components (inhibitors) in the competitive radioimmunoassay plotted against the inhibitor concentration. O^6-EtdGuo, O^6-ethyl-2'-deoxyguanosine; O^6-EtGuo, O^6-ethylguanosine; O^6-EtdGMP, O^6-ethyl-2'-deoxyguanosine 5'-monophosphate; O^6-EtGua, O^6-ethylguanine; O^6-MeGuo, O^6-methylguanosine; 7-EtdGuo, 7-ethyl-2'-deoxyguanosine; dGuo, 2'-deoxyguanosine; dAdo, 2'-deoxyadenosine; DNA-hydrolyzate, enzymic hydrolyzate of calf thymus DNA (deoxynucleosides). Reproduced from Müller and Rajewsky,[8] with permission of the publishers.

TABLE I

Inhibition of Tracer–Antibody Binding in
the Competitive Radioimmunoassay by
Various Alkylated and Natural Nucleic
Acid Components[a,b]

| Compound | Amount required for 50% inhibition of tracer–antibody binding | |
	pmol	Multiple of O^6-EtdGuo
O^6-EtdGuo	0.05	1
O^6-EtGuo	0.15	3
O^6-EtdGMP	0.7	14
O^6-EtGua	1.4	28
O^6-BuGuo	2.0	40
O^6-MeGuo	12	240
7-EtdGuo	170	3400
dGuo	$\sim 3 \times 10^{4c}$	$\sim 5 \times 10^6$
dAdo	$\sim 6 \times 10^{4c}$	$\sim 1 \times 10^7$
DNA hydrolyzate	$\sim 5 \times 10^{4c}$	$\sim 1 \times 10^7$

[a] Taken from Müller and Rajewsky,[8] with permission of the publishers.

[b] Antiserum was raised against O^6-EtGuo-KLH.

[c] Inhibition was 10% at this concentration.

methylguanosine (O^6-MeGuo)] as well as the position of the alkyl group [affinity for O^6-EtGuo > 7-ethylguanosine (7-EtGuo)]; (c) the antibodies recognize the sugar moiety of the nucleoside [affinity for O^6-EtdGuo > O^6-EtGuo > O^6-ethyl-2'-deoxyguanosine 5'-monophosphate (O^6-EtdGMP) > O^6-ethylguanine (O^6-EtGua)]; (d) the reactivity of the antibodies with naturally occurring deoxynucleosides is 6–7 orders of magnitude lower as compared to their reactivity with O^6-EtdGuo. Thus, using these antibodies in a competitive RIA, very low molar amounts of O^6-EtdGuo can be detected in DNA enzymatically hydrolyzed to deoxynucleosides.[7–10]

Calculation of K from Data Obtained by Competitive RIA: Mathematical Basis

The reaction of the radioactively labeled antigen (tracer) with the antibody binding sites is an equilibrium reaction:

$$[T] + [Ab] \rightleftharpoons [TAb]$$

where [T] and [Ab] are the concentrations of free tracer and antibody binding sites, respectively; and [TAb] is the concentration of the tracer–antibody complex.

According to the law of mass action:

$$K = \frac{[TAb]}{[T] [Ab]} \tag{1}$$

When b is the fraction of tracer bound by the antibodies, and $[T_t]$ is the total tracer concentration:

$$[TAb] = b[T_t] \tag{2}$$

and

$$[T] = [T_t] - [TAb] = [T_t] - b[T_t] = (1 - b) [T_t] \tag{3}$$

Substituting Eqs. (2) and (3) in Eq. (1):

$$K = \frac{b}{(1 - b)[Ab]} \quad \text{or} \quad [Ab] = \frac{b}{(1 - b)K} \tag{4}$$

The total concentration of antibody binding sites $[Ab_t]$ required for the binding of fraction b of the tracer is given by Eq. (5):

$$[Ab_t] = [TAb] + [Ab] \tag{5}$$

Substituting Eqs. (2) and (4) in Eq. (5):

$$[Ab_t] = b[T_t] + \frac{b}{(1 - b)K} \tag{6}$$

When tracer binding is 50% inhibited by unlabeled antigen (inhibitor), the fraction $0.5b$ of all antigen molecules (i.e., tracer + inhibitor) is bound:

$$[AgAb] = 0.5b[Ag_t] = 0.5b([T_t] + [I_t]) \tag{7}$$

where $[Ag_t]$ and $[I_t]$ are the total antigen and inhibitor concentrations, respectively, and [AgAb] is the concentration of the antigen (inhibitor + tracer)–antibody complex. Then the free antigen concentration is

$$[Ag] = [Ag_t] - [AgAb]$$
$$[Ag] = [T_t] + [I_t] - 0.5b([T_t] + [I_t])$$
$$= (1 - 0.5b)([T_t] + [I_t]) \tag{8}$$

The reaction of the total antigen (tracer + inhibitor) with the antibody binding sites is also an equilibrium reaction.

$$[Ag] + [Ab] \rightleftharpoons [AgAb]$$

According to the law of mass action:

$$K = \frac{[AgAb]}{[Ag][Ab]} \tag{9}$$

Substituting Eqs. (7) and (8) in Eq. (9):

$$K = \frac{0.5b}{(1 - 0.5b)[Ab]} \quad \text{or} \quad [Ab] = \frac{0.5b}{K - 0.5bK} \tag{10}$$

Make $[Ab_t] = [AgAb] + [Ab]$ and substitute $[Ab]$ by Eq. (10):

$$[Ab_t] = [AgAb] + \frac{0.5b}{K - 0.5bK} \tag{11}$$

Substituting Eqs. (6) and (7) in Eq. (11):

$$b[T_t] + \frac{b}{(1 - b)K} = 0.5b[T_t] + [I_t] + \frac{0.5b}{K - 0.5bK}$$

$$K = \frac{1}{([I_t] - [T_t])(1 - 1.5b + 0.5b^2)} \tag{12}$$

In case of 50% tracer binding ($b = 0.5$) in the absence of inhibitor:

$$K = \frac{8}{3([I_t] - [T_t])} \tag{13}$$

Thus, only the molar concentrations of the tracer and the inhibitor (at 50% inhibition of tracer–antibody binding) in the competitive RIA and the amount of tracer bound in the absence of inhibitor need be known in order to calculate the average antibody affinity constant K. The total concentration of antibody binding sites $[Ab_t]$ in the antiserum can then be calculated according to Eq. (6).

Another application of Eq. (12) is the calculation of the theoretical maximum sensitivity of a competitive RIA for a given antiserum. This sensitivity is reached when the specific radioactivity of the tracer is infinitely high; i.e., $[T_t]$ tends to 0. If K is known, the theoretical maximum sensitivity of the RIA (expressed as the inhibitor concentration required for 50% inhibition of tracer–antibody binding) may be calculated for $[T_t] = 0$ in Eq. (12):

$$[I_t] = \frac{1}{(1 - 1.5b + 0.5b^2)K}$$

Calculation of K from Data Obtained by Competitive RIA:
   Practical Application

   The average K values for dGuo, O^6-EtGuo, and O^6-EtdGuo in antisera containing antibodies of different affinity were determined using a competitive RIA. Constant amounts of labeled antigen (tracer) and varying amounts of unlabeled antigen (inhibitor) were allowed to react with the antisera under equilibrium conditions at dilutions giving 50% tracer–antibody binding in the absence of inhibitor ($b = 0.5$). After equilibration, the antibodies were precipitated with half-saturated ammonium sulfate. The inhibition of tracer–antibody binding was plotted against the respective inhibitor concentration in a probability grid (linearity over a large range), and the inhibitor concentrations giving 50% inhibition [I_t] were determined (Fig. 2). The average antibody affinity constants (2.7×10^5 to 1.8×10^{10} liters/mol) were calculated from the [I_t] values and the molar tracer concentrations in the assays from Eq. (13). They were found to be in good agreement with the K values obtained by the method of Steward and Petty[5,6] (see Table II).

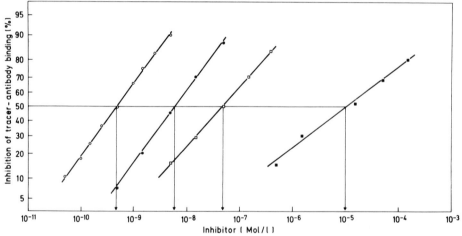

FIG. 2. Determination of the 50% inhibition values for dGuo, O^6-EtGuo, and O^6-EtdGuo in the competitive radioimmunoassay. The inhibition of tracer–antibody binding is plotted against the inhibitor concentration in a probability grid. ■, Antiserum A, tracer and inhibitor: dGuo; ●, antiserum C, tracer and inhibitor: O^6-EtdGuo; □, antiserum C, tracer and inhibitor: O^6-EtGuo; ○, antiserum B, tracer and inhibitor: O^6-EtdGuo. Antiserum A: anti-Guo-KLH; antiserum B: anti-O^6-EtGuo-KLH; antiserum C: anti-O^6-EtGuo-BSA. Nucleoside–protein conjugates were synthesized as described in references cited in footnotes 1, 7–10, 11. Taken from Müller,[1] with permission of the publishers.

TABLE II

COMPARISON OF K VALUES FOR dGuo, O^6-EtGuo, AND O^6-EtdGuo BY THE SCATCHARD PLOT ANALYSIS DESCRIBED BY STEWARD AND PETTY[5,6] AND BY THE TECHNIQUE DESCRIBED IN THIS CHAPTER FOR ANTISERA CONTAINING ANTIBODIES OF DIFFERENT AVERAGE AFFINITY[a,b]

Antiserum	Tracer	Inhibitor	$[I_t]$ (M)	$[T_t]$ (M)	K_R (liter/mol)	$K = 8/3([I_t] - [T_t])$ (l/mol)
A	[8-^{3}H]dGuo	dGuo	1.0×10^{-5}	4.2×10^{-9}	$2–5 \times 10^{5c}$	2.7×10^5
C	O^6-Et-[8-^{3}H]Guo	O^6-EtGuo	5.5×10^{-8}	2.4×10^{-8}	8×10^{7d}	8.6×10^7
C	O^6-Et-[8,5′-^{3}H]dGuo	O^6-EtdGuo	7.5×10^{-9}	4.0×10^{-9}	7×10^{8d}	7.6×10^8
B	O^6-Et-[8,5′-^{3}H]dGuo	O^6-EtdGuo	4.7×10^{-10}	3.2×10^{-10}	1.5×10^{10e}	1.8×10^{10}

[a] Taken from Müller;[1] with permission of the publishers.

[b] $[I_t]$, molar concentrations of inhibitor giving 50% inhibition of tracer–antibody binding in the competitive radioimmunoassay (RIA); K, average antibody affinity constant calculated from competitive RIA data; K_R, antibody affinity constant determined by Scatchard plot analysis[5,6]; $[T_t]$, total tracer concentration; $b = 0.5$ in all assays.

[c] Precise determination not possible, owing to lack of linearity of Scatchard plot.

[d] Mean of three determinations.

[e] Mean of two determinations.

The specific radioactivity of the tracer used in the RIA must be high enough to ensure that the expression $[I_t] - [T_t]$ in the denominator of Eq. (12) does not fall much below 0.5. Otherwise small deviations of the measured inhibition values would lead to large deviations of K. However, a specific radioactivity of 10 Ci/mmol, which may be obtained for almost all compounds, is sufficient to determine without difficulty K values of up to 10^{10} liters/mol.

Determination of Specific Radioactivity by RIA

The precision of K values calculated from data obtained by competitive RIA depends largely on a reliable value for the specific radioactivity of the tracer. If the molar amount of tracer available is limited (e.g., in the case of high specific radioactivity), determination of this parameter often proves to be difficult or even impossible. In those instances the specific radioactivity can be precisely determined by RIA in the following way. First, the molar concentration of an aliquot of tracer solution is raised by addition of the respective unlabeled compound (dilution of specific radioactivity) until a reliable determination of the molar concentration (e.g., spectrophotometrically) and thus of the specific radioactivity becomes possible. Then, applying the conditions described above in the section on competitive RIA (using an increased antibody concentration depending on the specific radioactivity of the diluted tracer; in our hands usually ~5- to 10-fold for a 10- to 20-fold dilution of tracer, i.e., ~1 Ci/mmol), a constant amount of antibodies is allowed to react with various amounts of both the diluted and the undiluted tracer. The relative amount (%) of tracer bound by the antibodies is calculated for each tracer concentration and plotted against the respective amount of radioactivity (dpm). The amount of tracer giving 50% tracer binding in the RIA is determined for both the diluted (dpm_d) and the undiluted (dpm_y) tracer. Since in both cases the molar amount of tracer bound by the antibodies is the same, the specific radioactivity of the undiluted tracer (Y) can then easily be calculated by the following formula:

$$Y = \frac{dpm_y}{dpm_d} \times D$$

where D is the specific radioactivity of the diluted tracer.

Determinants of Antibody Affinity and Specificity

Antibody affinity and specificity depend largely on the immunization procedure and the carrier protein to which the hapten is linked in order to exhibit immunogenic properties. Although it is probably impossible to judge whether a certain procedure is generally more advantageous than others, the findings of Müller and Rajewsky[8] may provide some guidelines in this direction. In this study, several carrier proteins and two different immunization procedures were used for the production of O^6-EtdGuo-specific antibodies in rabbits. The immunization procedures were as follows:

Procedure 1. The rabbits were immunized with 5 mg of nucleoside–protein conjugate per animal in 2.5 ml of PBS emulsified in 2.5 ml of complete Freund's adjuvant, by injections into the hind foot pads and about five other sites (intramuscular and subcutaneous). Four weeks later, the rabbits were boosted by intraveneous injection of 500 μg of conjugate per animal in PBS at 2-week intervals. Sera were collected at 10 days after the second booster.

Procedure 2. A mixture of 500 μg of nucleoside–protein conjugate in 0.5 ml of PBS and 0.5 ml of aluminum hydroxide slurry (Alugel S; Serva, Heidelberg, Germany) was stirred for 1 hr at 4° and emulsified in 1 ml of complete Freund's adjuvant. The rabbits were immunized by injection into the hind foot pads and into multiple intracutaneous sites (neck and axillae). Eight weeks later, the animals were boosted by the same procedure. After another 8-week interval, the rabbits received a second boost by intramuscular injection of 500 μg of conjugate in 1 ml of PBS emulsified in 1 ml of incomplete Freund's adjuvant. Sera were collected 2 weeks later.

Table III shows that the low-dose, slow immunogen-release (adsorption of the immunogen to aluminum hydroxide; intracutaneous injections; long intervals between boosters) procedure 2 was clearly superior to immunization procedure 1 in terms of the resulting antibody affinity and specificity to immunization procedure 1 leading to a stronger stimulation of the immune system, which is also reflected by the higher antibody concentration found in the antisera raised by this procedure. The superiority of procedure 2 could be explained by a preferential stimulation of a limited number of lymphocytes carrying high-affinity receptors. The antisera with the highest affinity and lowest cross-reactivity were elicited by using immunization procedure 2 and keyhole limpet hemocyanin (KLH) as a carrier protein (Table III). The strong immunogenic properties of KLH are presumably due to its large molecular weight and its phylogenetic distinction from the protein species of the immunized animals. All other carrier proteins studied were inferior with regard to the affinity and specificity of the antibodies obtained (KLH > horseshoe crab hemocyanin ≫ ovalbumin, bovine serum albumin, edestine ⋙ rabbit serum albumin).[8,10,12] Immunization procedure 2 and the use of KLH as a carrier protein proved to be suitable conditions also for the production of

---

[12] J. Adamkiewicz, R. Müller, and M. F. Rajewsky, unpublished results.

TABLE III

AFFINITY AND SPECIFICITY OF ANTI-O^6-EtdGuo SERA OBTAINED WITH THE USE OF DIFFERENT CARRIER PROTEINS FOR THE HAPTEN O^6-EtGuo AND BY DIFFERENT IMMUNIZATION PROCEDURES[a,b]

Immunogen	Immunization procedure	No. of rabbits immunized	Antibody titer	Concentration of specific antibodies (IgG) (mg/ml)	Antibody affinity constant, K (liter/mol)	Relative antibody reactivity	
						O^6-EtdGuo : dGuo	O^6-EtdGuo : dAdo
O^6-EtGuo-RSA	1	4	<1:3	ND[c]	ND	ND	ND
O^6-EtGuo-BSA	1	4	1:20,000–1:40,000	1.9–4.3	$6–8 \times 10^8$	$1.5 \times 10^6:1$	$3 \times 10^5:1$
O^6-EtGuo-BSA	2	2	1:60,000	1.1	$2–3 \times 10^9$	ND	ND
O^6-EtGuo-KLH	2	3	1:50,000–1:200,000	0.9–1.5	$1–2 \times 10^{10}$	$5 \times 10^6:1$	$1 \times 10^7:1$

[a] Reproduced from Müller and Rajewsky,[8] with permission of the publishers.

[b] Rabbits were immunized with conjugates of O^6-EtGuo and various carrier proteins (RSA, rabbit serum albumin; BSA, bovine serum albumin; KLH, keyhole limpet hemocyanin), using immunization procedure 1 or 2 (see text). Antibody titers in the antisera are expressed as the final serum dilutions giving 50% tracer binding in the radioimmunoassay (RIA) in the absence of inhibitor. Relative antibody reactivities for O^6-EtdGuo as compared to dGuo and dAdo were estimated from the respective inhibition values in the RIA.

[c] ND, not done.

high-affinity and specificity antibodies against other alkylated nucleosides in rabbits and other animal species (mouse, rat) used for the production of monoclonal antibodies.[9,10,12]

Acknowledgments

The work described in this chapter was supported by grants of the Deutsche For-schungsgemeinschaft (SFB 102/A9) and the Fritz Thyssen Stiftung (1980/2/41) to Professor M. F. Rajewsky. I am grateful to Drs. M. C. Poirier, M. F. Rajewsky, and J. Adamkiewicz for critical reading of the manuscript and for suggestions.

[44] Methods of Measuring Confidence Limits in Radioimmunoassay

By ROBERT C. BAXTER

In radioimmunoassay (RIA), as in other types of assay, it is impossible to achieve certainty in the estimation of an unknown analyte concentration. By calculating confidence limits, it is possible to define a region in which there is a high probability that the true answer will fall. Since the calculation of confidence limits of a concentration or potency estimate is based on a knowledge of within-assay errors, methods for calculating confidence limits are, in effect, methods for estimating within-assay variance.

For biological assays, Finney[1] has described the upper and lower limits as the greatest and smallest values for the true potency that would not be rejected by a significance test based on assay results. However, unlike bioassay, where unknown samples are commonly measured with many replicates at each of several dose levels, in RIA each sample is often measured in duplicate or triplicate at only a single dilution. An error estimate derived from such a measurement would be extremely inaccurate, being based only on 1 or 2 degrees of freedom. To overcome this problem it has become common to calculate the value for within-assay variance at any given dose level from an empirically derived function that describes the variance over the entire analytical range and is obtained by combining large numbers of variance measurements at different dose levels.

There are two somewhat different approaches to defining the relationship between dose estimates and their variance. The first, about which a

---

[1] D. J. Finney, "Statistical Methods in Biological Assay," 3rd ed. Griffin, London, 1978.

METHODS IN ENZYMOLOGY, VOL. 92

great deal has been written, involves determination of the errors in measuring the response variable, and then converting these response errors to the corresponding errors of dose estimate. The response variable measured may be raw count data or any of the many mathematical transformations of these data that have been used in processing RIA results: the fraction bound (B/T, or bound/total counts), the "bound/initial bound" ratio (B/B_0), etc.[1,2] The second approach, which appears the more straightforward, is to determine dose errors directly, regardless of the method used to obtain dose estimates. Since it has been demonstrated for several hormone radioimmunoassays that both methods provide appropriate values for confidence limits,[3] it is for the user to decide which is the more suitable in a particular situation.

Measurement of Errors in the Response Variable

Response errors in RIA have been the subject of a great number of investigations. A major contribution to knowledge on this topic has been made by Rodbard and his colleagues, and the reader is recommended to refer to several comprehensive articles by these authors,[2,4-8] and to the extensive reference lists, relating to earlier work in this area, contained in them. The particular interest in errors of the response variable is due not only to the importance of this information in RIA quality control, but also to its value in obtaining appropriate weights for each point of the standard curve when weighted least-squares methods are used for curve fitting.[8] It has become clear that the central problem in the estimation of response errors is the nonuniformity of variance, or heteroscedasticity, of the response variable across the dose-response curve.

When using experimental data to determine the relationship between the response variable Y and its variance Var(Y), several approaches may be taken. In the first method, only information from the standard curve is used. This assumes that sources of error leading to variation in the responses of standard curve points are identical to those affecting unknowns. It is of limited value to calculate the variance of the response variable from duplicate or triplicate measurements at each dose level in a

[2] D. Rodbard, P. L. Rayford, and G. T. Ross, *in* "Statistics in Endocrinology" (J. W. McArthur and T. Colton, eds.), p. 411. MIT Press, Cambridge, Massachusetts, 1970.

[3] R. C. Baxter, *Clin. Chem.* **26**, 763 (1980).

[4] D. Rodbard and G. R. Frazier, this series, Vol. 37, p. 3.

[5] D. Rodbard, *in* "Principles of Competitive Protein-Binding Assays" (W. D. Odell and W. H. Daughaday, eds.), p. 204. Lippincott, Philadelphia, Pennsylvania, 1971.

[6] D. Rodbard, *Clin. Chem.* **20**, 1255 (1974).

[7] D. Rodbard, *Radioimmunoassay Relat. Proced. Med., Proc. Int. Symp. 1977*, Vol. 2, p. 21 (1978).

[8] D. Rodbard, R. H. Lenox, H. L. Wray, and D. Ramseth, *Clin. Chem.* **22**, 350 (1976).

single standard curve, since these estimates of variance, based on only 1 or 2 degrees of freedom at each dose level, are very unreliable. Rodbard and Frazier[4] suggest running 10 replicates at each dose level and calculating the mean value and variance of Y at each point. Alternatively, combine data from a number of normal standard curves run in duplicate or triplicate and use analysis of variance to eliminate between-assay differences, giving an estimate of within-assay response variance at each dose level. Table I illustrates the calculations involved for a single standard dose, which should be repeated at each level.

TABLE I

CALCULATION OF WITHIN-ASSAY VARIANCE OF
THE RESPONSE VARIABLE $Y(B/B_0)$ BY ANALYSIS
OF VARIANCE[a]

Assay No.	Replicate response measurements		
	Y_1	Y_2	Y_3
1	0.362	0.368	0.362
2	0.391	0.389	0.399
3	0.362	0.378	0.348
4	0.345	0.352	0.366
5	0.396	0.420	0.392
6	0.353	0.354	0.338
7	0.372	0.374	0.364
8	0.367	0.381	0.337
9	0.366	0.370	0.349
10	0.356	0.378	0.363

[a] Number of replicates $(n) = 3$; number of assays $(N) = 10$; total number of values $(T) = 30$; mean $Y = 0.368$.

$$\sum_{N}^{N} \left(\sum_{n}^{n} (Y_i)^2 \right)$$

$$= \sum^{N} (Y_1^2 + Y_2^2 + Y_3^2) = S1 = 4.08210$$

$$\sum^{N} \left(\left(\sum^{n} Y_i \right)^2 / n \right)$$

$$= \sum^{N} (Y_1 + Y_2 + Y_3)^2/3 = S2 = 4.07916$$

Sum of squares for errors (SSE) $= S1 - S2 = 0.00294$. Mean square for errors $= \text{Var}(Y) = \text{SSE}/(T - N) = 0.000147$.

It is also possible to take advantage of response information from unknown specimens. Considerable information can be derived from a single assay, since each unknown measured in duplicate or triplicate contributes 1 or 2 degrees of freedom to an estimate of the response variance. A scattergram may be plotted relating the standard deviation or variance of Y to the mean value for each specimen, although it may be difficult to discern a trend in such a plot.[8] It is more useful to pool information into ranges of Y values. To achieve equal degrees of freedom for each point combine data for the 10 lowest values of Y, the next 10, and so on. However, if, as commonly happens, the data are not spread fairly evenly over the entire range of response values, it might be more useful to pool all points in the lowest 10% of the range, the next 10%, and so on (e.g., $B/B_0 = 0$–0.1, 0.1–0.2, etc.). Data are combined by the analysis of variance method described in Table I, allowing errors due to different specimens to be eliminated. Alternatively use the median variance for each range.

In the most complete treatment of response errors, all information from standards and unknowns is used, and data from many assays are combined. By pooling data from several thousand points, the relationship between Y and $\mathrm{Var}(Y)$ can be defined quite accurately.[8]

Models for the Response–Error Relationship

Whichever of the above methods is used to measure response variance, a graph of $\mathrm{Var}(Y)$ vs Y should be constructed. Three examples of such graphs are shown in Fig. 1. The relationship between Y and its variance (or standard deviation) is termed the "response-error relationship."[9] To define this relationship (that is, to fit the points on the graph), a number of different models have been suggested.

Quadratic model: $\mathrm{Var}(Y) = a_0 + a_1 Y + a_2 Y^2$ (1)

in which various authors have set to zero the coefficients a_2 (linear model), a_1 and a_2 (constant variance model), or a_0 and a_2 (Poisson-like model).

Exponential model: $\mathrm{Var}(Y) = a_0 Y^j$ (2)

which becomes the constant variance model or the Poisson-like model when $j = 0$ or 1, respectively, and the constant coefficient of variation model[10] when $j = 2$.

[9] R. P. Ekins, *Radioimmunoassay Relat. Proced. Med., Proc. Int. Symp. 1977*, Vol. 2, p. 39 (1978).

[10] G. M. Raab, R. Thompson, and I. McKenzie, *Comp. Prog. Biomed.* **12**, 111 (1980).

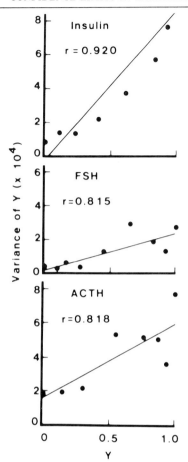

FIG. 1. The relationship between the response variable $Y(B/B_0)$ and its variance for insulin, follicle-stimulating hormone (FSH), and adrenocorticotropic hormone (ACTH) radioimmunoassays. Both B and B_0 are corrected for nonspecifically bound counts. Values are calculated by analysis of variance from response measurements for standards only, run in triplicate in 10 dose-response curves (see Table I). Lines represent least-squares fits to the data using the linear model, with correlation coefficients (r) as indicated.

Hyperbolic model: $\text{Var}(Y) = a_0 Y/(a_1 + a_2 Y)$ (3)

Log–log model: $\log \text{Var}(Y) = \log(a_0) + j \log Y$ (4)

a log-transformed version of the exponential model.

In one comparison of the performance of the linear, quadratic, and log–log models in fitting experimental data points, all three were found to give excellent fits.[8] The main point emerging from this study was that constancy of variance often is not found and should not be assumed.

Since it is mathematically simple to fit a straight line, there is an advantage in trying to use a linear fit to either untransformed or log-transformed data in preference to the nonlinear models. This has been done in the examples in Fig. 1, though it appears that a nonlinear model may have been more appropriate in some cases. This is clearly so when, as in the case of insulin, a linear model gives some negative values for Var(Y).

Conversion of Response Errors to Dose Errors

To calculate the standard deviation, s_X, of the dose estimate X for an unknown specimen from the response-error relationship, obtain an estimate of Var(Y) using the model most appropriate to the data, then divide s_Y, i.e. (Var(Y))$^{1/2}$, by the slope of the dose-response curve at the point corresponding to the mean response value. In a graphical treatment the curve may be assumed to be linear within the limits of s_Y, and the slope measured directly.[11] For a standard curve of the form $Y = 1/(aX + b) + c$, Brown et al.[12] have presented a method for approximating the slope of the tangent at the required point. Alternatively the slope must be calculated as the derivative dY/dX of the expression that defines the dose-response curve. Rodbard[7] has given several examples including the following, where y = counts and X = dose.

Example 1. For a polynomial fit, or segment of a spline curve, where $X' = \log_{10}X$:

$$y = a_0 + a_1X' + a_2(X')^2 + a_3(X')^3 + a_4(X')^4 \qquad (5)$$
$$dy/dX = [a_1 + 2a_2X' + 3a_3(X')^2 + 4a_4(X')^3]/2.303X \qquad (6)$$

Example 2. For a linear fit to logit-log transformed data, where $Y = B/B_0$, $Y' = \text{logit}(Y)$, $X' = \log_{10}X$, and β = the slope (dY'/dX'):

$$dY/dX = \beta(1 - Y)(Y)/2.303X \qquad (7)$$
$$dy/dX = (B_0)\beta(1 - Y)(Y)/2.303X \qquad (8)$$

Then for an unknown specimen, $s_X = s_Y/\text{slope}$; for example, using a linear model for response variance, $s_X = (a_0 + a_1 Y)^{1/2}/\text{slope}$. Confidence limits for the dose estimate can now be calculated simply: for a 95% confidence interval, the upper and lower limits are $X \pm t_{0.975}s_X$.

In an alternative treatment, Chang et al.[13] provide an equation, for use with logit-log transformed data, for the calculation of Var(Z) where $Z = \ln(X)$. Confidence limits for the dose estimate are then $\exp(Z \pm t_{0.975}s_Z)$

[11] R. P. Ekins, Radioimmunoassay Relat. Proced. Med., Proc. Int. Symp. 1977, Vol. 2, p. 6 (1978).

[12] M. Brown, M. Doron, and Z. Laron, Diabetologia 10, 23 (1974).

[13] P. C. Chang, R. T. Rubin, and M. Yu, Endocrinology 96, 973 (1975).

and are asymmetrical about the mean value. Yet another approach has been suggested by McHugh and Meinert,[14] who estimate upper and lower confidence limits for Y, then convert them to the corresponding limits for X using a formula that incorporates estimates of the concentration of antibody and tracer ligand, and the association constant of the binding reaction.

Direct Measurement of Errors in the Dose Estimate

As an alternative to determining errors in the response variable and converting them mathematically to the corresponding errors in dose estimate, dose errors may be determined directly. An advantage of this approach is that the treatment of errors is independent of the mathematical model used in obtaining dose estimates. Over a decade ago Rodbard *et al.*[2] noted that within-assay variation could be monitored by measuring the duplicate errors for a number of samples in each assay, pooling estimates for samples falling in a small dose range, and then combining data from several assays. This process can be repeated for a number of dose ranges across the entire analytical range, to give estimates of within-assay dose errors at different analyte levels. More recently the use of a number of quality control specimens, run in repeated assays to obtain error estimates with many degrees of freedom, has been suggested.[15] It becomes clear from such treatments that errors in dose estimate are not constant across the analytical range. As in the case of errors in the response variable, a major problem has been finding a simple relationship that would describe the change in dose errors across the entire dose-response curve. One commonly suggested relationship is that the variance is proportional to the square of the dose level (that is, the coefficient of variation is constant). However, this relationship can easily be seen not to hold at low dose levels, when errors become proportionately very high. In contrast, a quadratic relationship between the dose level and its standard deviation

$$s_X = a + bX + cX^2 \tag{9}$$

has been shown for several hormone radioimmunoassays to fit the experimental data satisfactorily.[3]

Estimation of Coefficients of the Quadratic Dose-Error Curve

To estimate the parameters a, b, and c, data can be combined, as previously described,[6] from samples falling into various dose regions

[14] R. B. McHugh and C. L. Meinert, *in* "Statistics in Endocrinology" (J. W. McArthur and T. Colton, eds.), p. 399. MIT Press, Cambridge, Massachusetts, 1970.

[15] B. F. McDonagh, P. J. Munson, and D. Rodbard, *Comp. Prog. Biomed.* **7**, 179 (1977).

across the dose-response curve. Another approach is to use error estimates from three plasma samples with low, medium, and high analyte levels, run in repeated assays as quality control specimens.[3] By using the same samples repeatedly, it is possible to accumulate many degrees of freedom for the estimation of within-assay error at the three dose levels. Once again, a simple analysis of variance is used to combine data, similar to that described in Table I. When the mean values (X_i) and within-assay standard deviations (s_{X_i}) are known for the three specimens, the data are combined easily by finding the unique quadratic curve passing through the points (X_1, s_{X_1}), (X_2, s_{X_2}) and (X_3, s_{X_3}). Table II illustrates the procedure, using the terminology from Table I. For this example, the dose-error relationship is $s_X = 0.354 + 0.0167 + 0.000349 \, X^2$. Thus at a dose level of 10 ng/ml, the estimated standard deviation is $0.354 + 0.167 + 0.035 = 0.556$ ng/ml. Then the 95% confidence limits are $10 \pm t_{0.975} (0.556)$, or 8.88–11.12 ng/ml.

The advantages of using the three routine quality control samples are (a) that no data must be accumulated above those normally used in between-assay quality control; and (b) that the fitting of the quadratic curve

TABLE II

CALCULATION OF THE DOSE-ERROR RELATIONSHIP FOR
PROLACTIN RADIOIMMUNOASSAY[a]

Parameter[b]	Quality control specimens		
	(1) Low	(2) Medium	(3) High
Number of assays (N)	40	40	40
Number of replicates (n)	2	2	2
Total number of values (T)	80	80	80
Degrees of freedom	40	40	40
Mean dose X (ng/ml)	4.25	17.1	52.1
SSE	7.44	21.96	188.4
MSE = SSE/$(T - N)$	0.186	0.549	4.71
$s_X = (\text{MSE})^{1/2}$	0.431	0.741	2.17

[a] For the relationship $s_X = a + bX + cX^2$

$$c = [(s_{X_1} - s_{X_2})/(X_1 - X_2) - (s_{X_2} - s_{X_3})/(X_2 - X_3)]/(X_1 - X_3) = 0.000349$$
$$b = (s_{X_1} - s_{X_2})/(X_1 - X_2) - c(X_1 + X_2) = 0.0167$$
$$a = s_{X_1} - b(X_1) - c(X_1)^2 = 0.354$$

[b] SSE, sum of squares for errors; MSE, mean square for errors.

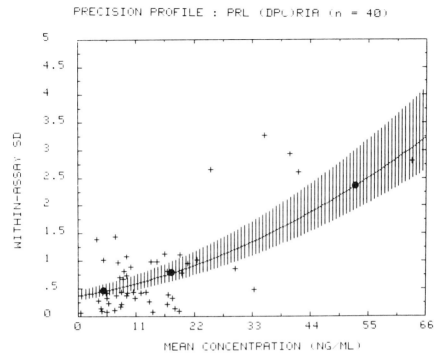

FIG. 2. Computer-derived (Hewlett-Packard 9845) dose-error relationship curve (precision profile) for the radioimmunoassay of prolactin. Filled circles represent the within-assay standard deviation (SD) of dose estimates of three quality-control specimens, run in duplicate in 40 assays. The solid line represents the quadratic curve joining the points (see Table II). The shaded area represents the 95% confidence envelope for the curve, calculated from chi-square tables. Crosses indicate the SD of duplicate dose estimates for individual unknown samples run in a single assay (1 degree of freedom per point).

is very simple. The main disadvantage is the relative instability of the fitted curve, due to the small number of points. This becomes less important as more assays are pooled for each estimate. Figure 2 shows a typical dose-error curve, based on the data in Table II, and its 95% confidence envelope, estimated from chi-square tables. It can be determined by reference to these tables that the confidence envelope would be almost 50% wider if calculations were based on only 20 degrees of freedom at each point.

Comparison of the Two Methods

To test the relative usefulness of response-error and dose-error methods in estimating confidence limits for a particular radioimmunoassay,

TABLE III
COMPARISON OF CONFIDENCE LIMITS FOR LUTEINIZING HORMONE
RADIOIMMUNOASSAY CALCULATED OF DOSE–ERROR METHOD AND
RESPONSE–ERROR METHOD

| Mean IU/liter | Observed[a] | 95% Confidence limits | |
| | | Calculated | |
		Dose-error method[b]	Response-error method[c]
2.3	1.3–3.3	1.7–2.9	1.7–3.1
5.9	5.0–6.8	5.2–6.6	5.2–6.7
10.9	9.5–12.3	10.0–11.8	10.0–11.9
18.0	16.6–19.3	16.7–19.2	16.8–19.2
23.9	22.1–25.6	22.4–25.4	22.4–25.4
30.8	29.0–32.6	29.0–32.6	28.9–32.8
41.0	37.9–44.1	38.6–43.4	38.4–43.8
55.9	51.0–60.8	52.3–59.5	52.0–60.1
68.5	62.9–74.0	64.5–72.5	63.3–74.2
82.8	75.1–90.6	77.9–87.7	75.9–90.3

[a] Samples were measured in 10 assays in triplicate, and within-assay errors were combined by analysis of variance (20 df per point). Confidence limits are means ± 2.09 standard deviations.

[b] Based on three quality-control specimens, each measured in duplicate in 20 assays. The dose-error relationship was

$$s_X = 0.239 + 0.01929\ X + 0.00008872\ X^2$$

[c] Based on 10 standard curves containing 6 standard doses run in triplicate. The response-error relationship was: $Var(Y) = 0.000036 + 0.000127\ Y$.

results obtained by the two methods can be compared to confidence limits determined directly by repeated measurements of plasma samples. Both methods should give confidence limits similar to the directly measured values. Table III illustrates such a comparison for the radioimmunoassay of luteinizing hormone. Ten plasma samples were run in triplicate in 10 successive assays, the within-assay variance of dose estimate was determined for each pool by analysis of variance, and 95% confidence limits were calculated as $X \pm ts_X$ ($t_{0.975}$, 20 df = 2.09). The dose-error method used three quality control specimens run in duplicate in 20 assays. The response-error method used standards only, run in triplicate in 10 assays. Good agreement between values calculated by the two methods and those obtained from repeated assay of the 10 samples suggests that either may be used in practice.

Author Index

Numbers in parentheses are reference numbers and indicate that an author's work is referred to although the name is not cited in the text.

T

Tacey, S. E., 257, 258(8), 265(8), 275(8)
Tada, N., 133
Tager, H. S., 334
Takács, B., 29
Talalay, P., 361
Tam, M. R., 162
Tamerius, J. D., 161, 162(16), 166(16)
Tanigaki, N., 133
Tao, Y., 377, 391, 392(2)
Taylor, C., 113, 136(23)
Taylor, D. M., 445
Taylor, G. F., 418
Teale, J. M., 48
Temperley, J. M., 295, 299(13)
ten Hoor, F., 40
Teramoto, Y. A., 162, 164(20)
Terhorst, C., 113, 133
Ternynck, T., 170, 257, 258(16), 394, 489, 492(10)
Thakur, A. K., 547, 551
Thierry, R., 493
Thompson, R., 604
Thorell, J. I., 310, 518
Thorpe, A., 330
Thron, F., 493
Thurmann, G. B., 34
Tirosh, N., 449
Tjian, R., 143, 145(10), 146
Tollaksen, S. L., 198, 202, 203(22)
Toms, E. J., 477
Tooze, J., 138
Tourtellote, W. W., 222, 223(7), 227(7)
Towbin, H., 139, 377
Tracy, R. P., 198, 199
Trayer, H. R., 417
Trayer, I. P., 417
Tregear, G. W., 404, 499
Treves, A., 12
Trinder, P., 415
Trisler, G. D., 194, 195(18)
Troll, V., 271, 272(48), 273(48)
Trosko, J. E., 445
Trowbridge, I., 90
Trucco, M. M., 110, 112, 130, 131
Truffa-Bachi, P., 175
Trujillo, J. M., 8
Tsang, V. C. W., 377, 391, 392(1, 2)
Tsumita, R., 275

Tucker, K. L., 438
Tung, A. S., 48, 50, 52, 57(18), 64
Turkewitz, A. P., 86, 95(10)
Turner, M. W., 158
Tuszynski, G. P., 99, 109(10)
Twine, M. E., 418
Tyrer, H., 81

U

Udenfriend, S., 96, 309
Ueda, Y., 300
Uehara, H., 86, 92(1)
Ui, N., 301
Ullman, E. F., 345, 360, 413, 426, 433, 458, 499, 518(4)
Umeda, Y., 345, 348(6), 350(8), 351(8), 353(8), 355(8), 360
Uterman, G., 126

V

Vaerman, J.-P., 89, 158, 159(23)
Valentine, R. C., 477
van Agthoven, A., 133
van Aken, W. G., 40
van Bavel, J. H., 577
Van Camp, L., 445
Van Dam, R. H., 89
Van Der Donk, J. A., 89
Vanderlaan, W. P., 459
van Es, 504
Van Hell, H., 359
van Mourik, J. A., 40
Van Raamsdonk, W., 377
Van Weemen, B. K., 55, 489
Veatch, W. R., 134, 135(65)
Velick, S. F., 589
Vitetta, E. S., 73, 86, 109
Voegtlé, D., 175, 180(4), 182(4)
Vogelhut, P. O., 413
Voller, A., 392, 393(9)

W

Wabl, M. R., 48
Wadley, F. M., 37

Subject Index

A

ABPC 48 cells, rosette assay of, 177

ACTH
high-performance liquid chromatography of, 295
pulse-chase studies on, 306–307

Actin
antibodies against, titration by lectin immunotest, 498
removal from monoclonal cell preparations, 134–135

Actinomyces sp., monoclonal antibody production against, 26

Affinity chromatography
of HLA monoclonal antibodies, 128–136
using monoclonal antibody, *see* Monoclonal antibody affinity chromatography

Affinity exclusion
affinity absorption of enzyme conjugate, 362–364
antigen enzyme preparation for, 362
enzyme immunoassay in, 364–366
glutathione estradiol absorbent for, 360–361
principle of, 360
use in enzyme immunoassay, 359–366

Agarose reagent, for replica plating assay, 189

AIDAs, rosette assays of, 180–182

Aldosterone, binding to A-6 epithelial cells, 562, 563

Alkaline phosphatase use in ELISA of MAb, 171

Aluminum hydroxide gel
as adjuvant for IgE induction, 49
preparation of, 50

Aminopterin
hybridoma sensitivity to, 10, 11
reagent, 21

Ammoniumperoxodisulfate reagent, for gel electrophoresis, 315

Ammonium sulfate, immunoglobulin precipitation by, 124

Ammonium thiocyanate, as eluting agent, 61–62

Angiotensins, high-performance liquid chromatography of, 295–296

Antibiotic-antimycotic mixture, for hybridization procedure, 21

Antibodies, *see also* Immunoglobulins

Antibodies
azophenyl hapten coupling to, 475
chains of, 206
enzyme-linked immunoelectrotransfer blot techniques (EITB) for, 377–391
labeling with β-D-galactosidase, 357–359
modification with SPDP, 480
multivalent, with dual specificity, 523–543
comparison with monovalent antibodies, 542
production by hydridomas, screening for, 15

Anti-bovine serum albumin antibodies, K-ELISA studies on, 400

Anti-Fab sera, use in MAb immunoassay, 156

Antigens
bound by monoclonal antibodies, 2-D gel analysis, 217
enzyme-linked immunoelectrotransfer blot techniques (EITB) for, 377–391
labeling with β-D-galactosidase, 357–359
mixtures of, for immunoadsorption, 217–218
plastic tubes coated with, for immunoassays, 403–413

Anti-hapten antibodies, determination of affinity of, by competitive radioimmunoassay, 589–601

J

JY cell line, HLA antigen purification from, 131, 134

K

K chain, secretion by HLK myeloma hybridomas, 149–156

Kappa chains, antisera to, for MAb quantitation, 156

Δ^5-3-Ketosteroid isomerase
as marker in enzyme immunoassay, 360
preparation of, 361

Keyhole limpet hemocyanin (KLH)
IgE hybridomas against, 49
as thymosin α_1 peptide carrier, 29, 34
use in hapten-sandwich applications, 479

Kinetic-dependent enzyme-linked immunosorbent assay, 391–403
applications of, 398–401
assay methods in, 393–394
error sources in, 401–403
instrumentation for, 396–397
procedure for, 397–398
solutions for, 394–396
theory of, 392–393

L

Lactoperoxidase
iodination with, 310–311
as surface label, 88

Lactoperoxidase iodination, of guinea pig anti-Ia antibodies, 73–75

Lambda chains, antisera to, for MAb quantitation, 156

Lectin-antibody conjugates for antigen quantitation, 489–498

Lectin immunotests, 489–498
applications of, 495–498
materials for, 491
methods, 492–495
principle of, 490
procedures, 493–495

Lens culinaris agglutinin, in lectin immunotests, 491, 494

Lentil lectin
in affinity chromatography of HLA monoclonal antibodies, 128–129
use in anti-Ia antibody purification, 73

Leucine-enkephalin, high-performance liquid chromatography of, 296–297

Levan, in rosette assay, 177

Lidex unit for noncentrifugation separation, 340
advantages of, 344

LIGAND
as computerized analysis of ligand binding data, 543–576
curve fitting by, 557

Ligand binding, mathematical model for, 575

Limulus polyphemus hemocyanin (Hy), IgE hybridomas against, 49

Lineweaver-Burk plot, for data analysis, 544, 571

Lipopolysaccharide
of erythrocyte-protein coupling agent, 274–276
lymphocyte stimulation by, 5–6

Lipopolysaccharide-stimulated spleen cells, MHC antigen purification from, 109

β-Lipotropin (β-LPH), pulse-chase studies on, 306–308

Liposomes, affinity targeting to, 479

Lithium sulfate, use in partition affinity ligand assay, 522

Luciferase, in bioluminescent assays, 427

Luteinizing hormone
antibodies to, binding characteristics, 249
radioimmunoassay of, confidence limits of, 610

Lymphocytes
cytotoxic, coated with hybrid antibody, 538–543
glycoproteins of, purification by antibody affinity columns, 135–136

Lymphocyte hybridomas, identification by electrophoresis of glucose-6-phosphate isomerase isozymes, 237–242

Lymphoid parental fusion partner cells, antigen-primed, 3–5